Bone Regeneration and Repair

Bone Regeneration and Repair

Biology and Clinical Applications

Edited by

Jay R. Lieberman, MD

Department of Orthopaedic Surgery
David Geffen School of Medicine
University of California, Los Angeles, CA

and

Gary E. Friedlaender, MD

Department of Orthopaedics and Rehabilitation
Yale University School of Medicine, New Haven, CT

HUMANA PRESS ✳ TOTOWA, NEW JERSEY

Library of Congress Cataloging in Publication Data

Bone regeneration and repair : biology and clinical applications / edited by Jay R. Lieberman and Gary E. Friedlander.
    p. ; cm.
  Includes bibliographical references and index.
  ISBN 0-89603-847-5 (alk. paper) e-ISBN 1-59259-863-3
  1. Bone regeneration. 2. Regeneration (Biology) 3. Transplantation of organs, tissues, etc.
  [DNLM: 1. Bone Regeneration—physiology. 2. Bone Regeneration—genetics. 3. Bone and Bones—physiology. WE 200 B71285 2005] I. Lieberman, Jay R. II. Friedlander, Gary E.
  RC930.B665 2005
  617.4'710592—dc22
                    2004010551

Preface

Bone is unique in its inherent capability to completely regenerate without scar tissue formation. This characteristic is central to skeletal homeostasis, fracture repair, as well as bone graft incorporation. However, in some circumstances the regenerative capacity of bone is altered or damaged in a manner that precludes such a special pattern of repair. Fracture nonunions, lost bone stock supporting total joint arthroplasties, and periodontal defects are frustrating examples of these difficult clinical challenges. Allogeneic bone and even autogenous bone grafts have not provided solutions for all these problems, at times related to limitations of their regenerative capacities and also when not used in a manner that respects their biological or biomechanical needs.

Over the past few decades, scientists and clinicians have been exploring the use of growth factors and bone graft substitutes to stimulate and augment the body's innate regenerative capabilities. The development of recombinant proteins and the application of gene therapy techniques could dramatically improve treatment for disorders of bone, cartilage and other skeletal tissues.

Bone Regeneration and Repair: Biology and Clinical Applications provides current information regarding the biology of bone formation and repair, reviews the basic science of autologous bone graft, skeletal allografts, bone graft substitutes, and growth factors, and explores the clinical applications of these exciting new technologies. An outstanding group of contributors has thoughtfully and skillfully provided current knowledge in this exciting area. This book should be of value to those in training, clinicians, and basic scientists interested in regeneration and repair of the musculoskeletal system.

Jay R. Lieberman, MD
Gary E. Friedlaender, MD

Contents

Color Illustrations

The following color illustrations are printed in the insert that follows page 212:

Chapter 1, Figure 3B, p. 10: Cutting cone in BMU.

Chapter 3, Figure 1, p. 47: Gene expression during mesenchymal cell condensation and cartilage development.

Chapter 3, Figure 2, p. 49: Gene expression during cartilage maturation, vascular invasion, and ossification.

Chapter 3, Figure 3, p. 52: Gene expression during early, intermediate, and late stages of nonstabilized fracture healing.

Chapter 5, Figure 2, p. 69: Mid-diaphysis of a stage 35 embryonic chick tibia.

Chapter 5, Figure 7, p. 78: Reactivity of antibodies SB-10 and SB-20 in longitudinal sections of developing human limbs.

Chapter 15, Figure 3, p. 296: Three-dimensional reconstructions of rat trabecular bone.

Chapter 17, Figure 1, p. 338: Four phases of fracture healing.

Contributors

SAM AKHAVAN, MD • *Department of Orthopaedic Surgery, Case Western Reserve University, Cleveland, OH*

JAMES ARONSON, MD • *Chief of Pediatric Orthopaedics, Arkansas Children's Hospital, Professor, Departments of Orthopaedics and Pediatrics, University of Arkansas for Medical Sciences, Little Rock, AR*

KODI AZARI, MD • *University of Pittsburgh School of Medicine, Pittsburgh, PA*

AXEL BALTZER, MD • *Praxis und Klink für Orthopaedie, Dusseldorf, Germany*

DANIEL J. BERRY, MD • *Orthopaedic Department, Mayo Clinic, Rochester, MN*

OLIVER BETZ, PhD • *Center for Molecular Orthopaedics, Harvard Medical School, Boston, MA*

K. CRAIG BOATRIGHT, MD • *The Emory Spine Center, Department of Orthopaedic Surgery, Emory University School of Medicine, Atlanta, GA*

SCOTT D. BODEN, MD • *The Emory Spine Center, Department of Orthopaedic Surgery, Emory University School of Medicine, Atlanta, GA*

MATHIAS BOSTROM, MD • *Hospital for Special Surgery, New York, NY*

SCOTT P. BRUDER, MD, PhD • *Department of Orthopaedics, Case Western Reserve University, Cleveland, OH; and DePuy Biologics, Raynham, MA*

BRUCE A. DOLL, DDS, PhD • *Department of Periodontics, University of Pittsburgh School of Dental Medicine, Pittsburgh, PA*

B. FRANK EAMES • *Department of Orthopaedic Surgery, University of California, San Francisco, CA*

THOMAS A. EINHORN, MD • *Professor and Chairman, Department of Orthopaedic Surgery, Boston Medical Center, Boson University School of Medicine, Boston, MA*

CHRISTOPHER H. EVANS, PhD • *Center for Molecular Orthopaedics, Harvard Medical School, Boston, MA*

GARY E. FRIEDLAENDER, MD • *Department of Orthopaedics and Rehabilitation, Yale University School of Medicine, New Haven, CT*

MARK C. GEBHARDT, MD • *Chief of Orthopaedics, Beth Israel Deaconess Hospital and Othopaedic Oncology Service, Children's Hospital, Boston, MA*

RICHARD S. GILBERT, MD • *Mount Sinai School of Medicine, New York, NY*

VICTOR M. GOLDBERG, MD • *Department of Orthopaedic Surgery, Case Western Reserve University, Cleveland, OH*

MOUSSA HAMADOUCHE, MD, PhD • *Department of Orthopaedic and Reconstructive Surgery (Service A), Centre Hospitalo-Universitaire Cochin-Port Royal, Paris, France*

JILL A. HELMS, DDS, PhD • *Department of Orthopaedic Surgery, University of California, San Francisco, CA*

SARAH HOLLAND, MD • *University of Pittsburgh School of Medicine, Pittsburgh, PA*

JEFFREY O. HOLLINGER, DDS, PhD • *Bone Tissue Engineering Center, Departments of Biomedical Engineering and Biological Sciences, Carnegie Mellon University, Pittsburgh, PA*

LAWRENCE HO • *University of Pittsburgh School of Medicine, Pittsburgh, PA*

FRANCIS J. HORNICEK, MD, PhD • *Orthopedic Oncology Service, Massachusetts General Hospital and Children's Hospital, Harvard Medical School, Boston, MA*

ELIZABETH JONESCHILD, MD • *Seattle Hand Surgery Group, Seattle, WA*

JAY R. LIEBERMAN, MD • *Department of Orthopaedic Surgery, David Geffen School of Medicine, University of California, Los Angeles, CA*

SAMUEL E. LYNCH, DMD, DMSc • *BioMimetic Pharmaceuticals Inc., Franklin, TN*

HENRY J. MANKIN, MD • *Orthopedic Oncology Service, Massachusetts General Hospital and Children's Hospital, Harvard Medical School, Boston, MA*

THEODORE MICLAU, MD • *Department of Orthopaedic Surgery, University of California, San Francisco, CA*

CALIN S. MOUCHA, MD • *Division of Adult Joint Replacement & Reconstruction, Department of Orthopedics, New Jersey Medical School, University of Medicine & Dentistry of New Jersey (UMDMJ), Newark, NJ*

DANIEL A. OAKES, MD • *Orthopaedic Department, Mayo Clinic, Rochester, MN*

PAUL D. ROBBINS, PhD • *Department of Molecular Genetics and Biochemistry, University of Pittsburgh School of Medicine, Pittsburgh, PA*

JAMES T. RYABY, PhD • *Senior Vice President, Research and Development, OrthoLogic Corp., Tempe, AZ*

RICHARD A. SCHNEIDER, PhD • *Department of Orthopaedic Surgery, University of California, San Francisco, CA*

TONY SCADUTO, MD • *Shriners Hospitals for Children, Los Angeles, CA*

CHARLES SFEIR, DDS, PhD • *University of Pittsburgh School of Dental Medicine, and Bone Tissue Engineering Center, Carnegie Mellon University, Pittsburgh, PA*

DOUG SUTHERLAND, MD • *Hospital for Special Surgery, New York, NY*

WILLIAM W. TOMFORD, MD • *Orthopedic Oncology Service, Massachusetts General Hospital and Children's Hospital, Harvard Medical School, Boston, MA*

STEPHEN B. TRIPPEL, MD • *Department of Orthopaedic Surgery, Indiana University School of Medicine, Indianapolis, IN*

JAMES R. URBANIAK, MD • *Division of Orthopaedic Surgery, Duke University Medical Center, Durham, NC*

MARK VRAHAS, MD • *Center for Molecular Orthopaedics, Harvard Medical School, Boston, MA*

SCOTT W. WOLFE, MD • *Professor of Orthopedic Surgery, Weill Medical College of Cornell University, and Attending Orthopedic Surgeon, Hospital for Special Surgery, New York, NY*

1

Bone Dynamics

Morphogenesis, Growth, Modeling, and Remodeling

Jeffrey O. Hollinger, DDS, PhD

INTRODUCTION

Morphogenesis, growth, and modeling of the skeletal system are dynamic processes, and the skeleton, once formed, is managed dynamically through remodeling. Attempts at consensus definitions and scripts for such processes will provoke heated debate, especially among the kinship of brothers who revere the skeleton. A review chapter of the likes conceived herein will provide grist for debate. This is good.

Considerable information is available in the literature on bone morphogenesis, growth, modeling, and remodeling. However, in preparing this review, it struck me that the line distinguishing growth, modeling, and remodeling, curiously, was sometimes gossamery. Fundamental and guiding building blocks from seminal publications of several distinguished workers helped focus my attention on key elements embodying definitions and principles necessary for a review chapter.

This chapter will provide a landscape of events embracing morphogenesis, growth, modeling, and remodeling. The benefits enjoyed by this author during the writing of this chapter are the sinew and power to inspire admiration and respect for the complexities and unity of form and function of the 206 bones of the skeleton *(1)*. I share this with you.

WORKING DEFINITIONS AND FOUNDATIONAL PRINCIPLES

Consensus definitions for knotty physiological processes can provide a sturdy platform for dialog. The underpinning for the chapter definitions was scoured from several sources, timeless epistles, consolidated, and reduced by the author. The curious reader can seek additional enlightenment and more detail in references provided.

Morphogenesis begets *growth*. *Morphogenesis* is a consummate series of events during *embryogenesis*, bringing cells together to permit inductive opportunities; the outcome is a three-dimensional structure, such as a bone *(2)*. The term *growth* embraces processes in endochondrally derived, tubular bones that increase length and girth prior to epiphyseal plate closure *(3)*. Intramembranous bone, not tubular in general form, but curved and platelike, without physes, enlarges in size under the aegis of a genetic script and then stops. In the cranium, the physis analog is the fontanelle. Fontanelles such as the bregmatic, frontal, occipital, mastoid, and sphenoidal provide linear space for growth (i.e., enlargement, increase in size). Heuristically, bone growth presupposes genetic controls prompting cell mitogenesis, differentiation, quantitative amplification, and enlargement (increase in cell mass and size).

Nononcologic cells have a built-in "governor" for cell divisions. For example, human fetal fibroblasts can undergo 80 cycles of cell divisions, whereas fibroblasts from an adult stop after about 40 divisions, and interestingly, embryonic mice fibroblasts stop at 30 divisions. Mechanisms controlling

From: *Bone Regeneration and Repair: Biology and Clinical Applications*
Edited by: J. R. Lieberman and G. E. Friedlaender © Humana Press Inc., Totowa, NJ

1

cell divisions are generally unknown; however, cyclin-dependent kinase inhibitor proteins, decrement in cyclin-dependent kinases, and cell contact-dependent cell–cell interactions have been implicated *(4, 5)*. From an embryological perspective, morphogenetic codes directing cell populations prompt inductive interactions for building three-dimensional structures *(2)*. A morphogenetic code could provide the guidelines ruling cell numbers, size, and growth. Therefore, growth may be perceived as dynamic events mentored by molecular cues.

The process that permits bone growth is *modeling,* an active pageantry of cells embraced in mysterious partnerships. Cells eagerly craft the growing 206 *(1)* bones using a three-dimensional blueprint that permits clinical recognition of a bone, whether it is the femur of the 6-month-old infant, a 3-year-old toddler, a 14-year old teenager, a 30-year-old surgery resident, or a 70-year-old professor emeritus. The preprogrammed architectural mold is a translation from an as yet to be deciphered genetic tome, hormonal directives (e.g., growth hormone), and mechanical cues: "modeling must alter both the size and architecture" *(6,7)*.

The final product of growth and modeling is a skeletal complex of 206 adult bones demanding continuous maintenance, which is accomplished by remodeling. *Remodeling* sustains structure and patches blemishes in the adult skeleton, while responding to homeostatic demands to ensure calcium and phosphate balance: "remodeling… [is] replacement of older by newer tissue in a way that need not alter its gross architecture or size" *(6,7)*.

In summary, as described in several recent reports *(8–14)* and stated succinctly by Frost; "*Growth* determines size. *Modeling* molds the growing shape. *Remodeling* then maintains functional competence" *(6,7)*.

MORPHOGENESIS AND GROWTH

For modeling to occur, there must be a structure to model. Fundamental questions need to be posed: (1) Why (and how) does a congregation of cells occur in a designated positional address? (2) Why (and how) do cells of that congregation produce a structure recognized as "a bone"? Molecular cues is the obvious answer. They drive cells, cells interact with other cells, and a structure, bone, takes shape. However, the response "molecular cues" spawns another query: Why are certain molecular cues expressed? *Morphogenesis* is the consummate porridge of molecular cues, and the inspiration for the cues is tangled in the genetic code. Morphogenesis begets growth, which begets modeling.

Morphogenesis is an epochal series of events during embryogenesis that brings cells together for inductive opportunities; the outcome is the skeletal system. Morphogenesis and bone are linked to a powerful family of cell morphogens: bone morphogenetic protein *(15,16)*. There are other key inductive morphogens that will be noted *(17)*.

Morphogenesis involves control centers with positional addresses in the developing embryo, where cells of that center regulate other cells through signaling factors. The signaling factors are proteins encoded by conserved multigene families; some multigene examples include *bone morphogenetic proteins (bmp), epidermal growth factors (egf), fibroblast growth factors (fgf)*, hedgehogs, and *Wnts (2,17–26)*.

The *hedgehog* family in vertebrates consists of three homologs of the *Drosophila melanogaster hedgehog* gene: *desert hedgehog, Indian hedgehog*, and *sonic hedgehog (shh)*. *Shh* may be the most important for the skeletal system, in that it mediates formation of the right–left axis (chicks) and initiates the anterior–posterior axis in limbs. *Shh* in limb bud formation induces *fgf4* expression, which acts with *Wnt7*. The name *Wnt* comes from fusing the *D. melanogaster* segment polarity gene *wingless* with the name of its vertebrate homolog *integrated*.

Signaling centers destined to be limb buds consist of aggregations of mesenchymal and epithelial cells and may be under the control of *fgf8, fgf10*, and *shh (27)* (reviewed in ref. 2). There are four axial levels where mesenchymal–ectodermal aggregates interact, called the *apical ectodermal ridge* (AER) *(28)*. Here, four limb buds form, and in the posterior zone of the AER, at the *zone of polarizing activity* (ZPA), *shh* acts as a mitogen for mesenchymal cells. Wisps of mesenchymal tissue stream in a centrifugal direction from the midline, and a further consolidation of cell phenotypes occurs, giving shape and form to a chondrogenic anlagen, where chondrocytes predominate and types II, IX, and XI collagens prevail (reviewed in ref. *13*).

Clusters of genes, *homeobox* genes *(Hox* genes), ensure limb bud location and limb constituents (reviewed in refs. *18* and *29*). In mice with abnormalities in expression of *Hox* genes, loss of digits can occur (associated with *Hoxa, Hoxd (30)*, and *Hoxd-13* may cause syndactyly in humans *(31)*.

The *shh* mediates anteroposterior patterning for metatarsal and metacarpals, as well as orchestrating expression of *bmps, fgfs,* and *Sox9* (the cartilage gene regulator for endochondral bone formation) (reviewed in ref. *17*). *Shh* prompts *fgf4* expression in ectoderm, *bmp2* expression in mesoderm *(32)*, and regulates anterior–posterior positioning and distal limb growth *(33)*.

With these cues flying around during morphogenesis, there is a potential for cells to get "confused." Through unidentified mechanisms, recklessness is not the rule, but rather, coordination and harmony among cells and cueing molecules propel growth. The process of growth and the dynamics of modeling (i.e., shaping growing bones) produce delicate digits, lovely shaped incus, maleus, and stapes, and the hulky femur. In addition to the signals for mitogenesis and differentiation, there are signals for programmed cell death: *apoptotic signals.*

As a symphony of life and death events, embryogenesis is a marvelous consortium of movements honed by a molecular tool kit that determines where congregations of cells will occur, the interactions among the cells, and the shape, size, and position of structures derived from that congregation, as well as the death of cells. Bundling of molecules in selectively positioned batches direct body position, form, cell, tissue, and organ development. This concept is underscored by the work reported by Storm and colleagues *(21,34)* on brachyopodism in mice (caused by a mutation in growth differentiation factors 5, 6, and 7) and by evidence from Kingsley on the *short-eared* mouse (associated with a corruption in the genetic coding for *bmp5) (19)*. The *short-ear* null mutation causes alterations in the size and shape of ears, sternum, and vertebrae that do not affect size and shape of limbs. In contrast, brachypodism null mutations reduce the length of limb bones and the number of segments in the digits but do not affect ears, sternum, ribs, or vertebrae. Explanation for the two phenotypes is that a mosaic for signaling centers exists, and during embryogenesis, some of the tiles in the mosaic become corrupted. The outcome is determined by the tiles corrupted.

During embryogenesis, controlling gates must be invoked to either stop or redirect events; extracellular stopping mechanisms broadly may include cell contact and extracellular inhibitory signals. *Bmps* are powerful, proactive inducers of events and must be tempered. A family of *anti-bmps* has been identified, and includes *noggin, fetuin, chordin, cerberus,* and *DAN* (reviewed in ref. *35*). *Noggin* antagonizes bmp-induced chondrocyte apoptosis *(36)*. (Chondrocyte apoptosis is required for joint formation *[37]*). When *noggin* expression is disrupted in mice, multiple skeletal defects occur, including short vertebrae, malformed ribs and limbs, and the absence of articulating joints. In terms of cartilage development and the growth of bone, growth differentiating factor-5 appears to be required in mice for joint cartilage, as long as cartilage-inducing signals from *bmp-7* are absent *(34)*.

Intracellular stopping mechanisms exist as well, and for bone may include the intracellular signaling molecules known as *smads* (the mammalian homolog to the *D. melanogaster* gene *Mothers against decapentaplegic) (38,39)*. *Smad* is a contraction of *D. melanogaster* Mothers against decapentaplegic —dpp—and *Caenorhabditis elegans* Sma. Bmps bind to serine–threonine transmembrane receptors, causing receptor phosphorylation, which activates a *smad* complex that transduces a signal to the cell nucleus and transcription ensues (**Fig. 1**) *(35)*. (More will be said about this in the section on osteoblasts.) Other *smad* complexes abrogate the process (reviewed in ref. *35*).

To this point, considerable information has been mentioned about cues, and nothing yet on the cellular craftsmen executing functions that result in growth and modeling. Pluripotential cells that can become chondrogenic, osteoblastic, and osteoclastic lineage cells will be mentioned next.

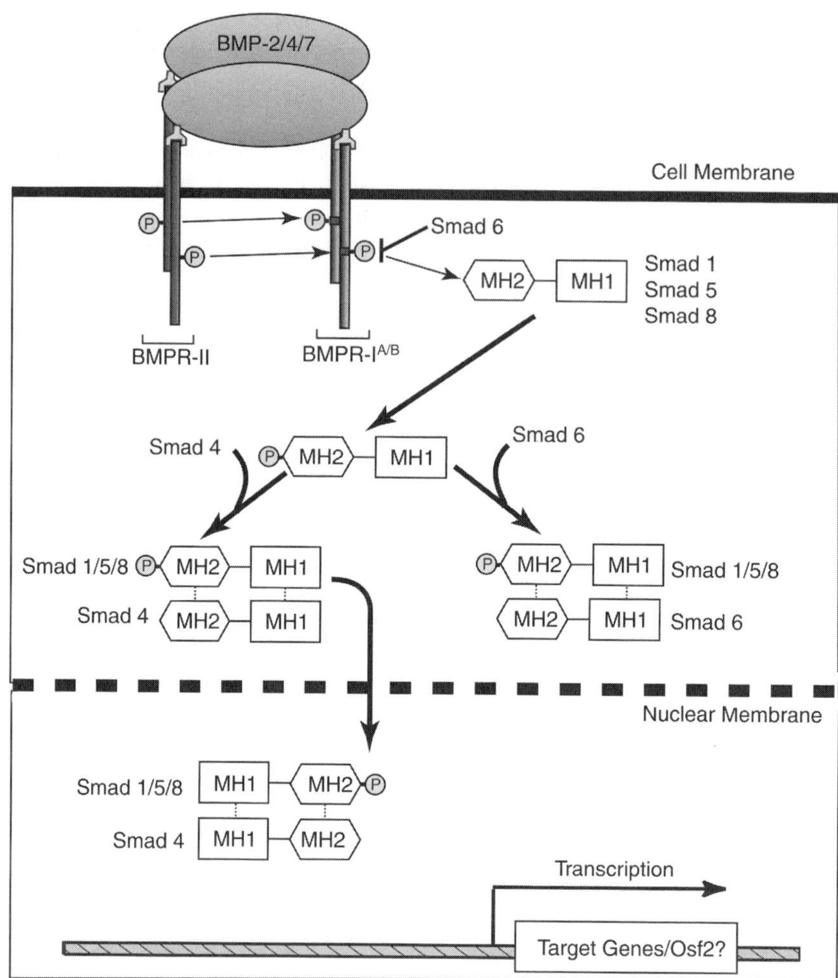

Fig. 1. BMP receptor binding and intracellular signal transduction. BMPs bind types I and II serine/threonine kinase receptors (BMPR-1[A/B] and BMPR-II) to form a heterodimer. Following binding, the type II receptors phosphorylate (P) the glycine/serine-rich domain of the type I receptor. The type I receptor phosphorylates the MH2 domain (Smad homology domain) of Smads 1, 5, and possibly 8. (Smad 6 may block the phosphorylation cascade by binding the type I receptor.) Following phosphorylation, the Smad1,5,8 complex either may bind to Smad 4 and translocate to the nucleus or may bind to Smad 6 and the signal is terminated. The Smad1, 5,8–Smad 4 complex translocated across the nuclear membrane can activate gene transcription either directly or indirectly through activation of the osteoblast-specific factor-2 (Osf2). (With permission from Schmitt, J. M., Hwang, K., Winn, S. R., and Hollinger, J. O. [1999] Bone morphogenetic proteins: An update on basic biology and clinical relevance. *J. Orthoped. Res.* **17,** 269–278.

Chondrocytes

Limb buds containing pluripotential mesenchymal cells destined to develop through endochondral bone formation express type IIb collagen, a chondrocyte-unique transcript of the *alpha1(II)* gene, type IX and type XI collagens, and matrix glutamic acid (gla) protein (reviewed in ref. *25*). Implicated in transcriptional control of chondrocyte differentiation has been *Sox9 (17,40)*. *Sox9* and *type II colla-*

gen are chondrocyte-specific genes coexpressed by chondrogenic lineage cells. Chondrocyte differentiation, maturation, and hypertrophy appear to be controlled by fibroblast growth factors, fibroblast growth factor receptors *(41)*, parathyroid hormone-related peptide (PTHrP) *(42,43)*, and the metalloproteinase gelatinase B *(44)*.

PTHrP controls the rate of differentiation of chondrocytes into hypertrophic chondrocytes. For example, bone explants exposed to elevated PTHrP have a delayed differentiation of hypertrophied chondrocytes; in PTHrP-deficient mice, there is premature differentiation of chondrocytes into hypertrophic chondrocytes (reviewed in ref. *25*). The upstream regulator for PTHrP is modulated by *Indian hedge hog* (Ihh), a gene product localized to the cartilage anlagen in endochondral bone *(45,46)*.

Until closure of the physes, long bones lengthen and increase in girth. Physeal energies for elongation are stimulated by growth hormone (GH), inspiring chondrocytes to express insulin-like growth factor-I (IGH-I). Acting in an *autocrine* manner, IGH-I "self-inspires" chondrocytes to express more IGH-I, proliferate, and, in a *paracrine* mode, incite other chondrocytes in a likewise fashion. Osteoblasts secrete IGF-I in response to PTH and GH; these factors are *osteoanabolic*, thus adding in expansion of girth (reviewed in ref. *47*).

Evidence suggests that fibroblast growth factor receptor 3 (fgfr3) negatively controls growth by limiting chondrocyte proliferation: absence of fgfr3 results in prolonged skeletal overgrowth (in mice) *(48)*. The metalloproteinase gelatinase B, a catalytic enzyme that is present in the extracellular matrix of cartilage, appears to control the final component of chondrocyte maturation, apoptosis, and vascularization *(44)*.

Vascularization of hypertrophic cartilage heralds calcification of the chondrocytes followed by programmed cell death (i.e., apoptosis). Streaming toward the calcified chondrocyte Cathedral are pluripotential mesenchymal cells destined to become chondroclasts, osteoblasts, and myeloid-derived cells, the osteoclast precursors.

Osteoblasts and Osteocyctes

During the complicated processes of embryogenesis, dorsal–ventral orientation, and limb bud development, a symphony of signaling cues (*bmps*, bmp-like molecules, *fgf, homeobox* gene products, *Ihh, shh*, TGF-β, and *Wnt*) weave a tapestry providing positional addresses for groups of pluripotential cells as well as fate-determining cues *(2,15,16,19,20,28,32,49–52)*. Cues for osteoblast lineage cell progression strongly suggest that the initiator is certain bmps, members of the TGF-β clan and *bmp*-like gene expression products (growth differentiation factor-5, gdf-5) *(53–59)*. (Certain bmps—except bmp-1—cause osteoblast differentiation; TGF-β stimulates proliferation and can inhibit differentiation *[60]*). The differentiation tempo is sustained through mediation with anti-bmps (e.g., *noggin, chordin, fetuin, DAN, cerberus*, reviewed in refs. *35, 61*, and *62*) that can short-circuit binding to cognate receptors, serine–threonine transmembrane receptor–ligand binding *(20,63–65)*, and transmembrane signal transduction through smads. The Smads shuttle signals received from receptor interaction with TGF-β and BMPs (i.e., the ligands) to the nucleus, where another set of signals begins.

Within the cell nucleus, an activated Smad complex can usher in the nuclear activities encoded by DNA (reviewed in ref. *35*) (**Fig. 2**). The process includes the nuclear transcriptional factor Runx-2 (a.k.a. core binding factor A: cbfa-1), which can stimulate expression of specific genes leading to differentiation of the osteoblast phenotype *(66–68)*. Runx-2 is a unique nuclear transcription factor for osteoblast differentiation (reviewed in refs. *35, 69*, and *70*). It is hypothesized that activation of Runx-2 results from a chain of events beginning with BMP-initiated receptor interaction, followed by intracellular Smad signaling.

The smad-activated complex transits the cell cytoplasm, crosses the nuclear membrane, and binds to DNA, where it induces a transcriptional response for *Runx-2*. *Runx-2* gene activation initiates expression of Runx-2 protein, which binds to the osteocalcin transcription promoter, heralding osteoblast differentiation *(23)*. Osteocalcin and Runx-2 are osteoblast icons.

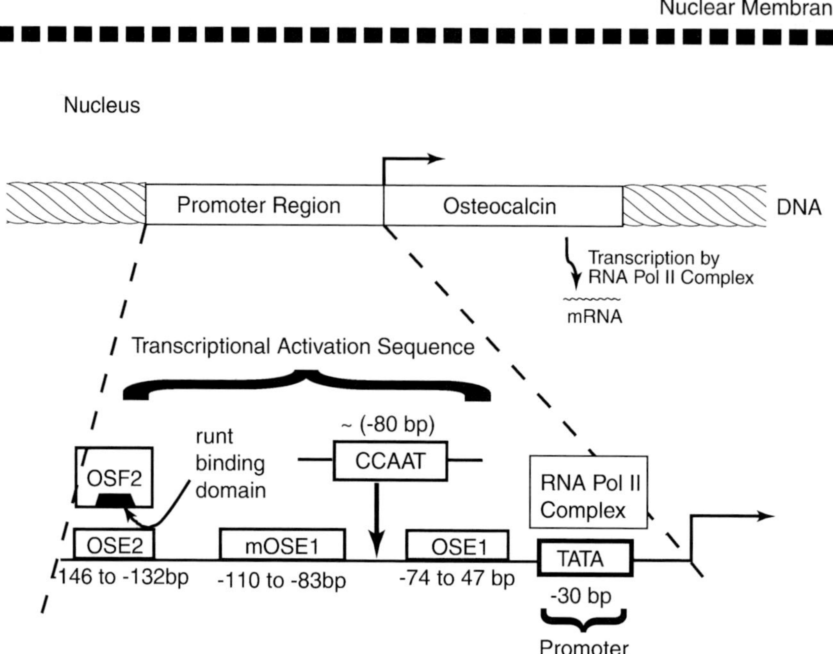

Fig. 2. The osteocalcin gene regulation is controlled by a promoter region where several specific nuclear proteins can activate gene transcription. Osteoblast-specific factor-2 (OSF-2) binds to the osteoblast-specific element-2 (OSE-2) by its runt domain. Following this action the TATA box, a nucleotide sequence with T–thymine nucleotide–and A–adenine nucleotide, binds RNA polymerase II (Pol II). This complex transcribes the osteocalcin genetic sequence into mRNA (messenger ribonucleic acid). The mRNA is translated into the osteocalcin protein on ribosomes. The illustration also shows that within the osteocalcin promoter region is the genetic sequences for mouse osteocalcin E-box sequence-1 (mOSE1) and osteoblast specific elelment-1 (OSE1). (bp stands for base pairs.) (Modified and with permission from Schmitt, J. M., Hwang, K., Winn, S. R., and Hollinger, J. O. [1999] Bone morphogenetic proteins: an update on basic biology and clinical relevance. *J. Orthoped. Res.* **17**, 269–278.

In homozygous deficient *Runx-2* mice, no osteoblasts form, and mice die postpartum due to intercostal muscle incompetency *(66)*. Heterozygously mutated mice for *Runx-2* have a phenotype consistent with cleidocranial dysostoses, the autosomal dominant disease characterized by hypoplastic clavicles, open fontanelles, supernumerary teeth, and short stature *(71)*.

The fate of the hard-working osteoblast can follow three pathways: programmed cell death (apoptosis), lining cells, and osteocytes. *Apoptosis* is the pathway most frequently taken, followed in order by *osteocytes* and *lining cells*. Osteocytes and lining cells are required to sustain bone viability and to respond to biomechanical signals. These two phenotypes will be addressed in more detail in the section on remodeling.

Osteoclasts

Balancing bone formation in the developing embryo and through the maturational period is the *osteoclast*, which is derived from the monocyte (reviewed in refs. *72* and *73*). It is generally concluded the osteoclast resorbs bone during growth, modeling, and remodeling.

Several factors have been associated with osteoclast formation, including PTH, PTHrP, vitamin D_3, interleukins-1, -6, and -11, tumor necrosis factor (TNF), leukemia inhibitory factor, ciliary neurotropic factor, prostaglandins, macrophage colony-stimulating factor (M-CSF), *c-fms, c-fos*, granulocyte colony-stimulating factor (reviewed in refs. *12* and *13*), and RANK *(74,75)*. A recently identified member of the TNF family, *osteoprotegerin*, has been shown to be an osteoclast inhibitor *(76,77)*.

A major extracellular differentiating factors for osteoclasts is RANK-L (rank ligand). RANK-L stimulates osteoclasts by a pathway mediated by osteoblasts. Osteoblast precursors express a unique molecule, TRANCE (also known as osteoclast differentiating factor), which activates osteoclast lineage cells by interacting with the RANK receptor *(74)*. Furthermore, osteoblast precursors also express osteoprotegerin *(62)* and, acting as a decoy receptor, can block TRANCE–RANK interaction, slamming the door on osteoclast formation (reviewed in ref. *14*).

Just as the osteoblast has a specific differentiation transcription factor (i.e., Runx-2), the factor for the osteoclast is *PU1 (78)*. *PU1*-deficient mice are osteopetrotic, lacking osteoclasts and macrophages *(78)*. Another transcription factor whose omission leads to osteopetrosis in mice is *c-fos (79)*.

Osteoclasts anchor to the surface of bone previously occupied by osteoblasts, and they do so through integrin extracellular matrix receptors: $\alpha v\beta 3$ (a vitronectin-type receptor), $\alpha 2\beta 1$ (a collagen receptor), and $\alpha v\beta 1$ *(80)*. In addition, osteopontin helps osteoclasts, as well as osteoblasts, stick to bone *(12)*.

Skeletal growth is a multidimensional, genetically coded process that destines size. A community of cells with a determined social hierarchy, bonded by signaling cues, sculpt growing bones, a process called modeling.

MODELING

Cells alter the shape and size of bone. Is this growth or modeling? Appendicular bones grow in length and girth. Physeal growth centers permit elongation, whereas the periosteal surface moves centrifugally, powered by osteoblastic deposition. Concurrently, endosteal growth proceeds centripetally, with a quanta of osteoclastic activity slowly enlarging the zone of bone marrow. The growth of appendicular bones maintains a gross morphology so the appearance of the pediatric "little" femur looks remarkably like the "adult" femur. In contrast, the axial and craniofacial skeletons do not possess physeal growth centers. Therefore, the axial growth for the vertebral bodies proceeds through a periosteal surface deposition titrated precisely with an endosteal deposition–resorption component. The adjective "drifts" *(6,7,81,82)* describe the waves of osteoblastic formation and osteoclastic resorption that move and mold bone in four dimensions: volume and time. This movement during growth is accomplished by the process of modeling.

The U-shaped mandible, mid-, and upper facial and cranial complexes may be viewed as plates of bones mortised together, with fontanelles in the cranial complex and formation and resorption drifts enabling expansion for brain growth. The skeletal complex of the cranium and upper face are often incorrectly described as "flat" bones. Studying a skull and midface, average freshman predental and premedical students would agree there is nothing "flat" in that area. Rather, gentle curves prevail and define the format. Therefore, bones of the craniofacial complex are correctly and accurately described as "curved" bones. Over 30 years ago, Enlow noted the intricate patterns of shaping, reshaping, resorption, and formation drifts of the growth of curved bones *(81,82)*.

Instructional guidance for growth (which includes *shape* and *size*) of osseous skeletal elements is controlled hormonally, and at pubescence the hormonal spigot is turned off, where GH, for example, is quenched—growth ceases.

Sustaining shape and size of bone in the adult skeleton is accomplished by the process of remodeling, where damaged bone is ceaselessly replaced by a tireless workforce of cells *(6,7,13)*. But does modeling really cease in the adult skeleton? The complexity of physiological issues and definitions often mire down rational dialog about the modeling and remodeling activities. Are the differences relative to timing? Relative to differences among processes? Frost proposes *(6,7)*, and Kimmel underscores *(8)*, that processes of macromodeling and minimodeling continue in the adult skeleton, where macromodeling increases the ability of bone to resist bending (by expanding periosteal and endosteal cortices) and minimodeling rearranges trabeculae to best adapt to functional challenges.

A recent review stated: "Remodeling of bone begins early in fetal life" *(14)*. The author then states: "Bone is initially formed by modeling, that is, the deposition of mineralized tissue at the developmentally determined sites" *(14)*. Another review noted that: "After modeling, the integrity of bone in the normal adult is maintained by the process of bone remodeling" *(13)*. Additional definitions are as follows: "Modeling is the process characterized by a change in bone shape or location of a bone structure... such as occurs during growth, fracture repair, or responses to altered biomechanical stress. As such, modeling processes occur in species with (open) growth plates (i.e., in immature animals) and in both immature and adults of species challenged with biomechanical stress (as in vigorous exercise) or fracture" *(83)*.

Simplification would be delightful: a simple definition for processes that shape bone, enable it to adapt to functional challenges, and meet physiological demands for homeostasis. While some of the cues for cells are likely to be different for bone with either open or closed physes, the active process could be conceived as the unity for simplicity. This notion could offer a broader base and a more simple platform for dialog between more clinicians and more bone scientists, and could be an inclusionary glue rather than an exclusionary barrier.

REMODELING

As reviewed by Frost *(7)*, remodeling of Haversian bone was first described in 1853, but the quintessential definition of the process, borrowed from Frost and modified, may be as follows: "Remodeling maintains functional (competence of bone)" *(7)*. Further, "remodeling serves the needs of replacement, maintenance, and homeostasis" *(6)*. Existing bone that becomes damaged is replaced with a like amount of bone without alteration in size and shape *(83)*. These definitions exclude fracture healing remodeling. Homeostatic remodeling and fracture healing remodeling are instigated by different promoters, sustained by different and similar stimulators, but common to both dynamic processes are the cellular craftsman. Fracture remodeling embraces a concept promulgated by Frost: regional area phenomenon (RAP) (reviewed in ref. *7*). The rest of this chapter will focus on homeostatic remodeling, and address RAP as needed.

A simple definition for complex physiological events is easy prey and elicits vocal challenges; retributions to the definer can be severe. Simple definitions for remodeling cloak a highly intricate, mysterious dance among cells and signaling factors (both soluble and mechanical). The appeal of simple definitions is that they can demystify the uninitiated, despite enraging the experts and inflicting angst on the definer. So, the challenges for the remainder of this chapter are to establish a silky passage to reason and understanding.

Understanding and explanation of the process of remodeling have been pursued with monklike fervor by a cadre of dedicated scientist/clinicians, and in the vanguard are Frost and Parfitt. They led; others followed. Reverence for their work provides inspiration and instructional guidance.

Remodeling sculpts what exists, making it bulkier, slimmer, redirecting trabecular struts, patching defects, and removing parts in response to homeostatic demands. Therefore, an enabling or activating signal must be evoked to jump-start the process. The signal can be either humoral (e.g., PTH) or biomechanical (e.g., strain), or both. The effector for the signal is a cell. The signal will activate the cell, the osteoblast. For cryptic reasons, the osteoblast vacates the bone surface, leaving behind a lure for osteoclasts *(14)*. The osteoclast arrives to the osteoblast-free surface, docks via integrinlike binding, resorbs a volume of bone (up to 5 µm/d *[84]*), and, for reasons to be determined, ceases activity, succumbs to programmed cell death, and detaches. Beckoned to the osteoclast-free site are osteoblasts. The lures are largely unknown. Osteoblasts attach to a remaining osteopontin-rich cement line, and in a sheetlike fashion spew forth an osteoid matrix that calcifies. Osteoid is produced at a rate of about 1–2 µm/d, and achieving a thickness of approximately 20 µm (after a maturation period of about 10 d), mineralizes at a rate of 1–2 µm/d *(85)*.

This truncated version of remodeling without regard at this instance for osseous location (i.e., periosteal, endosteal, Haversian, cortical, trabecular), is partitioned into activation–resorption–formation *(ARF) (86)*. (Osteoblasts are activated, vacate, are replaced by osteoclasts that resorb bone, vacate, and are replaced by osteoblasts that deposit bone. While some of the cues that turn on and off functional activity of the cells have been elucidated, many more must be discovered.) The gaggle of cell phenotypes (or cell packet *[3]*) responsible for remodeling is the basic multicellular unit (BMU), and the temporal duration (i.e., life span) of a BMU is called sigma (**Figs. 3A,B**) *(84)*.

Remodeling is a continuously active, dynamic activity driven by humoral and biofunctional cues, the outcome being that about 25% of trabecular bone and about 3% of cortical bone are removed and replaced each year (reviewed in ref. *9*). As we age, the balance between osteoblastic formation and osteoclastic resorption becomes asynchronous: bone loss occurs and results in the clinical disease osteoporosis *(87)*.

Trabecular bone (cancellous) and cortical bone remodel; the difference is that trabecular bone is trenched out by a BMU and cortical bone is burrowed out and the remnant is the cutting cone, which is eventually repaired with new bone. Regardless of the topographical differences that are BMU-crafted, the process begins on a surface populated by quiescent cells, lining cells, or preterminally differentiated osteoblasts. An ARF sequence for the remodeling BMU is invariant; duration of an active BMU (i.e., sigma) in either cancellous or cortical bone is about 2–8 mo *(9)*, whereas it can be prolonged from 2 to 10 yr in disease pathoses, such as osteoporosis and osteomalacia (reviewed in ref. *7*).

BMUs: Signals and Cells

Aggregates of osteoblast and osteoclast lineage cells and their end-stage phenotypes act continuously to replenish damaged bone (fatigue-damaged) with new bone *(88)*. Again, the consortium of cells wrapped in time (i.e., sigma) defines the BMU. Gearing up for the activity for a BMU requires an instigator, either humoral or biomechanical. The pageantry of bone physiology is too structured to enable arbitrariness; therefore, a skillful detector mechanism is necessary to determine a need for a response. What detects need? If biomechanical, what detects a biomechanical signal reflecting a need to respond? If humoral, the same question applies.

Origin of the Signals

The origin of the signals for remodeling relates to the three broad functional roles of the skeletal system—homeostasis, hematopoiesis, and mechanical (i.e., lever arms for muscles)—but not the fourth, protection. Homeostasis and the skeletal system are linked immutably to phosphate and calcium. Calcium concentration in the plasma averages about 9.4 mg/dL (varying between 9.0 and 10.0 mg/dL), whereas phosphate occurs predominantly in two anionic forms, divalent and univalent anions, at concentrations of 1.05 mmol/L and 0.26 mmol/L, respectively *(89)*. Titration of a precise calcium level is accomplished through feedback loops with participation of the liver and vitamin D_3; the kidney and 1,25-dihydroxy vitamin D_3 and PTH; and intestinal epithelium (where calcium is reabsorbed through binding to a calcium binding protein). Phosphate is a threshold ion, regulatable by the kidney, where increased secretion occurs as PTH expression elevates.

In response to homeostatic demands, systemic humoral cues for the cells of the BMU can include 1,25-dihydroxy vitamin D_3, androgen, calcitonin, estrogen, glucocorticoids, GH, PTH, and thyroid hormone (reviewed by several authors *[14,90–92]*). PTH and 1,25-dihydroxy vitamin D_3 stimulate resorption; they are countered by calcitonin, which inhibits resorption. Mechanisms for interactions are still not well known. The key systemic signal for bone is estrogen *(93)*: a decrease in this hormone can cause resorption to outstrip formation, bone mass falls, and the diagnosis for this disease is osteoporosis. Osteoporosis is not gender-specific. Estrogen is synthesized from testosterone (reviewed in ref. *14*). Advancing age is associated with an increased serum level of PTH and a decrease in estrogen, which

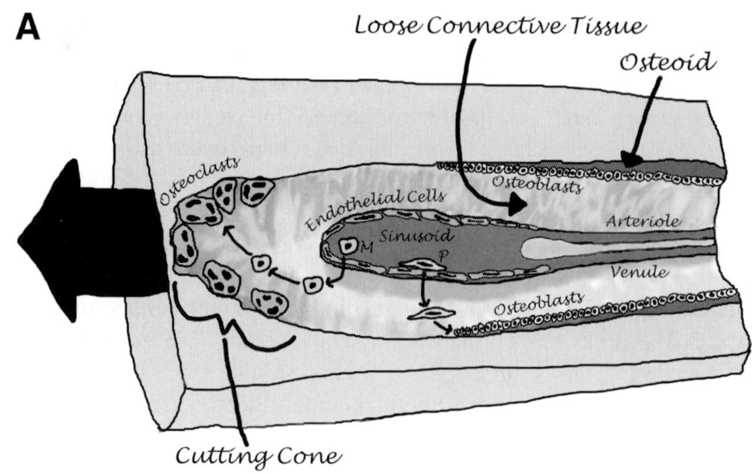
A
Loose Connective Tissue
Osteoid
Osteoclasts
Endothelial Cells
Osteoblasts
Arteriole
M
Sinusoid
P
Venule
Osteoblasts
Cutting Cone
3B1
3B2

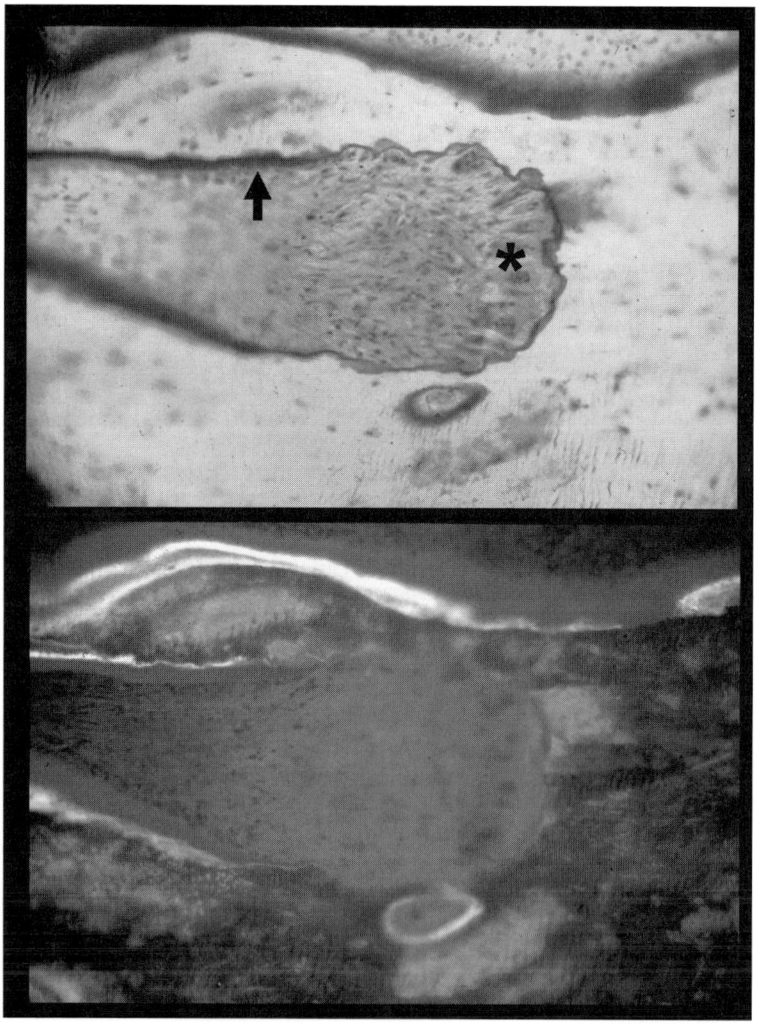

may evoke increased cytokine levels of IL-1, IL-6, TNF-α, and probably RANK-L *(93)*. Estrogen depletion provokes osteocyte apoptosis, and could cause bone loss *(94,95)*.

Local humoral cues can include BMPs, FGF, IGF, TGF-β, PDGF, PTHrP for formation and GM-CSF, ILs (1, 4, 6, 11, 13, 18), and M-CSF, leading to resorption (reviewed by several authors *[14,90–92]*). This dichotomy is *not* absolute; it is *general*. There is some controversy and contradictory data; for example, TGF-β can promote both resorption and formation.

Hematopoietic signals can include cytokines and lymphokines secreted in response to a regional phenomenon, such as inflammation. Examples can include IL-1, 6, 11, and TNF-α, as well as TGF-β and fgfs.

Marshaling signals for communication among cells in terms of mechanistic approaches to remodeling is daunting, the pitfalls many, and the data often confounding. The upregulation and downregulation modulators for cell responses to systemic and local soluble signals and biomechanical effects are often cloudy and speculative. Cell signaling molecules will decorate the landscape for discovery in the new millennium.

Detectors

Cells detect signals. Much needs to be learned about how cells detect signals and how they elicit a response.

A highly insightful article by Kimmel proposed a lucid argument for a paradigm that focuses on the mechanical function of the skeletal system that detects the mechanical need to remodel *(8)*. The signal is deformation of bone due to load: fatigue-damaged bone deforms, perhaps releasing cytokines (local humoral signals) *(60,96–98)*. The sensor for deformation is an osteocyte–bone lining cell complex *(99–101)*. It is unclear how messages are trafficked from the sensor to the effector (**Figs. 4** and **5**).

In addition to deformation, local biomechanical activity can provoke local release of extracellular matrix-containing arachidonic acid metabolites (e.g., prostaglandin E), leading to bone resorption *(26)*. What terminates biomechanical-induced resorption is likely estrogen *(14)*, and the local response to restore resorbed bone with deposition is likely prompted by androgens, BMPs, IGF, or TGF-β.

Osteoblasts that line bone surfaces to be remodeled must vacate that surface to permit osteoclasts to attach. This is part of the homeostatic partnership among cells. Nuances guiding this coupling process must be discovered. The stimuli for osteoblast-lining cells to depart the bone surface can include systemic factors (e.g., PTH) and local humoral factors (e.g., TGF-β) previously noted in this chapter. Collagenase digestion of the calcified surface by osteoblasts and their departure leaves exposed bone mineral and osteopontin, and establishes an enabling setting for osteoclast attachment. However, where are the osteoclasts? They must be recruited. Monocytes are the likely source for osteoclast lineage (reviewed in ref. *102*). ILs (1,6, and 11), PTH, PTHrP, TNF, prostaglandins, annexin-II, TGF-β, M-CSF, and RANK ligand stimulate osteoclast formation *(13,97,98)*. M-CSF and RANK ligand are suspected to be the strongest inducers for osteoclast formation, with RANK ligand promoting osteoclast formation

---

Fig. 3. *(Opposite page)* **(A)** An illustration of the basic multicellular unit (BMU) from human iliac bone where the movement is in the direction of the large arrow. The rate of travel for the osteoclast (OCl)-generated cutting cone is about 25 μm/d. The "cone" is about 500 μm in length and 200 μm wide. The zone between the osteoblasts (Ob) and sinusoid is lined by loose connective tissue stroma. M (monocytes) lured from the sinusoid can differentiate into the OCl phenotype. Pericytes (P) contiguous to sinusoidal endothelial cells can differentiate to osteoblasts (Ob). (Modified and with permission from Parfitt, A. [1998] Osteoclast precursors as leukocytes: importance of the area code. *Bone* **23(6),** 491–494. **(B1)** Cutting cone in BMU from human iliac crest biopsy. The OCls (*) are at the head of the resorption front and Obs (arrow) are following and depositing osteoid. (Villanueva Mineralized Bone Stain. 100×.) (Micrograph kindly provided by Antonio Villanueva, Ph.D). **(B2)** The same BMU as previous figure with fluorescent labels revealing osteoid mineralization. (Villanueva Mineralized Bone Stain. 100×.) (Micrograph kindly provided by Antonio Villanueva, Ph.D). (Color illustration of **B** in insert following p. 212.)

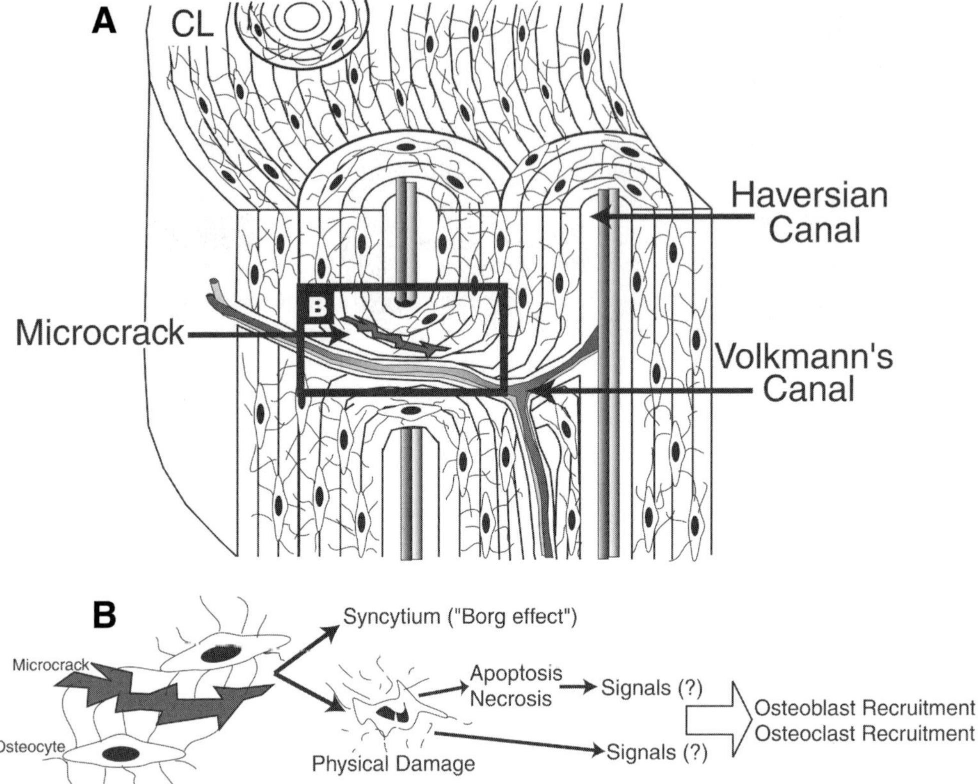

Fig. 4. The osteocyte syncytium (sensors) and osteoblast-osteoclast effectors work together to promote remodeling of fatigue-damaged bone. The osteocyte cytoplasmic process within canaliculi are disrupted due to a microcrack (fatigue damage). The cell–cell cytoplasmic junction sustaining community unity among osteocytes (the "Borg" effect) is abrogated. This can occur through the circumferential lamellar (CL), Haversian canal, and Volkmann canal. Physical damage and termination of cell–cell integration prompt regulatory signals that may cause ostecoyte necrosis or apoptosis, that in turn signal effectors: osteoclasts, osteoblasts.

by a signaling mechanism requiring expression of p50 and p52 subunits of a factor designated NF-κB *(103)*. M-CSF is necessary for osteoclast-committed mononuclear lineage cells, mediating its effects on osteoclastic bone resorption through a receptor kinase, the protooncogene c-fms *(73)*. It is unclear what the origin for some of these signaling molecules may be; however, TGF-β, IL-1, IL-6, and annexin-II have been shown to be expressed by osteoclasts *(98)*. The blizzard of remodeling activity requires continued renewal of osteoclasts, which undergo apoptosis after about 2 wk. Consequently, recruitment, differentiation, and activation must be unremitting throughout sigma.

Homing in of the osteoclast to the exposed mineral surface to be remodeled is controlled by factors that must be identified; the attachment to that surface by the osteoclast is essential for osteoclast activation. Activation of osteoclasts may be triggered by αvβ3 integrins *(104)* through a signal transduction pathway involving adhesion kinase PYK2 *(26)*. Attachment and activation are followed by resorption, where a team of osteoclasts scoop out furrows of bone along a 100–125-μm-thick by 2–3 mm in length trabeculum (of cancellous bone) about 30 μm in depth over a period of 2–4 wk *(105,106)*.

The process of remodeling in cortical bone is a bit different than the surface process in cancellous bone. The cutting cone is the hallmark of BMU cortical remodeling: approximately 2 mm in length, 0.2 mm wide, moving at a rate of about 20–40 μm/d, for a distance of 2–6 mm, and for a duration of

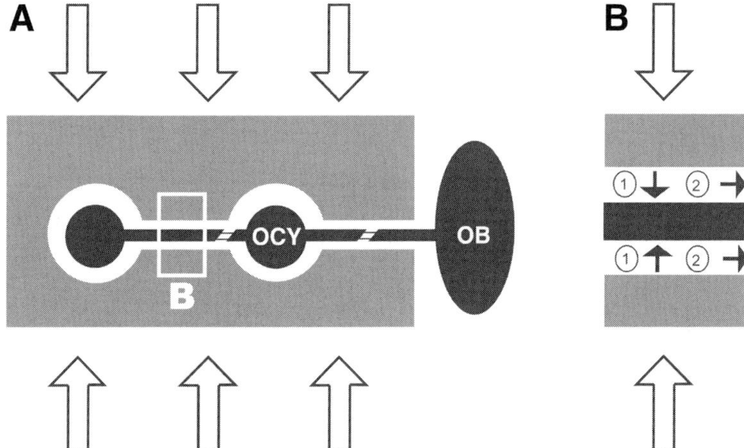

Fig. 5. Cytoplasmic stress–strain and fluid movement are possible operational mechanisms securing the osteocyte–osteoblast (OCy-Ob) interaction and may function as a mechanism for the transduction of mechanical strain to osteocytes in bone. **(A)** The Ocy-Ob cellular network on a section of bone under stress (large arrow). **(B)** Section "B" (from "A") depicts loading (large arrows) that causes straining of the cellular processes (1, vertical arrow heads) and fluid flow in canalicular extracellular matrix (2, horizontal arrow heads). (Modified and with permission from Klein-Nulend, J., van der Plas, A., Semeins, C. M., et al. (1995) Sensitivity of osteocytes to biomechanical stress in vitro. *FASEB J.* **9,** 441–445.

2–8 mo (sigma) *(6,7,9,106,107).* Osteoclasts are in the vanguard of the cutting cone, followed by a tessellation of osteoblasts. Osteoblasts have a life span of weeks to about 3 mo, and renewal through recruitment and differentiation is an unending directive. Pulsing through the center of each cutting cone is a blood vessel, providing transit for monocytic precursors and pluripotential cells (e.g., pericytes) that can differentiate to osteoclast and osteoblast phenotypes, respectively. It is still not clear what the signals are between the two phenotypes that couples their interactions. Neither is it clear what determines how much bone will be resorbed nor when new bone is deposited, or what stops deposition at a particular level.

What Happens When the Synchrony of Remodeling Gets Corrupted?

The osteoporotic condition mutes the capacity to sustain the homeostatic remodeling cycle, by which 25% of trabecular bone and 3% of cortical bone are resorbed and replaced each year *(9).* Instead, osteoclastic resorption proceeds without compensatory osteoblastic-mediated bone formation, and, consequently, in a lifetime, the aging process for women quietly steals up to 50% of their trabecular bone, while men lose about 25% of their bone *(108).* From age 20 to age 60, 25% of the cortical bone in men is depleted, and 35% in women *(108,109),* with a concomitant loss of 80–90% in bone strength *(110).* There is an overall risk of fracture of the hip, spine, and distal forearm that will afflict 40% of women and 13% of men 50 years of age and older *(111).* Moreover, it is projected that by 2005 there will be at least 25 million women in this country between 50 and 64 years old *(112),* and 33% of women older than 65 will experience at least one vertebral fracture *(113).*

The reduction of healthy marrow elements that occurs as a consequence of aging or disease (e.g., osteoporosis) is accompanied by a diminution of the cellular constituents, especially the osteogenic precursors *(114,115,129).* Moreover, the osteoporotic condition is plagued with a decrease in number and activity of osteoblasts *(116,117)* and a decrease in signaling molecules, such as estrogen, IGH, TGF-β, and calcitropic hormones *(118–123).* For postmenopausal women, osteoblast activity significantly decreases with estrogen depletion *(119,123,124)*. Furthermore, osteoblasts from "elderly" donors are less responsive to soluble signals than osteoblasts from "young" donors *(116,125)*. In addition, "old" osteoblast-like phenotypes in cell culture are three times less active than cells sourced from "young" stock *(115)*. Proliferation of human-derived cells of osteoblast phenotypes procured from donors of different ages revealed that osteogenic capacity decreased commensurately with increasing donor age *(126)*. In vivo, demineralized bone matrix from "young" donors is more osteoinductive than that derived from "old" donors, indicating a decrement of inductive factors in the matrix *(127)*. Significantly, there are irrefutable data that bone healing is delayed in the aged individual *(116,117,125,128– 130)*. In classic studies reported by Frost over 30 years ago, aging and osteoporosis were detailed clearly to retard remodeling dynamics *(131)*, and, using animal models, it was demonstrated that remodeling dynamics bog down with aging *(132–134)*.

What Happens When the Synchrony of Remodeling Is Accelerated?

Frost calls the general scenario of a regional noxious stimulus that evokes a series of events in an accelerated manner regional accelerated phenomenon (RAP) *(6)*. A fracture-healing site, a bone-graft bed, may be considered a place where a RAP will occur. Remodeling in such a zone, according to Frost, may be 50 times the normal, until form and function are restored *(6)*. Locally administered therapies to enhance fracture healing and regeneration of osseous deficits can boost RAP and ensure that healing deficits of aging are appropriately offset.

CONCLUSIONS

Embryogenesis, growth, modeling, and remodeling are dynamic processes, and the tool kit available to investigators is becoming more versatile and better packed, providing enabling technologies to understand and control these processes. The new millennium will be a jamboree of knowledge, spawning therapies to improve health care. This chapter identified many mysteries left in the 20th century by skeletal biologists that will usher in the 21st century. Our task and mission are to answer questions and find solutions to solve the mysteries, thereby improving lifestyle.

ACKNOWLEDGMENT

Support for this work was provided by the National Institutes of Health, NIDCR R01-DE13081 and RO1-DE11416.

GLOSSARY

Bone mass. The amount of bone tissue, often estimated by absorptiometry, preferably viewed as a volume minus the marrow cavity.

BMU. Basic multicellular unit of bone remodeling. In approximately 4 mo, and in a biologically coupled *activation → resorption → formation* (ARF) sequence, it turns over about 0.05 mm^3 of bone in humans. When it makes less bone than it resorbs (its disuse mode), this tends to remove bone, usually next to marrow. Adult humans may create about 3 million new BMUs annually, and about a million may function at any moment in the whole skeleton.

Modeling. Producing functionally purposeful sizes and shapes to bones. Mostly independent resorption and formation modeling drifts do it in bones and bone grafts. Modeling drifts mainly determine outside bone diameter, cortical thickness, and the upper limit of bone strength.

Remodeling. Turnover of bone in small packets by basic multicellular units. Literature published before 1964 did not distinguish between modeling and remodeling and lumped them together as remodeling. Some authors still do that, which can be confusing. However, while drifts and BMUs create and use what seem to be the same kinds of osteoblasts and osteoclasts to do their work, in different parts of the same bone at the same time the osteoblasts and osteoclasts in drifts and BMUs can act and respond differently and even oppositely to many influences. In remodeling disuse mode, BMU creations increase and completed BMUs make less bone than they resorb. In its conservation mode, BMU creations usually decrease and resorption and formation in completed BMUs tend to equalize.

Remodeling space. Each BMU makes a temporary hole in bone or on a bone surface. The sum of all such holes equals the remodeling space, which can vary from about 3% to occasionally more than 30% of a bone's volume. As a result of surface-to-volume ratio effects, its value in trabecular bone usually exceeds the value in compact bone.

Strain. The deformation or change in dimensions and/or shape caused by a load on any structure or structural material. Special gauges can measure bone strains in the laboratory and *in vivo*. Loads always cause strains, even if very small ones. In biomechanics, strain is often expressed in microstrain units, where 1000 microstrain in compression would shorten a bone by 0.1% of its original length, 10,000 microstrain would shorten it by 1% of that length, and 100,000 microstrain would shorten it by 10% of that length (and break it).

Stress. The elastic resistance of the intermolecular bonds in a material to being stretched by strains. Loads cause strains, which then cause stresses. Three principal strains and stresses include tension, compression, and shear. Stress cannot be measured directly but must be calculated from other information that often includes strain. The stress–strain curve of bone is not linear. The material is stiffer at small loads and strains than at larger ones.

Ultimate strength. The load or strain that, when applied once, usually fractures a bone. The fracture strength of normal lamellar bone is about 25,000 microstrain (CV about 0.3), which corresponds to a change in length of 2.5%, that is, from 100.0% to 97.5% of its original length under compression or to 102.5% of it under tension. That fracture strain corresponds to an ultimate or fracture stress of about 17,000 psi or about 120 MPa.

REFERENCES

1. Goss, C. (1965) *Gray's Anatomy of the Human Body,* 27th edition. Lea & Febiger. Philadelphia, p. 119.
2. Hogan, B. (1999) Morphogenesis. *Cell* **96,** 225–233.
3. Frost, H. (1983) Bone histomorphometry: choice of marking agent and labeling schedule, in *Bone Histomorphometry: Techniques and Interpretation* (Recker, R. R., eds.), CRC Press, Boca Raton, FL, pp. 37–52.
4. Alberts, B., Bray, D., Johnson, A., et al. (1998) *Essential Cell Biology. An Introduction to the Molecular Biology of the Cell.* Garland, New York, pp. 572–589.
5. Conlon, I. and Raff, M. (1999) Size control in animal development. *Cell* **96,** 235–244.
6. Frost, H. M. (1986) *Intermediary Organization of the Skeleton,* vol II. CRC Press, Boca Raton, FL, pp. 1–331.
7. Frost, H. M. (1986) *Intermediary Organization of the Skeleton,* vol I. CRC Press, Boca Raton, FL, pp. 1–365.
8. Kimmel, D. (1993) A paradigm for skeletal strength homeostasis. *J. Bone Joint Miner. Res.* **8(2),** 515–522.
9. Parfitt, A. M. (1994) Osteonal and hemi-osteonal remodeling: the spatial and temporal framework for signal traffic in adult human bone. *J. Cell Biochem.* **55,** 273–286.
10. Parfitt, A. M., Roodman, G. D., Hughes, D. E., and Boyce, B. F. (1996) A new model for the regulation of bone resorption, with particular reference to the effects of bisphosphonates. *J. Bone Miner. Res.* **11(2),** 150–159.
11. Rodan, G. (1996) Coupling of bone resorption and formation during bone remodeling, in *Osteoporosis* (Marcus, R., Feldman, D., and Kelsey, J., eds.), Academic Press, San Diego, CA, pp. 289–299.
12. Ng, W., Romas, E., Donnan, L., and Findlay, D. (1997) Bone biology. *Bailliére's Clin. Endocrinol. Metab.* **11(1),** 1–22.
13. Boyce, B., Hughes, D., Wright, K., Xing, L., and Dai, A. (1999) Recent advances in bone biology provide insight into the pathogenesis of bone diseases. *Lab. Invest.* **79(2),** 83–94.
14. Raisz, L. (1999) Physiology and pathophysiology of bone remodeling. *Clin. Chem.* **45(8B),** 1353–1358.
15. Hogan, L. B. (1996) Bone morphogenetic proteins in development. *Curr. Opin. Gen. Dev.* **6,** 432–438.
16. Hogan, B. L. M. (1996) Bone morphogenetic proteins: multifunctional regulators of vertebrate development. *Genes Dev.* **10,** 1580.
17. Ferguson, C., Miclau, T., Hu, D., Alpern, E., and Helms, J. (1998) Common molecular pathways in skeletal morphogenesis and repair. *Ann. Acad. NY Sci.* **3,** 121–131.
18. Duboule, D. (1994) How to make a limb? *Science* **266,** 575–576.
19. Kingsley, D. (1994) What do BMPs do in mammals? Clues from the mouse short-ear mutation. *Trends Genet.* **10(1),** 16–21.

20. Kingsley, D. M. (1994) The TGF-beta superfamily: new members, new receptors, and new genetic tests of function in different organisms. *Genes Dev.* **8,** 133–146.
21. Storm, E. E., Huynh, T. V., Copeland, N. G., Jenkins, N. A., Kingsley, D. M., and Lee, S. (1994) Limb alterations in brachypodism mice due to mutations in a new member of the TGF-beta superfamily. *Nature* **368,** 639–643.
22. Tickle, C. (1994) On making a skeleton. *Nature* **368,** 587–588.
23. Ducy, P., Zhang, R., Geoffroy, V., Ridall, A., and Karsenty, G. (1997) Osf2/Cbfa1: a transcriptional activator of osteoblast differentiation. *Cell* **89,** 747–754.
24. Rodan, G. and Harada, S.-I. (1997) The missing bone. *Cell* **89,** 677–680.
25. Ducy, P. and Karsenty, G. (1998) Genetic control of cell differentiation in the skeleton. *Curr. Opin. Cell Biol.* **10,** 614–619.
26. Rodan, G. (1998) Control of bone formation and resorption: biological and clinical perspective. *J. Cell. Biochem.* **Suppl. 30,** 55–61.
27. Martin, G. (1998) The roles of FGFs in the early development of vertebrate limbs. *Genes Dev.* **12,** 1571–1586.
28. Shubin, N., Tabin, C., and Carroll, S. (1997) Fossils, genes, and the evolution of animal limbs. *Nature* **388,** 639–648.
29. Duboule, D. (1992) The vertebrate limb: a model system to study the Hox/HOM gene network during development and evolution. *Bioessays* **14,** 375–384.
30. Kondo, T., Zakany, J., and Duboule, D. (1997) Of fingers, toes, and penises (letter). *Nature* **390,** 29.
31. Muragaki, Y., Mundlos, S., Upton, J., and Olsen, B. (1996) Altered growth and branching patterns in syndactyly caused by mutations in *HOXD13*. *Science* **272,** 548–551.
32. Laufer, E., Nelson, C. E., Johnson, R. L., Morgan, B. A., and Tabin, C. (1994) Sonic hedgehog and Fgf-4 act through a signaling cascade and feedback loop to integrate growth and patterning of the developing limb bud. *Cell* **79,** 993–1003.
33. Riddle, R. D., Johnson, R., Laufer, E., and Tabin, C. (1993) Sonic hedgehog mediates the polarizing activity of the ZPA. *Cell* **75,** 1401–1416.
34. Storm, E. and Kingsley, D. (1999) GDF5 coordinates bone and joint formation during digit development. *Dev. Biol.* **209,** 11–27.
35. Schmitt, J. M., Hwang, K., Winn, S. R., and Hollinger, J. O. (1999) Bone morphogenetic proteins: an update on basic biology and clinical relevance. *J. Orthop. Res.* **17(2),** 269–278.
36. Merino, R., Ganan, Y., Macias, D., Economides, A., Sampath, K., and Hurie, J. (1998) Morphogenesis of digits in the avian limb is controlled by FGFs, TGFbetas, and noggin through BMP signaling. *Dev. Biol.* **200,** 33–45.
37. Storm, E. and Kingsley, D. (1996) Joint patterning defects caused by single and double mutations in members of the bone morphogenetic protein (BMP) family. *Development* **233,** 3969–3979.
38. Sakou, T. (1998) Bone morphogenetic proteins: from basic studies to clinical approaches. *Bone* **22(6),** 591–603.
39. Sakou, T., Onishi, T., Yamamoto, T., Nagamine, T., Sampath, K., and Ten Dijke, P. (1999) Localization of Smads, the TGF-beta family intracellular signaling components during endochondral ossification. *J. Bone Miner. Res.* **14(7),** 1145–1152.
40. Ng, L., Wheatley, S., Muscat, G., et al. (1997) SOX9 binds DNA, activates transcription, and coexpresses with type II collagen during chondrogenesis in the mouse. *Dev. Biol.* **183,** 108–121.
41. Hurley, M. and Florkiewicz, R. (1996) Fibroblast growth factor and vascular endothelial cell growth factor families, in *Principles of Bone Biology* (Bilezikian, J., Raisz, L., and Rodan, G., eds.), Academic Press, San Diego, CA, pp. 627–645.
42. Martin, T., Findlay, D., and Mosley, J. (1996) Peptide hormones acting on bone, in *Osteoporosis* (Marcus, R., Feldman, D., and Kelsey, J., eds.), Academic Press, San Diego, CA, pp. 185–205.
43. Segre, A. (1996) Receptors for parathyroid hormone and parathyroid-related hormone, in *Principles of Bone Biology* (Bilezikian, J. P., Raisz, L. G., and Rodan, G. A., eds.), Academic Press, San Diego, CA, pp. 377–404.
44. Vu, T., Shipley, J., Bergers, G., et al. (1998) MMP-9/gelatinase B is a key regulator of growth plates angiogenesis and apoptosis of hypertrophic chondrocytes. *Cell* **93,** 411–422.
45. Bitgood, M. and McMahon, A. (1993) Hedgehog and Bmp genes are coexpressed at many diverse sites of cell-cell interaction in the mouse embryo. *Dev. Biol.* **172,** 126–138.
46. Vortcamp, A., Lee, K., Lanske, B., Segre, G., Kronenberg, H., and Tabin, C. (1996) Regulation of rate of cartilage differentiation by Indian hedgehog and PTH-related protein. *Science* **273,** 613–622.
47. Canalis, E. (1996) Skeletal growth factors, in *Osteoporosis* (Marcus, R., Feldman, D., and Kelsey, J., eds.), Academic Press, San Diego, CA, pp. 261–287.
48. Colvin, J., Bohne, B., Harding, G., and Ornitz, D. (1996) Skeletal overgrowth and deafness in mice lacking fibroblast growth factor receptor 3. *Nat. Genet.* **12,** 390–397.
49. McGinnes, W. and Krumlauf, R. (1992) Homeobox genes and axial patterning. *Cell* **68,** 283–302.
50. Hunt, P. and Krumlauf, R. (1992) *Hox* codes and positional specification in vertebrate embryonic axes. *Ann. Rev. Cell Biol.* **8,** 227–256.
51. Eriebacher, A., Filvaroff, E. H., Giteman, S. E., and Derynck, R. (1995) Toward molecular understanding of skeletal development. *Cell* **80,** 371–378.
52. Johnson, R. and Tabin, C. (1997) Molecular models for vertebrate limb development. *J. Biol. Chem.* **273,** 8799–8805.

53. Campbell, J. T. and Kaplan, F. (1992) The role of morphogens in endochondral ossification. *Calcif. Tissue Int.* **50,** 283–289.
54. Lyons, K., Pelton, R., and Hogan, B. (1989) Patterns of expression of murine Vgr-1 and BMP-2a RNA suggests that transforming growth factor-beta-like genes coordinately regulate aspects of embryonic development. *Genes Dev.* **3,** 1657–1668.
55. Lyons, K. M., Pelton, R. W., and Hogan, B. L. (1990) Organogenesis and pattern formation in the mouse: RNA distribution patterns suggest a role for bone morphogenetic protein-2A (BMP-2A). *Development* **109,** 833–844.
56. Macias, D., Ganan, Y., Sampath, K., Piedra, M., Ros, M., and Hurle, J. (1997) Role of BMP-2 and OP-1 (BMP-7) in programmed cell death and skeletogenesis during chick limb development. *Development* **124,** 1109–1117.
57. Ducy, P. and Karsenty, G. (2000) The family of bone morphogenetic proteins. *Kidney Int.* **57,** 2207–2214.
58. Ducy, P., Schinke, T., and Karsenty, G. (2000) The osteoblast: a sophisticated fibroblast under central surveillance. *Science* **289,** 1501–1504.
59. Yamaguchi, A., Komori, T., and Suda, T. (2000) Regulation of osteoblast differentiation mediated by bone morphogenetic proteins, hedgehogs, and cbfa1. *Endocrine Rev.* **21(4),** 393–411.
60. Mundy, G., Garrett, R., Harris, S., et al. (1999) Stimulation of bone formation *in vitro* and in rodents by statins. *Science* **286,** 1946–1949.
61. Schmitt, J. M., Winn, S. R., and Hollinger, J. O. (1998) Cellular and molecular events in bone repair, in *Bioceramics,* Vol. 11 (Sedel, L. and Rey, C., eds.), World Scientific Publishing Co. Ptc. Ltd. NY, NY, pp. 3–8.
62. Thirunavukkarasu, K., Halladay, D., Miles, R., et al. (2000) The osteoblast transcription factor Cbfa1 contributes to the expression of osteoprotegerin, a potent inhibitor of osteoclast differentiation and function. *J. Biol. Chem.* **33,** 25163–25172.
63. Bennett, N. T. and Schultz, G. S. (1993) Growth factors and wound healing. Part I. Biochemical properties of growth factors and their receptors. *Am. J. Surg.* **165,** 728–737.
64. Attisano, L., Wrana, J. L., Lopez-Castillas, F., and Massagué, J. (1994) Transforming growth factor-beta receptors and actions. *Biochim. Biophys. Acta* **1222,** 71–80.
65. ten Dijke, P., Yamashita, H., Sampath, K., et al. (1994) Identification of type I receptors for osteogenic protein-1 and bone morphogenetic protein-4. *J. Biol. Chem.* **269,** 16985–16988.
66. Komori, T. and Ksihimoto, T. (1998) Cbfa1 in bone development. *Curr. Opin. Gen. Dev.* **8,** 494–499.
67. Harada, H., Tagashira, S., Fujiwara, M., et al. (1999) *Cbfa1* isoforms exert functional differences in osteoblast differentiation. *J. Biol. Chem.* **274(11),** 6972–6978.
68. Karsenty, G. (2001) Transcriptional control of osteoblast differentiation. *Endocrinology* **14,** 2731–2733.
69. Reddi, A. H. (1998) Initiation of fracture repair by bone morphogenetic proteins. *Clin. Orthop. Rel. Res.* **355S,** S66–S72.
70. Ducy, P. (2000) Cbfa1: a molecular switch in osteoblast biology. *Dev. Dynam.* **219,** 461–471.
71. Mundlos, S., Otto, F., Mundlos, C., et al. (1997) Mutations involving the transcription factor CBFA1 cause cleidocranial dysplasia. *Cell* **89,** 773–779.
72. Suda, T., Udagawa, N., and Takahashi, N. (1996) Cells of bone: osteoclast generation, in *Principles of Bone Biology* (Bileziken, J. P., Raisz, L. G., and Rodan, G. A., eds.), Academic Press, New York, pp. 87–102.
73. Suda, T., Nakamura, I., Jimi, E., and Takahasi, N. (1997) Regulation of osteoclast formation. *J. Bone Miner. Res.* **12,** 869–879.
74. Yasuda, H., Shima, N., Nakagawa, N., et al. (1998) Osteoclast differentiation factor is a ligand for the osteoprotegerin/osteoclastogenesis-inhibitory factor and is identical to TRANCE/RANKL. *Proc. Natl. Acad. Sci. USA* **95,** 3597–3602.
75. Teitelbaum, S. (2000) Bone resorption by osteoclasts. *Science* **289,** 1504–1508.
76. Bucay, N., Sarosi, I., Dunstan, C., et al. (1998) Osteoprotegerin-deficient mice develop early onset osteoporosis and arterial calcification. *Genes Dev.* **12,** 1260–1268.
77. Simonet, W., Lacey, D., Dunstan, C., et al. (1997) Osteoprotegerin: a novel secreted protein involved in the regulation of bone density. *Cell* **89,** 309–319.
78. Tondravi, M., McKercher, S., Anderson, K., et al. (1997) Osteopetrosis in mice lacking haematopoietic transcription factor PU1. *Nature* **386,** 81–84.
79. Wang, Z., Ovitt, C., Grigoriades, A., Mohle-Steinlein, U., Ruther, U., and Wagner, E. (1992) Bone and haematopoietic defects in mice lacking c-fos. *Nature* **360,** 741–745.
80. Nesbit, S., Nesbit, A., Helfrich, M., and Hortone, M. (1993) Biochemical characterization of human osteoclast integrins. Osteoclasts express alpha v beta, alpha 2 beta, and alpha v beta 1 integrins. *J. Biol. Chem.* **268,** 16737–16745.
81. Enlow, D. H. (1963) *Principles of Bone Remodeling.* Charles C Thomas, IL.
82. Enlow, D. (1966) Comparative study of facial growth in *Homo* and *Macaca. Am. J. Phys. Anthropol.* **24,** 293–310.
83. Geddes, A. (1996) Animal models of bone disease, in *Principles of Bone Biology* (Bilezikian, J., Raisz, L., and Rodan, G., eds.), Academic Press, San Diego, CA, pp. 1343–1354.
84. Parfitt, M. A. (1983) The physiologic and clinical significance of bone histomorphometric data, in *Bone Histomorphometry: Techniques and Interpretation* (Recker, R. R., ed.), CRC Press, Boca Raton, FL, pp. 143–224.

2

Fracture Repair

Charles Sfeir, DDS, PhD, Lawrence Ho, Bruce A. Doll, DDS, PhD, Kodi Azari, MD, and Jeffrey O. Hollinger, DDS, PhD

INTRODUCTION

The economic impact of musculoskeletal conditions in the United States represents $126 billion. Bone fracture repairs are among the most commonly performed orthopedic procedures; about 6.8 million come to medical attention each year in the United States *(1)*. Advances through research and enhanced understanding of fracture repair have enabled orthopedic surgeons to provide patients with many treatment options and improved outcome. In this chapter we will review the current knowledge of fracture from both chronological and molecular biology aspects; we will then address bone healing in elderly patients and the different technologies used to enhance fracture repair.

Bone fracture healing is a very remarkable process because, unlike soft tissue healing, which leads to scar formation, the end result of normal healing is the regeneration of the anatomy of the bone and complete return to function. In general, fracture healing is completed by 6–8 wk after the initial injury. Fracture healing can be divided into two major categories: primary (direct, cortical) bone healing and secondary (indirect, spontaneous) bone healing, with the latter being discussed first because it is more common. Both of these are very complex processes that involve the coordination of a sequence of many biological events. With the recent advances made in molecular biology, the identification of various signaling molecules during specific phases of the healing process has been made possible.

SECONDARY BONE HEALING

Secondary fracture healing is characterized by spontaneous fracture healing in the absence of rigid fixation of the fracture site, and it is the more common method of bone healing as mentioned above. The complete process has been described as having three to five phases *(2–8)*. The biology of bone fracture repair is an organized pattern for repair and perhaps is best elucidated when viewed in histological sections *(2,9)*. Fracture repair can be easily divided into three phases, each characterized by the presence of different cellular features and extracellular matrix components. In temporal order, the events reflect an inflammatory phase; a reparative phase that includes intramembranous ossification; chondrogenesis, and endochondral ossification, and a remodeling phase *(2,10)*. The phases of secondary bone fracture repair are illustrated in **Fig. 1**. It is important to note that these three phases overlap one another and in effect form a continuous healing process.

Inflammatory Phase

An injury that fractures bone damages not only the cells, blood vessels, and bone matrix, but also the surrounding soft tissues, including muscles and nerves *(11)*. Immediately following the injury, an inflammatory response is elicited, which peaks in 48 h and disappears almost completely by 1 wk postfracture. This inflammatory reaction helps to immobilize the fracture in two ways: pain causes the individual

From: *Bone Regeneration and Repair: Biology and Clinical Applications*
Edited by: J. R. Lieberman and G. E. Friedlaender © Humana Press Inc., Totowa, NJ

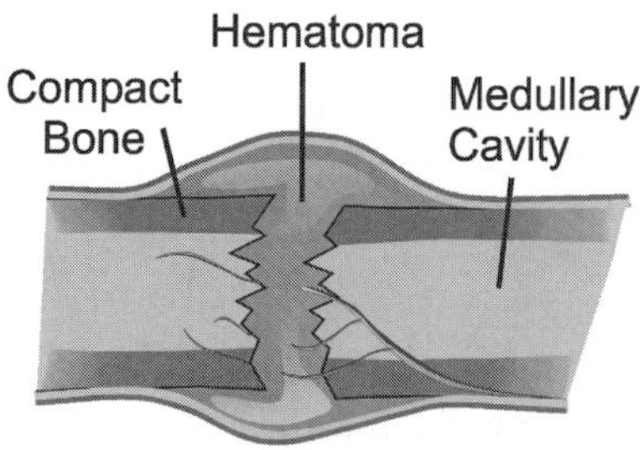

Inflammatory Phase- Vascular endothelial damage results in hematoma formation, complement cascade, and clotting cascade leading to the accumulation of PMNs, lymphocytes, platelets, blood monocytes, macrophages, neutrophils, osteoclasts, and undifferentiated cells. The cells express various genes including TGF-β, FGF-I, FGF-II, PDGF, IGF-I, IGF-II, BMP-2,-4,-7, osteonectin, BMPR I/II, IL-1, IL-6, and GMCSF.

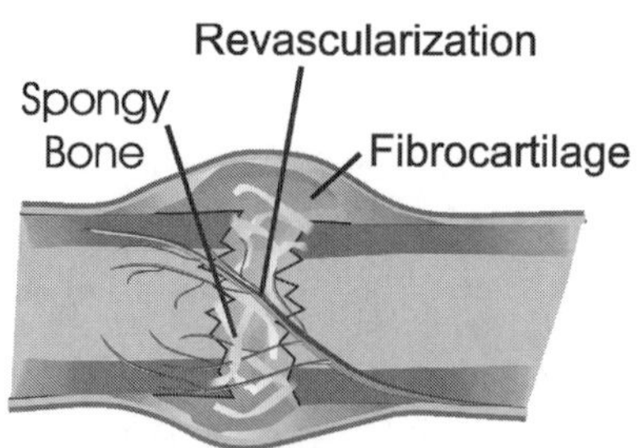

Reparative Phase (Fibrocartilage Callus Formation)- Fibrous tissue and new cartilage are beginning to form and revascularization is taking place. The appearance of macrophages, chondroblasts, chondrocytes, osteoclasts, fibroblasts, and endothelial cells is seen along with the expression of TGF-β, FGF-1, FGF-2, PDGF, IGF-I, IGF-II, BMP-2,-4,-7, osteonectin, fibronectin, BMPR I/II, smads 2,3,4, IL-1, IL-6, and collagens (II,III,IV,V,VI,IX,X).

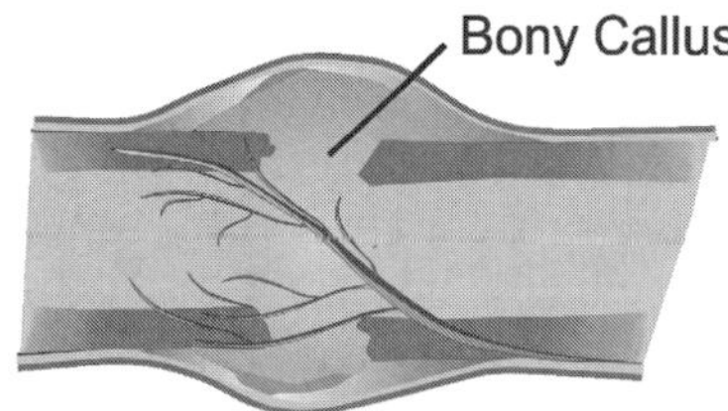

Reparative Phase (Bony Callus Formation)- Intramembranous and endochondral ossification is taking place to lay down new woven bone. The appearance of macrophages, chondroblasts, chondrocytes, osteoblasts, osteoclasts, and endothelial cells is seen along with the express ion of TGF-β, FGF-1, FGF-2, PDGF, IGF-I, IGF-II, BMP-2,-4,-7, osteonectin, osteopontin, osteocalcin, BMPR I/II, smads 2,3,4, IL-1, IL-6, and collagens (I, II, III, IV, V, VI, IX, X).

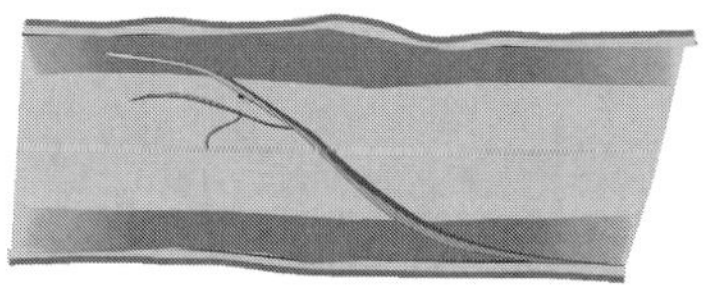

Remodeling Phase- Replacement of woven bone by lamellar bone and the resorption of excess callus. Gradual modification of the fracture region leads to restoration of normal bone architecture.

Fig. 1. Schematic representation of the three stages of fracture repair.

to protect the injury, and swelling hydrostatically keeps the fracture from moving *(3)*. At the injured site, vascular endothelial damage results in the activation of the complement cascade, platelet aggregation, and release of its α-granule contents. This platelet degranulation releases growth factors and triggers chemotactic signals. The conductors of the clotting cascade are the platelets, which have the duty of hemostasis and mediator signaling through the elaboration of chemoattractant growth factors. Polymorphonuclear leukocytes (PMNs), lymphocytes, blood monocytes, and tissue macrophages are attracted to the wound site and are activated to release cytokines that can stimulate angiogenesis *(12)*. The early fracture milieu is characteristically a hypoxic and acidic environment, which is optimal for the activities of PMNs and tissue macrophages *(13)*. The extravasated blood collection will clot. Hematoma accumulates within the medullary canal between the fracture ends and beneath elevated periosteum and muscle. Its formation serves as a hemostatic plug to limit further hemorrhage as well as becoming a fibrin network that provides pathways for cellular migration *(3,11,14,15)*. Recent evidence also suggests that the hematoma serves as a source of signaling molecules that initiate cellular events essential to fracture healing *(10)*. This whole process creates a reparative granuloma and is referred to as an external callus *(10)*.

Reparative Phase

The reparative phase occurs within the first few days, before the inflammatory phase subsides, and lasts for several weeks. The result of this phase will be the development of a reparative callus tissue in and around the fracture site, which will eventually be replaced by bone. The role of the callus is to enhance mechanical stability of the site by supporting it laterally. Osteocytes located at the fracture ends become deficient in nutrients and die, which is observed by the presence of empty lacunae extending for some distance away from the fracture *(5)*. Damaged periosteum and marrow as well as other surrounding soft tissues may also contribute necrotic tissue to the facture site *(3)*. While these tissues are being resorbed, pluripotent mesenchymal cells begin to form other cells such as fibroblasts, chondroblasts, and osteoblasts. These cells may originate in injured tissues, while others migrate to the site with the blood vessels. During this phase, the callus can be comprised of fibrous connective tissue, blood vessels, cartilage, woven bone, and osteoid. As repair progresses, the pH gradually becomes neutral and then slightly alkaline, which is optimal for alkaline phosphatase activity and its role in the mineralization of the callus *(11)*. It has been shown that the earliest bone forms from the cells in the cambium layer of the periosteum *(16)*. The composition of repair tissue and rate of repair may differ depending on where the fracture occurs in bone, the extent of soft tissue damage, and mechanical stability of the fracture site *(11)*. A closer look at the reparative phase focuses on intramembranous ossification, chondrogenesis, and endochondral ossification.

Intramembranous ossification begins within the first few days of fracture, but the proliferative activities appear to stop before 2 wk after the fracture. Histological evidence first shows osteoblast activity in the woven bone opposed to the cortex within a few millimeters from the fracture site *(7)*. Bone formation in this area occurs by the differentiation of osteoblasts directly from precursor cells, without the formation of cartilage as an intermediate step. The region of this type of bone formation occurring in the external callus is often referred to as the hard callus *(10)*.

While intramembranous ossification is taking place, chondrogenesis occurs in the periphery of the callus, where lower oxygen tension is present *(5)*. Mesenchymal or undifferentiated cells from the periosteum and adjacent external soft tissues are also seen in the granulation tissue over the fracture site *(7)*. These cells become larger, start to take on the appearance of cartilage, and begin to synthesize an avascular basophilic matrix much like what is seen in the proliferating zone of the growth plate. This region of fibrous tissue and new cartilage is referred to as the soft callus, and eventually the cartilage will replace all fibrous tissue *(10)*.

By the middle of the second week during fracture healing, there is abundant cartilage overlying the fracture site and calcification begins by the process of endochondral ossification *(7)*. This process is much like the one observed in the growth plate. Hypertrophic chondrocytes first secrete neutral proteoglycanases that degrade glycosaminoglycans, because high levels of glycosaminoglycans are shown to inhibit mineralization *(17)*. Then, these cells and later osteoblasts release membrane-derived vesicles that contain calcium phosphate complexes into the matrix *(18)*. They also carry neutral proteases and alkaline phosphatase enzymes that degrade the proteoglycan-rich matrix and hydrolyze high-energy phosphate esters in order to provide phosphate ions for precipitation with calcium *(11)*. As the mineralization process proceeds, the callus calcifies becoming more rigid and the fracture site is considered internally immobilized *(3)*. Capillaries from adjacent bone invade the calcified cartilage, increasing the oxygen tension. This is followed by invasion of osteoblasts, which form primary spongiosa consisting of both cartilage and woven bone *(10)*. Eventually the callus is composed of just-woven bone, which connects the two fracture ends, and the remodeling process begins.

Remodeling Phase

The remodeling phase is the final phase in fracture healing and begins with the replacement of woven bone by lamellar bone and the resorption of excess callus *(11,13)*. Although this phase represents the normal remodeling activity of bone, it may be accelerated in the fracture site for several years

(19). Remodeling of fracture repair after all woven bone is replaced consists of osteoclastic resorption of poorly located trabeculae and formation of new bone along lines of stress *(20)*. The result of the remodeling phase is a gradual modification of the fracture region under the influence of mechanical loads until optimal stability is achieved, where the bone cortex is typically similar to the architecture it had before the fracture occurred *(3)*.

PRIMARY BONE HEALING

Primary bone healing requires rigid stabilization with or without compression of the bone ends. Unlike secondary bone healing, this rigid stabilization suppresses the formation of a callus in either cancellous or cortical bone *(21–29)*. Because most fractures occurring worldwide either are untreated or are treated in a way that results in some degree of motion (sling or cast immobilization, external or intramedullary fixation), primary healing is rare *(7)*. Although some have considered this type of healing to be a goal of fracture repair, in many ways it is not shown to be advantageous over secondary bone healing *(30,31)*. The intermediate stages are weak, and it does not occur in an anaerobic environment *(3)*. Primary bone healing can be divided further into gap healing and contact healing, both of which are able to achieve bone union without external callus formation and any fibrous tissue or cartilage formation within the fracture gap.

Gap Healing

Gap healing occurs in two stages, starting with initial bone filling and followed by bone remodeling. In the first stage of gap healing, the width of the gap is filled by direct bone formation. An initial scaffold of woven bone is laid down, followed by formation of parallel-fibered and/or lamellar bone as support *(28,29)*. The orientation of the new bone formed in this first stage is transverse to that of the original lamellar bone orientation. There are no connective tissues or fibrocartilage within this gap preceding the production of bone. In the second stage of gap healing, which happens after several weeks, longitudinal haversian remodeling reconstructs the necrotic fracture ends and the newly formed bone such that the fracture site is replaced with osteons of the original orientation *(32)*. The end result of normal gap healing is the return of the bone structure to the way it was before the fracture.

Contact Healing

In contrast to gap healing, contact healing occurs where fragments are in direct apposition and osteons actually are able to grow across the fracture site, parallel to the long axis of the bone, without being preceded by the process of transverse bone formation between fracture ends *(23,26,28,29)*. Under these conditions, osteoclasts on one side of the fracture undergo a tunneling resorptive response, forming cutting cones that cross the fracture line. This resorptive cavity that develops allows the penetration of capillary loops and eventually the establishment of new haversian systems. These blood vessels are then accompanied by endothelial cells and osteoprogenitor cells for osteoblasts leading to the production of osteons across the fracture line *(7)*. The result of normal contact healing will also eventually lead to regeneration of the normal bone architecture.

The biology of bone fracture repair is a very complex process that leads to the regeneration of normal bone architecture. Primary bone healing occurs when there is rigid stabilization of the fracture site and the fracture callus is inhibited. Gap healing and contact healing are both considered to be primary bone healing processes. Secondary bone healing occurs when there is no rigid fixation of the fractured bone ends, which leads to the development of a fracture callus. This process is a little more complicated and consists of an inflammatory phase, a reparative phase, and a remodeling phase. Normal fracture repair is orchestrated through the expression of many different genes, which are turned on and off at very specific times throughout healing. Important gene expression includes TGF-β, FGF, PDGF, IGF, BMP, osteonectin, osteocalcin, osteopontin, fibronectin, BMPR, Smads, IL-1, IL-6, GMCSF, MCSF, and various collagen isotypes. The well-regulated expression of these genes enables the cellular interactions to take place that are responsible for restoring bone morphology and function.

GENE EXPRESSION DURING FRACTURE REPAIR

As described above, the process of fracture repair can be divided into three distinct phases: inflammation, reparative, and remodeling. During these phases, interactions among many different cells via various growth factors, cytokines, receptors, and intermediate signaling molecules take place. With recent advances in molecular biology, the identification and characterization of many of these interactions can now be elucidated. Although several growth factors and extracellular matrix proteins are involved in the repair process, **Table 1** and the following section summarizes the most investigated ones. The temporal and spatial expression of these growth factors and extracellular matrix proteins during different phases of bone repair is described below.

Transforming Growth Factor-β (TGF-β)

TGF-β is produced in the fracture site by platelets, inflammatory cells (monocytes, macrophages), osteoblasts, osteoclasts, and chondrocytes *(10)*. It is extracellularly present in the hematoma (fracture site and periosteum) during the immediate injury response (within 24 h). In the inflammatory phase, the mRNA of TGF-β is weakly expressed in proliferating mesenchymal cells and endothelial cells. It is strongly expressed in proliferating osteoblasts during intramembranous ossification, and strongly expressed in proliferating chondrocytes, not hypertrophic chondrocytes, during the chondrogenesis and endochondral ossification phases *(33)*. It exists first as an inactive precursor peptide that is activated by the acidic conditions of the callus or proteases and becomes the most potent chemoattractant identified for macrophages *(34–37)*. TGF-β also has many other roles, including promoting angiogenesis, which is essential for orderly fracture repair *(10)*; stimulating bone formation by inducing differentiation of periosteal mesenchymal cells into chondroblasts and osteoblasts *(38–40)*; regulating cartilage matrix calcification; and stimulating osteoblast activity and intraosseus wound regeneration *(13, 41,42)*. Other actions include inhibiting osteoblast differentiation and mineralization *(43,44)*, inhibiting osteoclast activity and the formation of osteoclasts *(45)*, and also increasing the production of other bone and cartilage components such as types I, II, III, IV, VI, and X collagen, fibronectin, osteopontin, osteonectin, thrombospondin, proteoglycans, and alkaline phosphatase *(40,46,47)*.

Fibroblast Growth Factors (FGFs)

FGFs are produced by inflammatory cells, osteoblasts, and chondrocytes within the fracture callus. There are two forms of FGF, designated FGF-I and FGF-II. FGF-I is expressed in macrophages and periosteal cells in the inflammatory phase of fracture. It is then expressed in osteoblasts during intramembranous ossification, followed by maximum expression in immature chondrocytes during chondrogenesis. During endochondral ossification, FGF-I is expressed only in osteoblasts. FGF-II has similar expression throughout repair, without any peaks. It is present in macrophages during the inflammatory phase, in osteoblasts during intramembranous ossification, in chondrocytes during chondrogenesis, and in hypertrophic chondrocytes and osteoblasts during endochondral ossification *(10)*. FGFs promote blood vessel formation *(48)*, has autocrine, intracellular functions, and stimulates type 4 collagenase *(10)*. FGF-II also serves as a chemoattractant and mitogen for chondrocytes and regulates differentiation of growth plate chondrocytes *(49,50)*.

Platelet-Derived Growth Factors (PDGFs)

PDGFs are produced by platelets, monocytes, activated tissue macrophages, and endothelial cells in the fracture callus. After being weakly expressed in the inflammatory phase, PDGF expression rises and remains constant throughout repair *(10)*. PDGF has many roles including having receptor tyrosine

Table 1
Gene Expression during Fracture Repair

Gene expression	Function	Temporal and spatial expression
Transforming growth factor-β (TGF-β)	–Most potent chemoattractant for macrophages *(34–37)* –Promotes angiogenesis *(10)* –Induces differentiation of periosteal mesenchymal cells into chondroblasts and osteoblasts *(38–40)* –Regulates cartilage matrix calcification and stimulates osteoblast activity *(13,41,42)* –Increases production of types I, II, III, IV, VI, and X collagen, fibronectin, osteopontin, osteonectin, thrombospondin, proteoglycans, and alkaline phosphatase *(40,46,47)*	–Produced by platelets, inflammatory cells (monocytes, macrophages), osteoblasts, osteoclasts, mesenchymal cells, endothelial cells, and chondrocytes *(10,33)* –Weakly expressed in proliferating mesenchymal cells and endothelial cells in the inflammatory phase, strongly expressed in proliferating osteoblasts during intramembranous ossification, and strongly expressed in proliferating chondrocytes during chondrogenesis and endochondral ossification *(33)*
Fibroblast growth factor-I (FGF-I)	–Promotes blood vessel formation *(48)*, has autocrine, intracellular functions, and stimulates type 4 collagenase *(10)*	Expressed in macrophages and periosteal cells in inflammatory phase, in osteoblasts during intramembranous ossification, maximum expression occurs in immature chondrocytes during chondrogenesis, and it is expressed in osteoblasts during endochondral ossification *(10)*
Fibroblast growth factor-II (FGF-II)	–Promotes blood vessel formation *(48)*, has autocrine, intracellular functions, and stimulates type 4 collagenase *(10)* –A chemoattractant and mitogen for chondrocytes and regulates differentiation of growth plate chondrocytes *(49,50)*	Constant expression throughout repair in macrophages during the inflammatory phase, in osteoblasts during intramembranous ossification, in chondrocytes during chondrogenesis, and in hypertrophic chondrocytes and osteoblasts during endochondral ossification *(10)*
Platelet-derived growth factor (PDGF)	–Has receptor tyrosine kinase activity, stimulates mesenchymal cell proliferation, helps form cartilage and intramembranous bone, and initiates callus formation *(10)* –Potent mitogen for connective tissue cells, stimulates bone cell DNA and protein synthesis, and promotes resorption via prostaglandin synthesis *(51)* –Enables cells to respond to other biologic mediators, increases type I collagen *in vitro*, modulates blood flow *(13,52,53)* –Increases expression of *c-myc* and *c-fos* protooncogenes *(40,54)*	Constant expression in platelets, monocytes, activated tissue macrophages, and endothelial cells in the fracture callus after being weakly expressed in the inflammatory phase *(10)*
Insulin-like growth factor-I (IGF-I)	–Increases collagen synthesis and decreases collagen degradation *(40,62)* –Stimulates clonal expansion of chondrocytes in proliferative zone *(57)*	–In osteoblasts during the intramembranous ossification phase and present in prehypertrophic chondrocytes *(55)* –mRNA peaks at 8 d postfracture *(56)*

Table 1 (Continued)

Gene expression	Function	Temporal and spatial expression
Insulin-like growth factor-I (IGF-I) (*continued*)	–Stimulates replication of preosteo-blastic cells *(51)* –Increases osteoclast formation from mouse osteoclast precursors *(59,60)*	–IGF-I in callus extracts increased at 13 wk after fracture *(58)*
Insulin-like growth factor-II (IGF-II)	–Increases collagen synthesis and decreases collagen degradation *(40,62)* –Increases osteoblast precursor cell proliferation during resorption *(37)* –Promotes cartilage matrix synthesis *(13)* –Modulates osteoclast function leading to bone remodeling *(33)*	–IGF-II mRNA is in fetal rat precarti-laginous condensations, perichondrium, and proliferating chondrocytes *(61)* –IGF-II mRNA is detected in some osteoclasts next to osteoblasts that also expressed IGF-II, whereas most other osteoblasts in bone remodeling were negative for IGF-II *(55)*
Bone morphogenetic proteins (BMP-2, BMP-4, BMP-7)	–BMP-2 increases rat osteoblast IGF-I and IGF-II expression *(69)* –BMP-2 increases TGF-β and IL-6 expression in HOBIT cells *(70)* –BMP-4 stimulates TGF-β expression in monocytes *(71)* –BMP-4 binds to type IV collagen, type I collagen, and heparin *(74)*, and may explain in part the role of vasculogenesis and angiogenesis in fracture healing *(74,75)* –BMP-7 induces expression of Osf2/Cbfa1, a transcription factor associated with early osteoblast differentiation *(76)* –BMP-7 or osteogenic protein-1 (OP-1) *(72)*, increases IGF type 2 receptor expression *(73)*	–Produced by primitive mesenchymal and osteoprogenitor cells, fibroblasts, and proliferating chondrocytes *(66–68)* –Present in newly formed trabecular bone and multinucleated osteoclast-like cells *(68)* –Strongly present in undifferentiated mesenchymal cells during the inflam-matory phase *(33,68)* –Strongly present in the proliferating osteoblasts in intramembranous ossifi-cation *(33,68)* –During chondrogenesis and enochondral ossification, BMP-2 and -4 are in pro-liferating chondrocytes, weakly in mature and hypertrophic chondrocytes, and strongly in osteoblasts near endochondral ossification front, BMP-7 is in proliferating chondrocytes and weakly in mature chondrocytes *(33,68)*
Osteonectin	–Most abundant noncollagenous organic component of bone and serves to bind calcium *(82)* –May regulate tissue morphogenesis *(7)*	–mRNA is found throughout the healing process *(83,84)* –Expression peaks in the soft callus on d 9 and a prolonged peak in expression in the hard callus observed from d 9 to d 15 *(85)* –In d 4–7, the osteonectin signal is found to be strongest in the osteoblastic cells where intramembranous ossification was occurring *(7)* –By d 10, osteonectin signal diminishes, is detected only at the endochondral ossification front, and only weakly in proliferative chondroctyes *(7,84)*
Osteocalcin	–Participates in regulation of hydroxy-apatite crystal growth *(40)*	–Thought to be osteoblast-specific *(7)* –Osteocalcin was not detected in the soft callus but was in the hard callus, and initiation of osteocalcin occurred between d 9 and d 11, with peak expression at about d 15 *(85)*

(continued)

Table 1 (Continued)

Gene expression	Function	Temporal and spatial expression
Osteopontin	–Interacts with CD-44, which is a cell-surface glycoprotein that binds hyaluronic acid, type I collagen, and fibronectin *(88)* –Mediates cell–cell interaction in bone repair and remodeling *(7)* –Helps anchor osteoclasts to bone through vitronectin receptors *(91)*	Detected in osteocytes and osteoprogenitor cells in the subperiosteal hard callus, and by d 7 after fracture it is found in the junction between the hard and soft callus *(7,89,90)*
Fibronectin	–Helps in adhesion and cell migration *(7)* –Plays an important role in the establishment of provisional fibers in cartilaginous matrices *(7)*	–Produced by fibroblasts, osteoblasts, and chondrocytes and is detected in the hematoma within the first 3 d after fracture and in the fibrous portions of the provisional matrices *(7)* –Low levels of fibronectin mRNA in intact bone and marked expression in the soft callus within 3 d after fracture that reaches peak level at d 14 *(92)*
Bone morphogenetic protein receptors (BMPR-I, -II)	–Findings suggest an association of the receptors with the differentiation of mesenchymal cells into chondroblastic and osteoblastic lineages *(33)*	–Strongly present in undifferentiated mesenchymal cells during the inflammatory phase, in proliferating osteoblasts during intramembranous ossification, and are found in proliferating chondrocytes, weakly in mature and hypertrophic chondrocytes, and strongly in osteoblasts near the endochondral ossification front during chondrogenesis and endochondral ossification *(33,93)*
smads (2, 3, 4)	–Components of the intracellular signaling cascade that starts with BMPs *(94,95)* –smad 2 and smad 3 help to mediate TGF-β signaling *(94)* –smad 4 forms a heterodimeric complex with other pathway restricted smads and translocates into the nucleus to modulate important BMP response genes *(96)*	–In the inflammatory phase, the mRNA for smads 2, 3, 4 are not expressed, and in chondrogenesis and endochondral ossification, the mRNA for smads 2, 3, 4 are upregulated and the smad 2 protein is present in chondroblasts and chondrocytes *(33)*
Interleukin-1 (IL-1)	–Induces the secretion of IL-6, GMCSF, and MCSF *(98)* –May stimulate activities of neutral proteases to selectively degrade callus tissue *(17,99)* –May increase fibroblastic collagen synthesis, collagen cross-linking, and stimulate angiogenesis *(98,100–103)*	–Produced by macrophages and is expressed at low constitutive levels throughout fracture healing but can be induced to high activities in the early inflammatory phase (d 3) *(97)*
Interleukin-6 (IL-6)	–Very sensitive to IL-1 stimulation *(106)* –May be a stimulator of bone resorption *(107–109)*	–Produced by osteoblasts during fracture repair *(104,105)* –Shows a high constitutive activity early in the healing process *(97)*
Granulocyte-macrophage colony-stimulating factor (GMCSF)	–May stimulate formation of osteoclasts, increase the proliferation of T-lymphocytes, and stimulate cytokine secretion *(102,111–114)*	–Produced by T-lymphocytes during the fracture healing process and is expressed at early time points after fracture *(97)*

Table 1 (Continued)

Gene expression	Function	Temporal and spatial expression
Granulocyte-macrophage colony-stimulating factor (GMCSF) *(continued)*	–Associated with increased fibroblast migration and collagen synthesis *(115–117)* –Associated with the proliferation and differentiation of granulocytic and monocyte/macrophage lineages *(118)* –May suppress the expression of receptors for other cytokines in different cell types *(97,111)*	–May be produced from osteoblasts *(102,111–114)*
Macrophage colony-stimulating factor (MCSF)	–An important growth factor for development of macrophage colonies by hematopoietic tissues *(121)*	–Lack of expression in the fracture callus may be due to complex interactions between immune, hematopoietic and musculoskeletal systems not yet understood *(97)* –Constitutive secretion by osteoblast-like cells in culture is observed *(119,120)*
Collagens (types I, II, III, IV, V, VI, IX, X, XI)	–Type I collagen aids in developing cross-linkages which produce collagen fibrils that mature to collagen fibers, creating regions allowing for the deposition and growth of hydroxy-apatite crystals *(13)* –Aberrations in type III collagen production may lead to delayed union or nonunion *(124)* –Type IV (and types I and X) may aid in converting mesenchymal lineage cells into osteoblasts *(128)* –Type V and XI may regulate the growth and orientation of type I and type II collagen in cartilaginous and noncartilagenous tissues *(129,130)* –Type V collagen has been associated with blood vessels in granulation tissue *(124)* –Type IX may mediate interactions between collagen fibrils and proteo-glycans in cartilage *(40,132)* –Type X collagen may play a role in the mineralization of cartilage *(40)*	–Type I is associated with bone, type II with cartilage, types III and V with granulation tissue, types IV and VI with the endothelial matrix, and type X with hypertrophic cartilage *(123)* –Mechanically stable fractures have predominately type I collagen along with types II and V *(124)* –Mechanically unstable fractures are characterized by initial production of types III and V collagen which is replaced by types II and IX collagen and very little type I collagen *(122)* –Type II collagen mRNA is detectable as early as d 5 postfracture in cells that have chondrocytic phenotype, has a peak expression approximately 9 d after fracture in the mouse and rat, and by d 14 after fracture the expression of mRNA for type II chain becomes absent *(7,85,125,126)* –Type III collagen mRNA increases rapidly during the first week of fracture healing *(127)* –Type V collagen is expressed through out healing process with the highest accumulation of type V collagen in the subperiosteal callus *(89)* –Expression of type IX collagen and aggrecan coincides with expression of type II collagen *(40,132)* –Expression of type X collagen occurs later than that of other cartilage specific genes *(40)*

kinase activity, stimulating mesenchymal cell proliferation, initiating fracture repair, helping to form cartilage and intramembranous bone, and initiating callus formation *(10)*. They are released from the α-granules of platelets and become potent mitogens for connective tissue cells, stimulate bone cell DNA and protein synthesis, and promote resorption via prostaglandin synthesis *(51)*. PDGF also serves as a competence factor that enables cells to respond to other biological mediators; increase type I collagen *in vitro*; modulate blood flow, which has a positive impact on wound healing *(13,52,53)*; and are shown to increase expression of *c-myc* and *c-fos* protooncogenes, which encode nuclear proteins involved in regulating cell proliferation, growth, and differentiation *(40,54)*.

Insulin-Like Growth Factors (IGFs)

IGFs are also often referred to as somatomedins or sulfation factors. IGF expression is high in cells of the developing periosteum and growth plate, healing fracture callus tissue, and developing ectopic bone tissue induced by DBM *(40,47,55,56)*. IGFs produced by bone cells not only act as autocrine and paracrine regulators, but also become incorporated into bone matrix and are later released during resorption, which increases osteoblast precursor cell proliferation *(37)*. IGFs may also become secreted by chondrocytes and respond in an autocrine manner to promote cartilage matrix synthesis *(13)*. However, IGFs may not only contribute to bone formation, they may modulate osteoclast function, leading to bone remodeling during fracture repair *(33)*.

IGF-I mRNA is not expressed in the inflammatory phase of repair. However, mRNA expression is seen in osteoblasts during the intramembranous ossification phase and are also present in prehypertrophic chondrocytes *(55)*. Actually, the level of mRNA peaks at 8 d postfracture *(56)*. IGF-I may stimulate clonal expansion of chondrocytes in proliferative zone through an autocrine mechanism, much like in the chondrogenesis stage of fracture repair *(57)*. IGF-I also stimulates replication of preosteoblastic cells and induces collagen production by differentiated osteoblasts *(51)*. It should be noted that IGF-I in callus extracts increased at 13 wk after fracture *(58)*, and has been shown to increase osteoclast formation from mouse osteoclast precursors, which suggests some involvement during remodeling *(59,60)*. In addition, IGF-II mRNA is observed in fetal rat precartilaginous condensations, perichondrium, and proliferating chondrocytes *(61)*. IGF-II mRNA is detected in some osteoclasts in the fracture healing model next to osteoblasts that also expressed IGF-II, whereas most other osteoblasts in bone remodeling were negative for IGF-II *(55)*. IGF-I and IGF-II have been observed to increase collagen synthesis and decrease collagen degradation *(40,62)*.

Bone Morphogenetic Proteins (BMPs)

BMPs are members of the TGF-β superfamily and were discovered as the noncollagenous and water-soluble substances in bone matrix that have osteoinductive activity *(63–65)*. In general, recent studies reveal increased presentation of BMP-2, -4, and -7 in the primitive mesenchymal and osteoprogenitor cells, fibroblasts, and proliferating chondrocytes present at the fracture site *(66–68)*. In a rat model, mesenchymal cells that had migrated into the fracture gap and had begun to proliferate showed increased statement of BMP-2 and -4 *(66)*. In a similar rat fracture healing model, it was confirmed that BMP-2, -4, and -7 were present in newly formed trabecular bone and multinucleated osteoclast-like cells *(68)*. More specifically, when the expression is broken down into the phases of healing, BMP-2, -4, and -7 are strongly present in undifferentiated mesenchymal cells during the inflammatory phase. During intramembranous ossification, these BMPs are strongly present in the proliferating osteoblasts. In chondrogenesis and endochondral ossification, BMP-2 and -4 are found in proliferating chondrocytes, weakly in mature and hypertrophic chondrocytes, and strongly in osteoblasts near endochondral ossification front. In these later stages of healing, BMP-7 is found in proliferating chondrocytes and weakly in mature chondrocytes *(33,68)*.

BMPs affect expression of other growth factors that may function to mediate the effects of BMPs on bone formation *(37)*. BMP-2 increased rat osteoblast IGF-I and IGF-II expression *(69)*, and increased

TGF-β and IL-6 expression in HOBIT cells *(70)*. BMP-4 stimulated TGF-β expression in monocytes *(71)*. BMP-7 or osteogenic protein-1 (OP-1) *(72)* is shown to increase IGF type 2 receptor expression *(73)*.

BMPs also have other roles in fracture repair. BMP-4 binds to type IV collagen, type I collagen, and heparin *(74)*. The interaction of BMP-4 with type IV collagen and heparin may explain in part the role of vasculogenesis and angiogenesis in bone development such as in fracture healing *(74,75)*. BMP-7 also stimulates normal human osteoblast proliferation by inducing expression of Osf2/Cbfa1, a transcription factor associated with early osteoblast differentiation *(76)*. It should be noted that although they were identified and named because of their osteoinductive activity *(77,78)*, the BMPs play many diverse roles during embryonic and postembryonic development as signaling molecules in a wide range of tissues *(79,80)*. In conclusion, a number of findings suggest that BMP-2, -4, and -7 work to promote fracture healing and bone regeneration *(81)*.

Osteonectin is one of many extracellular matrix proteins involved with bone repair and regeneration. In fact, osteonectin is the most abundant noncollagenous organic component of bone and serves to bind calcium *(82)*. Osteonectin mRNA is found throughout the healing process *(83,84)*. Its expression peaks in the soft callus on d 9, and a prolonged peak in expression in the hard callus is observed from d 9 to d 15 *(85)*. During d 4–7, the osteonectin signal is found to be strongest in osteoblastic cells where intramembranous ossification was occurring *(7)*. By d 10, this signal diminished and the signal was detected only at the endochondral ossification front. No osteonectin was detected in hypertrophic chondrocytes and only weakly in proliferative chondroctyes *(7,84)*. Incidentally, type I and V collagen followed similar expression patterns, which suggests that osteonectin may regulate tissue morphogenesis *(7)*.

Osteocalcin, an osteoblast-specific protein, contains three γ-carboxyglutamic acid residues, which provide it with calcium-binding properties. Osteocalcin has been suggested to participate in regulation of hydroxyapatite crystal growth *(40)*, and may possess other functions, as it is also expressed in human fetal tissues *(86)*. In one study, osteocalcin was not detected in the soft callus but was detected in the hard callus. Initiation of osteocalcin occurred between d 9 and d 11, and peak expression was at about d 15 *(85)*. Osteocalcin levels in plasma depend on the formation of new bone, and the concentration may be an indicator of the activity of osteoblasts *(87)*.

Osteopontin, an extracellular matrix protein known to be important in cellular attachment, interacts with CD-44, which is a cell-surface glycoprotein that binds hyaluronic acid, type I collagen, and fibronectin *(88)*. *In situ* studies have shown that this protein is detected in osteocytes and osteoprogenitor cells in subperiosteal hard callus; however, little is seen in cuboid osteoblasts and by d 7 after fracture. Osteopontin is found in the junction between the hard and soft callus *(7,89,90)*. The coexistence of CD-44 and osteopontin in osteocytes and osteoclasts implies the presence of an osteopontin/CD-44 mediated cell–cell interaction in bone repair *(7)*. Another theory suggests that osteopontin helps anchor osteoclasts to bone through vitronectin receptors, helping in the resorption process *(91)*.

Fibronectin is a protein that helps in adhesion and cell migration, making it important in the repair process. In the fracture callus, this protein is produced by fibroblasts, osteoblasts, and chondrocytes. It is detected in the hematoma within the first 3 d after fracture and in the fibrous portions of the provisional matrices and less in the cartilage matrix *(7)*. There was no evidence of this protein in the periosteum, in osteoblasts, or osteocytes of periosteal woven bone using *in situ* hybridization. Northern hybridization showed low levels of fibronectin mRNA in intact bone and marked expression in the soft callus within 3 d after fracture, reaching a peak level at d 14 *(92)*. Because fibronectin production appears to be greatest in the earlier stages of repair, it is thought that it plays an important role in the establishment of provisional fibers in cartilaginous matrices *(7)*.

Bone Morphogenetic Protein Receptors (BMPRs)

The receptors for BMPs are strongly present in undifferentiated mesenchymal cells during the inflammatory phase. Then, they are strongly present in proliferating osteoblasts of intramembranous ossification. BMPR I/II are found in proliferating chondrocytes, weakly in mature and hypertrophic chondrocytes, and strongly in osteoblasts near the endochondral ossification front during chondrogenesis and endochondral ossification *(93)*. The association of these receptors with the differentiation of mesenchymal cells into chondroblastic and osteoblastic lineages has been suggested *(33)*.

Smads are essential components of the complex intracellular signaling cascade that starts with BMPs *(94,95)*. During the inflammatory phase, the mRNA for smads 2, 3, 4 are not expressed, and smad 2 protein is not present. During the intramembranous ossification phase, smad 2 is still not present yet. In chondrogenesis and endochondral ossification, the mRNA for smads 2, 3, 4 are upregulated and the smad 2 protein is present in chondroblasts and chondrocytes *(33)*. Smad 2 and smad 3 help to mediate TGF-β signaling *(94)*. Smad 4 forms a heterodimeric complex with other pathway-restricted smads and translocates into the nucleus in order to modulate important BMP response genes *(96)*.

Interleukin-1 (IL-1)

IL-1 is an important cytokine produced by macrophages and is expressed at low constitutive levels throughout fracture healing but can be induced to high activities in the early inflammatory phase (d 3) *(97)*. It induces the secretion of IL-6, GMCSF, and MCSF, which means that the early expression of IL-1 may indicate a triggering mechanism that initiates a cascade of events that regulate repair and remodeling *(98)*. IL-1 may stimulate activities of neutral proteases to selectively degrade callus tissue *(17,99)*. The action of macrophages, which include increasing fibroblastic collagen synthesis, increasing collagen crosslinking, stimulating angiogenesis, and improving wound breaking strength, may also be attributed to IL-1 production *(98,100–103)*.

Interleukin-6 (IL-6) is an important cytokine that is produced by osteoblasts during fracture repair *(104,105)*. It is very sensitive to IL-1 stimulation *(106)*, and shows a high constitutive activity early in the healing process *(97)*. Several lines of evidence suggest that it is a stimulator of bone resorption *(107–109)*.

Granulocyte-Macrophage Colony-Stimulating Factor (GMCSF)

T-lymphocytes have been identified morphologically in fracture calluses and may be a part of the healing process *(110)*. GMCSF is produced by T-lymphocytes during the fracture healing process and is expressed at early time points after fracture but then gradually declines *(97)*. It is also suggested that GMCSF may be produced from osteoblasts to stimulate formation of osteoclasts, increases the proliferation of T-lymphocytes, and stimulates cytokine secretion *(102,111–114)*. This cytokine activity has been associated with increased fibroblast migration and collagen synthesis *(115–117)*, and the proliferation and differentiation of granulocytic and monocyte/macrophage lineages *(118)*. GMCSF may also suppress the expression of receptors for other cytokines in different cell types *(97,111)*.

Macrophage Colony-Stimulating Factor (MCSF) was not detected in the fracture callus according to one study *(97)*; however, constitutive secretion by osteoblast-like cells in culture is observed *(119, 120)*. It has been shown to be an important growth factor for development of macrophage colonies by hematopoietic tissues *(121)*. The lack of expression in the fracture callus may be due to the complex interactions among immune, hematopoietic, and musculoskeletal systems as a result of injury, which are not yet understood *(97)*.

Collagens

The overall quantity and type of collagen influences callus formation and fracture healing and the expression of these extracellular matrix proteins has also been documented *(122)*. There are at least 18 isotypes of collagens: type I is associated with bone, type II with cartilage, types III and V with granulation tissue, types IV and VI with the endothelial matrix, and type X with hypertrophic cartilage *(123)*. Mechanically stable fractures have predominately type I collagen, along with types II and V *(124)*. Mechanically unstable fractures are characterized by initial production of types III and V collagen, which is replaced by types II and IX collagen and very little type I collagen *(122)*.

Type I collagen, which is the main collagen type in bone, aids in developing cross-linkages. These linkages produce collagen fibrils that mature to collagen fibers, creating regions allowing for the deposition and growth of hydroxyapatite crystals about 10 d postfracture *(13)*. Type II collagen is a major structural protein of cartilage and has a peak expression approx 9 d after fracture in the mouse and rat. Pro-α-2 collagen mRNA is seen in the proliferative chondrocytes. By d 14 after fracture, expression of mRNA for type II collagen becomes absent. Almost all chondrocytes are hypertrophied, and there is no expression of type 2 procollagen chain. Type II mRNA is detectable as early as d 5 postfracture *(7,85,125,126)*. Type III collagen mRNA increases rapidly during the first week of fracture healing *(127)*, particularly in bone, and aberrations in its production may lead to delayed union or nonunion *(124)*. Type IV (and types I and X) may aid in converting mesenchymal lineage cells into osteoblasts *(128)*. Types V and XI have a closely related structures it has been suggested that they regulate the growth and orientation of type I and type II collagen in cartilaginous and noncartilagenous tissues *(129,130)*. Type V collagen is expressed in both soft and hard callus throughout the healing process. The highest accumulation of type V collagen was detected in the subperiosteal callus, where intramembranous ossification was taking place *(89)*. Type V collagen has also been associated with blood vessels in granulation tissue *(124)*. Type XI collagen is found in cartilage and is a minor component of collagen fibrils, but expression of this collagen type is not restricted to cartilage *(40,131)*. The expression of type IX collagen and aggrecan coincides with expression of type II. Type IX collagen is seen in cartilage and may mediate interactions between collagen fibrils and proteoglycans *(40,132)*. The expression of type X collagen, a marker for hypertrophic chondrocytes during endochondral ossification, occurs later than that of other cartilage-specific genes and may play a role in the mineralization of cartilage *(40)*.

As our understanding of bone repair at a molecular level increases, we will be able to engineer comprehensive bone regenerative therapies. This knowledge will guide us to better design delivery systems that are biology driven; for example, if multiple growth factors are being delivered to a fractured bone site, one might imagine that different growth factors could be released at different times to optimize the healing cascade. Another area of research that will also influence our therapy design is the bone healing related to age; research indicates that bone repair is different between young and elderly patients. This topic is discussed in the following section.

FRACTURE HEALING IN THE ELDERLY

It has been established that bone formation during bone remodelling and fracture healing in the elderly patient appears to be reduced. Causes include a reduced number of recruited osteoblast precursors, a decline in proliferative activity of osteogenic precursor cells, and a reduced maturation of osteoblast precursors. Advanced age-related changes occur in the bone mineral, bone matrix *(133)*, and osteogenic cells *(134,135)*. Common clinical experience indicates that fractures heal faster in children than in adults *(136)*. Mechanisms causing these alterations are unclear. The observations have been attributed to slow wound healing, reflecting a general functional decline in the homeostatic mechanisms during aging and senescence. Furthermore, differences in fracture healing in the elderly population can be caused by local or systemic changes in hormonal and growth factor secretion and altered receptor levels, or changes in the extracellular matrix composition.

Several publications deal with the delicate relationship between bone resorption and bone formation and its imbalances, leading to osteopenia and osteoporosis. Presently, less information is obtainable as to similarities and changes in the process of fracture healing in the elderly patient in comparison to the physiological process of bone healing in children and young adults. In addition, the data obtained in animal fracture healing models (rat, rabbit) are difficult to transfer to the human physiological fracture repair process in the elderly patient.

General cellular and biochemical processes of fracture repair in the elderly, healthy (nonosteoporotic) patient receive less focus. Demographic changes and with an overaging population, steadily increasing fracture numbers in the elderly population will mandate more emphasis as a means to enhance the process.

In vitro evidence of age-related changes in cell behavior indicate a reduced proliferative capacity. Christiansen et al. have demonstrated that serially passaged cultures of human trabecular osteoblasts exhibit limited proliferative activity and undergo cellular aging. They reported a number of changes during serial passaging of human trabecular osteoblasts, which include alterations in morphology and cytoskeleton organization; an increase in cell size and higher levels of senescence-associated β-galactosidase activity. They studied changes of topoisomerase I levels during cellular aging of human trabecular osteoblasts. They reported an age-related progressive and significant decline in steady-state mRNA levels of this gene in human bone cells undergoing cellular aging *in vitro (137)*. Taken together, these observations facilitate a further understanding of reduced osteoblast functions during cellular aging. These results concur with previous former findings of a correlation between donor age and the impairment of osteoblastic functions such as production of Col I, OC, and other extracellular matrix components in *in vitro* culture of human mature osteoblasts *(138–140)*.

Martinez et al. examined the cell proliferation rate and the secretion of C-terminal type I procollagen and alkaline phosphatase (ALP). They noted a lower proliferation rate and osteocalcin secretion in osteoblastic cells from the older donors than in those from younger subjects. They also found significant differences of these parameters in relation to the skeletal site of origin *(141)*. Theoretical basis of these experiments and their importance for the understanding of the process of bone aging and bone healing in the elderly patient is the consideration as a useful tool for evaluating osteoblastic alterations associated with bone pathology and aging *(142)*. Other groups have shown that human bone-derived cells show a dramatic decrease in their proliferative capacity with donor age. Studying the gender and age-related changes in iliac crest cortical bone and serum osteocalcin in humans subjects, Vanderschueren et al. *(143)* also detected a significant age-related decline of bone and serum osteocalcin content with age in vivo. Furthermore, a parallel decrease in age-matched groups revealed a generally higher concentration of bone and serum osteocalcin in men.

With advancing age, the membrane-like arrangement of the osteogenic cells in the periosteum is lost, leaving a reduced number of precursor cells to draw from *(134)*. These electron microscopy-based results were confirmed by an organ culture model investigating the relationship between chondrogenic potential of periosteum and aging. In this model, periosteal explants from the medial tibiae of rabbits (age range between 2 wk and 2 yr) were cultured in agarose suspension conditions conductive for chondrogenesis. A significant decline of chondrogenic potential of periosteum with increasing age was apparent. Furthermore, a significant decrease of proliferative activity was found by ^{3}H-thymidin incorporation *(144)*.

Enhancing Fracture Healing

The goal is to accelerate or to assure the healing of a fracture, which is likely not able to heal without invasive or noninvasive intervention. Several methods could be used to enhance bone fracture healing. The approaches could be biological or mechanical and biophysical enhancement *(145–147)*. In this section we will focus on the biological approaches.

The local methods for fracture enhancements involve the use of biological bone grafts, synthetic grafts, and delivery of growth factors. The autologous cancellous bone graft is considered the gold standard and has been extensively used in orthopedics. This type of grafting material will provide some living bone-producing cells, inductive growth factors, and hydroxyapatite mineral. The disadvantages are morbidity at the donor site, scarring and risk of infection, and most often the graft volume needed is greater than what is available. Thus, the need for alternative graft material has been sought, but none yet provide all the qualities of autologous cancellous bone. Different categories of grafting materials are available and are summarized in **Table 2**.

Table 2
Alternative Grafts Used to Enhance Fracture Healing

Absorbable	Nonabsorbable
Natural	Synthetic polymers
• Allogeneic bone	• Polytetrafluoroethlene
• Collagen	• Synthetic composite
• Collagen-GAG	• Bioactive glasses
• Fibrin	• Calcium-based ceramic grafts
• Hyaluranic acid	Hydroxyapatite
Natural mineral	Composite
• Hydroxyapatite	• Calcium-collagen composite
• Xenogeneic derivatives (anorganic bone)	
Synthetic	
• Polylactic acid	
• Polyglycolic acid	
• Tri-calcium phosphate	
• Calcium sulfate	
Cellular grafts	
• Autogenous bone marrow grafts	
• Autogenous bone grafts	

In addition to grafts, bone marrow has been shown to contain a population of mesenchymal stem cells that are capable of differentiating into osteoblasts and form bone as well as other connective tissues. Connolly et al. reported that injectable bone marrow cells could stimulate osteogenic repair. They developed techniques for clinical application by harvesting autologous bone marrow, centrifuging, and concentrating the osteogenic marrow prior to implantation. Garg et al. *(148)* also reported the successful use of autogenous bone marrow as an osteogenic graft. Seventeen of the 20 ununited long bone fractures healed according to clinical and radiographic criteria.

Extensive research has been carried out and in progress aimed at isolating, purifying and expanding marrow-derived mesenchymal cells *(149–152)*. Once these cells are isolated, they may be expanded (not differentiated) in a specialized medium and ultimately yield a source of cells that are highly osteogenic. These cells could then be delivered to enhance bone repair *(150,153,154)*.

Other attempts to enhance bone healing are the use of osteoinductive factors such as recombinant growth factors. This osteoinductive therapy induces mitogenesis of undifferentiated perivascular mesenchymal cells and leads to the formation of osteoprogenitor cells with the capacity to form bone. Several growth factors are potentially beneficial for bone and cartilage healing, such as TGF-β, fibroblast growth factor (FGF), platelet-derived growth factor (PDGF), and the BMPs. Since these factors have been shown to be produced during fracture repair and to participate in the regulation of the healing process, it was logical to administer some of these factors exogenously at the site of injury. Extensive research has been carried to enhance bone healing in different animal models; we summarize these advances in **Table 3**.

Although there is increasing evidence supporting the use of growth factors to enhance fracture healing, the clinical data have been hindered by the selection of optimal carrier and dosage. Only three peer-reviewed clinical studies using rhBMP have been published *(183–185)*, and BMP doses suggesting efficacy ranged from 1.7 to 3.4 mg. These results mute clinical enthusiasm. To overcome difficulties using growth factors, alternatives have been investigated. Such alternatives are gene therapy for fracture healing.

Table 3
Growth Factors and Delivery Systems Used in Different Animal Models to Enhance Bone Healing

Growth	Carrier	Animal	Tissue regenerated	References
TGF-β1	Gelatin	Rabbit	Skull bone	*(155)*
	PLGA	Rat	Skull bone	*(156)*
	Collagen	Mouse	Dermis	*(157)*
FGF-1	Demineralized bone matrix (DBM)	Rabbit	Long bone	*(158)*
FGF-2	Alginate	Mouse	Angiogenesis	*(159)*
FGF-2	Agarose/heparin	Mouse, pig	Angiogenesis	*(160,161)*
FGF-2	Gelatin	Mouse	Angiogenesis	*(162)*
FGF-2	Gelatin	Rabbit, monkey	Skull bone	*(162,163)*
FGF-2	Fibrin gel	Rat	Long bone	*(164)*
FGF-2	Collagen minipellet	Rabbit	Long bone	*(165)*
FGF-2	Collagen	Mouse	Cartilage	*(166)*
RhBMP2	PLA	Dog	Maxilla	*(167)*
BMP	PLA	Dog	Long bone	*(168)*
rhBMP2	PLA (porous)	Dog	Vertebrae	*(169)*
rhBMP2	PLA-coating gelatin sponge	Dog	Long bone, maxilla	*(170)*
rhBMP7	Collagen	Dog	Vertebrae	*(171)*
rhBMP7	Collagen	Monkey	Long bone	*(171)*
rhBMP2	Porous HA	Monkey	Skull	*(172)*
rhBMP2	PLA/PGA	Rabbit	Long bone	*(173)*
rhBMP2	Porous HA	Rabbit	Skull	*(174)*
rhBMP2	PLA	Rabbit	Long bone	*(175)*
rhBMP2	Injection into intervertebral disk	Rabbit	Vertebrae	*(176)*
rhBMP2	Gelatin	Rabbit	Skull	*(177)*
rhBMP2	PLGA	Rat	Long bone	*(178)*
rhBMP2	PLA	Rat	Skull bone	*(179)*
rhBMP2	Collagen sponge	Rat	Skull	*(173)*
rhBMP2	PLA-PEG copolymer	Rat	Long bone	*(180)*
rhBMP2	Inactive bone matrix	Sheep	Long bone	*(181)*
rhBMP2	PLGA	Sheep	Long bone	*(182)*

Fracture Enhancement via Gene Therapy

Gene-based delivery systems offer the potential to deliver and produce proteins locally at therapeutic levels and in a sustained fashion within the fracture site. To transfer genes into a cell, two main choices have to be made. The first is to determine the gene delivery vehicle, known as the vector. The second is to determine if the genes should be introduced into the cell in vivo or ex vivo.

To introduce exogenous DNA into the cell and more specifically into the nucleus where the transcriptional machinery resides, vectors must be used. These vectors could be viral or nonviral. Each system has its advantages and disadvantages. Naked DNA delivery is usually achieved by direct local injection; more recently, combining the DNA with cationic liposomes or other transfecting agents or a biodegradable polymer improved the transfection efficiency. Although transfection efficiency in general was lower than with viral vectors, gene expression from delivered plasmid DNA was sufficient to promote osteogenesis *(186,187)* and angiogenesis *(188–190)*. The main advantages of plasmid DNA are cost, safety, transient expression, and less antigenicity than viral vectors.

Viral vectors have been developed from various viruses. The most widely used viruses are derived from retroviruses, adenoviruses, adeno-associated, and herpes simplex viruses. **Table 4** summarizes the clinical research conducted so far in orthopedics using these various viruses.

With continuing advances in gene technology, gene therapy will likely become increasingly important in healing both acute and chronic wounds. As our understanding of the physiology of bone fracture

Table 4
Summary of Gene Therapy to Bone

Virus type/gene delivered	Tissue targeted	References
Retroviral		
• lacZ marker gene, hBMP-7	Periosteal cells/rabbit femoral osteochondral defects	*(191)*
• Collagen alpha 1	*In vitro* expression in bone marrow stromal cells	*(192)*
• LacZ marker gene	Human osteoprogenitors bone marrow fibroblast were transduced with retrovirus-LacZ and implanted in calvariae of SCID mouse	*(193)*
• BMP-2 and BMP-4	Ectopical expression in developing chick limbs	*(194)*
Adenoviruses		
• LacZ	Rabbit femur (diaphysis)	*(195)*
• BMP-2	Rabbit femur	*(196)*
• FGF		
• BMP-7	Adeno-CMV-BMP-7 virus particles mixed with bovine bone-derived collagen carrier and was implanted into mouse muscle and dermal pouches	*(197)*
• BMP-7	*Ex vivo* transduction of human gingival fibroblasts or rat dermal fibroblasts. The transduced cells were then implanted in critical size skeletal defects in rat calvariae	*(198)*
• LacZ	Rat mandibular osteotomy model,	*(199)*
• BMP-9	Injection of 7.5×10^8 pfu of a BMP-9 adenoviral vector in the lumbar paraspinal musculature.	*(200)*
• Human TGF-β1	Rabbit lumbar intervertebral disks	*(201)*
• BMP-2	Athymic nude rats were injected with Ad-BMP-2 in the thigh musculature	*(202,203)*
• LacZ	Direct injection into the temporomandibular joints of Hartley guinea pigs	*(204)*
• BMP-2	Intramuscular direct injection	*(205)*
Adeno-associated viruses (AAVs)		
• Murine IL-4	Synovial tissues	*(206)*

• To the best of our knowledge, no AAV vectors have been used to enhance bone fracture repair. The difficulty in preparing and purifying this viral vector in large quantities remains a major obstacle for evaluating AAV vectors in clinical trials. Recently, methods for producing a high titer *(207)* and purification *(208)* were published. These advances will allow further studies using AAV vectors.

Herpex simplex virus type 1 (HSV-1)

• Has not been used in bone fracture healing models. The HSV-1 amplicon vector is a very promising genetic vehicle for *in vivo* gene delivery. The HSV-1 amplicon vectors consists of a plasmid containing a transgene(s) and the HSV-1 origin of DNA replication and packaging sequence, packaged in a HSV-1 virion free of HSV-1 helper virus.

repair and the role of the various repair regulators at the molecular level increases, this will ultimately accelerate the progress of gene therapy. In addition, the transfection efficiency and the safety of the delivery systems is expected to improve, providing a therapy with fewer hurdles to overcome in order to become an accepted therapy.

In summary, newly developed comprehensive therapies based on biological understanding, using either recombinant proteins or their genes, will enhance bone regeneration. The challenging task of tissue engineering bone is being tackled by many multidisciplinary research groups involving engineers, biologists, and polymer chemists. This effort should yield optimization of current therapies or the development of therapies that will enhance clinical treatment outcomes.

REFERENCES

1. Praemer, A, Furner, S., et al. (1992) Musculoskeletal conditions in the United States. American Academy of Orthopaedic Surgeons, Park Ridge, IL.
2. Cruess, R. L. and Dumont, J. (1975) Fracture healing. *Can. J. Surg.* **18(5),** 403–413.
3. Madison, M. and Martin, R. B. (1993) Fracture healing, in *Operative Orthopaedics* (Chapman, M. W., ed.), Lippincott, Philadelphia, pp. 221–228.
4. Frost, H. M. (1989) The biology of fracture healing. An overview for clinicians. Part I. *Clin. Orthop.* **248,** 283–293.
5. McKibbin, B. (1978) The biology of fracture healing in long bones. *J. Bone Joint Surg. Br.* **60B(2),** 150–162.
6. White, A. A. I., Punjabi, M. M., and Southwick, W. O. (1977) The four biomechanical stages of fracture healing. *J. Bone Joint Surg.* **59A,** 188.
7. Thomas, T. A. (1998) The cell and molecular biology of fracture healing. *Clinical Orthopaedics & Related Research. Fracture Healing Enhancement* **1(355 Suppl),** S7–S21.
8. Marsh, D. R. and Li, G. (1999) The biology of fracture healing: optimising outcome. *Br. Med. Bull.* **55(4),** 856–869.
9. Nemeth, G. G., Bolander, M. E., and Martin, G. R. (1988) Growth factors and their role in wound and fracture healing. *Prog. Clin. Biol. Res.* **266,** 1–17.
10. Bolander, M. E. (1992) Regulation of fracture repair by growth factors. *Proc. Soc. Exp. Biol. Med.* **200(2),** 165–170.
11. Buckwalter, J. A., et al. (1996) Healing of the musculoskeletal tissues, in *Rockwood and Green's Fractures in Adults* (Rockwood, C. A., et al., ed.), Lippincott-Raven, Philadelphia, pp. 261–304.
12. Glowacki, J. (1998) Angiogenesis in fracture repair. *Clin. Orthop.* **355(Suppl),** S82–S89.
13. Hollinger, J. and Wong, M. E. (1996) The integrated processes of hard tissue regeneration with special emphasis on fracture healing. *Oral Surg. Oral Med. Oral Pathol. Oral Radiol. Endodont.* **82(6),** 594–606.
14. Grundnes, O. and Reikeras, O. (1993) The role of hematoma and periosteal sealing for fracture healing in rats. *Acta Orthop. Scand.* **64(1),** 47–49.
15. Grundnes, O. and Reikeras, O. (1993) The importance of the hematoma for fracture healing in rats. *Acta Orthop. Scand.* **64(3),** 340–342.
16. Tonna, E. A. and Cronkite, E. P. (1963) The periosteum: autoradiographic studies on cellular proliferation and transformation utilizing tritiated thymidine. *Clin. Orthop.* **30,** 218–233.
17. Einhorn, T. A., et al. (1989) Neutral protein-degrading enzymes in experimental fracture callus: a preliminary report. *J. Orthop. Res.* **7(6),** 792–805.
18. Brighton, C. T. and Hunt, R. M. (1986) Histochemical localization of calcium in the fracture callus with potassium pyroantimonate. Possible role of chondrocyte mitochondrial calcium in callus calcification. *J. Bone Joint Surg. Am.* **68(5),** 703–715.
19. Wendeberg, B. (1961) Mineral metabolism of fractures of the tibia in man studied with external counting of strontium 85. *Acta Orthop. Scand. Suppl.* **52,** 1–79.
20. Frost, H. M. (1980) Skeletal physiology and bone remodeling, in *Fundamental and Clinical Bone Physiology* (Urist, M. R. ed.), Lippincott, Philadelphia, pp. 208–241.
21. Allgower, M. and Spiegel, P. G. (1979) Internal fixation of fractures: evolution of concepts. *Clin. Orthop.* **138,** 26–29.
22. Muller, M. E., Allgower, M., and Willenegger, H. (1965) *Technique of internal fixation of fractures.* Springer-Verlag, New York.
23. Perren, S. M. (1979) Physical and biological aspects of fracture healing with special reference to internal fixation. *Clin. Orthop.* **138,** 175–196.
24. Perren, S. M. (1989) The biomechanics and biology of internal fixation using plates and nails. *Orthopedics* **12(1),** 21–34.
25. Perren, S. M., Cordey, J., and Gautier, E. (1987) Rigid internal fixation using plates: terminology, principle and early problems, in *Fracture Healing* (Lang, J. M., ed.), Churchill Livingstone, New York, pp. 139–151.
26. Perren, S. M., Huggler, A., and Russenberger, S. (1969) Cortical Bone Healing. *Acta Orthop. Scand. Suppl.* 125.
27. Schatzker, J., Waddell, J., and Stoll, J. E. (1989) The effects of motion on the healing of cancellous bone. *Clin. Orthop.* **245,** 282–287.
28. Schenk, R. (1987) Cytodynamics and histodynamics of primary bone repair, in *Fracture Healing* (Lang, J. M., ed.), Churchill Livingstone, New York, pp. 23–32.
29. Schenk, R. and Willenegger, H. (1967) Morphological findings in primary fracture healing: callus formation. *Simp. Biol. Hungar.* **7,** 75–80.
30. Rahn, B. A., et al. (1971) Primary bone healing. An experimental study in the rabbit. *J. Bone Joint Surg. Am.* **53(4),** 783–786.
31. Schenk, R. and Willenegger, H. (1963) Histologie der primaren Knochenheilung. *Arch. Klin. Chir.* **19,** 593.
32. DeLacure, M. D. (1994) Physiology of bone healing and bone grafts. *Otolaryngol. Clin. N. Am.* **27(5),** 859–874.
33. Liu, Z., et al. (1999) Molecular signaling in bone fracture healing and distraction osteogenesis. *Histol. Histopathol.* **14(2),** 587–595.

34. Bonewald, L. F. and Dallas, S. L. (1994) Role of active and latent transforming growth factor beta in bone formation. *J. Cell Biochem.* **55(3)**, 350–357.
35. Dallas, S. L., et al. (1994) Characterization and autoregulation of latent transforming growth factor beta (TGF beta) complexes in osteoblast-like cell lines. Production of a latent complex lacking the latent TGF beta-binding protein. *J. Biol. Chem.* **269(9)**, 6815–6821.
36. Miyazono, K., Ichijo, H., and Heldin, C. H. (1993) Transforming growth factor-beta: latent forms, binding proteins and receptors. *Growth Factors* **8(1)**, 11–22.
37. Linkhart, T. A., Mohan, S., and Baylink, D. J. (1996) Growth factors for bone growth and repair: IGF, TGF beta and BMP. *Bone* **19(1 Suppl)**, 1S–12S.
38. Joyce, M. E., et al. (1990) Transforming growth factor-beta and the initiation of chondrogenesis and osteogenesis in the rat femur. *J. Cell Biol.* **110(6)**, 2195–2207.
39. Noda, M. and Camilliere, J. J. (1989) *In vivo* stimulation of bone formation by transforming growth factor-beta. *Endocrinology* **124(6)**, 2991–2994.
40. Sandberg, M. M., Aro, H. T., and Vuorio, E. I. (1993) Gene expression during bone repair. *Clin. Orthop.* **289**, 292–312.
41. Mundy, G. R. (1993) Cytokines and growth factors in the regulation of bone remodeling. *J. Bone Miner. Res.* **8(Suppl 2)**, S505–S510.
42. Beck, L. S., et al. (1991) *In vivo* induction of bone by recombinant human transforming growth factor beta 1. *J. Bone Miner. Res.* **6(9)**, 961–968.
43. Talley Ronsholdt, D., Lajiness, E., and Nagodawithana, K. (1995) Transforming growth factor-beta inhibition of mineralization by neonatal rat osteoblasts in monolayer and collagen gel culture. *In Vitro Cell Dev. Biol. Anim.* **31**, 274–282.
44. Cheifetz, S., et al. (1996) Influence of osteogenic protein-1 (OP-1;BMP-7) and transforming growth factor-beta 1 on bone formation in vitro. *Connect. Tissue Res.* **35(1–4)**, 71–78.
45. Pfeilschifter, J., Seyedin, S. M., and Mundy, G. R. (1988) Transforming growth factor beta inhibits bone resorption in fetal rat long bone cultures. *J. Clin. Invest.* **82(2)**, 680–685.
46. Massague, J. (1990) The transforming growth factor-beta family. *Annu. Rev. Cell Biol.* **6**, 597–641.
47. Prisell, P. T., et al. (1993) Expression of insulin-like growth factors during bone induction in rat. *Calcif. Tissue Int.* 1993. **53(3)**, 201–205.
48. Eppley, B. L., et al. (1988) Enhancement of angiogenesis by bFGF in mandibular bone graft healing in the rabbit. *J. Oral. Maxillofac. Surg.* **46(5)**, 391–398.
49. Trippel, S. B., et al. (1992) Interaction of basic fibroblast growth factor with bovine growth plate chondrocytes. *J. Orthop. Res.* **10(5)**, 638–646.
50. Chintala, S. K., Miller, R. R., and McDevitt, C. A. (1994) Basic fibroblast growth factor binds to heparan sulfate in the extracellular matrix of rat growth plate chondrocytes. *Arch. Biochem. Biophys.* **310(1)**, 180–186.
51. Canalis, E., McCarthy, T., and Centrella, M. (1988) Growth factors and the regulation of bone remodeling. *J. Clin. Invest.* **81(2)**, 277–281.
52. Centrella, M., et al. (1992) Isoform-specific regulation of platelet-derived growth factor activity and binding in osteoblast-enriched cultures from fetal rat bone. *J. Clin. Invest.* **89(4)**, 1076–1084.
53. Robson, M. C., et al. (1992) Platelet-derived growth factor BB for the treatment of chronic pressure ulcers. *Lancet* **339(8784)**, 23–25.
54. Muller, R., et al. (1984) Induction of c-fos gene and protein by growth factors precedes activation of c-myc. *Nature* **312(5996)**, 716–720.
55. Andrew, J., et al. (1993) Insuling like growth factor gene expression in human fracture callus. *Calcif. Tissue Int.* **53**, 97–102.
56. Edwall, D., et al. (1992) Expression of insulin-like growth factor I messenger ribonucleic acid in regenerating bone after fracture: influence of indomethacin. *J. Bone Miner. Res.* **7(2)**, 207–213.
57. Nilsson, A., et al. (1986) Regulation by growth hormone of number of chondrocytes containing IGF-I in rat growth plate. *Science* **233(4763)**, 571–574.
58. Lammens, J., et al. (1998) Distraction bone healing versus osteotomy healing: a comparative biochemical analysis. *J. Bone Miner. Res.* **13(2)**, 279–286.
59. Hill, P. A., Reynolds, J. J., and Meikle, M. C. (1995) Osteoblasts mediate insulin-like growth factor-I and -II stimulation of osteoclast formation and function. *Endocrinology* **136(1)**, 124–131.
60. Mochizuki, H., et al. (1992) Insulin-like growth factor-I supports formation and activation of osteoclasts. *Endocrinology* **131(3)**, 1075–1080.
61. Beck, F., et al. (1987) Histochemical localization of IGF-I and -II mRNA in the developing rat embryo. *Development* **101(1)**, 175–184.
62. McCarthy, T. L., Centrella, M., and Canalis, E. (1989) Regulatory effects of insulin-like growth factors I and II on bone collagen synthesis in rat calvarial cultures. *Endocrinology* **124(1)**, 301–309.
63. Urist, M. R. and Strates, B. S. (1971) Bone morphogenetic protein. *J. Dent. Res.* **50(6)**, 1392–1406.

64. Urist, M. R., Iwata, H., and Strates, B. S. (1972) Bone morphogenetic protein and proteinase in the guinea pig. *Clin. Orthop.* **85,** 275–290.
65. Urist, M. R., et al. (1983) Human bone morphogenetic protein (hBMP). *Proc. Soc. Exp. Biol. Med.* **173(2),** 194–199.
66. Bostrom, M. P., et al. (1995) Immunolocalization and expression of bone morphogenetic proteins 2 and 4 in fracture healing. *J. Orthop. Res.* **13(3),** 357–367.
67. Nakase, T., et al. (1994) Transient and localized expression of bone morphogenetic protein 4 messenger RNA during fracture healing. *J. Bone Miner. Res.* **9(5),** 651–659.
68. Onishi, T., et al. (1998) Distinct and overlapping patterns of localization of bone morphogenetic protein (BMP) family members and a BMP type II receptor during fracture healing in rats. *Bone* **22(6),** 605–612.
69. Canalis, E. and Gabbitas, B. (1994) Bone morphogenetic protein 2 increases insulin-like growth factor I and II transcripts and polypeptide levels in bone cell cultures. *J. Bone Miner. Res.* **9(12),** 1999–2005.
70. Zheng, M. H., et al. (1994) Recombinant human bone morphogenetic protein-2 enhances expression of interleukin-6 and transforming growth factor-beta 1 genes in normal human osteoblast-like cells. *J. Cell Physiol.* **159(1),** 76–82.
71. Cunningham, N. S., Paralkar, V., and Reddi, A. H. (1992) Osteogenin and recombinant bone morphogenetic protein 2B are chemotactic for human monocytes and stimulate transforming growth factor beta 1 mRNA expression. *Proc. Natl. Acad. Sci. USA* **89(24),** 11740–11744.
72. Ozkaynak, E., et al. (1990) OP-1 cDNA encodes an osteogenic protein in the TGF-beta family. *EMBO J.* **9(7),** 2085–2093.
73. Knutsen, R., et al. (1995) Regulation of insulin-like growth factor system components by osteogenic protein-1 in human bone cells. *Endocrinology* **136(3),** 857–865.
74. Paralkar, V. M., et al. (1990) Interaction of osteogenin, a heparin binding bone morphogenetic protein, with type IV collagen. *J. Biol. Chem.* **265(28),** 17281–17284.
75. Wozney, J. M. (1992) The bone morphogenetic protein family and osteogenesis. *Mol. Reprod. Dev.* **32(2),** 160–167.
76. Ducy, P., et al. (1997) Osf2/Cbfa1: a transcriptional activator of osteoblast differentiation. *Cell* **89(5),** 747–754.
77. Celeste, A. J., et al. (1990) Identification of transforming growth factor beta family members present in bone-inductive protein purified from bovine bone. *Proc. Natl. Acad. Sci. USA* **87(24),** 9843–9847.
78. Wozney, J. M. (1989) Bone morphogenetic proteins. *Prog. Growth Factor Res.* **1(4),** 267–280.
79. Hogan, B. L. (1996) Bone morphogenetic proteins in development. *Curr. Opin. Genet. Dev.* **6(4),** 432–438.
80. Hogan, B. L. (1996) Bone morphogenetic proteins: multifunctional regulators of vertebrate development. *Genes Dev.* **10(13),** 1580–1594.
81. Sakou, T. (1998) Bone morphogenetic proteins: from basic studies to clinical approaches. *Bone* **22(6),** 591–603.
82. Bolander, M. E., et al. (1988) Osteonectin cDNA sequence reveals potential binding regions for calcium and hydroxyapatite and shows homologies with both a basement membrane protein (SPARC) and a serine proteinase inhibitor (ovomucoid). *Proc. Natl. Acad. Sci. USA* **85(9),** 2919–2923.
83. Hiltunen, A., et al. (1994) Retarded chondrogenesis in transgenic mice with a type II collagen defect results in fracture healing abnormalities. *Dev. Dyn.* **200(4),** 340–349.
84. Yamazaki, M., Majeska, R. J., and Moriya, H. (1997) Role of osteonectin during fracture healing. *Trans. Orthop. Res. Soc.* **22,** 254.
85. Jingushi, S., Joyce, M. E., and Bolander, M. E. (1992) Genetic expression of extracellular matrix proteins correlates with histologic changes during fracture repair. *J. Bone Miner. Res.* **7(9),** 1045–1055.
86. Metsaranta, M., et al. (1989) Localization of osteonectin expression in human fetal skeletal tissues by *in situ* hybridization. *Calcif. Tissue Int.* **45(3),** 146–152.
87. Price, P. A., Williamson, M. K., and Lothringer, J. W. (1981) Origin of the vitamin K-dependent bone protein found in plasma and its clearance by kidney and bone. *J. Biol. Chem.* **256(24),** 12760–12766.
88. Nakamura, H., et al. (1995) Localization of CD44, the hyaluronate receptor, on the plasma membrane of osteocytes and osteoclasts in rat tibiae. *Cell Tissue Res.* **280(2),** 225–233.
89. Yamazaki, M., et al. (1997) Spatial and temporal expression of fibril-forming minor collagen genes (types V and XI) during fracture healing. *J. Orthop. Res.* **15(5),** 757–764.
90. Hirakawa, K., et al. (1994) Localization of the mRNA for bone matrix proteins during fracture healing as determined by *in situ* hybridization. *J. Bone Miner. Res.* **9(10),** 1551–1557.
91. Reinholt, F. P., et al. (1990) Osteopontin—a possible anchor of osteoclasts to bone. *Proc. Natl. Acad. Sci. USA* **87(12),** 4473–4475.
92. Bourghol, A., et al. (1996) Fibronectin expression during fracture healing in the rat. *Trans. Orthop. Res. Soc.* **21,** 618.
93. Ishidou, Y., et al. (1995) Enhanced expression of type I receptors for bone morphogenetic proteins during bone formation. *J. Bone Miner. Res.* **10(11),** 1651–1659.
94. Baker, J. C. and Harland, R. M. (1997) From receptor to nucleus: the Smad pathway. *Curr. Opin. Genet Dev.* **7(4),** 467–473.
95. Heldin, C. H., Miyazono, K., and Dijke, P. (1997) TGF[beta] signaling from cell membrane to nucleus through Smad proteins. *Nature* **390,** 465–471.

96. Reddi, A. H. (1998) Initiation of fracture repair by bone morphogenetic proteins. *Clin. Orthop.* **355(Suppl)**, S66–S72.

97. Einhorn, T. A., et al. (1995) The expression of cytokine activity by fracture callus. *J. Bone Miner. Res.* **10(8)**, 1272–1281.

98. Beuscher, H. U., et al. (1992) Transition from interleukin 1 beta (IL-1 beta) to IL-1 alpha production during maturation of inflammatory macrophages *in vivo*. *J. Exp. Med.* **175(6)**, 1793–1797.

99. Brown, C. C., Hembry, R. M., and Reynolds, J. J. (1989) Immunolocalization of metalloproteinases and their inhibitor in the rabbit growth plate. *J. Bone Joint Surg. Am.* **71(4)**, 580–593.

100. Leibovich, S. J. and Ross, R. (1975) The role of the macrophage in wound repair. A study with hydrocortisone and antimacrophage serum. *Am. J. Pathol.* **78(1)**, 71–100.

101. Polverini, P. J., et al. (1977) Activated macrophages induce vascular proliferation. *Nature* **269(5631)**, 804–806.

102. Zlotnik, A., Daine, B., and Smith, C. A. (1985) Activation of an interleukin-1-responsive T-cell lymphoma by fixed P388D1 macrophages and an antibody against the Ag:MHC T-cell receptor. *Cell. Immunol.* **94(2)**, 447–453.

103. Browder, W., et al. (1988) Effect of enhanced macrophage function on early wound healing. *Surgery* **104(2)**, 224–230.

104. Feyen, J., et al. (1989) Interleukin-6 is produced by bone and modulated by parathyroid hormone. *J. Bone Miner. Res.* **4**, 633–638.

105. Horowitz, M. C., et al. (1990) Il-6 secretion by osteoblasts. *Trans. Orthop. Res. Soc.* **15**, 379.

106. Helle, M., et al. (1988) Interleukin 6 is involved in interleukin 1-induced activities. *Eur. J. Immunol.* **18(6)**, 957–959.

107. Lowik, C., et al. (1989) Parathyroid hormone (PTH) and PTH-like protein (PLP) stimulate interleukin-6 production by osteogenic cells: a possible role of interleukin-6 in osteoclastogenesis. *Biochem. Biophys. Res. Commun.* **162**, 1546–1552.

108. Roodman, G. (1992) Interleukin-6: An osteoptropic factor? *J. Bone Miner. Res.* **7**, 475–478.

109. Jilka, R. L., et al. (1992) Increased osteoclast development after estrogen loss: mediation by interleukin-6. *Science* **257(5066)**, 88–91.

110. Hulth, A., et al. (1985) Appearance of T-cells and Ia-expressing cells in fracture healing in rats. *Int. J. Immunother.* **1**, 103–106.

111. Gliniak, B. C. and Rohrschneider, L. R. (1990) Expression of the M-CSF receptor is controlled posttranscriptionally by the dominant actions of GM-CSF or multi-CSF. *Cell* **63(5)**, 1073–1083.

112. Horowitz, M. C., et al. (1989) Parathyroid hormone and lipopolysaccharide induce murine osteoblast-like cells to secrete a cytokine indistinguishable from granulocyte-macrophage colony-stimulating factor. *J. Clin. Invest.* **83(1)**, 149–157.

113. Horowitz, M. C., et al. (1989) Osteotropic agents induce the differential secretion of granulocyte-macrophage colony-stimulating factor by the osteoblast cell line MC3T3-E1. *J. Bone Miner. Res.* **4**, 911–921.

114. Horowitz, M. C., Friedlaender, G. E., and Qian, H.-T. (1994) T cell activation and the immune response to bone allografts. *Trans. Orthop. Res. Soc.* **19**, 180.

115. Barbul, A., et al. (1986) Interleukin 2 enhances wound healing in rats. *J. Surg. Res.* **40(4)**, 315–319.

116. Dinarello, C. A. and Mier, J. W. (1987) Lymphokines. *N. Engl. J. Med.* **317(15)**, 940–945.

117. Turck, C. W., Dohlman, J. G., and Goetzl, E. J. (1987) Immunological mediators of wound healing and fibrosis. *J. Cell Physiol. Suppl.* **Suppl(5)**, 89–93.

118. Nicola, N. A. (1989) Hemopoietic cell growth factors and their receptors. *Annu. Rev. Biochem.* **58**, 45–77.

119. Felix, R., Fleisch, H., and Elford, P. R. (1989) Bone-resorbing cytokines enhance release of macrophage colony-stimulating activity by the osteoblastic cell MC3T3-E1. *Calcif. Tissue Int.* **44(5)**, 356–360.

120. Horowitz, M. C., et al. (1989) Functional and molecular changes in colony stimulating factor secretion by osteoblasts. *Connect Tissue Res.* **20(1–4)**, 159–168.

121. Stanley, E. R., et al. (1976) Factors regulating macrophage production and growth: identity of colony-stimulating factor and macrophage growth factor. *J. Exp. Med.* **143(3)**, 631–647.

122. Hayda, R. A., Brighton, C. T., and Esterhai, J. L. Jr. (1998) Pathophysiology of delayed healing. *Clin. Orthop.* **355(Suppl)**, S31–S40.

123. Pan, W. and Einhorn, T. A. (1992) The biochemistry of fracture healing. *Curr. Orthop.* **6**, 207–213.

124. Page, M., Hogg, J., and Ashhurst, D. E. (1986) The effects of mechanical stability on the macromolecules of the connective tissue matrices produced during fracture healing. I. The collagens. *Histochem. J.* **18(5)**, 251–265.

125. Hiltunen, A., Aro, H. T., and Vuorio, E. (1993) Regulation of extracellular matrix genes during fracture healing in mice. *Clin. Orthop.* **297**, 23–27.

126. Sandberg, M., et al. (1989) *In situ* localization of collagen production by chondrocytes and osteoblasts in fracture callus. *J. Bone Joint Surg. Am.* **71(1)**, 69–77.

127. Multimaki, P., Aro, H., and Vuorio, E. (1987) Differential expression of fibrillar collagen genes during callus formation. *Biochem. Biophys. Res. Commun.* **142(2)**, 536–541.

128. Paralkar, V. M., Vukicevic, S., and Reddi, A. H. (1991) Transforming growth factor beta type 1 binds to collagen IV of basement membrane matrix: implications for development. *Dev. Biol.* **143(2)**, 303–308.

129. Birke, D., et al. (1990) Collagen fibrillogenesis *in vitro*: interaction of types I and V collagen regulates fibril diameter. *J. Cell Sci.* **95**, 649–657.

130. Mendler, M., et al. (1989) Cartilage contains mixed fibrils of collagen types II, IX, and XI. *J. Cell Biol.* **108(1)**, 191–197.
131. Mayne, R. and Burgeson, R. E., eds. (1987) *Structure and Function of Collagen Types.* Academic Press, Orlando, FL.
132. Shimokomaki, M., et al. (1990) The structure and macromolecular organization of type IX collagen in cartilage. *Ann. NY Acad. Sci.* **580**, 1–7.
133. Tonna, E. A. (1958) Histochemical and antoradiograhic studies on the effects of aging of the mucopolysaccharides of the periosteum. *J. Biophys. Biochem. Cytol.* **6**, 171–175.
134. Tonna, E. A. (1978) Electron microscopic study of bone surface changes during aging. The loss of cellular control and biofeedback. *J. Gerontol.* **33(2)**, 163–177.
135. Syftestad, G.T. and Urist, M.R. (1982) Bone aging. *Clin. Orthop.* **162**, 288–297.
136. Mears, D.C. (1999) Surgical treatment of acetabular fractures in elderly patients with osteoporotic bone. *J. Am. Acad. Orthop. Surg.* **7(2)**, 128–141.
137. Christiansen, M., et al. (2000) CBFA1 and topoisomerase I mRNA levels decline during cellular aging of human trabecular osteoblasts. *J. Gerontol. A Biol. Sci. Med. Sci.* **55(4)**, B194–B200.
138. Ankersen, L.(1994) Aging of humantravecular osteoblasts in culture. *Arch. Gerontol. Geriatr.* **4(Suppl)**, 5–12.
139. Kassem, M., et al. (1997) Demonstration of cellular aging and senescence in serially passaged long-term cultures of human trabecular osteoblasts. *Osteopor. Int.* **7(6)**, 514–524.
140. Chavassieux, P. M., et al. (1990) Influence of experimental conditions on osteoblast activity in human primary bone cell cultures. *J. Bone Miner. Res.* **5(4)**, 337–343.
141. Martinez, M. E., et al. (1999) Influence of skeletal site of origin and donor age on osteoblastic cell growth and differentiation. *Calcif. Tissue Int.* **64(4)**, 280–286.
142. Marie, P. J. (1995) Human endosteal osteoblastic cells: relationship with bone formation. *Calcif. Tissue Int.* **56(Suppl 1)**, S13–S16.
143. Vanderschueren, D., et al. (1990) Sex- and age-related changes in bone and serum osteocalcin. *Calcif. Tissue Int.* **46(3)**, 179–182.
144. O'Driscoll, S. W., et al. (2001) The chondrogenic potential of periosteum decreases with age. *J. Orthop. Res.* **19(1)**, 95–103.
145. Evans, R. D., Foltz, D., and Foltz, K. (2001) Electrical stimulation with bone and wound healing. *Clin. Podiatr. Med. Surg.* **18(1)**, 79–95, vi.
146. Spadaro, J. A. (1997) Mechanical and electrical interactions in bone remodeling. *Bioelectromagnetics* **18(3)**, 193–202.
147. Goodman, S. and Aspenberg, P. (1993) Effects of mechanical stimulation on the differentiation of hard tissues. *Biomaterials* **14(8)**, 563–569.
148. Garg, N. K., Gaur, S., and Sharma, S. (1993) Percutaneous autogenous bone marrow grafting in 20 cases of ununited fracture. *Acta Orthop. Scand.* **64(6)**, 671–672.
149. Caplan, A. I. (1994) The mesengenic process. *Clin. Plast. Surg.* **21(3)**, 429–435.
150. Bruder, S. P., Fink, D. J., and Caplan, A. I. (1994) Mesenchymal stem cells in bone development, bone repair, and skeletal regeneration therapy. *J. Cell Biochem.* **56(3)**, 283–294.
151. Goshima, J., Goldberg, V. M., and Caplan, A. I. (1991) The osteogenic potential of culture-expanded rat marrow mesenchymal cells assayed *in vivo* in calcium phosphate ceramic blocks. *Clin. Orthop.* **262**, 298–311.
152. Caplan, A. I. (1991) Mesenchymal stem cells. *J. Orthop. Res.* **9(5)**, 641–650.
153. Goshima, J., Goldberg, V. M., and Caplan, A. I. (1991) The origin of bone formed in composite grafts of porous calcium phosphate ceramic loaded with marrow cells. *Clin. Orthop.* **269**, 274–283.
154. Dennis, J. E. and Caplan, A. I. (1993) Porous ceramic vehicles for rat-marrow-derived (Rattus norvegicus) osteogenic cell delivery: effects of pre-treatment with fibronectin or laminin. *J. Oral Implantol.* **19(2)**, 106–115.
155. Hong, L., et al. (2000) Promoted bone healing at a rabbit skull gap between autologous bone fragment and the surrounding intact bone with biodegradable microspheres containing transforming growth factor-beta1. *Tissue Eng.* **6(4)**, 331–340.
156. Gombotz, W. R., et al. (1993) Controlled release of TGF-beta 1 from a biodegradable matrix for bone regeneration. *J. Biomater. Sci. Polym. Ed.* **5(1–2)**, 49–63.
157. Mustoe, T. A., et al. (1987) Accelerated healing of incisional wounds in rats induced by transforming growth factor-beta. *Science* **237(4820)**, 1333–1336.
158. Mackenzie, D. J., et al. (2001) Recombinant human acidic fibroblast growth factor and fibrin carrier regenerates bone. *Plast. Reconstr. Surg.* **107(4)**, 989–996.
159. Downs, E. C., et al. (1992) Calcium alginate beads as a slow-release system for delivering angiogenic molecules *in vivo* and *in vitro. J. Cell Physiol.* **152(2)**, 422–429.
160. Edelman, E. R., et al. (1991) Controlled and modulated release of basic fibroblast growth factor. *Biomaterials* **12(7)**, 619–626.
161. Harada, K., et al. (1994) Basic fibroblast growth factor improves myocardial function in chronically ischemic porcine hearts. *J. Clin. Invest.* **94(2)**, 623–630.

162. Tabata, Y., et al. (1999) Skull bone regeneration in primates in response to basic fibroblast growth factor. *J. Neurosurg.* **91(5)**, 851–856.

163. Tabata, Y., et al. (1998) Bone regeneration by basic fibroblast growth factor complexed with biodegradable hydrogels. *Biomaterials* **19(7–9)**, 807–815.

164. Kawaguchi, H., et al. (1994) Stimulation of fracture repair by recombinant human basic fibroblast growth factor in normal and streptozotocin-diabetic rats. *Endocrinology* **135(2)**, 774–781.

165. Inui, K., et al. (1998) Local application of basic fibroblast growth factor minipellet induces the healing of segmental bony defects in rabbits. *Calcif. Tissue Int.* **63(6)**, 490–495.

166. Fujisato, T., et al. (1996) Effect of basic fibroblast growth factor on cartilage regeneration in chondrocyte-seeded collagen sponge scaffold. *Biomaterials* **17(2)**, 155–162.

167. Hollinger, J. and Wong, M. E. (1996) The integrated processes of hard tissue regeneration with special emphasis on fracture healing. *Oral Surg. Oral Med. Oral Pathol. Oral Radiol. Endodont.* **82(6)**, 594–606.

168. Heckman, J. D., et al. (1991) The use of bone morphogenetic protein in the treatment of non-union in a canine model. *J. Bone Joint Surg. Am.* **73(5)**, 750–764.

169. Muschler, G. F., et al. (1994) Evaluation of human bone morphogenetic protein 2 in a canine spinal fusion model. *Clin. Orthop.* **308**, 229–240.

170. Miyamoto, S., et al. (1992) Evaluation of polylactic acid homopolymers as carriers for bone morphogenetic protein. *Clin. Orthop.* **278**, 274–285.

171. Cook, S. D., et al. (1994) *In vivo* evaluation of recombinant human osteogenic protein (rhOP-1) implants as a bone graft substitute for spinal fusions. *Spine* **19(15)**, 1655–1663.

172. Ripamonti, U., et al. (1992) Osteogenin, a bone morphogenetic protein, adsorbed on porous hydroxyapatite substrata, induces rapid bone differentiation in calvarial defects of adult primates. *Plast. Reconstr. Surg.* **90(3)**, 382–393.

173. Zellin, G. and Linde, A. (1997) Treatment of segmental defects in long bones using osteopromotive membranes and recombinant human bone morphogenetic protein-2. An experimental study in rabbits. *Scand. J. Plast. Reconstr. Surg. Hand. Surg.* **31(2)**, 97–104.

174. Ono, I., et al. (1992) A study on bone induction in hydroxyapatite combined with bone morphogenetic protein. *Plast. Reconstr. Surg.* **90(5)**, 870–879.

175. Zegzula, H. D., et al. (1997) Bone formation with use of rhBMP-2 (recombinant human bone morphogenetic protein-2). *J. Bone Joint Surg. Am.* **79(12)**, 1778–1790.

176. Muschik, M., et al. (2000) Experimental anterior spine fusion using bovine bone morphogenetic protein: a study in rabbits. *J. Orthop. Sci.* **5(2)**, 165–170.

177. Hong, L., et al. (1998) Comparison of bone regeneration in a rabbit skull defect by recombinant human BMP-2 incorporated in biodegradable hydrogel and in solution. *J. Biomater. Sci. Polymer. Ed.* **9(9)**, 1001–1014.

178. Lee, S. C., et al. (1994) Healing of large segmental defects in rat femurs is aided by RhBMP-2 in PLGA matrix. *J. Biomed. Mater. Res.* **28(10)**, 1149–1156.

179. Kenley, R., et al. (1994) Osseous regeneration in the rat calvarium using novel delivery systems for recombinant human bone morphogenetic protein-2 (rhBMP-2). *J. Biomed. Mater. Res.* **28(10)**, 1139–1147.

180. Miyamoto, S., et al. (1993) Polylactic acid-polyethylene glycol block copolymer. A new biodegradable synthetic carrier for bone morphogenetic protein. *Clin. Orthop.* **294**, 333–343.

181. Gerhart, T. N., et al. (1993) Healing segmental femoral defects in sheep using recombinant human bone morphogenetic protein. *Clin. Orthop.* **293**, 317–326.

182. Kirker-Head, C. A., et al. (1998) Healing bone using recombinant human bone morphogenetic protein 2 and copolymer. *Clin. Orthop.* **349**, 205–217.

183. Boden, S. D., et al. (2000) The use of rhBMP-2 in interbody fusion cages. Definitive evidence of osteoinduction in humans: a preliminary report. *Spine* **25(3)**, 376–381.

184. Boyne, P. J., et al. (1997) A feasibility study evaluating rhBMP-2/absorbable collagen sponge for maxillary sinus floor augmentation. *Int. J. Periodont. Restor. Dent.* **17(1)**, 11–25.

185. Cochran, D. L., et al. (2000) Evaluation of recombinant human bone morphogenetic protein-2 in oral applications including the use of endosseous implants: 3-year results of a pilot study in humans. *J. Periodontol.* **71(8)**, 1241–1257.

186. Fang, J., et al. (1996) Stimulation of new bone formation by direct transfer of osteogenic plasmid genes. *Proc. Natl. Acad. Sci. USA* **93(12)**, 5753–5758.

187. Bonadio, J., et al. (1999) Localized, direct plasmid gene delivery *in vivo*: prolonged therapy results in reproducible tissue regeneration. *Nat. Med.* **5(7)**, 753–759.

188. Baumgartner, I., et al. (1998) Constitutive expression of phVEGF165 after intramuscular gene transfer promotes collateral vessel development in patients with critical limb ischemia. *Circulation* **97(12)**, 1114–1123.

189. Symes, J. F., et al. (1999) Gene therapy with vascular endothelial growth factor for inoperable coronary artery disease. *Ann. Thorac. Surg.* **68(3)**, 830–836; discussion 836–837.

190. Losordo, D. W., et al. (1998) Gene therapy for myocardial angiogenesis: initial clinical results with direct myocardial injection of phVEGF165 as sole therapy for myocardial ischemia. *Circulation* **98(25)**, 2800–2804.

191. Mason, J. M., et al. (1998) Expression of human bone morphogenic protein 7 in primary rabbit periosteal cells: potential utility in gene therapy for osteochondral repair. *Gene Ther.* **5(8),** 1098–1104.

192. Stover, M. L., et al. (2001) Bone-directed expression of Col1a1 promoter-driven self-inactivating retroviral vector in bone marrow cells and transgenic mice. *Mol. Ther.* **3(4),** 543–550.

193. Oreffo, R. O., Virdi, A. S., and Triffitt, J. T. (2001) Retroviral marking of human bone marrow fibroblasts: *in vitro* expansion and localization in calvarial sites after subcutaneous transplantation *in vivo. J. Cell Physiol.* **186(2),** 201–209.

194. Duprez, D., et al. (1996) Overexpression of BMP-2 and BMP-4 alters the size and shape of developing skeletal elements in the chick limb. *Mech. Dev.* **57(2),** 145–157.

195. Baltzer, A. W., et al. (1999) A gene therapy approach to accelerating bone healing. Evaluation of gene expression in a New Zealand white rabbit model. *Knee Surg. Sports Traumatol. Arthrosc.* **7(3),** 197–202.

196. Baltzer, A. W., et al. (2000) Genetic enhancement of fracture repair: healing of an experimental segmental defect by adenoviral transfer of the BMP-2 gene. *Gene Ther.* **7(9),** 734–739.

197. Franceschi, R. T., et al. (2000) Gene therapy for bone formation: *in vitro* and *in vivo* osteogenic activity of an adenovirus expressing BMP7. *J. Cell Biochem.* **78(3),** 476–486.

198. Krebsbach, P. H., et al. (2000) Gene therapy-directed osteogenesis: BMP-7-transduced human fibroblasts form bone *in vivo. Hum. Gene Ther.* **11(8),** 1201–1210.

199. Spector, J. A., et al. (2000) Expression of adenovirally delivered gene products in healing osseous tissues. *Ann. Plast. Surg.* **44(5),** 522–528.

200. Helm, G. A., et al. (2000) Use of bone morphogenetic protein-9 gene therapy to induce spinal arthrodesis in the rodent. *J. Neurosurg.* **92(2 Suppl),** 191–196.

201. Nishida, K., et al. (1999) Modulation of the biologic activity of the rabbit intervertebral disc by gene therapy: an *in vivo* study of adenovirus-mediated transfer of the human transforming growth factor beta 1 encoding gene. *Spine* **24(23),** 2419–2425.

202. Alden, T. D., et al. (1999) *In vivo* endochondral bone formation using a bone morphogenetic protein 2 adenoviral vector. *Hum. Gene Ther.* **10(13),** 2245–2253.

203. Alden, T. D., et al. (1999) Percutaneous spinal fusion using bone morphogenetic protein-2 gene therapy. *J. Neurosurg.* **90(1 Suppl),** 109–114.

204. Kuboki, T., et al. (1999) Direct adenovirus-mediated gene delivery to the temporomandibular joint in guinea-pigs. *Arch. Oral. Biol.* **44(9),** 701–709.

205. Musgrave, D. S., et al. (1999) Adenovirus-mediated direct gene therapy with bone morphogenetic protein-2 produces bone. *Bone* **24(6),** 541–547.

206. Watanabe, S., et al. (2000) Adeno-associated virus mediates long-term gene transfer and delivery of chondroprotective IL-4 to murine synovium. *Mol. Ther.* **2(2),** 147–152.

207. Xiao, W., et al. (1999) Gene therapy vectors based on adeno-associated virus type 1. *J. Virol.* **73(5),** 3994–4003.

208. Gao, G., et al. (2000) Purification of recombinant adeno-associated virus vectors by column chromatography and its performance *in vivo. Hum. Gene Ther.* **11(15),** 2079–2091.

Common Molecular Mechanisms Regulating Fetal Bone Formation and Adult Fracture Repair

Theodore Miclau, MD, Richard A. Schneider, PhD,
B. Frank Eames, and Jill A. Helms, DDS, PhD

INTRODUCTION

Skeletal formation involves synchronized integration of genetic programs governing the specification, proliferation, differentiation, and programmed death of cells, remodeling of the extracellular matrix, and vasculogenesis. These same cellular and extracellular events occur during adult bone repair, leading us and others to propose that the molecular machinery responsible for fetal skeletogenesis also plays a role in the process of skeletal repair *(1–5)*. The goal of this review is to highlight recent advances in understanding molecular and cellular mechanisms regulating fetal skeletal development and adult fracture repair. We are optimistic that these advances will ultimately facilitate the manipulation of molecular programs in order to prevent bone disease and treat traumatic injury.

BONE FORMATION DURING DEVELOPMENT

The skeleton can be divided into three parts based on anatomical location and embryonic origin. The axial skeleton arises from condensations of paraxial mesoderm that form adjacent to the embryonic notochord and that comprise the future vertebral column. The appendicular skeleton is derived from localized proliferation of lateral plate mesoderm in the trunk and, along with the axial skeleton, forms bone through endochondral ossification. The skeleton of the head has a far more complex developmental history, being derived from paraxial mesoderm as well as the cranial neural crest. Cranial skeletal tissues form bone through both endochondral and intramembranous ossification. Despite these differences in embryonic origin, cartilages and bones in the head are histologically indistinguishable from those tissues found elsewhere in the body. For the sake of simplicity in this review, we will focus the remaining discussion on development of the appendicular skeleton. However, two issues should be kept in mind. First, mechanisms initiating and controlling skeletal development in the head may be qualitatively different from those regulating appendicular or axial skeletogenesis. Second, these differences may be reflected in the mechanisms by which these tissues undergo repair and/or regeneration.

Appendicular skeletal development begins shortly after the onset of limb bud outgrowth, at a time when the limb primordia consist only of mesenchymal cells sheathed in an ectodermal jacket. Histologically, the mesenchymal cells in these early limb buds may appear identical to one another, but a "molecular map" of the limb field belies this fact. *Sonic hedgehog (Shh)*, which encodes a secreted protein involved in patterning and growth in a number of systems *(6)*, is expressed in a localized region of the posterior limb mesenchyme *(7)*. Shh directly or indirectly regulates the expression of a wide variety of growth and transcription factors, including members of the Bone Morphogenic Protein (BMP) and fibroblast growth factor (FGF) families *(8)*. At this early stage of appendicular skeletal development, all of the mesenchymal cells in the limb are competent to adopt a chondrogenic fate *(9,10)*.

From: *Bone Regeneration and Repair: Biology and Clinical Applications*
Edited by: J. R. Lieberman and G. E. Friedlaender © Humana Press Inc., Totowa, NJ

expression, which in turn regulates the cell adhesion molecule N-CAM *(17,29)*. This alteration in cell–ECM contact is a prerequisite for condensation. Another member of the TGF-β superfamily, growth and differentiation factor-5 (GDF-5), affects condensation size by increasing cell adhesion, which is a critical determinant of condensation *(30)*. Later in development, GDF-5 stimulates the proliferation of chondrocytes. However, mice carrying deletions in GDF-5 exhibit only subtle alterations in skeletal development, specifically a loss or abnormal development of some joints *(31)*.

Chondrogenesis

During condensation, mesenchymal cells begin to alter their phenotype from small, fibroblast-like cells to rounded, enlarged cells (**Fig. 2**). At the same time, there is a shift from the production of a mesenchymal matrix, characterized by collagen types I and III, to the production of a cartilaginous matrix, typified by the expression of collagen types II, IX, and XI. The transition from an undifferentiated mesenchymal cell to a differentiated, mature chondrocyte is incremental. Apparently, cells must continue to express Sox9 and Col2α1α1 before becoming irrevocably committed to a chondrogenic lineage. In the head, for example, mesenchymal cells that contribute to the cranial vault express Col2α1α1, yet these cells do not progress to form a mature cartilage *(19)*. After their initiation into chondrogenesis, however, cells must downregulate *Sox9* in order to mature *(11,32,33)*.

Shortly after the induction of Col2α1α1, the secreted factor Indian hedgehog (Ihh) is expressed in mesenchymal cells in the central region of the condensation *(34–37)*. Ihh binds to a cell-surface receptor complex encoded by *Patched (Ptc)* and *Smoothened (6)*. Ihh expression persists throughout fetal chondrogenesis and postnatal growth, and then disappears around the time of puberty *(35)*. Mice carrying deletions in *Ihh* develop condensations, yet they have a delay in chondrocyte maturation *(38)*. Ihh appears to regulate the rate of chondrocyte maturation through a feedback loop involving parathyroid hormone-related protein (PTHRP) and its receptors *(37)*. Ihh appears to regulate angiogenesis as well

---

Fig. 2. (*Opposite page*) Gene expression during cartilage maturation, vascular invasion, and ossification. **(A)** By e18, bone formation has begun in the forelimbs. Mature (mc) and hypertrophic chondrocytes (hc) border the primary ossification center (b), which is evident within the center of the distal ulna. The periosteum (p) has formed a bony collar and the perichondrium is visible as a thickened epithelium adjacent to the mature chondrocytes. Cbfa1 is expressed in the perichondrium and periosteum, in mature and hypertrophic chondrocytes, and in bone (b). On an adjacent section, Osteocalcin is expressed in both the perichondrium and periosteum, and in the primary ossification center. The Ihh, BMP-6 and Col-10 expression domains overlap with Cbfa1. **(B)** Nuclear Hoechst stain illustrates the cellular outline of the primary ossification center at e18.25. Note that the periosteum (p) has formed around the periphery of the skeletal element, and bone (b) is forming in the central region, surrounded on either side by hypertrophic chondrocytes (hc). Cbfa1 (yellow) and MMP-13 (aqua) signals are superimposed to show the extent of overlap between the two transcripts in hypertrophic chondrocytes. Note the absence of MMP-13 in the periosteum, where intramembranous ossification is occurring. Cbfa-1 (yellow signal) and Osteocalcin (red) are co-expressed in areas of new bone formation, including the periosteum and in the primary ossification center. VEGF is expressed strongly in hypertrophic chondrocytes and weakly in bone, where it overlaps with Osteocalcin. **(C)** Higher magnification shows that MMP-13 transcripts are limited to the hypertrophic and terminally differentiated chondrocytes, similar to VEGF. Cbfa1 is detected in chondrocytes, bone, and periosteum, coincident with Osteocalcin. **(D)** In the tibial growth plate of a 10-d-old mouse, there is an orderly progression of chondrocytes from a proliferative (pc) to a hypertrophic state (hc). New bone formation is evident distal to the hypertrophic zone (b). In addition, the secondary ossification center (2°) is evident; the arrow indicates the location of hypertrophic chondrocytes in this center. At this stage, Cbfa1 is expressed in mature and hypertrophic chondrocytes in both the growth plate and secondary ossification center. Cbfa1 is expressed in regions of new bone formation. Osteocalcin transcripts are detected throughout the trabecular bone of the growth plate. Ihh is restricted to mature and early hypertrophic chondrocytes of the growth plate, with very low levels detected in the secondary ossification center. The Col10 expression domain overlaps with that of Cbfa1 in the secondary ossification center (arrow) and, to a lesser extent, in hypertrophic chondrocytes of the growth plate. (From ref. 2 with permission.) (Color illustration in insert following p. 212.)

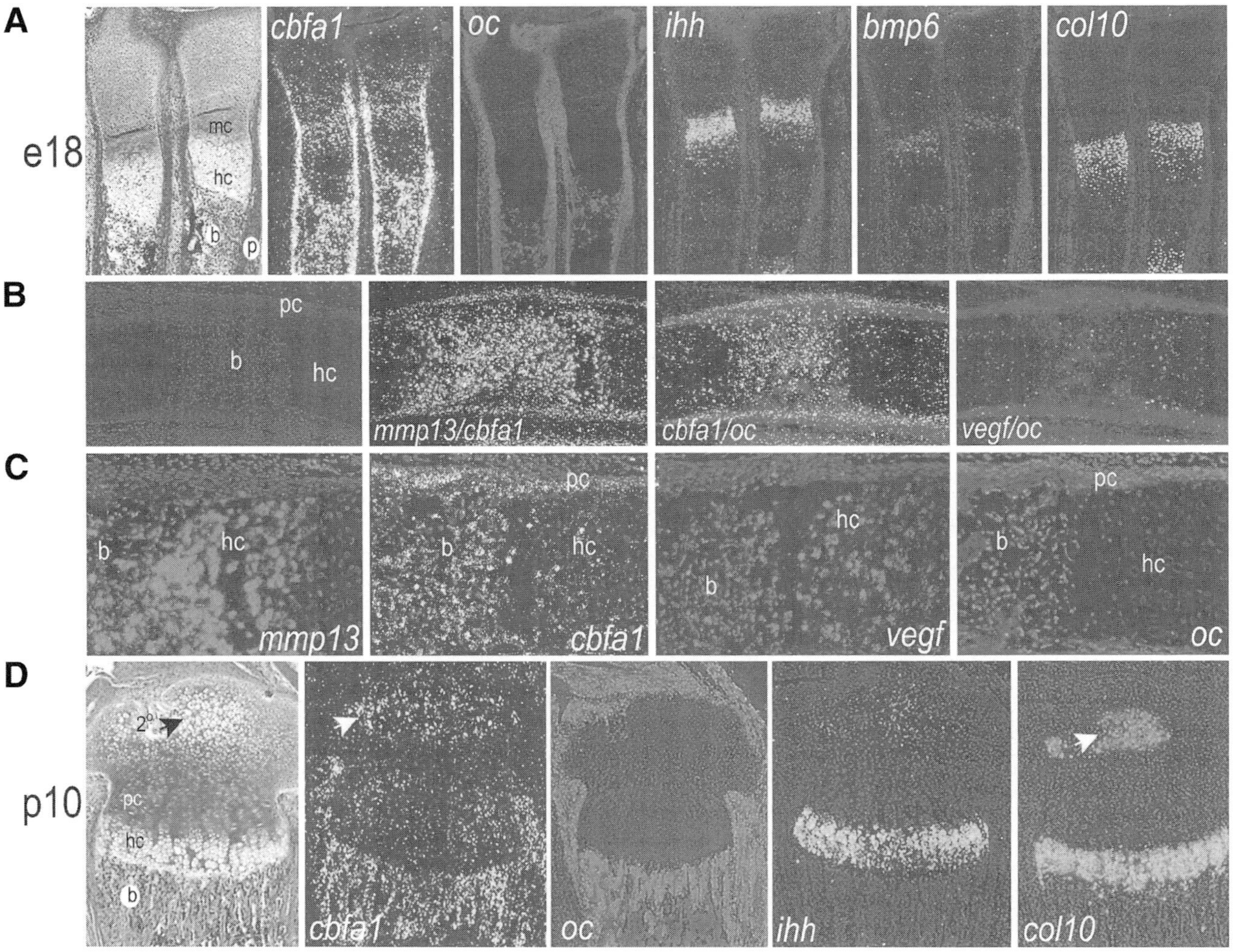

A
e18
mc
hc
b
p
cbfa1
oc
ihh
bmp6
col10
B
pc
b
hc
mmp13/cbfa1
cbfa1/oc
vegf/oc
C
b
hc
pc
mmp13
cbfa1
vegf
oc
D
p10
2°
pc
hc
b
cbfa1
oc
ihh
col10

(39), which may account, in part, for the delay in ossification seen in the *Ihh*-null mutant. Curiously, the ability of mesenchymal cells to undergo condensation and initial chondrogenesis is apparently unaffected in the *Ihh*-null mutant *(38)*. Either *Ihh* does not participate in the programs of condensation and initial chondrogenesis, or it does not play an essential role. However, some clues about Ihh function come from the expression of PTC, which is found in the perichondrial mesenchyme surrounding the skeletal condensations. Ihh and PTC expression patterns are complementary, strongly suggesting that even at this early stage of skeletogenesis, cell communication has been established between the future chondrocytes and cells of the perichondrium.

Another transcription factor that plays an essential role in chondrogenesis is Runx2 (previously termed Cbfa1, Aml3, Pebp2αA, or Osf2). Runx2 is expressed in chondrogenic condensations of the limb and osteogenic condensations *(40,41)*. Mice with null mutations in *Runx2* form mesenchymal cell aggregations in the limb, but later have an arrest in chondrocyte maturation and osteoblast differentiation *(41–43)*. Runx2 directly regulates *Osteopontin* and *Osteocalcin*, two genes associated with hypertrophic and terminally differentiated chondrocytes and osteoblasts *(40,44)*. One function of *Osteopontin* is to mediate the attachment of cells, such as osteoclasts, to the extracellular matrix *(45–47)*. In addition, Runx2 directly induces at least one matrix metalloproteinase, collagenase-3 *(48)*, which is also expressed by hypertropic chondrocytes. The loss of *Runx2* may therefore result in a misregulation of genes associated with the degradation of the hypertrophic cartilage matrix. Accordingly, *Runx2–/–* cartilage fails to undergo vascular invasion *(43)*. This mouse mutant demonstrates an important connection between the differentiation of chondrocytes, remodeling of the extracellular matrix, angiogenesis, and bone formation.

Angiogenesis and Osteogenesis

Vascular invasion is essential for the formation of bone during both intramembranous and endochondral ossification. In intramembranous ossification, the endothelial cells are incorporated into growing mesenchymal cell condensations and provide a blood supply for subsequent ossification. In endochondral ossification, chondrocytes undergo hypertrophy, terminal differentiation, and apoptosis. The hypertrophic cartilage matrix simultaneously is degraded by matrix metalloproteinases, such as MMP9, and invaded by blood vessels. The molecular regulation of new blood vessel formation during endochondral ossification is beginning to be understood, and a number of key angiogenic regulators have been identified. These molecules include members of the FGF, insulin-like growth factor (IGF), TGF-β, and vascular endothelial growth factor (VEGF) families *(49,50)*. VEGF is of particular importance to the vascularization of the cartilaginous skeleton. Several forms of VEGF bind to tyrosine-kinase receptors, Flt-1 and Flk-1, and the coreceptors Neuropilin-1 and Neuropilin-2 *(51,52)*. *Flt-1* and *Flk-1* are expressed in endothelial cells. VEGF is essential for embryonic development, and even the loss of a single *VEGF* allele in restricted embryonic domains causes embryonic death *(53)*. VEGF induces endothelial cell proliferation, stimulates cell migration, and inhibits programmed cell death. Whether apoptosis of hypertrophic chondrocytes is the stimulus for vascular invasion, or conversely, whether blood vessel recruitment is the trigger for cell death, is not clear. However, the coordination of the two steps is essential for osteogenesis.

VEGF-mediated angiogenesis is critical for coupling the resorption of cartilage with the deposition of bone *(50)*. One possible mechanism by which this is achieved is that VEGF is produced by hypertrophic chondrocytes but is only active in or adjacent to those cells that also express *MMP9*. *MMP9* may function to release VEGF from the extracellular matrix and initiate a series of signaling cascades that induce endothelial cell proliferation and invasion, the result of which would be the introduction of osteoprogenitor cells from marrow or endothelial pericytes. As opposed to this indirect mechanism, VEGF may have direct effects on Flt-1-expressing osteoblasts *(50)*. VEGF-mediated angiogenesis is an essential step in the replacement of cartilage by bone during development. As is becoming clear from recent experiments in our laboratory, the same events are important in both proper skeletal development and healing (Colnot et al., unpublished observations).

BONE FORMATION DURING FRACTURE REPAIR

Histologically, bone formed during skeletogenesis has much in common with bone formed during fracture repair *(1,2,35)* (**Fig. 3**). In response to injury, mesenchymal cells from surrounding tissues invade the wound site, where they proliferate, condense, and differentiate into cartilage or bone, much like that seen during development. The similarities and differences between fetal skeletal development and adult fracture repair will be outlined in the following sections.

Mesenchymal Cell Aggregation Following Injury

Immediately after tissue injury, vascular and inflammatory processes trigger a cascade of signaling events that coordinate the invasion of macrophages and other inflammatory cells to the site of injury. Unlike skeletal development, the inflammatory process is a component of adult fracture repair, although its precise contribution remains unclear. One important element of the inflammatory response is the local increase in the number of macrophages, which release molecular signals regulating differentiation *(54)*. Precisely which cells respond to these cues is not entirely clear, but one possibility is mesenchymal stem cells. These cells may be capable of differentiating into chondrocytes or osteoblasts, as suggested by analogy with mesenchymal cells present during the initial stages of skeletal development. Our own analyses of cells populating a fracture site during this inflammatory stage of healing support this hypothesis. Mesenchymal cells populating the site of injury in a closed murine fracture express genes such as *Runx2*, *Sox9*, and *Col2α1*, similar to mesenchymal cells in the developing limb (Miclau et al., unpublished observations).

A clear difference between fetal development and adult repair is the influence of the mechanical environment on the differentiation of mesenchymal stem cells into chondrogenic or osteogenic fates *(55,56)*. In an unstable mechanical environment, mesenchymal cells differentiate into cartilage, whereas a stable environment favors the differentiation of these cells into osteoblasts. How this is achieved remains largely unknown, but some biomechanical and molecular data suggest at least one possible mechanism. Fracture site instability may prevent formation of an intact vasculature. In such a scenario, the fracture site would develop a low oxygen tension, permit the formation of an avascular tissue such as cartilage, and inhibit the generation of a highly vascular tissue such as bone. In a sense, the cartilage stabilizes the fracture site for intact vascularity. As avascular chondrocytes differentiate and eventually hypertrophy, they express angiogenic factors such as VEGF that induce new blood vessel formation. In addition, conditions such as lower oxygen tension lead to the induction of hypoxia-inducible factor, which directly regulates the expression of VEGF *(57)*. This mechanism may account for the formation of cartilage in an avascular situation, but whether an unstable mechanical environment actually disrupts angiogenesis is still an untested hypothesis. Although stabilizing the fracture results in an intramembranous form of healing *(58)*, this scenario additionally does not explain how mesenchymal cells can sense their mechanical environment. These are areas that remain open to inquiry and will likely yield important clues about how bone healing can be stimulated in different mechanical environments such as a stabilized fracture or distraction osteogenesis *(59)*.

Chondrogenesis, Osteogenesis, and Angiogenesis

The maturation of a cartilage callus following a fracture closely parallels cartilage maturation during development. Similar growth and transcription factors expressed during development are also detected during the soft and hard callus phases of fracture repair *(2,5)*. For example, with the conversion of the cartilage callus to woven bone, *Runx2*, *BMP-6*, *Ihh*, and *Col2α1α1* are expressed in chondrocytes of the callus. Similarly, BMP-6, Gli-3, osteocalcin, and collagen type X are detected as the cartilage is replaced by bone. Therefore, the cellular and molecular programs for endochondral and intramembranous ossification during adult fracture healing may recapitulate those operating during development. Recent findings suggest that, as during development, angiogenesis is a key regulator of the conversion of cartilage to bone during fracture repair (Colnot et al., unpublished observations), and also

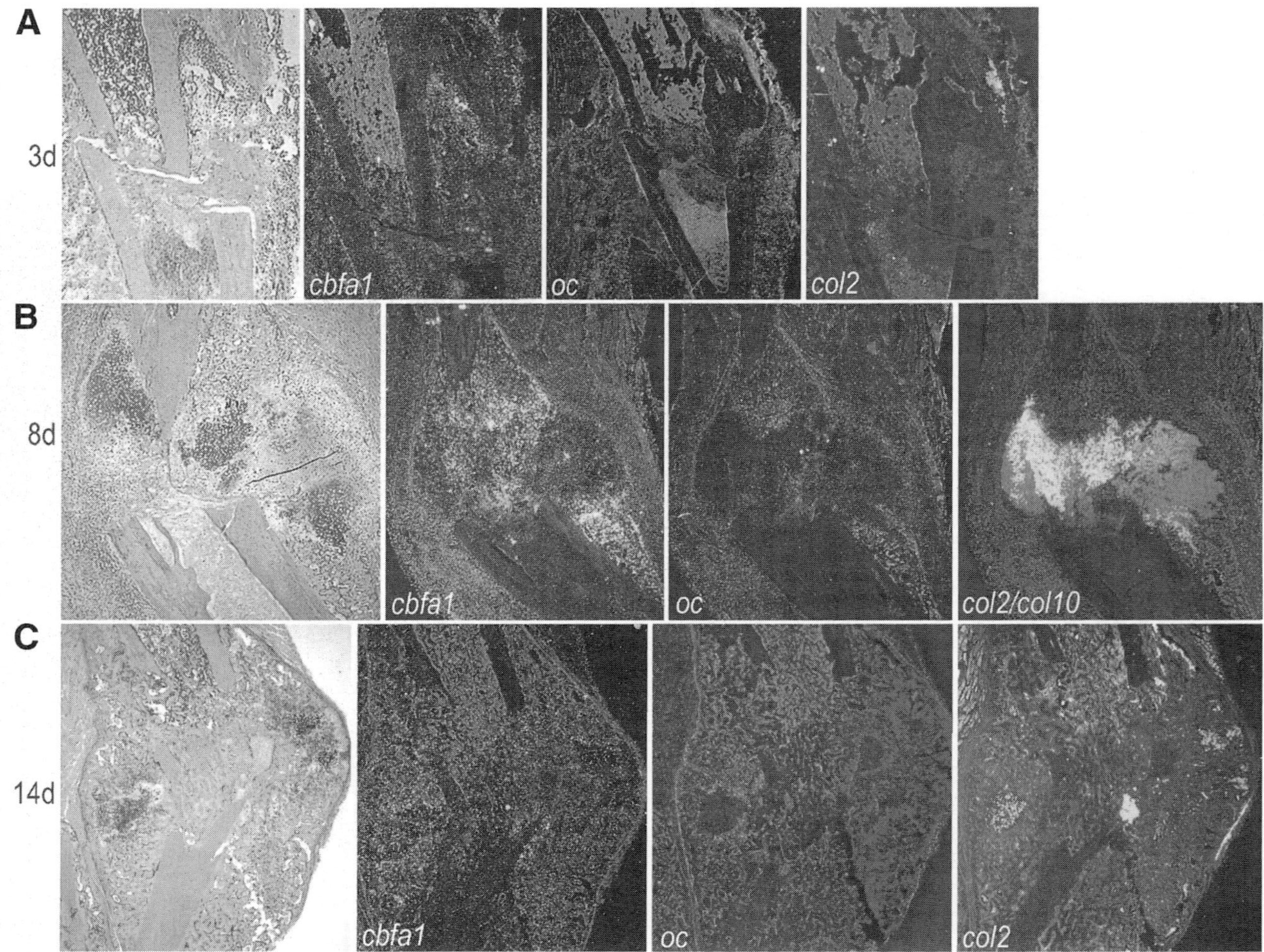
A
3d
cbfa1
oc
col2
B
8d
cbfa1
oc
col2/col10
C
14d
cbfa1
oc
col2

demonstrate that similar to the process of embryonic development, the breakdown of the extracellular matrix is important for the vascularization of the fracture callus.

CONCLUSIONS

During both fetal skeletal development and adult fracture repair, the creation of bone requires a precise coordination of genetic programs that mediate chondrogenesis, osteogenesis, angiogenesis, and bone remodeling. Substantial advances have been made in identifying some of the key molecules and mechanisms that regulate the processes of skeletal development and repair. Collectively, this work indicates that there are remarkable similarities between the cellular and molecular programs for bone formation that function in both embryos and adults. Whether during fetal skeletogenesis or adult healing, bone formation clearly involves a series of discrete phases that are highly coordinated to produce a complete, intact skeleton. Future studies focusing on the molecular and cellular regulation of skeletal morphogenesis and the development of new models of bone repair will undoubtedly provide the foundation for novel therapies to treat bone diseases and traumatic injuries.

REFERENCES

1. Schneider, R. A. and Helms, J. A. (1998) Development and regeneration of the musculoskeletal system. *Curr. Opin. Orthop.* **9**, 20–24.
2. Ferguson, C., Alpern, E., Miclau, T., and Helms, J. A. (1999) Does adult fracture repair recapitulate embryonic skeletal formation? *Mech. Dev.* **87(1–2)**, 57–66.
3. Iwasaki, M., Le, A. X., and Helms, J. A. (1997) Expression of indian hedgehog, bone morphogenetic protein 6 and gli during skeletal morphogenesis. *Mech. Dev.* **69(1–2)**, 197–202.
4. Sandberg, M. M., Aro, H. T., and Vuorio, E. I. (1993) Gene expression during bone repair. *Clin. Orthop.* **289**, 292–312.
5. Vortkamp, A., Pathi, S., Peretti, G. M., Caruso, E. M., Zaleske, D. J., and Tabin, C. J. (1998) Recapitulation of signals regulating embryonic bone formation during postnatal growth and in fracture repair. *Mech. Dev.* **71(1–2)**, 65–76.
6. Hammerschmidt, M., Brook, A., and McMahon, A. P. (1997) The world according to hedgehog. *Trends Genet.* **13(1)**, 14–21.
7. Riddle, R. D., Johnson, R. L., Laufer, E., and Tabin, C. (1993) Sonic hedgehog mediates the polarizing activity of the ZPA. *Cell* **75(7)**, 1401–1416.
8. Laufer, E., Nelson, C. E., Johnson, R. L., Morgan, B. A., and Tabin, C. (1994) Sonic hedgehog and FGF-4 act through a signaling cascade and feedback loop to integrate growth and patterning of the developing limb bud. *Cell* **79(6)**, 993–1003.
9. Hurle, J. M., Ganan, Y., and Macias, D. (1989) Experimental analysis of the *in vivo* chondrogenic potential of the interdigital mesenchyme of the chick leg bud subjected to local ectodermal removal. *Dev. Biol.* **132(2)**, 368–374.
10. Kawakami, Y., Ishikawa, T., Shimabara, M., et al. (1996) BMP signaling during bone pattern determination in the developing limb. *Development* **122(11)**, 3557–3566.
11. Akiyama, H., Chaboissier, M. C., Martin, J. F., Schedl, A., and de Crombrugghe, B. (2002) The transcription factor Sox9 has essential roles in successive steps of the chondrocyte differentiation pathway and is required for expression of Sox5 and Sox6. *Genes Dev.* **16(21)**, 2813–2828.

Fig. 3. (*Opposite page*) Gene expression during the early, intermediate, and late stages of nonstabilized fracture healing. (**A**) Three days after fracture, the periosteum near the fracture site appears thickened, with no cartilage detected by Safranin O/Fast Green staining. Cbfa1 transcripts are present in the thickened periosteal cells. Coincident with Cbfa1 expression, Osteocalcin is detected in the periosteum as well as in the periphery of the callus. Col2 is limited to a small region near the periosteum of the proximal fracture fragment. (**B**) By 8 d postfracture, abundant cartilage appears within the callus. Col-2 and ColX are detected throughout the callus. Osteocalcin is expressed on the periphery of the cartilaginous portion of the callus and within newly formed woven bone, overlapping with Cbfa1. (**C**) At 14 d postfracture, extensive new bone formation is evident, with small islands of residual cartilage persisting. Cbfa1 is detected at low levels in areas of ossification and within the cartilage islands, but is not detected in the periosteum. Osteocalcin is evident throughout areas of bone formation, but is excluded from the cartilage islands. Col2 shows a reciprocal expression pattern, being restricted to the small islands of cartilage in the callus. (Data from ref. *2*.) (Color illustration in insert following p. 212.)

12. Bi, W., Deng, J. M., Zhang, Z., Behringer, R. R., and de Crombrugghe, B. (1999) Sox9 is required for cartilage formation. *Nat. Genet.* **22(1)**, 85–89.
13. Zákány, J. and Duboule, D. (1999) Hox genes in digit development and evolution. *Cell Tissue Res.* **296(1)**, 19–25.
14. Peichel, C. L., Prabhakaran, B., and Vogt, T. F. (1997) The mouse *Ulnaless* mutation deregulates posterior HoxD gene expression and alters appendicular patterning. *Development* **124(18)**, 3481–3492.
15. Goff, D. J. and Tabin, C. J. (1997) Analysis of *Hoxd-13* and *Hoxd-11* misexpression in chick limb buds reveals that *Hox* genes affect both bone condensation and growth. *Development* **124(3)**, 627–636.
16. Capdevila, J., Tsukui, T., Rodríquez Esteban, C., Zappavigna, V., and Izpisúa Belmonte, J. C. (1999) Control of vertebrate limb outgrowth by the proximal factor Meis2 and distal antagonism of BMPs by Gremlin. *Mol. Cell* **4(5)**, 839–849.
17. Hall, B. K. and Miyake, T. (2000) All for one and one for all: condensations and the initiation of skeletal development. *Bioessays* **22(2)**, 138–147.
18. Merino, R., Rodriguez-Leon, J., Macias, D., Gañan, Y., Economides, A. N., and Hurle, J. M. (1999) The BMP antagonist Gremlin regulates outgrowth, chondrogenesis and programmed cell death in the developing limb. *Development* **126(23)**, 5515–5522.
19. Hall, B. K. and Miyake, T. (1995) Divide, accumulate, differentiate: cell condensation in skeletal development revisited. *Int. J. Dev. Biol.* **39(6)**, 881–893.
20. Wright, E., Hargrave, M. R., Christiansen, J., et al. (1995) The *Sry*-related gene *Sox9* is expressed during chondrogenesis in mouse embryos. *Nat. Genet.* **9(1)**, 15–20.
21. Ng, L. J., Wheatley, S., Muscat, G. E., et al. (1997) SOX9 binds DNA, activates transcription, and coexpresses with type II collagen during chondrogenesis in the mouse. *Dev. Biol.* **183(1)**, 108–121.
22. Lefebvre, V., Huang, W., Harley, V. R., Goodfellow, P. N., and de Crombrugghe, B. (1997) SOX9 is a potent activator of the chondrocyte-specific enhancer of the pro alpha1(II) collagen gene. *Mol. Cell Biol.* **17(4)**, 2336–2346.
23. Zhang, P., Jimenez, S. A., and Stokes, D. G. (2003) Regulation of human COL9A1 gene expression. Activation of the proximal promoter region by sox9. *J. Biol. Chem.* **278(1)**, 117–123.
24. Liu, Y., Li, H., Tanaka, K., Tsumaki, N., and Yamada, Y. (2000) Identification of an enhancer sequence within the first intron required for cartilage-specific transcription of the alpha2(XI) collagen gene. *J. Biol. Chem.* **275(17)**, 12712–12718.
25. Healy, C., Uwanogho, D., and Sharpe, P. T. (1999) Regulation and role of Sox9 in cartilage formation. *Dev. Dyn.* **215(1)**, 69–78.
26. Brunet, L. J., McMahon, J. A., McMahon, A. P., and Harland, R. M. (1998) Noggin, cartilage morphogenesis, and joint formation in the mammalian skeleton. *Science* **280(5368)**, 1455–1457.
27. Zimmerman, L. B., De Jesús-Escobar, J. M., and Harland, R. M. (1996) The Spemann organizer signal noggin binds and inactivates bone morphogenetic protein 4. *Cell* **86(4)**, 599–606.
28. Duprez, D., Bell, E. J., Richardson, M. K., et al. (1996) Overexpression of BMP-2 and BMP-4 alters the size and shape of developing skeletal elements in the chick limb. *Mech. Dev.* **57(2)**, 145–157.
29. Hall, B. K. and Miyake, T. (1992) The membranous skeleton: the role of cell condensations in vertebrate skeletogenesis. *Anat. Embryol. (Berl.)* **186(2)**, 107–124.
30. Francis-West, P. H., Abdelfattah, A., Chen, P., et al. (1999) Mechanisms of GDF-5 action during skeletal development. *Development* **126(6)**, 1305–1315.
31. Storm, E. and Kingsley, D. (1996) Joint patterning defects caused by single and double mutations in members of the bone morphogenetic protein BMP family. *Development* **122(12)**, 3969–3979.
32. Zhao, Q., Eberspaecher, H., Lefebvre, V., and De Crombrugghe, B. (1997) Parallel expression of Sox9 and Col2a1 in cells undergoing chondrogenesis. *Dev. Dyn.* **209(4)**, 377–386.
33. Bi, W., Huang, W., Whitworth, D. J., et al. (2001) Haploinsufficiency of Sox9 results in defective cartilage primordia and premature skeletal mineralization. *Proc. Natl. Acad. Sci. USA* **98(12)**, 6698–6703.
34. Bitgood, M. J. and McMahon, A. P. (1995) Hedgehog and BMP genes are coexpressed at many diverse sites of cell–cell interaction in the mouse embryo. *Dev. Biol.* **172(1)**, 126–138.
35. Ferguson, C. M., Miclau, T., Hu, D., Alpern, E., and Helms, J. A. (1998) Common molecular pathways in skeletal morphogenesis and repair. *Ann. NY Acad. Sci.* **857(5)**, 33–42.
36. Helms, J. A., Alpern, E. J., and Ferguson, C. (1998) Indian hedgehog signaling in embryonic skeletal morphogenesis and adult repair. Orthopaedic Research and Education Foundation Grant.
37. Vortkamp, A., Lee, K., Lanske, B., Segre, G. V., Kronenberg, H. M., and Tabin, C. J. (1996) Regulation of rate of cartilage differentiation by Indian hedgehog and PTH-related protein. *Science* **273(5275)**, 613–622.
38. St-Jacques, B., Hammerschmidt, M., and McMahon, A. P. (1999) Indian hedgehog signaling regulates proliferation and differentiation of chondrocytes and is essential for bone formation. *Genes Dev.* **13(16)**, 2072–2086.
39. Byrd, N., Becker, S., Maye, P., et al. (2002) Hedgehog is required for murine yolk sac angiogenesis. *Development* **129(2)**, 361–372.
40. Ducy, P., Zhang, R., Geoffroy, V., Ridall, A. L., and Karsenty, G. (1997) Osf2/Cbfa1: a transcriptional activator of osteoblast differentiation. *Cell* **89(5)**, 747–754.

41. Otto, F., Thornell, A. P., Crompton, T., et al. (1997) Cbfa1, a candidate gene for cleidocranial dysplasia syndrome, is essential for osteoblast differentiation and bone development. *Cell* **89(5)**, 765–771.
42. Mundlos, S., Otto, F., Mundlos, C., et al. (1997) Mutations involving the transcription factor CBFA1 cause cleidocranial dysplasia. *Cell* **89(5)**, 773–779.
43. Komori, T., Yagi, H., Nomura, S., et al. (1997) Targeted disruption of Cbfa1 results in a complete lack of bone formation owing to maturational arrest of osteoblasts. *Cell* **89(5)**, 755–764.
44. Sato, M., Morii, E., Komori, T., et al. (1998) Transcriptional regulation of osteopontin gene *in vivo* by PEBP2alphaA/CBFA1 and ETS1 in the skeletal tissues. *Oncogene* **17(12)**, 1517–1525.
45. Somerman, M. J., Prince, C. W., Sauk, J. J., Foster, R. A., and Butler, W. T. (1987) Mechanism of fibroblast attachment to bone extracellular matrix: role of a 44 kilodalton bone phosphoprotein. *J. Bone Miner. Res.* **2(3)**, 259–265.
46. Reinholt, F. P., Hultenby, K., Oldberg, A., and Heinegård, D. (1990) Osteopontin—a possible anchor of osteoclasts to bone. *Proc. Natl. Acad. Sci. USA* **87(12)**, 4473–4475.
47. Miyauchi, A., Alvarez, J., Greenfield, E. M., et al. (1991) Recognition of osteopontin and related peptides by an alpha v beta 3 integrin stimulates immediate cell signals in osteoclasts. *J. Biol. Chem.* **266(30)**, 20369–20374.
48. Jiménez, M. J., Balbín, M., López, J. M., Alvarez, J., Komori, T., and López-Otín, C. (1999) Collagenase 3 is a target of Cbfa1, a transcription factor of the runt gene family involved in bone formation. *Mol. Cell. Biol.* **19(6)**, 4431–4442.
49. Solheim, E. (1998) Growth factors in bone. *Int. Orthop.* **22(6)**, 410–416.
50. Gerber, H., Vu, T., Ryan, A. M., Kowalski, J., Werb, Z., and Ferrara, N. (1999) VEGF couples hypertrophic cartilage remodeling, ossification and angiogenesis during endochondral bone formation. *Nat. Med.* **5**, 623–628.
51. Risau, W. (1997) Mechanisms of angiogenesis. *Nature* **386(6626)**, 671–674.
52. Neufeld, G. (1999) Vascular endothelial growth factor (VEGF) and its receptors. *FASEB J.* **13**, 9–22.
53. Haigh, J. J., Gerber, H. P., Ferrara, N., and Wagner, E. F. (2000) Conditional inactivation of VEGF-A in areas of collagen2a1 expression results in embryonic lethality in the heterozygous state. *Development* **127(7)**, 1445–1453.
54. Probst, A. and Spiegel, H. U. (1997) Cellular mechanisms of bone repair. *J. Invest. Surg.* **10(3)**, 77–86.
55. Carter, D. C. and Giori, N. J. (1991) Effect of mechanical stress on tissue differentiation in the bony implant bed. University of Toronto Press. Toronto.
56. Carter, D. R., Beaupré, G. S., Giori, N. J., and Helms, J. A. (1998) Mechanobiology of skeletal regeneration. *Clin. Orthop. Related Res.* **82(355 Suppl)**, S41–S55.
57. Minet, E., Michel, G., Remacle, J., and Michiels, C. (2000) Role of HIF-1 as a transcription factor involved in embryonic development, cancer progression and apoptosis (review). *Int. J. Mol. Med.* **5(3)**, 253–259.
58. Thompson, Z., Miclau, T., Hu, D., and Heims, J. A. (2002) A model for intramembranous bone healing during fracture repair. *J. Invest. Med.* **50(1)**, 83A.
59. Tay, B. K., Le, A. X., Gould, S. E., and Helms, J. A. (1998) Histochemical and molecular analyses of distraction osteogenesis in a mouse model. *J. Orthop. Res.* **16(5)**, 636–642.
60. Sandell, L. J., Sugai, J. V., and Trippel, S. B. (1994) Expression of collagens I, II, X, and XI and aggrecan mRNAs by bovine growth plate chondrocytes *in situ*. *J. Orthop. Res.* **12(1)**, 1–14.

4

Biology of Bone Grafts

Victor M. Goldberg, MD and Sam Akhavan, MD

INTRODUCTION

The need for bone grafting to replace skeletal defects or augment bony reconstruction has become more prevalent recently because of enhanced capability to salvage major bone loss. There are many bone graft options available for the surgeon, including autografts or allografts, either of a cortical or cancellous structure, each of which has specific biological and mechanical properties. Some grafts are more dependent on the host bed for successful incorporation, such as freeze-dried allografts, while others, such as vascularized autografts, are capable of incorporating into the host bone under adverse physiological conditions. An understanding of the biological events and biomechanical aspects of autografts and allografts is important in understanding the processes that influence the incorporation of the bone graft into the host skeleton.

DEFINITION OF BONE GRAFT TERMS

The first level of describing bone grafts refers to the origin of the graft *(1,2)*. A graft transplanted from one site to another within the same individual is called an *autograft*. *Allografts* are tissues transferred from two genetically different individuals of the same species. *Xenografts* are transplanted from one species to a member of a different species. An *isograft* is transferred from one monozygotic twin to the other. This is usually described for laboratory experiments when tissue is transferred from inbred, genetically identical strains of animals. The anatomical placement of the graft is an important descriptor of bone transplantation. A graft transplanted to an anatomically appropriate site is defined as *orthotopic*, whereas if it is transplanted to an anatomically dissimilar site, it is termed *heterotopic*. Additionally, the graft may be described as *cortical, cancellous, corticocancellous,* or *osteochondral*. Fresh grafts are transferred directly from the donor to the recipient site. These grafts are usually autografts because of the immunogenetic potential of fresh allografts. The graft may be vascularized with its own blood supply or it may be nonvascularized. Allografts are usually modified or preserved to reduce immunogenicity before transplantation. These modification processes include freezing, freeze-drying, irradiation, or chemomodification *(1)*.

BONE GRAFT FUNCTION

The biological activity of bone graft is a result of two functions: osteogenesis and mechanical support (**Table 1**). Bone regeneration usually requires three processes: osteoinduction, osteochonduction, and osteogenic cells *(3)*. Osteogenesis is the physiological process whereby new bone is synthesized by cells of the graft or cells of host origin. Surface cells that survive transplantation of either cortical or cancellous grafts can produce new bone *(4–7)*. This new bone may initially be important for the development of callous during the early phase of bone graft incorporation. Cancellous bone, because

From: *Bone Regeneration and Repair: Biology and Clinical Applications*
Edited by: J. R. Lieberman and G. E. Friedlaender © Humana Press Inc., Totowa, NJ

Table 1
Comparative Properties of Bone Grafts

Bone Grafts	Mechanical properties	Osteoconduction	Inductiveness	Osteogenesis graft derived
Autograft				
Cancellous	±	+++	+++	+++
Cortical	+++	++	++	+
Allograft				
Cancellous				
Frozen	±	++	+	− −
Freeze-dried	−	++	+	− −
Cortical				
Frozen	+++	+	±	− −
Freeze-dried	+	+	±	− −
Demineralized	− −	+	++	− −
Allogenic				
Cancellous chips				

of its large surface area, has a greater potential for forming new bone than does cortical bone. Osteo-induction provides osteogenic potential by inducing the host bed to synthesize new bone. This is achieved by the recruitment of mesenchymal cells that differentiate initially into cartilage and then bone-forming cells. This much-studied process is achieved through the recruitment of graft-derived proteins that drive this physiological process. The most completely studied of these low-molecular-weight peptides are the bone morphogenic proteins (BMPs). A number of BMPs have been identified, and some are already in clinical use. The most active BMPs in bone include BMP-2, -4, and -7 *(8,9)*. These proteins play an important role in the differentiation of stem cells into osteoblasts and are also important in fracture healing and bone remodeling *(10)*. The presence of BMP in bone has been demonstrated both experimentally and clinically when the matrix has been demineralized and sequentially extracted to remove any antigenic materials *(3,11)*. This activity does not require viable graft cells because it is property of bone matrix. It is present in all bones, whether autografts, or allografts that have been preserved using a method that does not destroy the BMP, such as autoclaving. Osteoconduction is provided by all grafts as well as by biomaterials such as ceramics. This graft function provides the three-dimensional configuration for the ingrowth into the graft of host capillaries, perivascular tissue, and osteoprogenitor cells from the recipient.

Bone graft incorporation requires an interaction of osteoinduction and osteoconduction, described as creeping substitution. This ultimately leads to the replacement of the graft by host bone in a predictable pattern under the influences of load bearing *(3)*. Bone graft incorporation is a sequence of well-balanced processes between the graft and the host bed. Under most circumstances, all of the functions described above are in play. The initial inflammatory response results in the migration of inflammatory cells and fibroblasts into the graft. The hematoma formation that occurs enhances the release of both cytokines and growth factors. Osteoinduction drives chemotaxis, mitosis, and differentiation of the host osteoprogenitor cells. By d 5, chondrocytes are usually recognizable; and osteoblasts can be seen by the 10th posttransplantation day. Host blood vessels quickly invade the graft through existing haversian and Volkmann canals and also provide the osteoclasts that resorb the surfaces of the graft. Both intramembranous and endochondral bone formation usually occurs on graft surfaces. Osteo-conduction proceeds in large cortical or cancellous grafts for many years and ultimately results in the resorption of the original graft tissue and replacement with new host bone. The remodeling that results is a response to weight bearing. A successful outcome depends on a balance between revascularization and osteogenesis and the graft's response to applied loads. A biological balance must be achieved between the graft and the host bed to ensure successful incorporation. Clearly, bone graft incorporation is a dynamic interplay of the biological function of the bone graft, the graft environment, and the host–graft mechanical interactions.

BIOLOGY OF BONE GRAFTS

Bone graft incorporation is a prolonged process that involves a sequence of complex steps involving the interrelationship of the graft and host *(12–13)*. Autografts in general are more rapidly and completely integrated into the host than allografts. However, because of the morbidity associated with autograft harvesting and inadequate material, allografts have been utilized *(14)*. Allografts, although potentially functional, may remain an admixture of graft and host bone for many years. Because of the prolonged remodeling activity of the graft, it is usually difficulty to define the absolute end point of incorporation. However, one can consider a bone graft functional when it can withstand the normal loads of activities of daily living. Although there have been many approaches to understanding the incorporation of bone grafts such as radiographic, histological, and biomechanical methods, it is the histological process of bone graft incorporation that reflects the biological events of this process.

BIOLOGY OF AUTOGRAFTS

Cancellous Autografts

Hematoma formation and inflammation are rapidly seen in the early phases of bone autograft incorporation. Surface osteocytes may survive and are important in synthesizing new bone during the initial phases of incorporation *(15)*. The inflammatory infiltrate is minimal and consists mainly of small mononuclear cells. Revascularization of the graft occurs rapidly and is characterized by considerable capillary ingrowth *(16)*. Both osteoclasts and osteoblast precursors are seen early. The graft-derived BMPs play an important and central role in inducing host mesenchymal stem cells to migrate into the graft and differentiate into osteoblasts. There is an early, dynamic equilibrium established among inflammation, revascularization, osteoinduction, and osteoconduction, that by 4 wk provides active bone resorption and formation throughout the graft. Following the ingrowth of capillaries, osteoconduction proceeds rapidly. Osteoblasts line the edges of the dead trabecular during this stage, and new bone formation takes the form of immature woven bone on the surfaces of the original graft trabeculae. Seams of osteoid tissues surround the core of necrotic bone of the graft. Ultimately, hematopoietic cells accumulate within the transplanted bone and form a viable new bone marrow. The graft is well underway to complete resorption and new bone formation by 6 mo after transplantation. The remodeling and complete replacement of the nonviable graft bone is directed by Wolff's law. By 1 yr, the process of incorporation is usually complete and the graft is completely resorbed and replaced with new viable bone **(Fig. 1)**.

Cortical Autografts

Cortical grafts differ from cancellous grafts mainly in their rate of revascularization, their mechanism of repair, and the completeness of the repair *(1,13,17)*. The overall process of incorporation proceeds in a manner similar to that of cancellous autograft incorporation; however, because of the density of cortical bone, a cortical graft has a decreased rate of revascularization and remodeling. Vascular penetration of the graft by host tissue occurs only after resorption of the dense cortical surface by osteoclasts is initiated. The slower revascularization of these grafts in contrast to cancellous grafts usually results in the cortical bone becoming more radiolucent and significantly weaker than normal bone. This reduction in strength may last from months to years after transplantation, depending on the graft size and the implantation site. Unlike cancellous grafts, bone incorporation of cortical grafts occurs

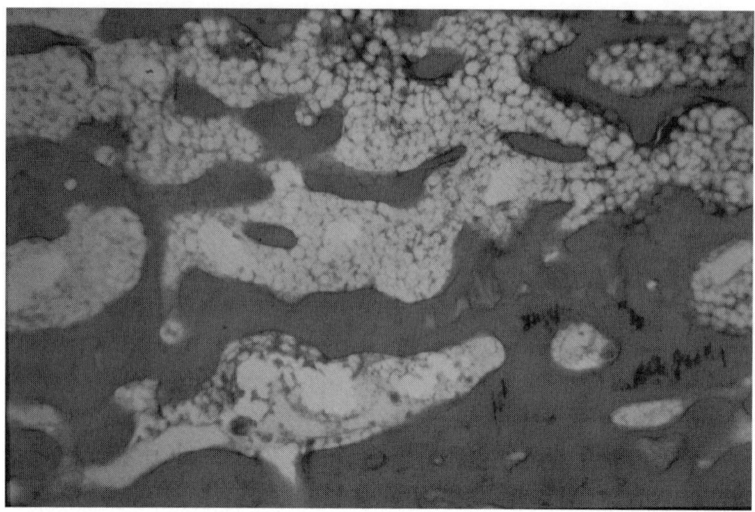

Fig. 1. Photomicrograph of a cancellous autograft 1 yr after transplantation to a canine ulnar defect, demonstrating complete replacement of the graft with viable host bone (hematoxylin and eosin ×25).

by appositional bone growth over a necrotic core *(18–22)*. Cortical grafts may be involved in load bearing sooner then cancellous grafts if the graft–host junction is stabilized by adequate fixation and heals rapidly to the host *(20)*.

Initially, following transplantation there is little histological differences between autologous cortical and cancellous grafts. There is a similar rapid inflammatory response and hematoma formation. The pattern of revascularization that follows slowly behind osteoclast resorption follows preexisting haversian canals from the periphery to the interior *(1,21,22)*. By 2 wk, widespread resorption of the grafts is well underway, and it increases during the initial 6 mo. As discussed earlier, the result of this process is a generalized decrease in the strength of cortical autografts. There is a fine balance between graft resorption by osteoclasts and new bone formation by osteoblasts. There is, however, a normal physiological uncoupling of bone resorption and bone formation as a result of the differential activity of osteoblasts and osteoclasts. Osteoclasts can resorb bone at a rate of 50 μm/d, whereas osteoblasts can synthesize new bone tissue at a rate of only 1 μm/d. This uncoupling may cause cortical bone grafts to fail, even under the best circumstances. Any additional significant uncoupling of this process will cause delay in the incorporation process. Lack of immediate blood supply in nonvascularized cortical autograft leads to the death of most of the graft's osteocytes. This may reduce the osteoinductive potential of these grafts, although osteoconduction is present throughout the process of incorporation. Normal marrow may appear in the remodeled bone graft by 9 mo following implantation.

Because nonvascularized cortical autografts result in osteocyte death and the reduction in the efficiency of function, vascularized cortical autografts have been used to reconstruct deficient bone. Vascularized cortical autografts provide an immediate blood supply and experience only a transient ischemia. Many studies have demonstrated the superiority of vascularized grafts in the immediate, short- and long-term postoperative period *(23–25)*. Clinical studies have demonstrated a 50% reduction in failure rate when these grafts are used to reconstruct defects larger then 6 cm in length when compared to nonvascularized autografts. Over 90% of the donor osteocytes may survive the transplantation. Rapid healing is seen at the graft–host interface, and the uncoupling of bone resorption and formation is diminished. Vascularized cortical grafts are therefore not usually weakened by resorption. In general, these grafts during the first 3 mo are stronger, stiffer, and less porotic then nonvascularized autografts.

Because vascularized cortical autografts may be incorporated independently of the host bed, they may function in biologically deficient host environments. Remodeling of cortical autografts is a complex process that requires graded controlled loading. During the initial phases of incorporation, protection of the graft–host interface with the use of internal or external fixation is crucial to prevent the formation of fractures and failure of the grafts.

BIOLOGY OF ALLOGRAFTS

Bone allografts have been used effectively in clinical practice because of the inadequate supply and donor site morbidity associated with autogenous bone grafts. Bone allografts have been shown to be immunogenic and to demonstrate a higher failure rate when compared to autografts. The immunogenicity of these grafts appears to play an important role in the successful incorporation of bone allografts *(26–32)*. The antigens most responsible for the recognition of the graft by the host are those of the major histocompatibility complex (MHC) *(32)*. The MHC antigens most important in this process are those of the Class I and Class II molecules. These major alloantigens are recognized by the host responding T-lymphocytes. The cells of all musculoskeletal tissues display Class I MHC antigens, and some cells may display Class II antigens. Minor histocompatibility antigens may also be present on the surface of cells and can be important in the late rejection of bone allografts. In general, the mechanism of immune rejection of bone grafts is similar to that of parenchymal organs. These mechanisms may include cell-mediated toxicity, antibody-mediated toxicity, and antibody-dependent cell-mediated cytotoxicity. Many experimental studies have shown that bone allografts may invoke all of these responses in vivo *(32)*. Furthermore, these studies have shown that when histocompatibility differences are reduced by either matching tissue types or modifying allografts to reduce immunogenicity, allograft acceptance is improved. However, the exact mechanism and importance of the immune response in bone allograft incorporation is unclear. These studies suggest that the immune response delays and may destroy the initial osteoinduction phase of the bone graft, and that any blood vessels present in the allograft are quickly surrounded by inflammatory cells, occluded, and result in rapid necrosis of marrow cells and osteoblasts. Additionally, animal studies have demonstrated that bone allografts do induce graft-specific antibodies *(30,31,33)*. Mismatched fresh grafts appear to invoke humoral responses to Class I antigens more than when frozen allografts are utilized *(34)*. Other studies have demonstrated that T-cells may be activated by MHC-mismatched allogeneic bone *(31)*. Studies have demonstrated that bone marrow cells may be the primary means of inducing the immune response, while other studies suggest that cells within the cortex are also capable of activating allogeneic T-cells *(32)*. Taken together, these experimental studies provide data that allografts can induce an immune reaction in the host. However, the clinical significance of these responses is still unclear. Notwithstanding this controversy, it is the capability of fresh bone allografts to evoke a rapid immune response that results in destruction of the graft that has led to the use of preserved modified allografts *(3)*.

Cancellous Allografts

Preservation of cancellous allografts reduces the immune response and does improve graft acceptance. Preservation of allograft using freezing or freeze drying has been demonstrated to improve incorporation of the graft *(3,35)*. Revascularization and remodeling of these processed cancellous allografts are delayed compared with fresh autografts, but osteoinduction and osteoconduction are generally preserved and incorporation of the graft can be complete. The overall process is slower, although a similar sequence of events occurs when compared to autografts. Initially, hemorrhage and a cellular infiltrate is seen. It usually reaches its peak during the first 2 wk. The processes of resorption of the grafts, osteoinduction and osteoconduction, proceeds until viable bone is present that can withstand load bearing. The clinical incorporation of massive frozen allografts is improved when rigid internal fixation is provided. Other preservation techniques such as decalcification and demineralization, although somewhat less effective than freezing or freeze drying grafts in regenerating bone, do provide to some extent osteoinduction and osteoconduction. Demineralization of grafts results in the loss of their inherent mechanical strength, while morselized cancellous frozen or freeze-dried material may retain some resistance to compression. Because processed cancellous allografts remain an admixture of necrotic graft bone and viable host tissue for a prolonged period of time, their clinical use has been confined to filler material for cavitary skeletal defects.

Cortical Allografts

Because fresh cortical allografts also invoke an immune response, clinical use of cortical allografts is generally confined to processed allografts that have been either frozen or freeze-dried *(3)*. The processing of these allografts also plays an important role in ensuring the safety of bone allografts. The chance of transmission of communicable diseases—mainly human immune virus and hepatitis B and C—has been dramatically reduced by appropriate donor screening and sterilization techniques *(36)*. The transplantation of cortical allografts free of marrow and blood products has resulted in safe outcomes. When allografts are modified and stabilized to the host with internal fixation devices, the biological process of incorporation proceeds in orderly steps. Osteoclastic resorption is the initial event and provides the means for vascular invasion of the haversian and Volkmann's canals by host capillaries and osteoprogenitor cells. The process is delayed compared to fresh autografts, but new appositional bone growth occurs by osteoconduction. Cortical allografts may be substantially weaker than autografts for as long as 2 yr after surgery *(20,37)*. However, results from retrieved human bone allografts demonstrate that bony union occurs at the host graft cortical–cortical junction by means of bridging external callous that originates from the periosteum of the host bone and extends for up to 3 cm on the surface of the allograft *(20,38)*. Junctional discontinuities are filled with fibrovascular tissue that progresses to woven bone, and eventually haversian bone under the influence of Wolff's law is formed. Central to successful bone allograft functioning is the stability of the graft–host junction. Even in experimental models using fresh allografts, when interfaces were successfully stabilized, fresh allograft host junctions healed *(31)*. When intimate host–graft contact was not achieved or the union was not stable, these interfaces invariably failed.

Although freeze drying reduces the inflammatory response to bone graft, it also reduces the mechanical properties of the graft and results in a significantly weaker graft *(39)*. Sterilization of bone by irradiation of more than 30 kGy may destroy any osteoinductive function. Chemosterilized, autolyzed, antigen-extracted allogeneic bone, although providing inductive capabilities, has little strength *(3)*.

In an effort to improve incorporation of cortical allografts, new methodologies have been developed. Perforation of the graft from a biological standpoint increases the available surface area for ingrowth and ongrowth of new bone *(40,41)*. Additionally, it provides easier access to the intramedullary canal. Several studies have demonstrated that perforated grafts indeed have more new bone ingrowth when compared to similar nonperforated grafts *(40,41)*. These grafts are also more porous in the 6 mo following surgery because of the increased area availability to osteoclasts for bone resorption. This method may also help revascularization of the allograft by providing channels for ingrowth of host blood vessels. However, the overall repair process is not that different from that in standard cortical grafts. Perforation of the graft has raised some concern in the past about the possibility of stress rises at the perforation sites *(42,43)*. Studies have shown, however, that the strength of the bone immediately following drilling is not diminished significantly in either compression or bending *(17,40)*. There is a decrease in the overall strength after 4 wk, but this was associated with the increased porosity of the graft rather than drill holes. By 6 wk after transplantation, the strength of the graft returned to that of the nonperforated grafts. Recent studies have combined partial demineralization with perforation of the graft and have demonstrated a positive effect on overall osteoinduction while preserving some of the biomechanical properties of this graft *(41)*. Overall cortical allograft incorporation is a complex process that has significant variables which influence the ultimate incorporation and function of the graft.

BIOMECHANICS OF BONE GRAFTS

Mechanical performance of a bone graft *in vivo* as has been discussed is a function of the intrinsic property of the graft and the properties of the graft–host interface *(44)*. The intrinsic properties of the graft are a function of its geometry as well as its composition and includes properties such as fracture toughness, yield strength, and its elastic modulus. If the graft possesses the same mechanical properties and geometry as the host bone, it may function in a clinical setting almost immediately *(45)*. In a situation where the mechanical properties of the graft are inferior to that of the host, additional graft material should be used or the construct may be augmented with internal fixation until remodeling occurs and the graft can provide load-bearing function. A bone graft must be biologically incorporated into the host in order to function successfully in load bearing. Incorporation of the graft is related directly to the mechanical and biological properties of the graft–host junction. A well-incorporated graft such as a cortical graft bridging a femoral defect shares some of the load of the femur and remodels to the requirement of the host. If the same graft does not heal to the host, aberrant loads may be experienced and failure of the construct usually occurs. In a large segmental defect, it may be necessary to augment the graft–host junction with either internal or external fixation in order to protect the graft while it is being remodeled. Studies have shown that a dominant parameter in determining the material properties of bone is volume fracture of tissue in any given sample *(44,46)*. This value, also known as porosity, is related directly to the stiffness of the tissue (as a third power of porosity) and yield strength (as a second power of porosity) *(46)*. As a result, small changes in porosity result in large changes in the material properties of bone. Cortical bone grafts initially may have as little as 5–10% porosity. The biological sequence of events proceeds and, as graft incorporation occurs, large increases in porosity may result. This significantly reduces the strength of the bone graft. This critical period may last for as long as 2 yr and ultimately, through the biological processes described earlier, result in the successful incorporation of the bone graft. However, if the process becomes uncoupled and bone resorption significantly outstrips bone accretion, and the graft–host interface is inadequate, rapid failure of the bone graft may occur.

Cortical and cancellous grafts have different biomechanical properties because of the different biology of each graft type. Porosity of cancellous grafts may be as high as 70–80%, leading to material strength roughly equivalent to 4% of that of cortical bone *(44)*. The strength of cancellous bone grafts increases as new bone is laid down on the preexisting trabecula. However, until the new graft is successfully integrated into the host, it is critical that fixation methods sustain a significant portion of the load. In both cancellous and cortical grafts, the remodeling process is driven by functional loading under the influence of Wolff's law. It is important to achieve a balance between appropriately protecting the graft during the remodeling phase while allowing the bone graft to experience physiological loads necessary for remodeling to occur. Internal and external systems have been developed that provide protection for the graft while allowing some loading by the patient. As has been discussed previously, bone allografts do evoke an immune response. In order to reduce the immune reaction, modification of the graft has become a common practice as a method of preservation and sterilization. These modifications, however, have a profound effect on the biomechanical properties of the graft. Freezing has been shown to have minimal effects on the biomechanical properties of the graft, while freeze drying significantly reduces both the yield strength and stiffness of the bone graft *(47)*. Other methods have varying effects. Autoclaving has been shown to produce a dose-dependent decrease in both strength and stiffness of bone *(48)*. Irradiation, although effective in destroying bacteria at relatively low doses (<20 kGy), does not usually destroy viruses at this dose. However, virucidal doses greater the 30 kGy significantly reduce the material properties of the bone graft. Complete demineralization of the bone graft, although significantly reducing the immunogenicity of the bone, results in loss of all of virtually its mechanical properties. The ultimate success of any bone graft, however, requires a balance between its biological functions and biomechanical properties.

CONCLUSIONS

A successful clinical outcome for a bone grafting procedure requires an understanding of the biological and mechanical environment into which the graft will be placed. Although the biological aspects of bone graft incorporation are critical in determining this outcome, the technical aspects of the surgery are as important. A clean, well-vascularized host bed is critical in providing the satisfactory host environment. Wide excision of scar tissue, treatment of infection, protection of the blood supply, and satisfactory soft tissue coverage is mandatory. The selection of appropriate graft material for the desired clinical function will also help determine the clinical outcome of bone grafting. Central, however, in the successful incorporation of the bone graft is a stable fixation and contact between the host bone and the graft. Experimental studies have demonstrated that when the host–graft interfaces are tightly apposed and fixed with internal fixation, the interfaces healed whether the grafts were autogenous, allogeneic, fresh, or frozen. Even under stable conditions, but without closely apposed host bone, graft tissue retrieval studies have demonstrated that interfaces did not heal and had a profound effect on the biological characteristics of the graft *(38)*. When no apposition or stability was provided at this interface, bone grafts have uniformly failed. It is therefore important to provide intrinsic and stable graft–host fixation and satisfactory soft tissue coverage. The bone graft must also be protected from full weight bearing until remodeling enables it to function fully in a loaded environment. When the appropriate bone graft is selected and the surgical technique is synergistic, bone grafts do incorporate both biologically and functionally and provide clinically functional load bearing.

REFERENCES

1. Burchardt, H. (1987) Biology of bone transplantation. *Orthop. Clin. N. Am.* **18(2)**, 187–196.
2. Goldberg, V. M. (2001) Bone grafts and their substitutes: facts, fiction and failures. *Orthopaedics* **24**, 875–876.
3. Urist, M. (1980) Bone transplants and implants, in *Fundamental and Clinical Bone Physiology* (Urist, M., ed.), Lippincott, Philadelphia, pp. 331–368.
4. Gray, J. and Elves, M. (1982) Donor cells' contribution to osteogenesis in experimental cancellous bone grafts. *Clin. Orthop.* **163**, 261.
5. Gray, J. and Elves, M. (1979) Early osteogenesis in compact isografts: a quantitative study of contributions of the different graft cells. *Calcif. Tissue Int.* **29**, 225–237.
6. Bonfiglio, M. (1958) Repair of bone-transplant fractures. *J. Bone Joint Surg.* **40A**, 446–456.
7. Bassett, C. (1972) Clinical implications of cell function in bone grafting. *Clin. Orthop.* **87**, 49–59.
8. Muschler, G., Hyodo, A., and Manning, T. (1994) Evaluation of human bone morphogenic protein 2 in a canine fusion model. *Clin. Orthop.* **308**, 229–240.
9. Cook, S., et al. (1995) Effect of recombinant human osteogenic protein-1 on healing of segmental defects in non-human primates. *J. Bone Joint Surg.* **77a**, 734–750.
10. Long, M., et al. (1995) Regulation of human bone marrow-derived osteoprogenitor cells by osteogenic growth factors. *J. Clin. Invest.* **95**, 881–887.
11. Mizutani, H. and Urist, M. (1982) The nature of bone morphogenic protein fractions derived from bovine bone matrix gelatin. *Clin. Orthop.* **171**, 213.
12. Stevenson, S., Emery, S., and Goldberg, V. (1996) Factors affecting bone graft incorporation. *Clin. Orthop.* **323**, 66–74.
13. Stevenson, S., et al. (1997) Critical biological determinants of incorporation of non-vascularized cortical bone grafts. *J. Bone Joint Surg.* **79A**, 1–6.
14. Younger, E. and Chapman, M. (1989) Morbidity of bone graft donor sites. *J. Orthop. Trauma* **3(3)**, 192–195.
15. Axhausen, W. (1956) The osteogenesitic phase of regeneration of bone. *J. Bone Joint Surg.* **38**, 593–600.
16. Goldberg, V. and Lance, E. (1972) Revascularization and accretion in transplantation. *J. Bone Joint Surg.* **54A**, 807–816.
17. Burchardt, H., Busbee, G., and Enneking, W. (1975) Repair of experimental autologous grafts of cortical bone. *J. Bone Joint Surg.* **57A**, 814.
18. Heiple, K., Chase, S., and Herndon, C. (1963) A comparative study of the healing process following different types of bone transplantation. *J. Bone Joint Surg.* **45A**, 1593–1616.
19. Chase, S. and Herndon, C. (1955) The fate of autogenous and homogenous bone grafts. A historical review. *J. Bone Joint Surg.* **37A**, 809–841.
20. Enneking, W. and Campanacci, D. (2001) Retrieved human allografts. a clinicopathological study. *J. Bone Joint Surg.* **83A(7)**, 971–986.

21. Abbott, L., et al. (1947) The evaluation of cortical and cancellous bone as grafting material: a clinical and experimental study. *J. Bone Joint Surg.* **29A**, 391.

22. Enneking, W., et al. (1975) Physical and biological aspects of repair in dog cortical-bone transplants. *J. Bone Joint Surg.* **57A**, 237–252.

23. Gazdag, A., et al. (1995) Alternatives to autogenous graft: efficacy and indications. *J. Am. Acad. Orthop. Surg.* **3**, 1–8.

24. Doi, K., Tominaga, S., and Shibata, T. (1977) Bone grafts with microvascular anastomoses of vascular pedicles. *J. Bone Joint Surg.* **59A**, 809–815.

25. Goldberg, V., et al. (1990) Biological and physical properties of autogenous vascularized fibular grafts in dogs. *J. Bone Joint Surg.* **72A**, 801–810.

26. Bos, G., et al. (1983) The effect of histocompatibility matching on canine frozen bone allograft. *J. Bone Joint Surg.* **65A**, 89–96.

27. Goldberg, V., et al. (1984) Improved acceptance of frozen bone allografts in genetically mismatched dogs by immuno-suppression. *J. Bone Joint Surg.* **66**, 937–950.

28. Goldberg, V., et al. (1985) Bone grafting: role of histocompatibility in transplantation. *J. Orthop. Res.* **3(4)**, 389–404.

29. Bos, D., et al. (1985) The long-term fate of fresh and frozen orthotopic bone allografts in genetically defined rats. *Clin. Orthop.* **197**, 245–254.

30. Stevenson, S. (1987) The immune response to osteochondral allografts in dogs. *J. Bone Joint Surg.* **69A**, 573–582.

31. Stevenson, S, Li, S., Martin, B. (1991) The fate of cancellous and cortical bone after transplantation of fresh and frozen tissue-antigen-matched and mismatched osteochondral allografts in dogs. *J. Bone Joint Surg.* **73A**, 1143–1156.

32. Stevenson, S. and Horowitz, M. (1992) Current concepts review: the response to bone allografts. *J. Bone Joint Surg. Am.* **74A**, 939–950.

33. Friedlaender, G., Strong, D., and Sell, K. (1976) Studies on the antigenicity of bone. *J. Bone Joint Surg.* **58A**, 854–858.

34. Stevenson, S., Shaffer, J., and Goldberg, V. (1996) The humoral response to vascular and nonvascular allografts of bone. *Clin. Orthop.* **323**, 86–95.

35. Burchardt, H., et al. (1978) Freeze-dried allogeneic segmental cortical-bone grafts in dogs. *J. Bone Joint Surg.* **60A**, 1082–1090.

36. Tomford, W. and Mankin, H. (1999) Bone banking—update in methods and materials. *Orthop. Clin. N. Am.* **30(4)**, 565–570.

37. Burchardt, H., Glowczewskie, F., and Enneking, W. (1977) Allogenic segmental fibular transplants in azathioprine-immunosuppressed dogs. *J. Bone Joint Surg.* **59A**, 881.

38. Enneking, W. and Mindell, E. (1991) Observations on massive retrieved human allografts. *J. Bone Joint Surg.* **73A**, 1123–1142.

39. Pelker, R. and Friedlaender, G. (1987) Biomechanical aspsects of bone autografts and allografts. *Orthop. Clin. N. Am.* **18**, 235–239.

40. Delloye, C., et al. (2002) Perforations of cortical bone allografts improve their incorporation. *Clin. Orthop.* **396**, 240–247.

41. Lewandrowski, K., et al. (1997) Improved osteoinduction of cortical bone allografts: a study of the effects of laser perforation and partial demineralization. *J. Orthop. Res.* **15(5)**, 748–756.

42. Brooks, D., Burstein, A., and Frankel, V. (1970) The biomechanics of torsional fractures: the stress concentration effect of a drill hole. *J. Bone Joint Surg.* **52A**, 507–514.

43. Burstein, A., et al. (1972) Bone strength: the effect of screw holes. *J. Bone Joint Surg.* **54A**, 1143–1156.

44. Davy, D. (1999) Biomechanical issues in bone transplantation. *Orthop. Clin. N. Am.* **30(4)**, 553–563.

45. Goldberg, V. (2000) Selection of bone grafts for revision total hip arthroplasty. *Clin. Orthop.* **381**, 68–76.

46. Carter, D. and Hayes, W. (1977) The compressive behavior of bone as a two-phase porous structure. *J. Bone Joint Surg.* **59A**, 954.

47. Pelker, R., Friedlaneder, G., and Markham, T. (1983) Biomechanical properties of bone allograft. *Clin. Orthop.* **174**, 54.

48. Taguchi, R., et al. (1995) Autoclaved autograft bone combined with vascularized bone and bone marrow. *Clin. Orthop.* **320**, 220.

5

Cell-Based Strategies for Bone Regeneration

From Developmental Biology to Clinical Therapy

Scott P. Bruder, MD, PhD and Tony Scaduto, MD

INTRODUCTION

The cellular and molecular events of skeletal morphogenesis, remodeling, and repair have their basis in the embryonic production of bone *(1–7)*. For this reason, detailed analysis of developing osseous tissue in experimental models can provide the framework for understanding the cellular processes that occur in skeletal tissue as an organism grows and matures. This chapter is not intended to provide a comprehensive review of all cell-based therapies directed at bone regeneration. Instead, it aims to characterize the cellular events of osteoblastic differentiation, help clarify the lineage relationships between cells at various stages of maturity, and demonstrate how this cellular information has been used to design effective therapeutic strategies for bone repair and regeneration. While the bulk of the experimental data illustrated herein is derived from the authors' laboratory, we provide these as examples against which results from additional investigators can be compared and contrasted.

HISTOGENESIS OF EMBRYONIC BONE

Detailed investigations of the morphological changes that occur during embryonic bone development highlight the existence of several key features common among various species including aves, rodents, and man *(2)*. Using the embryonic chick, we articulated a model for describing limb development, which has as its central tenets the following observations: (1) chondrogenic and osteogenic cell commitment occur early in limb formation, (2) a layer of osteoprogenitor cells resides in a collar surrounding a prechondrogenic core of cells, (3) expression of the osteoblast phenotype governs the boundaries of cartilage development, (4) cartilage is replaced by vascular and marrow elements, and (5) vasculature plays a key role in the patterning of bone formation *(8,9)*.

At the cellular level, osteoprogenitors derived from lateral plate mesoderm are observed in a stacked configuration around a prechondrogenic core in the limb. These stacked cells are fated to become the periosteum, which is eventually composed of an outer fibrous layer and an inner cambium layer. Newly committed osteogenic cells emerge from the stacked cell layer to form the osteoblastic portion of the periosteum, as noted by their alkaline phosphatase activity. Subsequently these cells secrete and mineralize osteoid matrix toward the cartilage core, and are followed spatially and functionally by a continuous wave of newly differentiated osteoblasts arising from the stacked cell layer. Capillaries penetrate the periosteum and align themselves above the first elements of mineralized bone as an anastamosing network that orients the secretion of resident osteoblasts (**Figs. 1** and **2A,B**). The process of mineralization in avian bone is associated with the elaboration of various phosphoproteins, including an approx

From: *Bone Regeneration and Repair: Biology and Clinical Applications*
Edited by: J. R. Lieberman and G. E. Friedlaender © Humana Press Inc., Totowa, NJ

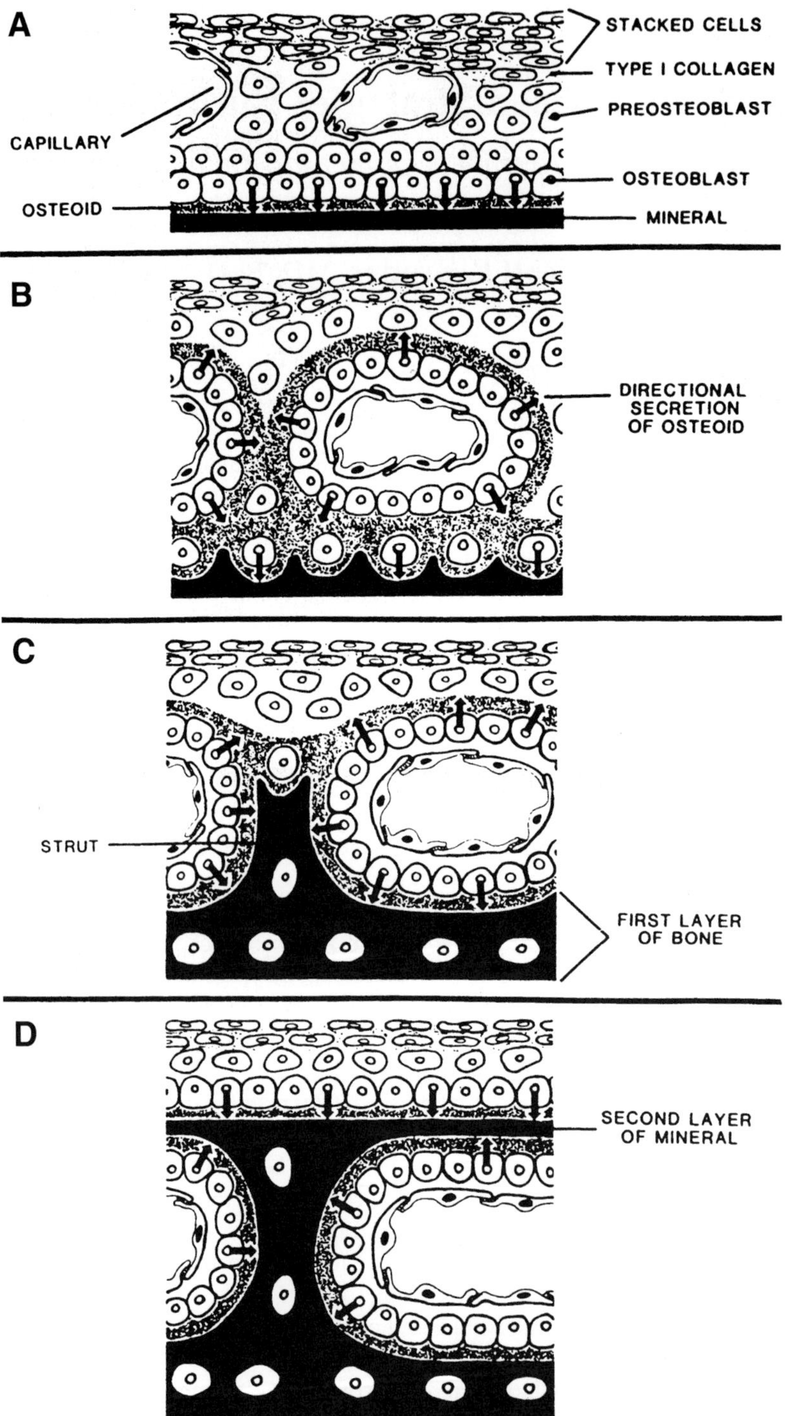

Fig. 1. This schematic drawing depicts the association of vasculature with the fabrication of struts and rings which comprise the developing chick bone. (**A**) The orientation of osteoblast secretion is in the direction of the arrows, with the cell's back against the vasculature, and its front toward the region of fresh osteoid. In this model, osteoblasts secrete osteoid away from the vasculature (**B**), causing the formation of a strut (**C**) and eventually the second layer of mineralized bone.

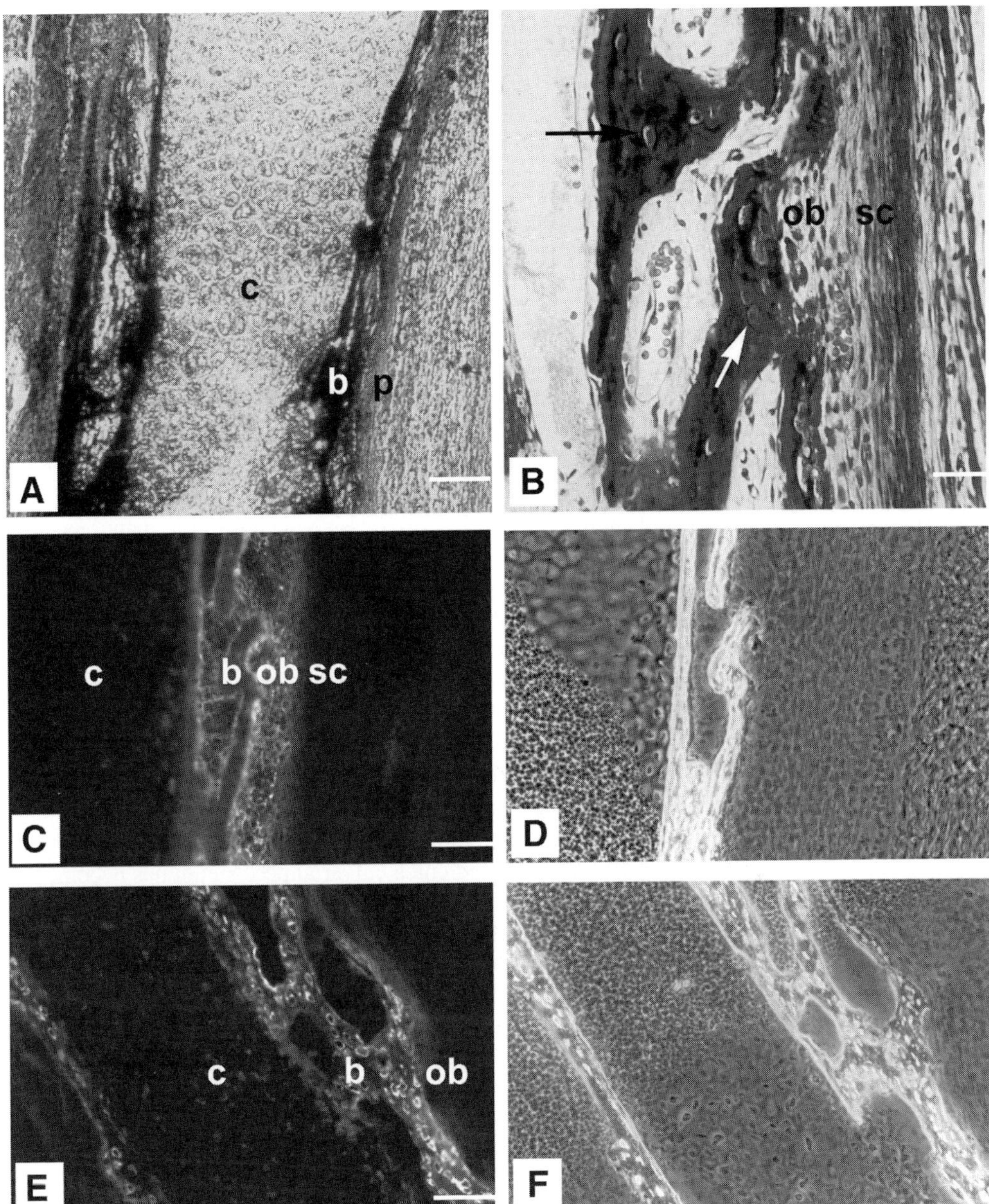

Fig. 2. (A) The mid-diaphysis of a stage 35 embryonic chick tibia is presented in longitudinal section. The cartilage core (c) anlagen is surrounded by a von Kossa/H&E-stained mineralized collar of bone (b), which in turn is circumscribed in a bilayered periosteum (p). **(B)** High-power magnification of a Mallory-stained osteogenic collar illustrates the outer layer of stacked cells (sc), the inner layer of osteoblastic cells (ob), and the osteocytes (arrows) embedded in the bone matrix. Immunofluorescent **(C)** staining with SB-1 demonstrates a broad band of osteoblastic cells within the periosteum and on the surface of the bone, which is apparent in the phase-contrast image **(D)**. **(E)** Osteocytes encased within the bone matrix are stained with SB-5. **(F)** Phase-contrast microscopy illustrates the cartilage core, the bone matrix, and the surrounding periosteum in this osteocyte-rich specimen. Bar in **A**, 120 μm. Bar in **B**, 40 μm. Bar in **C** and **E**, 60 μm. (Color illustration in insert following p. 212.)

66-kDa homolog of mammalian osteopontin *(10)*. As the vasculature penetrates the first collar of bone formed along the diaphysis, phagocytic cells remove the hypertrophic cartilage core and allow its replacement by stromal and hematopoietic marrow elements. In this way, the cartilage core precisely defines the geometric boundaries of the eventual marrow cavity.

CELL LINEAGE AND THE ORIGIN OF OSTEOBLASTS

Differentiation of the totipotent zygote into developmentally restricted pluripotent stem cell populations, and the subsequent commitment and expression of specific cellular phenotypes, is believed to be regulated by a variety of factors including inherent genomic potential, cell–cell interactions, and environmental cues. In considering the mechanisms involved, the concept of cell lineage is fundamentally relevant. Our logic, in part, is based on the cellular relationships proven to exist in the hematopoietic lineage pathways. As it is generally understood, the term *lineage* refers to the progression of particular cell precursors as they mature and give rise to differentiated cells, tissues, and organs. The accurate description of such a cell lineage depends on the ability to identify a particular feature, or collection of features, which can be traced from the differentiated cell type back through its precursors. Because the formation of specialized tissues is a progressive phenomenon, many generations of cells lie between the stem cell and the fully differentiated phenotype, which forms the mature tissue.

Our hypothesis was that a discrete series of steps, or transitions, exists between osteoprogenitor cells and the fully expressive osteoblast, comparable to that documented during hematopoiesis *(11)* or development of the nematode *Caenorhabditus elegans (12)*. Analysis of these lineage steps is, paradoxically, both facilitated and complicated by the variety of tissues containing osteoblast progenitors. Embryonic limb bud mesenchyme, developing and mature periosteum, and bone marrow all contain these precursors. In addition, calvarial tissue, which is derived from neural crest, possesses a repository of progenitor cells. Fortunately, experimental systems for analyzing each of these tissues have been developed. In addition to dynamic studies of limb development *in situ*, conditions to demonstrate differentiation of osteoblasts from isolated periosteum in vitro *(13,14)* and in vivo *(15,16)* have been established. For example, organ culture of folded chick calvarial periosteum has become a useful model for studies of bone cell differentiation. In addition, inoculating marrow cell suspensions into diffusion chambers and implanting these chambers into athymic mouse hosts served to provide the first evidence for osteochondral progenitors in bone marrow *(17,18)*. While host-derived cells are prevented from entering the chamber, nutrients and growth factors may pass freely through its pores. In this setting, bone forms along the inner surface of the porous membrane, adjacent to external vasculature, and cartilage forms within the center of the chamber.

Although these anatomically distinct sources of progenitor cells all give rise to bone, the precise sequence of cellular transitions that occurs during maturation has not been appreciated fully. That is, do marrow-derived and periosteal osteoblasts proceed through the same developmental pathway? Does embryonic limb bud mesenchyme give rise to osteoblasts through a different sequence than ectodermal neural crest? And finally, how do these cellular transitions compare between embryonic and adult sources of osteoprogenitors, both in vivo and in vitro? As a basis for answering these questions, one must first understand that many developing eukaryotic systems have been studied using biochemical and immunological techniques aimed at demonstrating alterations in the surface architecture of cells as a function of stage-specific morphologies, activities, and requirements. In recent years, investigators have used monoclonal antibody technology to generate probes that detect these alterations. This is especially clear in the study of hematopoiesis, which now boasts over 160 cell-surface cluster designation (CD) antigens. These probes also provide the means by which to purify antigens or cells, and in some cases, determine the function of the molecules for which they select. As an extension of this successful logic, we sought to generate a battery of monoclonal antibody probes selective for surface antigens on osteogenic cells at various stages of differentiation.

THE GENERATION OF MONOCLONAL
ANTIBODIES AGAINST OSTEOGENIC CELL SURFACES

We immunized mice with a heterogeneous population of chick embryonic bone cells obtained from the first bony collar formed in the tibia, and subsequently generated and selected for monoclonal antibodies against osteogenic cell surface determinants. Supernatants from growing hybridomas were definitively screened immunohistochemically against frozen sections of developing tibiae. Four unique cell lines were cloned, referred to as SB-1, SB-2, SB-3, and SB-5, each of which reacts with a distinct set of cells in the developing bone *(19–22)*. Detailed morphologic analyses of the dynamic changes during bone histogenesis document the restricted expression of specific antigens during embryogenesis. Progenitor cells in the stacked cell layer are not stained by any of these antibodies; however, a broad layer of cells between the surface of newly formed bone and the overlying inner cambium layer react with SB-1 (**Fig. 2C,D**). By contrast, SB-3 and SB-2 appear sequentially during the maturation of cells as they begin to secrete osteoid matrix and initiate mineralization. As a subset of these cells begins to surround themselves with bone matrix, the SB-1 and SB-3 antigens are lost. The resulting SB-2-positive cells then express the SB-5 antigen, which is restricted to nascent and mature osteocytes. The subsequent loss of SB-2 reactivity and the formation of characteristic stellate processes that react with SB-5 and extend through the bone matrix define this terminal differentiation step (**Fig. 2E,F**). By carefully tracking the reactivity of discrete cell populations, these experiments not only establish the existence of an osteoblastic lineage, but also indicate that osteocytes are derived directly from cells expressing the SB-1, -2, and -3 antigens.

A natural progression of this effort was to identify the antigens recognized by these antibodies. One antibody, SB-1, was observed to react with a family of cells in bone, liver, kidney, and intestine that were identically stained by the histochemical substrate for alkaline phosphatase (APase) *(20)*. Partial purification of intestinal or bone APase on a Sepharose CL-6B column results in the co-elution of enzyme activity and high-affinity antibody-binding material. Western immunoblots of bone extract show that SB-1 reacts with a single approx 155-kDa band, which also is stained in the sodium dodecyl sulfate (SDS)-polyacrylamide gel by APase substrate. In a similar set of immunoblot experiments, SB-1 reacts with an intestinal APase isoenzyme whose molecular weight is approx 185 kDa. The reactive epitope was found to be stable to SDS denaturation, not associated with the active site of the enzyme, and dependent on disulfide bonds that impart secondary structure to the protein *(23)*. Efforts to identify the antigens recognized by the other antibodies have met with only limited success. Preliminary immunoblot data indicate that SB-5 reacts with an approx 37-kDa protein extracted from freshly isolated osteocyte membranes; however, neither we nor Nijweide and colleagues *(5,24)* have yet identified a specific antigen present on avian osteocytes. Nevertheless, it is important to emphasize that the identity of the antigens need not be known in order for these probes to remain as useful markers for characterizing the lineage of osteogenic cells.

OSTEOPROGENITOR CELLS
FROM ISOLATED PERIOSTEUM AND BONE MARROW

Unlike traditional culture methods using collagenase-liberated osteoblastic cells, calvarial periosteal explants form a mineralized bone tissue in 4–6 d that is virtually identical to the in vivo counterpart *(14)*. Examination of fresh explants confirmed that no mature osteoblastic cells were present, although a discontinuous layer of SB-1-reactive preosteoblasts was evident in some regions. The inner (cambial) surface of the tissue was folded onto itself, and the explant was then cultured at the air–fluid interface in the presence of dexamethasone, a synthetic glucocorticoid capable of stimulating osteoprogenitor cell differentiation. As the wave of differentiation swept through the cultured tissue, antibody SB-1 reacted with the surface of a large family of cells associated with the developing bone. SB-3 and SB-2 reacted with progressively smaller subsets of cells, namely, those in successively closer

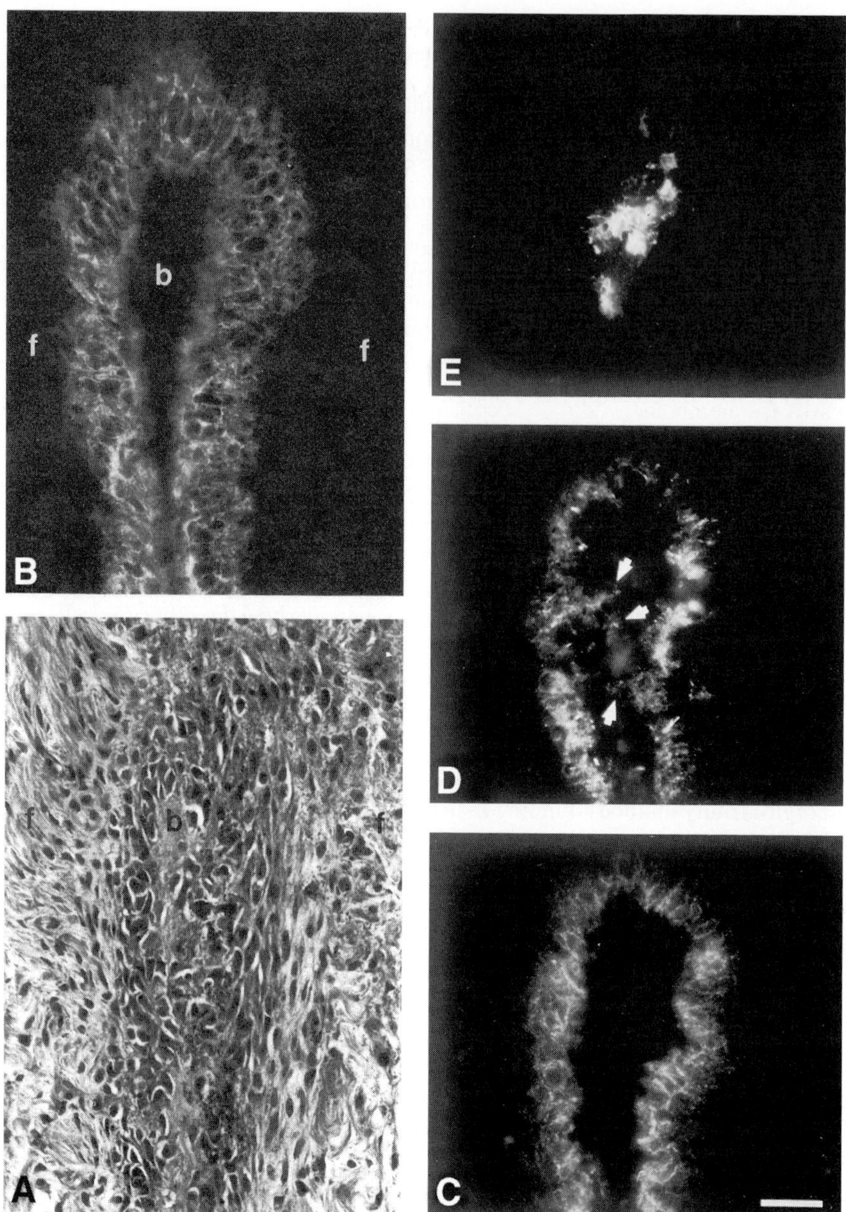

Fig. 3. Expression of osteogenic cell surface antigens in a 4-d-old periosteal culture. A H&E-stained section from one end of the tissue fold is illustrated in **(A)**. Bone matrix (b) containing osteocytes is evident, as is the fibrous tissue (f) in the outer region of the explant. A broad band of cells are reactive with antibody SB-1 **(B)**, while a restricted population of cells reacts with SB-3 **(C)**. A further subset of the SB-3-reactive cells is stained by SB-2 **(D)**, along with some cells that were recently encased in bone matrix (arrows). Morphologically recognizable osteocytes are stained with SB-5 **(E)**. Bar, 25 µm.

physical association with the newly formed and mineralizing bone (**Fig. 3**). Since the early events of osteogenesis are extended over a 4-d period in this culture system, folded periosteal explants provide an exaggerated model useful for the study of individual lineage steps. Specifically, this system allows further dissection of the transitory stages associated with sequential acquisition of the SB-3 and SB-2 antigens. Furthermore, the relatively high cellularity of the bone matrix accentuates the brief stage of SB-2 and SB-5 co-expression prior to terminal differentiation of SB-5-positive osteocytes *(25)*. Additional studies document that in the absence of β-glycerophosphate, which is necessary for mineralization in vitro, the SB-5 antigen is not expressed despite the normal morphological appearance of osteocytes in the developing bone *(26,27)*. These experiments support the conclusion that expression of the SB-5 antigen is an inducible process, is associated with bone mineralization, and that such mineralization is obligatory to the terminal differentiation of osteogenic cells.

The emergence of osteogenic cell-surface molecules by avian marrow–derived osteochondral progenitors was similarly evaluated in diffusion chamber cultures in vivo. Fresh marrow cells from young chick tibiae were implanted intraperitoneally in athymic mice and harvested at multiple time points out to 60 d. Although first noted in other species *(17,18,28,29)*, these marrow-derived avian cells also gave rise to bone and cartilage within the chambers *(30)*. Type I collagen was observed adjacent to the inner surface of the membrane, and type II collagen was elaborated by chondrogenic cells in the central portion of the chamber, where access to vascular-derived nutrients and cues was relatively reduced (**Fig. 4**). Immunostaining with SB-1 revealed the expression APase-positive cells 12 d after implantation. As development progressed, the staining intensity and number of SB-1-positive cells increased. By 20 d after implantation, antibodies SB-3 and SB-2 were observed to react with cells associated with the developing bone. Finally, cells within the type I collagen matrix reacted with the osteocyte-specific antibody SB-5 (**Fig. 4**). The morphology of these cells, with their slender pseudopodia-like processes entering matrix-free canaliculi, is identical to that seen in embryonic chick bone and periosteal explant cultures.

THE FIRST OSTEOGENIC CELL LINEAGE MODEL

The above investigations led to the creation of a lineage paradigm presented diagrammatically in **Fig. 5**. The key aspects of this model describe the differentiation of APase-positive preosteoblasts from undifferentiated progenitor cells. These preosteoblasts undergo a series of transitory osteoblast stages, defined in part by their sequential SB-3 and SB-2 immunoreactivity, before becoming secretory osteoblasts. A fraction of these cells surround themselves in matrix as SB-2/SB-5-positive osteocytic osteoblasts, and terminal differentiation into an osteocyte is characterized by loss of the SB-2 antigen. That osteocytes are derived directly from secretory osteoblastic cells is now clear; however, whether incorporation of cells into the matrix is a random event or specifically programmed to a subset of cells is not yet known. Importantly, these molecular probes document that the cellular transitions of the osteogenic lineage are shared by embryonic limb bud mesenchyme, by periosteal cells from the long bone or calvarium, and by postnatal stromal cells from the marrow.

With such a lineage in mind, we have used the antibodies to isolate and purify cells at key stages along their pathway. We employed antibody-coated magnetic bead techniques, as well as complement-mediated cell lysis, to purify preosteoblastic populations and follow their subsequent expression of mature phenotypic markers in vitro *(31)*. We have also used fluorescent-activated cell sorting (FACS) to isolate SB-5-positive osteocytes for further in vitro characterization *(32)*. In addition, collaborators have used these probes to establish statistical models for evaluating the effect of various hormones on cells at specific lineage stages *(33–35)*. Finally, these antibodies have been used to describe the differentiation of scleral ossicles in the avian eye *(7,36)*, and during osteogenesis of isolated periosteal cells in diffusion chambers *(16)*, on tissue culture plastic *(13)*, and in subcutaneous implantations in athymic mice *(15)*.

IDENTIFICATION OF HUMAN OSTEOBLASTIC PROGENITORS

Studies of animal bone marrow–mediated osteogenesis in diffusion chambers *(17,18)* and ectopic implants *(37–39)* served as the foundation for isolating analogous progenitor cells from humans. Haynesworth and his colleagues first reported the isolation, cultivation, and characterization of human

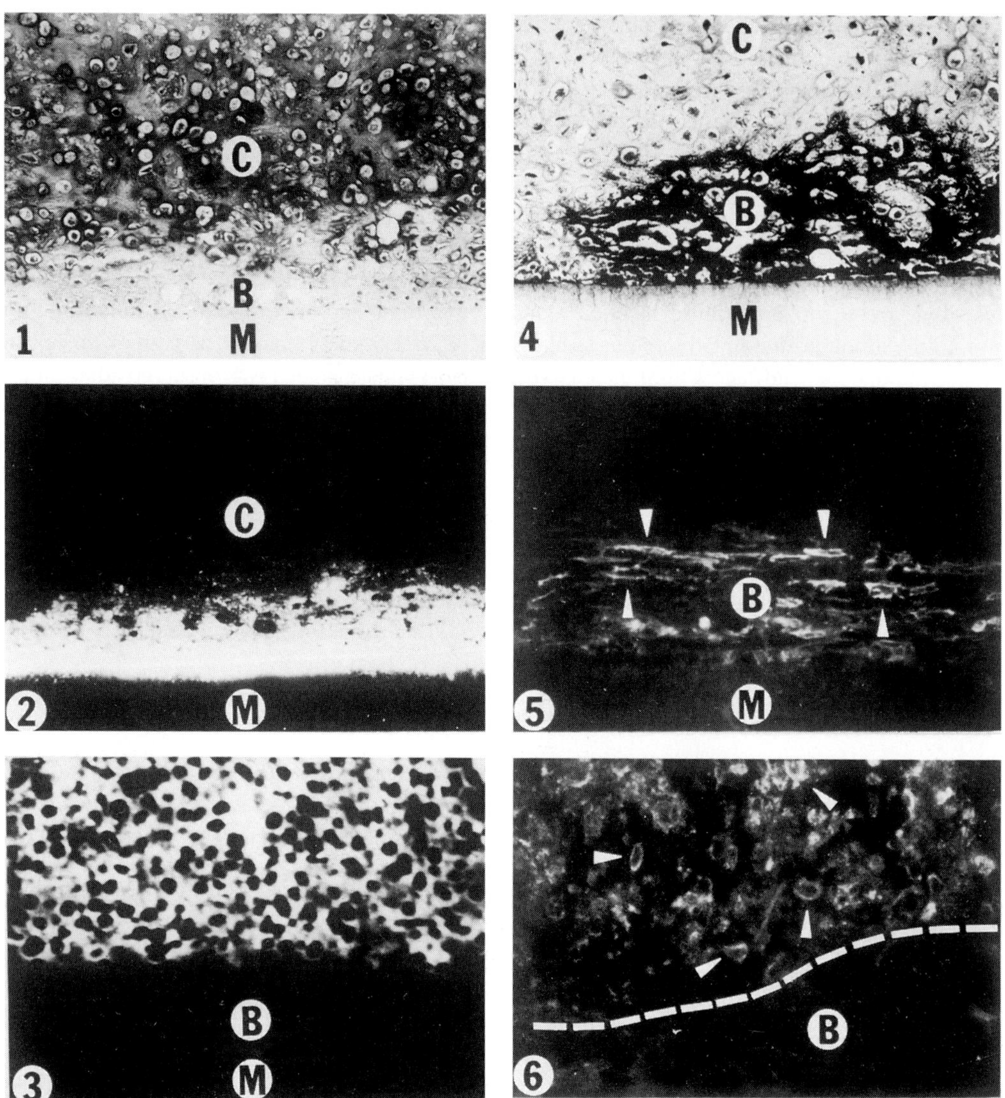

Fig. 4. (**1**) Toluidine-blue staining of membranous bone (B) and hyaline cartilage (C) in a diffusion chamber inoculated with fresh chick marrow and intraperitoneally incubated for 21 d. Bone is formed predominantly along the inner face of the membrane filter (M). (**2**) Type I collagen immunofluorescence shows reactivity within the bone, and type II collagen immunofluorescence (**3**) resides exclusively within the cartilage. (**4**) Von Kossa-stained bone (B) along the inner surface of the membrane is filled with SB-5-positive osteocytes in this 40-d sample (**5**), while adjacent polygonal osteoblasts are stained by SB-1 along their surface (**6**). Note that SB-1 and SB-5 staining is mutually exclusive. Magnification in 1–3, ×125. Magnification in 4–6, ×300.

marrow–derived progenitor cells with osteochondral potential *(40,41)*. By loading small porous hydroxy-apatite/tricalcium phosphate (HA/TCP) cubes (3 mm per side) with culture-expanded cells, and implanting the construct into athymic mice, Haynesworth demonstrated that bone and cartilage would form in the pores of the ceramic. These cells are now known as mesenchymal stem cells (MSCs) *(42)*, because they have a high replicative capacity and give rise to multiple mesenchymal tissue types including bone, cartilage, tendon, muscle, fat, and marrow stroma *(43–48)*. We have extended these observations

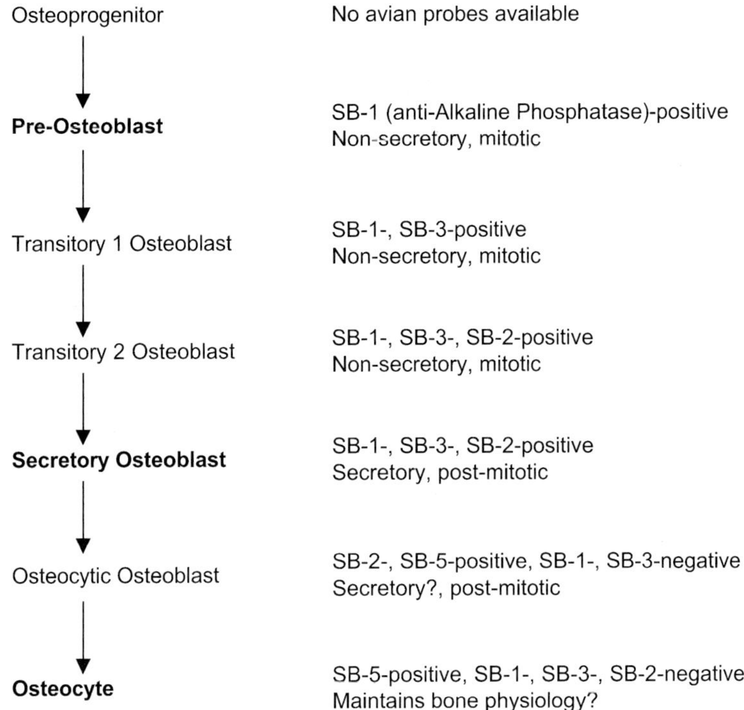

Fig. 5. Diagrammatic representation of discrete cell stages that comprise the avian osteogenic cell lineage.

to provide a detailed analysis of the surface molecules that characterize culture-expanded human MSCs (**Table 1**) *(49)*. This work stems from our effort to document the changes that occur in cell-surface architecture as a function of lineage progression. The profile of cell adhesion molecules, growth factor and cytokine receptors, and miscellaneous antigens serves to establish the unique phenotype of these cells, and provides a basis for exploring the function of selected molecules during osteogenic, and other, lineage progression.

Because MSCs are understood to be the source of osteoblastic cells during the processes of normal bone growth, remodeling, and fracture repair in humans *(1,4,6)*, we have used them as a model to study aspects of osteogenic differentiation. When cultivated in the presence of osteogenic supplements (OSs) (dexamethasone, ascorbic acid, and β-glycerophosphate), purified MSCs undergo a developmental cascade defined by the acquisition of cuboidal osteoblastic morphology, transient induction of APase activity, and deposition of a hydroxyapatite-mineralized extracellular matrix *(50,51)* (**Fig. 6A–C**). Gene expression studies illustrate that APase is transiently increased, type I collagen is downregulated during the late phase of osteogenesis, and osteopontin is upregulated at the late phase *(49)* (**Fig. 6D**). Similarly, bone sialoprotein and osteocalcin *(51)* are upregulated late in the differentiation cascade, while osteonectin is constitutively expressed. Additional studies detail the growth kinetics and high replicative capacity of these cells, which do not lose their osteogenic potential following a 1 billion-fold expansion and/or cryopreservation *(52,53)*. We have documented that OS-treated MSCs secrete a small-molecular-weight osteoinductive factor into their conditioned medium, which is capable of stimulating osteogenesis in naïve cultures *(54)*, similar to that reported for rat marrow stromal cells directed into the osteogenic lineage *(55)*. We have completed a comprehensive series of pulse-chase and transient exposure experiments using dexamethasone to determine which steps of the lineage pathway are dependent on exogenous factors, and which are supported by either (1) paracrine/autocrine fac-

Table 1
The Cell Surface Molecular Profile of Human MSCs

Molecules present	Molecules absent
Integrins	
α1, α2, α3, α5, α6, αv, β1, β3, β4	β2, α4, αL
Growth factor and cytokine receptors	
bFGFR, PDGFR, IL-1R, IL-3R, IL-4R, IL-6R, IL-7R, IFN-γR, TNFIR and TNFIIR, TGFβIR and TGFβIIR	EGFR-3, IL-2R
Cell adhesion molecules	
ICAM-1 and -2, VCAM-1, L-selectin, LFA-3, ALCAM	ICAM-3, cadherin-5, E-selectin, P-selectin, PECAM-1
Miscellaneous antigens	
Transferrin receptor, CD9, Thy-1, SH-2, SH-3, SH-4, SB-20, SB-21	CD4, CD14, CD34, CD45, von Willebrand factor

bFGFR = basic fibroblast growth factor receptor; PDGFR = platelet-derived growth factor receptor; IL-#R = inter-leukin-# receptor; IFN-γR = interferon gamma-receptor; TNFIR = tumor necrosis factor I receptor; TNFIIR = tumor necrosis factor II receptor; TGFβIR = transforming growth factor beta I receptor; TGFβIIR = transforming growth factor beta II receptor; EGFR-3 = epidermal growth factor receptor 3; ICAM = intercellular adhesion molecule; VCAM = vascular cell adhesion molecule; LFA-3 = lymphocyte function-related antigen-3; ALCAM = activated leukocyte cell adhesion molecule; PECAM = platelet endothelial cell adhesion molecule; CD = cluster designation.

tors in culture or (2) sustained lineage progression events following brief exposure to dexamethasone *(56,57)*. A diagrammatic representation of these results is presented in **Fig. 6E**.

Additional collaborations have led to insights regarding the role of BMP receptors and downstream signaling events in osteogenesis *(58–60)*. Recent studies of the MAP and JUN kinase signal transduction pathways establish pivotal roles for these family members in the differential commitment of human MSCs to either the osteogenic or adipogenic lineage *(61,62)*. Finally, studies using two-dimensional electrophoresis of culture samples at specific time points have led to the identification of molecules, such as α-B crystalline, that are differentially regulated during osteogenesis *(63,64)*.

MONOCLONAL ANTIBODIES AGAINST HUMAN OSTEOGENIC CELLS

As part of characterizing the dynamic events of the differentiation process, we have generated a number of monoclonal antibodies that react specifically with the surface of human cells during discrete stages of osteogenesis. As was the case for avian-specific antibodies SB-1 through SB-5, new probes known as SB-10, SB-20, and SB-21 have been used to localize MSCs and their progeny during development of the fetal human skeleton *(65,66)*. Antibody SB-10 recognizes a family of osteoprogenitor cells present exclusively in the outer stacked cell region of the periosteum, while SB-20 and SB-21 react with a subset of inner cambium cells expressing APase on their surface (**Fig. 7**). By tracking bone-related markers during the developmental process, we have refined our understanding of the specific lineage transitions in osteogenesis. These data serve as a basis for our belief that sequential acquisition and loss of specific surface molecules can be used to define positions of individual cells within the osteogenic lineage (**Fig. 8**).

The antigen recognized by one of these antibodies, SB-10, was identified following its immunopurification from MSC plasma membrane preparations. Western blots initially demonstrated a single approx 99-kDa-reactive band *(67)*, which upon immunoprecipitation, purification, and peptide fragment sequencing, was determined to be a cell-surface molecule known as ALCAM *(68)*, a member of the immunoglobulin superfamily of cell adhesion molecules *(69)* (**Fig. 9A–C**). Molecular cloning of a full-length cDNA from a human MSC expression library confirmed nucleotide sequence identity with ALCAM (Activated Leukocyte Cell Adhesion Molecule), and allowed us to discover homologs present

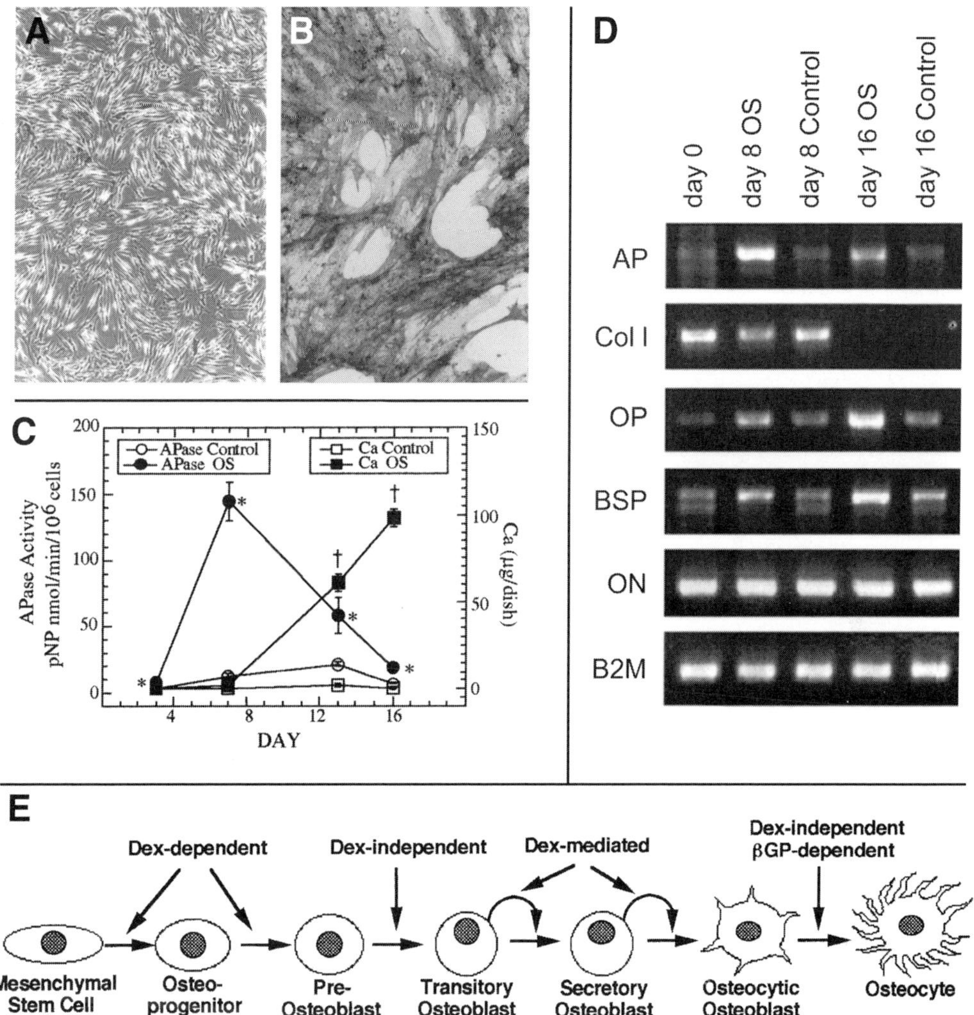

Fig. 6. Osteogenic differentiation of human MSCs in vitro. Phase-contrast photomicrographs of: **(A)** human MSC cultures under growth conditions display characteristic spindle-shaped morphology and uniform distribution (unstained ×18); **(B)** human MSC cultures grown in the presence of osteoinductive supplements (OS) for 16 d and stained for APase and mineralized matrix. APase staining appears gray in these micrographs (originally red) and mineralized matrix appears dark (APase and von Kossa histochemistry, ×45). **(C)** Mean APase activity and calcium deposition of MSC cultures grown in control or OS medium and harvested on d 3, 7, 13, and 16 ($n = 3$). The vertical bars indicate standard deviations. $*p < 0.05$, $p < 0.005$ (compared to control). **(D)** Expression of characteristic osteoblast mRNAs during in vitro osteogenesis. Reverse transcriptase-polymerase chain reactions using oligonucleotide primers specific for selected bone-related proteins were performed with RNA isolated at the indicated times. **(E)** Diagrammatic representation of the stages of dexamethasone-induced osteogenic differentiation of MSCs in vitro.

in rat, rabbit, and canine MSCs *(68)* (**Fig. 9D–F**). The addition of antibody SB-10 F_{ab} fragments to MSCs undergoing osteogenic differentiation in vitro accelerated the process, thereby implicating a role for ALCAM during bone morphogenesis and including ALCAM in the group of cell adhesion molecules involved in osteogenesis. Together, these results provide evidence that ALCAM plays a critical role in the differentiation of mesenchymal tissues in multiple species across the phylogenetic tree.

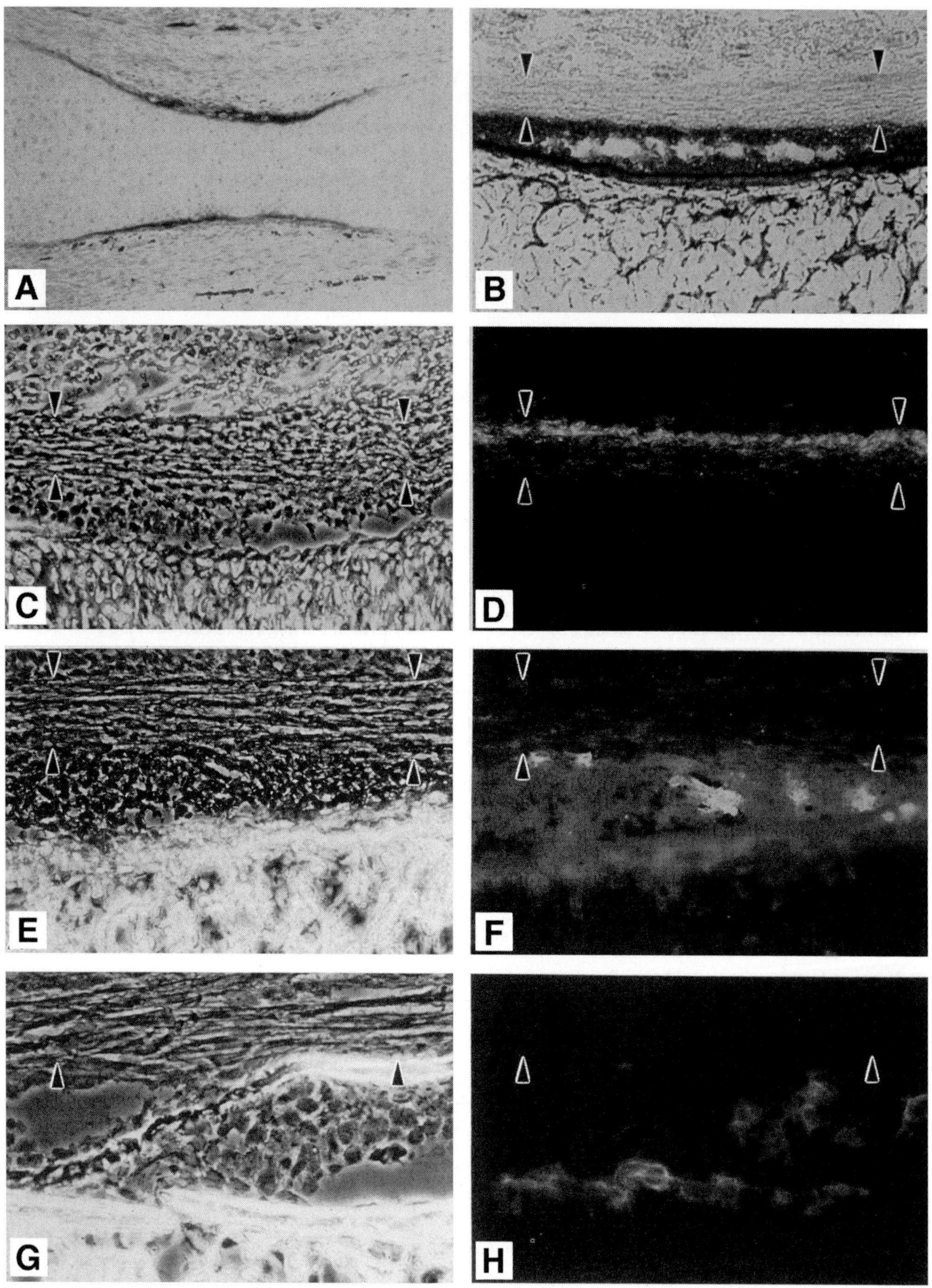

Fig. 7. Reactivity of antibodies SB-10 and SB-20 in longitudinal sections of developing human limbs. **(A)** A 55-d embryonic tibia illustrates the cartilaginous core that is surrounded by a primary collar of diaphyseal bone and a rudimentary periosteum. (Mallory-Heidenhain, ×30). **(B)** High-power view of the periosteum, first layer of bone, and underlying cartilage stained histochemically for APase (red). While the inner cambium layer of the periosteum is intensely stained, the outer stacked cell layer (arrowheads) is free of APase activity.

DEVELOPMENTAL BIOLOGY APPLIED TO CLINICAL THERAPY

We have extended our basic science investigations to examine not only the role that cells of the osteogenic lineage play in normal bone homeostasis, but also the therapeutic potential of these cells in clinical situations requiring bone repair or bone augmentation. While autologous cancellous bone is the current gold standard for bone grafting, a variety of problems are associated with its acquisition including donor site morbidity, loss of function, and a limited supply *(70,71)*. These problems have inspired the development of alternative strategies for the repair of clinically significant bone defects. Some of these tactics have tried to mimic portions of the natural biological sequence that occur following a fracture. This cascade, however, is composed of a complex series of steps including inflammation, chemotaxis of progenitor cells (MSCs) to the injured site, local proliferation of MSCs to form a repair blastema, and eventually differentiation of these cells into bone or cartilage, depending on the mechanical stability of the site. Our approach has been to develop techniques that directly provide the cellular machinery, namely MSCs, to the site in need of bone augmentation *(1,3,49)*. This approach can circumvent the early steps of bone repair, and may be particularly attractive for patients who have fractures that are difficult to heal, or patients who have a decline in their MSC repository as a result of age, osteoporosis, or other metabolic derangement *(72–77)*.

Our initial efforts to design cell-based implants focused on the evaluation of a variety of delivery vehicles. We have used a standard rat femoral gap model *(78,79)* to screen myriad cell-matrix combinations thus far. Selection of the ideal carrier for repair of such local defects is based on several criteria: (1) the material should foster uniform cell loading and retention; (2) the scaffold should support rapid vascular invasion; (3) the matrix should be designed to orient the formation of new bone in anatomically relevant shapes; (4) the composition of materials should be resorbed and replaced by new bone as it is formed; (5) the material should be radiolucent to allow the new bone to be distinguished radiographically from the original implant; (6) the formulation should encourage osteoconduction with host bone; and (7) it should possess desirable handling properties for the specific clinical indication *(1,3,49)*.

PRECLINICAL ANIMAL MODELS OF BONE REGENERATION

One of the original preclinical studies showed that culture-expanded, syngeneic rat MSCs loaded onto a porous HA/TCP cylinder are able to regenerate bone in a critical-sized segmental femoral defect *(80)*. In samples loaded with MSCs, bone formed rapidly throughout the biomatrix as a result of the osteoblastic differentiation of the implanted cells (**Fig. 10A,B**). Quantitative histologic assessment of these MSC-loaded implants demonstrated that as early as 4 wk postoperatively, bone had filled 20% of the available pore space of the biomatrix, and by 8 wk, over 40% bone fill was achieved (**Table 2**). Cell-free samples did not exceed 10% fill (osteoconduction), and even samples loaded with fresh marrow derived from one entire femur were not significantly better at 17% fill. Our results compare favorably with other approaches described in the literature, and are strikingly better than those reported for purified BMP in the same HA/TCP carrier *(81)*.

Fig. 7. *(Continued)* Phase-contrast (**C**) and SB-10 immunostaining (**D**) of a serial section to that presented in Panel **B** show that the outer stacked cell layer is strongly reactive with SB-10, while the inner APase-positive layer is negative. Panels **B** and **D** represent reciprocal staining patterns with regard to the periosteum. (Original magnification in **B–D** ×150.) (**E**) Phase-contrast image of the mid-diaphysis of a 62-d tibia histochemically stained for APase activity. The stacked cell layer (arrowheads) is negative, while the inner cambium layer and isolated chondrocytes are positive. (**F**) The section in panel **E** was also stained by antibody SB-20. Double exposure demonstrates selected osteoblastic cells stained by SB-20 (yellow), which are a subset of the APase-positive (red) cells in the periosteum. The stacked cell layer is not immunoreactive with SB-20. (Original magnification ×150.) (**G, H**) At high power, cell surface staining on a subset of cells within the inner periosteum is apparent in this 62-d embryonic femur (original magnification in **E–H** ×300). (Color illustration in insert following p. 212.)

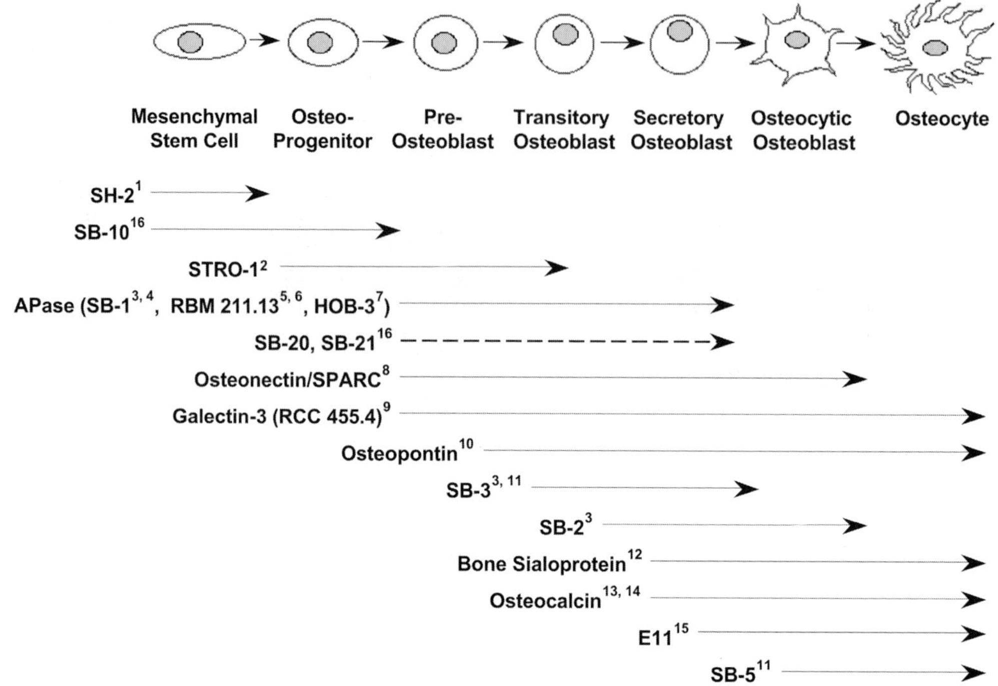

Fig. 8. Comprehensive description of the osteogenic cell lineage. Expression of selected cell surface and extracellular matrix molecules, reported by various investigators using either monoclonal or polyclonal antibodies on sections of developing bone, was used to generate this model. The beginning of each arrow reflects the stage of differentiation when expression is first detected, while the arrowhead notes the point when expression is no longer detected. The data presented in this figure represent a collection of studies performed on several species, including chick, pig, rat, and human. The dashed line used for antibodies SB-20 and SB-21 indicates that only a subset of cells within these stages of differentiation is immunoreactive. See original references for additional details. 1, ref. *40*; 2, ref. *104*; 3, ref. *20*; 4, ref. *23*; 5, ref. *105*; 6, ref. *106*; 7, ref. *107*; 8, ref. *108*; 9, ref. *109*; 10, ref. *110*; 11, ref. *25*; 12, ref. *111*; 13, ref. *112*; 14, ref. *113*; 15, ref. *114*; 16, ref. *65*.

To determine the ability of purified *human* MSCs to heal a clinically significant bone defect, culture-expanded cells were loaded onto a HA/TCP cylinder and implanted into a segmental defect in the femur of adult athymic (HSD:Rh-*rnu/rnu*) rats *(82,83)*. Healing of bone defects was compared on the basis of high-resolution radiography, immunohistochemistry, quantitative histomorphometry, and biomechanical testing. The percentage bone fill with human MSCs in this study was equivalent to that seen in euthymic rats who received syngeneic MSCs (**Table 2**). Immunohistochemical evaluation using antibodies to distinguish human cells from rat cells demonstrated that tissue within the pores of the implant during the early phase of repair was derived from donor (human) MSCs. The biomechanical data demonstrate that torsional strength and stiffness, as measured through the implant and adjoining diaphyseal shaft at 12 wk, were approx 40% that of intact control limbs, which is more than twice that observed with the cell-free carrier (**Fig. 10D**). This result also compares favorably with the mechanical outcome achieved in a similar study of bone repair in a primate long bone defect model, where autogenous bone produced only 23% of the strength of intact contralateral limbs 20 wk after implantation *(84)*. These studies confirm that purified, culture-expanded *human* MSCs can be used to regenerate bone in a clinically significant osseous defect.

Subsequent investigations focused on advancing this technology into large animal models, and developing prototype procedures for shipping marrow, MSCs, and autologous implants to and from

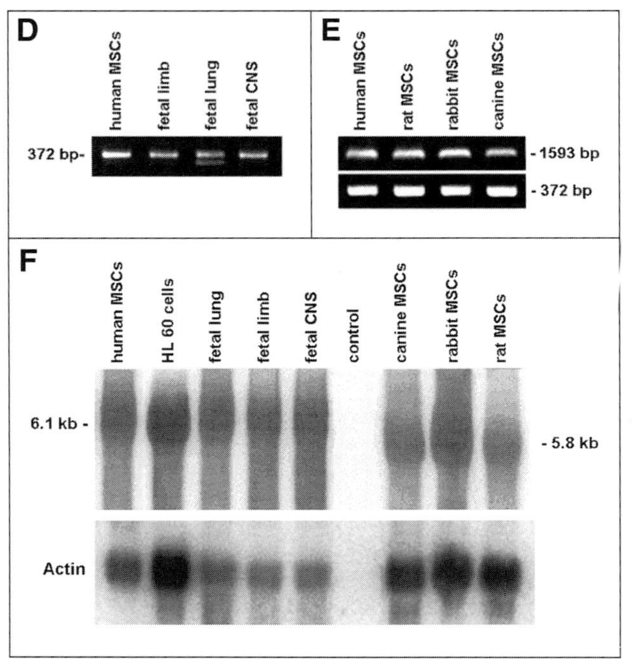
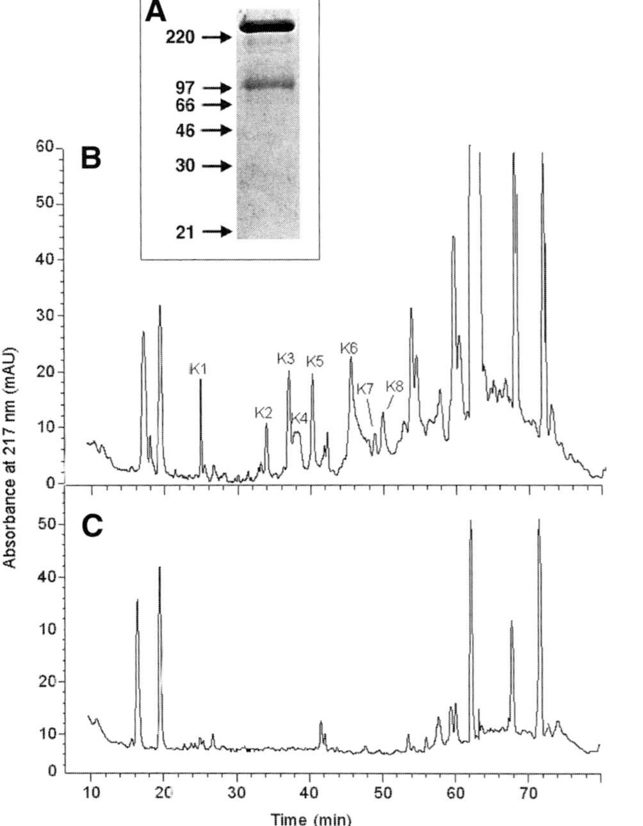

Fig. 9. Identification of the SB-10 surface antigen. (**A**) The SB-10 antigen was immunoprecipitated, and excised from a polyacrylamide gel for lysine C-endoproteinase digestion. (**B**) Recovered peptides were separated by reverse-phase high-performance liquid chromatography (HPLC). Collected peaks referred to as K1 through K8 were subjected to N-terminal sequence analysis and found to correspond to ALCAM. (**C**) Control digest of a blank piece of polyacrylamide excised from the same gel. (**D**) Polymerase chain reaction (PCR) amplification of an ALCAM-specific fragment in cultured human MSCs, fetal limb, and other tissues known to express ALCAM. (**E**) PCR amplification of ALCAM fragments in cultures of human, rat, rabbit, and canine MSCs. (**F**) Northern blot analysis of human MSCs shows a single mRNA species approximately 6.1 kb in size, while animal ALCAM has an approximate mRNA size of 5.8 kb.

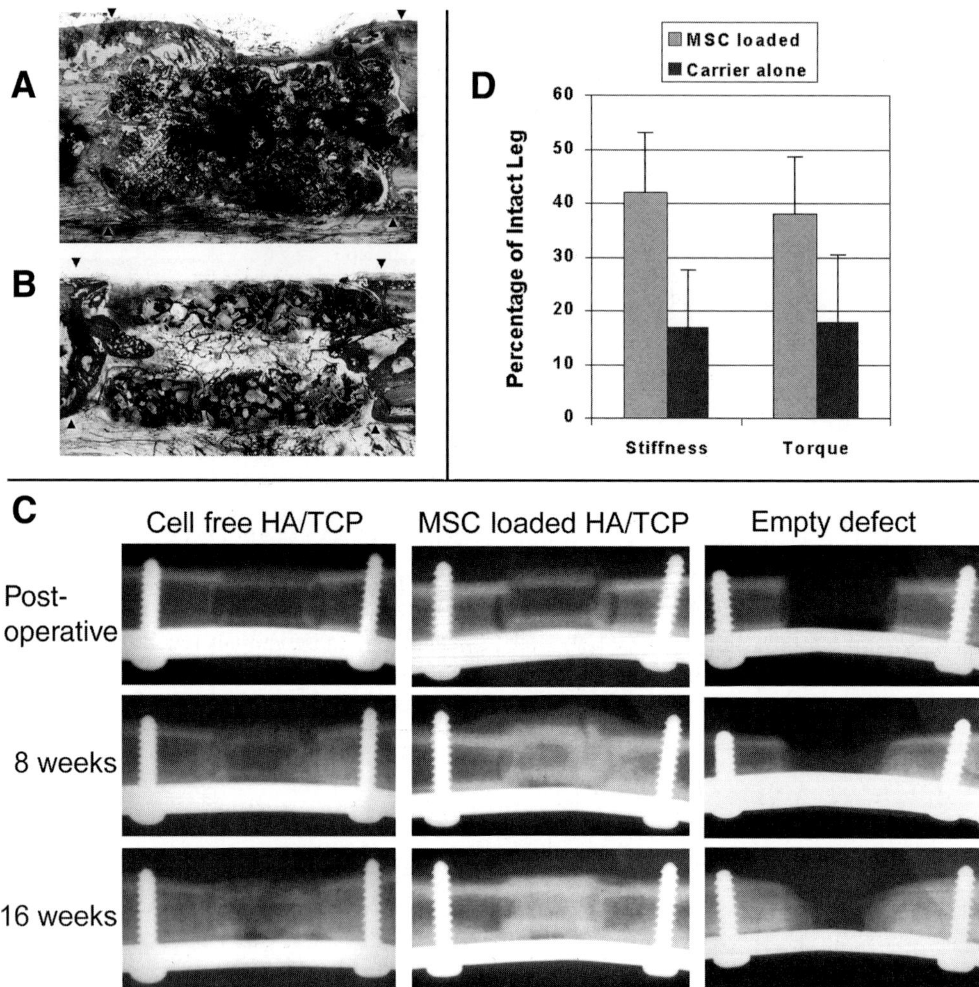

Fig. 10. MSC-mediated bone regeneration in preclinical animal studies of segmental femoral defect repair. **(A)** Rat defects fitted with a MSC-loaded HA/TCP carrier form a solid osseous union with the host, and contain substantial new bone throughout the pores by 8 wk. **(B)** Defects fitted with a cell-free HA/TCP carrier do not contain bone within the pores of the implant, nor is there significant union at the interfaces, noted by the arrowheads. (Toluidine blue-O, ×16.) **(C)** Radiographic appearance of bone healing in a 21-mm canine femoral gap defect. Animals that did not receive an implant established a fibrous nonunion by 16 wk. Animals that received an MSC-loaded HA/TCP cylinder regenerated a substantial amount of bone at the defect site, including a peri-implant callus that remodeled to the size of the original bone by 16 wk. Those animals receiving cell-free implants did not successfully heal their defects, as noted by the lack of new bone and the multiple fractures throughout the body of the implant material. **(D)** Graphic results of biomechanical torsion testing performed on athymic rat femora 12 wk following implantation with human MSC-loaded ceramics. (*$p < 0.05$ compared to carrier alone.)

distant clinical sites. Using a standardized strategy for the isolation of marrow-derived MSCs *(85)*, we identified conditions for effective cultivation and in vivo osteogenic differentiation of canine cells *(86)*. We then established a critical-sized femoral gap defect model to determine the efficacy of MSC-based bone regeneration therapy in large dogs *(87)*. As was done in the rodent studies, a ceramic cylinder was used to deliver autologous MSCs back to the site of a 21-mm-long osteoperiosteal segmental

Table 2
Quantitative Histomorphometry of Bone Fill as a Percentage
of Available Space in Selected Models of Segmental Bone Defect Repair

Implant type	Canines (implanted with autologous MSCs)	Athymic rats (implanted with human MSCs)	Fischer rats (implanted with syngeneic MSCs)
Cell-free HA/TCP	24.0 ± 15.5	29.5 ± 8.9	10.4 ± 2.4
MSC-loaded HA/TCP	39.9 ± 6.1*	46.6 ± 14.8*	43.2 ± 7.7*

*Indicates $p < 0.05$ compared to cell-free controls.

femoral defect, which was stabilized by a stainless-steel internal fixation plate with bicortical screws. Radiographic (**Fig. 10C**) and histological evidence reveal an impressive periimplant callus of bone, as well as bone throughout the pores of the entire implant by 16 wk *(88,89)*. We attribute the formation of this large callus to the combined action of cells delivered on the surface of the ceramic material and the secretion of osteoinductive factors by these cells during the process of differentiation *(54)*. Such combined osteogenic and osteoinductive activity serve to create a mass of new bone that is derived from the implanted cells, as well as host-derived cells that are competent to respond to secreted bone morphogens. Importantly, none of the empty defects healed, and those animals receiving cell-free ceramics did not possess any periimplant callus or bone in the center of the implant region. **Table 2** demonstrates the similarity of bone fill between the canine studies outlined here and the previous efforts using rat or human MSCs in rodent hosts.

PRECLINICAL ANIMAL MODELS
OF BONE MARROW-BASED BONE REGENERATION

Culture expansion of MSCs can provide an abundant supply of osteogenic cells for repair and defect healing, but the steps necessary for expansion, and the delay between harvest and implantation are challenging to integrate into a clinical setting. An intraoperative technique that eliminates the steps of culture expansion but provides an enriched population of osteoprogenitor cells to the graft site may be effective in many clinical conditions.

Osteoprogenitor cells present in bone marrow are obtained by simple aspiration. We initially focused on optimizing the osteogenic capacity of fresh, intraoperatively manipulated bone marrow. Employing our standard rat femoral defect model, we evaluated a variety of matrix carriers including ceramics, synthetic polymers, and natural polymers. When bone marrow was combined with a porcine-derived gelatin product (Gelfoam Upjohn, Kalamazoo, MI) and peripheral blood, the femoral defects healed successfully; however, such defects did not heal when the same amount of marrow was implanted on a synthetic matrix, or when a reduced amount of marrow was delivered using Gelfoam (see **Fig. 11** for details). When using a similar combination of fresh bone marrow with Gelfoam in a large animal model of bone repair, excellent results were observed in several animals, though the uniformity of the outcome was not ideal—only six of nine animals had a solid bony bridge spanning the defect *(90)*. This line of investigation also highlights two important issues: (1) that there are non-MSC components in the marrow that are important to the healing response; and (2) that the delivery matrix is critical to eventual success. We conclude this based on the observation that even the large number of purified MSCs required to heal a long bone defect on HA/TCP is not capable of healing the defect when delivered on Gelfoam. However, successful healing is observed on Gelfoam when as little as 500 times fewer MSCs are delivered in conjunction with other endogenous marrow-derived cells and factors. We refer to these other non-MSC, marrow-derived cells as *accessory cells*. Whether accessory cell function is paracrine in nature or mediated by cell-to-cell contact remains to be evaluated.

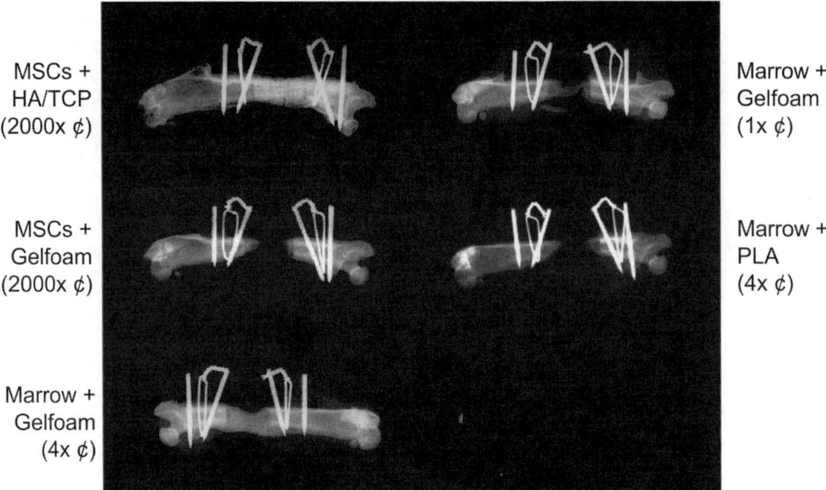

Fig. 11. Fresh marrow-based bone regeneration in preclinical animal studies of segmental femoral defect repair. As in **Fig. 10**, rat segmental defects were fitted with various implants containing either culture-expanded, purified MSCs, or fresh marrow obtained from 1–4 diaphyseal segments of syngeneic femora. MSCs on a HA/TCP cylinder reproducibly heal defects, and such implants contain approximately 2000 times the number of MSCs when compared to the same volume of fresh marrow from one diaphyseal segment. Purified MSCs delivered on a porcine gelatin sponge (Gelfoam) exhibit no healing, but when the same carrier is used to deliver whole marrow from four femoral segments, solid bone bridging ensues. A dose–response effect is observed when marrow from only one femur is delivered in Gelfoam; these animals show modest bone formation. Animals implanted with synthetic polylactic acid (PLA) carriers and marrow from four femora similarly showed poor bone formation. Together, these data provide insight into the importance of proper carriers, proper cellular dosing, and the benefit of accessory cells in fresh marrow that reduce the need for large numbers of purified MSCs.

Osteoprogenitors constitute significantly less than 1% of the nucleated cells in the marrow of a healthy adult *(41,53)*. Because these are the cells that go on to synthesize new bone, one possible way to improve the efficacy of a bone marrow aspirate is by concentrating the endogenous osteoprogenitor cells *(91)*. Using simple centrifugation of fresh whole marrow, Connolly reported successful treatment of 18 of 20 tibial nonunions via percutaneous injection of bone marrow concentrate with and without intramedullary nailing *(92)*.

Recent work by several investigators has focused on developing a means to intraoperatively concentrate osteoprogenitor cells while optimizing their clinical delivery and local retention. Ideally, this process would combine cells participating in bone formation with a directly implantable substrate that enhances their activity. Bone marrow cells have been shown to possess a high affinity for certain solid substrates. For example, osteoprogenitor cells are selectively retained when marrow is filtered through specific porous configurations of calcium phosphate or bone matrix. Using demineralized bone matrix to capture osteoprogenitors cells, and then implant the composite graft directly, Takigami et al. reproducibly obtained spine fusion in a canine model *(93)*. The results of the cell-enriched graft were significantly better than allograft alone or allograft mixed with whole marrow. Kapur and colleagues *(94)* have similarly shown, in a canine long bone defect model, the beneficial effect of selective retention on graft performance (see **Fig. 12**). In this study, the bone grafts were created using a matrix consisting of a mixture of allogeneic demineralized bone fibers and undemineralized cancellous bone chips (DBM-CC). The experimental group contained grafts prepared by flowing marrow through the matrix under controlled conditions that selectively retain the osteoprogenitors. As part of

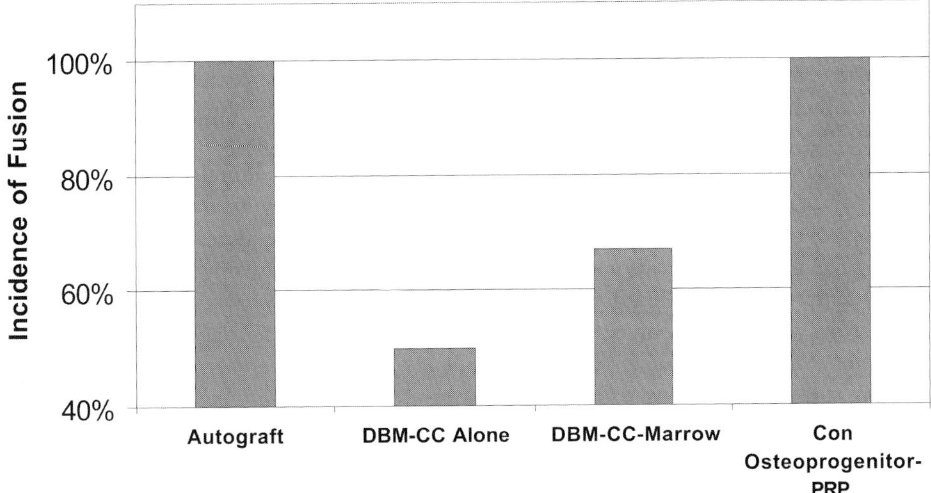

Fig. 12. Summary of canine femoral gap study. Radiographic fusion was observed in all animals treated with autograft or concentrated osteoprogenitor cells and platelet-rich plasma. The benefit of selective retention compared to DBM-CC plus fresh marrow or DBM-CC alone is apparent. Each group contained at least five animals, all sacrificed at 16 wk.

the final step of graft preparation, the concentrated osteoprogenitor-graft was clotted together with autologous platelet-rich plasma (PRP) (Con Osteoprogenitor-PRP). The control groups consisted of an iliac crest bone graft, allogeneic DBM-CC mixture alone, or the DBM-CC mixture loaded with whole marrow (DBM-CC-Marrow). The rate and incidence of union was assessed by radiographic analysis, including plain films every 4 wk and CT scan upon sacrifice at 16 wk. In the Autograft and Con-Osteoprogenitor-PRP groups, fusion was achieved in all animals (**Fig. 12**). In contrast, when the allogeneic DBM-CC mixture was used alone or in combination with native bone marrow, there was an unsatisfactory healing response, with approximately half of the canines going onto fusion.

Although the above results are promising, maximizing cell capture and concentration does not necessarily guarantee optimal conditions for bone formation. To better approximate the ideal biological milieu for bone formation, conditions must aim to optimize cell interaction and supply the cytokines and growth factors involved in bone formation. Using the selective retention technique, Muschler et al. demonstrated improved graft performance when a bone marrow clot was added to the enriched cell matrix in a canine spine fusion model *(95)*. Interestingly, a cellular composite that contained twice the number of osteogenic cells was inferior to a graft containing fewer progenitors but included the clot environment. They hypothesized that the fibrin clot may provide additional mechanical stability, deliver osteotropic and angiogenic growth factors, and possibly replace cells that contribute to bone formation that are excluded by the selective retention process. Toward this overall goal, a disposable, single-use kit for the preparation of osteoprogenitor cell-enriched bone graft materials has recently been cleared for use by the US Food and Drug Administration. The initial clinical study results in both long bone repair *(96)* and spine fusion *(97)* are encouraging.

THE FUTURE OF CELL-BASED THERAPY

In summary, these studies establish the existence of an osteogenic cell lineage, which can be defined by the sequential expression of specific cell surface and extracellular matrix molecules. In an effort to refine our understanding of the specific transition steps that constitute development of the osteoblast phenotype, we have generated a battery of specific monoclonal antibody probes against cell sur-

It may also be possible to further expedite the healing process by directing culture-expanded MSCs to enter the osteogenic lineage prior to implantation, thus decreasing the *in situ* interval between surgical delivery and their biosynthetic activity as secretory osteoblasts. Yoshikawa *(101)* and Ishuag-Riley *(102)* have set the stage for this approach by showing that, following implantation in syngeneic rat hosts, rat marrow stromal cells directed into the osteogenic lineage in vitro form a greater amount of bone faster than undifferentiated stromal cells. With this in mind, other investigators have demonstrated that modifications of the ceramic carrier itself can also induce osteogenic differention of cultured MSCs *(103)*.

Continuing studies of the regulatory pathways and transitions comprising osteogenic lineage progression will serve to guide our cellular treatment protocols and define the precise stage of cells that are implanted in patients for therapeutic purposes.

ACKNOWLEDGMENTS

Portions of this work were funded by research grants from the National Institutes of Health, The Arthritis Foundation, The Orthopaedic Research and Education Foundation, Case Western Reserve University, Osiris Therapeutics, Inc., and DePuy, Inc.

REFERENCES

1. Bruder, S. P., Fink, D. J., and Caplan, A. I. (1994) Mesenchymal stem cells in bone formation, bone repair, and skeletal regeneration therapy. *J. Cell Biochem.* **56(3),** 283–294.
2. Caplan, A. I. and Pechak, D. (1987) The cellular and molecular biology of bone formation, in *Bone and Mineral Research,* vol. 5 (Peck, W. A., ed.), Elsevier, New York, pp. 117–183.
3. Caplan, A. I. and Bruder, S. P. (1997) Cell and molecular engineering of bone regeneration, in *Textbook of Tissue Engineering* (Lanza, R., Langer, R., and Chick, W., eds.), R.G. Landes, Georgetown, pp. 603–618.
4. Hollinger, J., Buck, D. C., and Bruder, S. P. (1998) Biology of bone healing: its impact on clinical therapy, in *Applications of Tissue Engineering in Dentistry* (Lynch, S., ed.), Quintessence, Chicago, pp. 17–53.
5. Nijweide, P. J., Van der Plas, A., and Olthof, A. A. (1988) Osteoblastic differentiation, in *Cell and Molecular Biology of Vertebrate Hard Tissues,* CIBA Foundation Symposium 136 (Evered, D. and Harnett, S., eds.), John Wiley and Sons, New York, pp. 61–72.
6. Owen, M. (1985) Lineage of osteogenic cells and their relationship to the stromal system, in *Bone and Mineral,* 3 (Peck, W. A., ed.), Elsevier, Amsterdam, pp. 1–25.
7. Watanabe, K., Bruder, S. P., and Caplan, A. I. (1994) Transient expression of type II collagen and tissue mobilization during development of the scleral ossicle, a membranous bone, in the chick embryo. *Dev. Dyn.* **200,** 212–226.
8. Bruder, S. P. (1987) Rethinking embryonic bone formation. *JAMA* **257(17),** Pulse 3–7.
9. Bruder, S. P. and Caplan, A. I. (1989) Cellular and molecular events during embryonic bone development. *Conn. Tiss. Res.* **20,** 65–71.
10. Bruder, S. P., Caplan, A. I., Gotoh, Y., Gerstenfeld, L. C., and Glimcher, M. J. (1991) Immunohistochemical localization of a ~66kD glycosylated phosphoprotein during development of the embryonic chick tibia. *Calcif. Tissue Int.* **48,** 429–437.
11. Ogawa, M., Porter, P., and Nakahata, T. (1983) Renewal and commitment to differentiation of hemopoietic stem cells. *Blood* **61,** 823–829.
12. Sulston, J., Schierenberg, E., White, J., and Thomson, J. (1983) The embryonic cell lineage of the nematode *Caenorhabditis elegans. Dev. Biol.* **110,** 64–119.
13. Nakahara, H., Dennis, J. E., Bruder, S. P., Haynesworth, S. E., Lennon, D. P., and Caplan, A. I. (1991) In vitro differentiation of bone and hypertrophic cartilage from periosteal-derived cells. *Exp. Cell Res.* **195,** 492–503.
14. Tenenbaum, H. C. and Heersche, J. N. M. (1985) Dexamethasone stimulates osteogenesis in chick periosteum in vitro. *Endocrinology* **117,** 2211–2217.
15. Nakahara, H., Bruder, S. P., Goldberg, V. M., and Caplan, A. I. (1990) In vivo osteo-chondrogenic potential of cultured cells derived from the periosteum. *Clin. Orthop. Rel. Res.* **259,** 223–232.
16. Nakahara, H., Bruder, S. P., Haynesworth, S. E., et al. (1990) Bone and cartilage formation in diffusion chambers by subcultured cells derived from the periosteum. *Bone* **11,** 181–188.
17. Ashton, B., Allen, T., Howlett, C., Eaglesome, C., Hattori, A., and Owen, M. (1980) Formation of bone and cartilage by bone marrow stromal cells in diffusion chambers in vivo. *Clin. Orthop. Rel. Res.* **151,** 294–307.
18. Friedenstein, A. J. (1976) Precursor cells of mechanocytes. *Int. Rev. Cytol.* **47,** 327–355.
19. Bruder, S. P. and Caplan, A. I. (1989) Discrete stages within the osteogenic lineage are revealed by alterations in the cell surface architecture of embryonic bone cells. *Connect. Tissue Res.* **20,** 73–79.

20. Bruder, S. P. and Caplan, A. I. (1989) First bone formation and the dissection of an osteogenic lineage in the embryonic chick tibia is revealed by monoclonal antibodies against osteoblasts. *Bone* **10**, 359–375.

21. Bruder, S. P. and Caplan, A. I. (1989) Monoclonal antibodies against osteoblasts reveal cell surface alterations correlating to discrete stages of differentiation. *Trans. Orthop. Res. Soc.* **14**, 61.

22. Bruder, S. P. and Caplan, A. I. (1990) Terminal differentiation of osteogenic cells in the embryonic chick tibia is revealed by a monoclonal antibody against osteocytes. *Bone* **11**, 189–198.

23. Bruder, S. P. and Caplan, A. I. (1990) A monoclonal antibody against the surface of osteoblasts recognizes alkaline phosphatase in bone, liver, kidney, and intestine. *Bone* **11**, 133–139.

24. Nijweide, P. and Mulder, R. (1986) Identification of osteocytes in osteoblast-like cell cultures using a monoclonal antibody specifically directed against osteocytes. *Histochemistry* **84**, 342–347.

25. Bruder, S. P. and Caplan, A. I. (1990) Osteogenic cell lineage analysis is facilitated by organ cultures of embryonic chick periosteum. *Dev. Biol.* **141**, 319–329.

26. Bruder, S. P., Nakamura, O., and Caplan, A. I. (1991) Terminal osteogenic cell differentiation in culture requires beta-glycerol phosphate. *Trans. Orthop. Res. Soc.* **16**, 58.

27. Chung, C.-H., Bruder, S. P., and Shapiro, I. M. (1997) Beta-glycerophosphate induces osteocyte differentiation in chick osteoblast cultures. *Dent. Res.* **76**, 341 (#2621).

28. Bab, I., Passi-Even, L., Gazit, D., et al. (1988) Osteogenesis in *in vivo* diffusion chamber cultures of human marrow cells. *Bone Miner.* **4**, 373–386.

29. Johnson, K. A., Howlett, C. R., Bellenge, C. R., and Armati-Guslon, P. (1988) Osteogenesis by canine and rabbit bone marrow in diffusion chambers. *Calcif. Tissue Int.* **42**, 113–118.

30. Bruder, S. P., Gazit, D., Passi-Even, L., Bab, I., and Caplan, A. I. (1990) Osteochondral differentiation and the emergence of osteogenic cell-specific markers by bone marrow cells in diffusion chambers. *Bone Miner.* **11**, 141–151.

31. Bruder, S. P. and Caplan, A. I. (1989) Isolation of osteogenic cells at distinct stages of differentiation is facilitated by monoclonal antibodies against osteogenic cells. *J. Bone Miner. Res.* **4**, S128.

32. Kaysinger, K. K., Ramp, W. K., and Bruder, S. P. (1996) Low pH stimulates the activity of an osteocyte-enriched population. *Trans. Orthop. Res. Soc.* **21**, 105.

33. Broess, M., Schaffer, J. L., Davidovitch, M., et al. (1995) Differential phenotypic expression and responsiveness during osteoblast lineage progression. *Trans. Orthop. Res. Soc.* **20**, 487.

34. Gerstenfeld, L. C., Broess, M., Bruder, S. P., Caplan, A. I., and Landis, W. J. (1992) Regulation of osteoblast extracellular matrix formation: post-translational regulation of extracellular matrix deposition and relationship between embryonic development and hormonal response, in *The Chemistry and Biology of Mineralized Tissues* (Slavkin, H. and Price, P., eds), Elsevier Science, New York, pp. 287–295.

35. Gerstenfeld, L. C., Zurakowski, D., Schaffer, J. L., et al. (1996) Variable hormone responsiveness of osteoblast populations solated at different stages of embryogenesis and its relationship to the osteogenic lineage. *Endocrinology* **137(9)**, 3957–3968.

36. Bruder, S. P., Watanabe, K., and Caplan, A. I. (1994) Bone cell differentiation and matrix morphogenesis in the scleral ossicle occurs via transient epithelial interaction and expression of type II collagen. *J. Bone Miner. Res.* **9**, S-236.

37. Dennis, J. E., Haynesworth, S. E., Young, R. G., and Caplan, A. I. (1992) Osteogenesis in marrow-derived mesenchymal cell porous ceramic composites transplanted subcutaneously: effect of fibronectin and laminin on cell retention and rate of osteogenic expression. *Cell Transplant* **1**, 23–32.

38. Goshima, J., Goldberg, V. M., and Caplan, A. I. (1991) The origin of bone formed in composite grafts of porous calcium phosphate ceramic loaded with marrow cells. *Clin. Orthop. Rel. Res.* **269**, 274–283.

39. Goshima, J., Goldberg, V. M., and Caplan, A. I. (1991) The osteogenic potential of culture-expanded rat marrow mesenchymal cells assayed in vivo in calcium phosphate ceramic blocks. *Clin. Orthop. Rel. Res.* **262**, 298–311.

40. Haynesworth, S. E., Baber, M. A., and Caplan, A. I. (1992) Cell surface antigens on human marrow-derived mesenchymal cells are detected by monoclonal antibodies. *Bone* **13**, 69–80.

41. Haynesworth, S. E., Goshima, J., Goldberg, V. M., and Caplan, A. I. (1992) Characterization of cells with osteogenic potential from human marrow. *Bone* **13**, 81–88.

42. Caplan, A. I. (1991) Mesenchymal stem cells. *J. Orthop. Res.* **9**, 641–650.

43. Johnstone, B., Hering, T. M., Caplan, A. I., Goldberg, V. M., and Yoo, J. U. (1998) In vitro chondrogenesis of bone marrow-derived mesenchymal progenitor cells. *Exp. Cell Res.* **238**, 265–272.

44. Majumdar, M. K., Thiede, M. A., Mosca, J. D., Moorman, M. K., and Gerson, S. L. (1998) Phenotypic and functional comparison of cultures of marrow-derived mesenchymal stem cells (MSCs) and stromal cells. *J. Cell Physiol.* **176**, 57–66.

45. Pittenger, M. F., Mackay, A. M., Beck, S. C., et al. (1999) Multilineage potential of adult human mesenchymal stem cells. *Science* **284**, 143–147.

46. Saito, T., Dennis, J. E., Lennon, D. P., Young, R. G., and Caplan, A. I. (1995) Myogenic expression of mesenchymal stem cells within myotubes of mdx mice in vitro and in vivo. *Tissue Eng.* **1(4)**, 327–343.

47. Wakitani, S., Saito, T., and Caplan, A. I. (1995) Myogenic cells derived from rat bone marrow mesenchymal stem cells exposed to 5-azacytidine. *Muscle Nerve* **18**, 1417 1426.

48. Young, R. G., Butler, D. L., Weber, W., Caplan, A. I., Gordon, S. L., and Fink, D. J. (1998) The use of mesenchymal stem cells in a collagen matrix for Achilles tendon repair. *J. Orthop. Res.* **16**, 406–413.

49. Bruder, S. P., Jaiswal, N., Ricalton, N. S., Mosca, J. D., Kraus, K. H., and Kadiyala, S. (1998) Mesenchymal stem cells in osteobiology and applied bone regeneration. *Clin. Orthop. Rel. Res.* **355S**, 247–256.

50. Bruder, S. P., Eames, B. F., and Haynesworth, S. E. (1995) Osteogenic induction of purified human mesenchymal stem cells *in vitro*: quantitative assessment of the osteoblastic phenotype. *Trans. Orthop. Res. Soc.* **20**, 464.

51. Jaiswal, N., Haynesworth, S. E., Caplan, A. I., and Bruder, S. P. (1997) Osteogenic differentiation of purified, culture-expanded human mesenchymal stem cells in vitro. *J. Cell. Biochem.* **64(2)**, 295–312.

52. Bruder, S. P. and Jaiswal, N. (1996) The osteogenic potential of human mesenchymal stem cells is not diminished after one billion-fold expansion in vitro. *Trans. Orthop. Res. Soc.* **21**, 580.

53. Bruder, S. P., Jaiswal, N., and Haynesworth, S. E. (1997) Growth kinetics, self-renewal and the osteogenic potential of purified human mesenchymal stem cells during extensive subcultivation and following cryopreservation. *J. Cell Biochem.* **64(2)**, 278–294.

54. Jaiswal, N. and Bruder, S. P. (1997) Human osteoblastic cells secrete paracrine factors which regulate differentiation of osteogenic precursors in marrow. *Trans. Orthop. Res. Soc.* **22**, 524.

55. Hughes, F. and McCulloch, C. A. G. (1991) Stimulation of the differentiation of osteogenic rat bone marrow stromal cells by osteoblast cultures. *Lab. Invest.* **64**, 617–622.

56. Bruder, S. P. and Jaiswal, N. (1995) Transient exposure of human mesenchymal stem cells to dexamethasone is capable of inducing sustained osteogenic lineage progression. *J. Bone Miner. Res.* **10**, S-41.

57. Jaiswal, N. and Bruder, S. P. (1996) The pleiotropic effects of dexamethasone on osteoblast differentiation depend on the developmental state of the responding cells. *J. Bone Miner. Res.* **11**, S-259.

58. Connolly, T., Bruder, S. P., and Kerrigan, L. (1996) Dexamethasone induces the expression of a type I BMP receptor in pluripotential mesenchymal progenitor cells. *Trans. Orthop. Res. Soc.* **21**, 180.

59. Kerrigan, L., Murphy, J. M., Connolly, T. J., and Bruder, S. P. Human mesenchymal stem cells are responsive to BMP receptor-mediated signaling. Submitted.

60. Kerrigan, L. A., Murphy, J. M., Bruder, S. P., and Connolly, T. J. (1997) Bone morphogenic protein (BMP) receptor expression in human mesenchymal stem cells. *Trans. Orthop. Res. Soc.* **22**, 225.

61. Jaiswal, R. K., Jaiswal, N., Bruder, S. P., Marshak, D. R., and Pittenger, M. F. (2000) Inverse regulation of osteogenic and adipogenic differentiation of human mesenchymal stem cells by MAP kinases. *J. Biol. Chem.* **275(13)**, 9645–9652.

62. Jaiswal, R. K., Jaiswal, N., Bruder, S. P., Marshak, D. R., and Pittenger, M. F. (1997) Differential regulation of MAP kinases during osteogenic differentiation of human mesenchymal stem cells. *Mol. Biol. Cell* **8S**, 247a (1432).

63. Pittenger, M. F., Beck, S., Bruder, S. P., Mackay, A. M., and Kobayashi, R. (1996) Alpha B crystalline is expressed during osteogenic differentiation of human mesenchymal stem cells. *Mol. Biol. Cell* **7S**, 583a (3390).

64. Pittenger, M. F., Bruder, S. P., and Kobayashi, R. (1996) Identification of proteins involved in osteogenesis of human mesenchymal stem cells by high resolution 2D gel electrophoresis. *Trans. Orthop. Res. Soc.* **21**, 111.

65. Bruder, S. P., Horowitz, M. C., Mosca, J. D., and Haynesworth, S. E. (1997) Monoclonal antibodies reactive with human osteogenic cell surface antigens. *Bone* **21(3)**, 225–235.

66. Bruder, S. P., Lawrence, E. G., and Haynesworth, S. E. (1995) The generation of monoclonal antibodies against human osteogenic cells reveals embryonic bone formation in vivo and differentiation of purified mesenchymal stem cells in vitro. *Trans. Orthop. Res. Soc.* **20**, 8.

67. Bruder, S. P., Barry, F. P., and Lawrence, E. G. (1996) Identification and characterization of a cell surface differentiation antigen on human osteoprogenitor cells. *Trans. Orthop. Res. Soc.* **21**, 574.

68. Bruder, S. P., Ricalton, N. S., Boynton, R., et al. (1998) Mesenchymal stem cell surface antigen SB-10 corresponds to activated leukocyte cell adhesion molecule and is involved in osteogenic differentiation. *J. Bone Miner. Res.* **13(4)**, 655–663.

69. Bowen, M. A., Patel, D. D., Li, X., et al. (1995) Cloning, mapping, and characterization of activated leukocyte-cell adhesion molecule (ALCAM), a CD6 ligand. *J. Exp. Med.* **181**, 2213–2220.

70. Laurie, S. W. S., Kaban, L. B., Mulliken, J. B., and Murray, J. E. (1984) Donor-site morbidity after harvesting rib and iliac bone. *Plast. Reconstr. Surg.* **73**, 33–938.

71. Younger, E. M. and Chapman, M. W. (1989) Morbidity at bone graft donor sites. *J. Orthop. Trauma* **3**, 192–195.

72. Bergman, R. J., Gazit, D., Kahn, A. J., Gruber, H., McDougall, S., and Hahn, T. J. (1996) Age-related changes in osteogenic stem cell in mice. *J. Bone Miner. Res.* **11**, 568–577.

73. Inoue, K., Ohgushi, H., Yoshikawa, T., et al. (1997) The effect of aging on bone formation in porous hydroxyapatite: biochemical and histological analysis. *J. Bone Miner. Res.* **12(6)**, 989–994.

74. Kahn, A., Gibbons, R., Perkins, S., and Gazit, D. (1995) Age-related bone loss: a hypothesis and initial assessment in mice. *Clin. Orthop. Rel. Res.* **313**, 69–75.

75. Quarto, R., Thomas, D., and Liang, T. (1995) Bone progenitor cell deficits and the age-associated decline in bone repair capacity. *Calcif. Tissue Int.* **56**, 123–129.

76. Tabuchi, C., Simmon, D. J., Fausto, A., Russell, J., Binderman, I., and Avioli, L. (1986) Bone deficit in ovariectomized rats. *J. Clin. Invest.* **78**, 637–642.

77. Tsuji, T., Hughes, F. J., McCulloch, C. A., and Melcher, A. H. (1990) Effect of donor age on osteogenic cells of rat bone marrow in vitro. *Mech. Ageing Dev.* **51,** 121–132.
78. Einhorn, T. A., Lane, J. M., Burstein, A. H., Kopman, C. R., and Vigorita, V. J. (1984) The healing of segmental bone defects induced by demineralized bone matrix. *J. Bone Joint Surg.* **66,** 274–279.
79. Takagi, K. and Urist, M. R. (1982) The role of bone marrow in bone morphogenetic protein-induced repair of femoral massive diaphyseal defects. *Clin. Orthop. Rel. Res.* **171,** 224–231.
80. Kadiyala, S., Jaiswal, N., and Bruder, S. P. (1997) Culture-expanded, bone marrow-derived mesenchymal stem cells can regenerate a critical-sized segmental bone defect. *Tissue Eng.* **3,** 173–185.
81. Stevenson, S., Cunningham, N., Toth, J., Davy, D., and Reddi, A. H. (1994) The effect of osteogenin (a bone morphogenetic protein) on the formation of bone in orthotopic segmental defects in rats. *J. Bone Joint Surg.* **76,** 1676–1687.
82. Bruder, S. P., Kurth, A. A., Shea, M., Hayes, W. C., and Kadiyala, S. (1997) Quantitative parameters of human mesenchymal stem cell-mediated bone regeneration in an orthotopic site. *Trans. Orthop. Res. Soc.* **22,** 250.
83. Bruder, S. P., Kurth, A. A., Shea, M., Hayes, W. C., Jaiswal, N., and Kadiyala, S. (1998) Bone regeneration by implantation of purified, culture-expanded human mesenchymal stem cells. *J. Orthop. Res.* **16(2),** 155–162.
84. Cook, S. D., Wolfe, M. W., Salkeld, S. L., and Rueger, D. C. (1995) Effect of recombinant human osteogenic protein-1 on healing of segmental defects in non-human primates. *J. Bone Joint Surg.* **77A,** 734–750.
85. Lennon, D. P., Haynesworth, S. E., Bruder, S. P., Jaiswal, N., and Caplan, A. I. (1996) Development of a serum screen for mesenchymal progenitor cells from bone marrow. *In Vitro Cell Dev. Biol.* **32,** 602–611.
86. Kadiyala, S., Young, R. G., Thiede, M. A., and Bruder, S. P. (1997) Culture-expanded canine mesenchymal stem cells possess osteochondrogenic potential in vivo and in vitro. *Cell Transplant* **6(2),** 125–134.
87. Kraus, K. H., Kadiyala, S., Wotton, H. M., et al. (1999) Critically sized osteo-periosteal femoral defects: a dog model. *J. Invest. Surg.* **12,** 115–124.
88. Bruder, S. P., Kraus, K. H., Goldberg, V. M., and Kadiyala, S. (1998) Critical-sized canine segmental femoral defects are healed by autologous mesenchymal stem cell therapy. *Trans. Orthop. Res. Soc.* **23,** 147.
89. Bruder, S. P., Kraus, K. H., Goldberg, V. M., and Kadiyala, S. (1998) The effect of implants loaded with autologous mesenchymal stem cells on the healing of canine segmental bone defects. *J. Bone Joint Surg.* **80A,** 985–996.
90. Bruder, S. P. (2004) A bench to bedside case study—the development pathway of an intraoperative, autologous progenitor cell preparation kit, in *Tissue Engineering in Musculoskeletal Clinical Practice* (Sandell, L. J. and Grodzinsky, A. J., eds.), AAOS Press, Chicago, IL, pp. 377–387.
91. Connolly, J. F., Guse, R., Lippiello, L., and Dehne, R. (1989) Development of an osteogenic bone-marrow preparation. *J. Bone Joint Surg.* **71A,** 684–691.
92. Connolly, J. F., Guse, R., Lippiello, L., and Dehne, R. (1991) Autologous marrow injection as a substitute for operative grafting of tibial nonunions. *Clin. Orthop. Rel. Res.* **266,** 259–270.
93. Takigami, H., Muschler, G. F., Matsukura, Y., et al. (2002) Spine fusion using allograft bone matrix enriched in bone marrow cells and connective tissue progenitors. *J. Orthop. Res. Soc.* **27,** 807.
94. Kapur, T., Kadiyala, S., Urbahns, D. J., and Bruder, S. P. (2003) Autologous growth factors and progenitor cells as effective components in bone grafting products for spine, in *Advances in Spinal Fusion: Molecular Science, Biomechanics, and Clinical Management* (Lewandrowski, K. U., Wise, D. L., Trantolo, D. J., Yaszemski, M. J., and White, A. A., eds.), Cambridge Scientific, Cambridge, MA, pp. 381–395.
95. Muschler, G. F., Nitto, H., Matsukura, Y., et al. (2003) Spine fusion using cell matrix composites enriched in bone marrow-derived cells. *Clin. Orthop. Rel. Res.* **407,** 102–118.
96. Takigami, H., Matsukura, Y., and Muschler, G. F. (2002) Retrospective comparison of bone grafting techniques using cellular and acellular grafts. *Assoc. Bone Joint Surg. Annu. Meet.* **54,** 74–75.
97. Lieberman, I. (2003) Clinical use of bone marrow-derived osteoprogenitor cells in orthopaedic surgery. *Trans. Orthop. Res. Soc.* **28,** 28.
98. Quarto, R., Mastrogiacomo, M., Cancedda, R., et al. (2001) Repair of large bone defects with the use of autologous bone marrow stromal cells. *N. Engl. J. Med.* **344,** 385–386.
99. Arinzeh, T. L., Peter, S. J., Archambault, M. P., et al. (2003) Allogeneic mesenchymal stem cells regenerate bone in a critical-sized canine segmental defect. *J. Bone Joint Surg.* **85,** 1927–1935.
100. Tsuchida, H., Hashimoto, J., Crawford, E., Manske, P., and Lou, J. (2003) Engineered allogeneic mesenchymal stem cells repair femoral segmental defect in rats. *J. Orthop. Res.* **21,** 44–53.
101. Yoshikawa, T., Ohgushi, H., and Tamai, S. (1996) Immediate bone forming capability of prefabricated osteogenic hydroxyapatite. *J. Biomed. Mater. Res.* **32,** 481–492.
102. Ishuag-Riley, S., Crane, G. M., Gurlek, A., et al. (1997) Ectopic bone formation by marrow stromal osteoblast transplantation using poly (DL-lactic-coglycolic acid) foams implanted into the rat mesentery. *J. Biomed. Mater. Res.* **36,** 1–8.
103. Ikeuchi, M., Ito, A., Dohi, Y., et al. (2003) Osteogenic differentiation of cultured rat and human bone marrow cells on the surface of zinc-releasing calcium phosphate ceramics. *J. Biomed. Mater. Res.* **67A(4),** 1115–1122.
104. Stewart, K., Screen, J., Jefferiss, C. M., Walsh, S., and Beresford, J. (1996) Co-expression of the STRO-1 antigen and alkaline phosphatase in cultures of human bone and marrow cells. *J. Bone Miner. Res.* **11,** S142 (abstr.).

105. Turksen, K. and Aubin, J. E. (1991) Positive and negative immunoselection for enrichment of two classes of osteo-progenitor cells. *J. Cell Biol.* **114,** 373–384.
106. Turksen, K., Bhargava, U., Moe, H. K., and Aubin, J. E. (1992) Isolation of monoclonal antibodies recognizing rat bone-associated molecules in vitro and in vivo. *J. Histochem. Cytochem.* **40,** 1339–1352.
107. Yamaguchi, A. and Kahn, A. J. (1993) Monoclonal antibodies that recognize antigens in human osteosarcoma cells and normal fetal osteoblasts. *Bone Miner.* **22,** 165–175.
108. Bianco, P., Silvestrini, G., Termine, J. D., and Bonucci, E. (1988) Immunohistochemical localization of osteonectin in developing human and calf bone using monoclonal antibodies. *Calcif. Tissue Int.* **43,** 155–161.
109. Aubin, J. E., Gupta, A. K., Bhargava, U., and Turksen, K. (1996) Expression and regulation of galectin 3 in rat osteo-blastic cells. *J. Cell Physiol.* **169,** 468–480.
110. Chen, J. K., Zhang, Q., McCulloch, A. G., and Sodek, J. (1991) Immunohistochemical localization of bone sialo-protein (BSP) in fetal porcine bone tissues: comparisons with secreted phosphoprotein 1 (SPP1, osteopontin) and SPARC (osteonectin). *Histochem. J.* **23,** 281–289.
111. Mark, M. P., Prince, C. W., Oosawa, T., Gay, S., Bronckers, A. L., and Butler, W. T. (1987) Immunohistochemical demonstration of a 44 kDa phosphoprotein in developing rat bones. *J. Histochem. Cytochem.* **35,** 707–715.
112. Bronckers, A. L., Gay, S., Finkelman, R. D., and Butler, W. T. (1987) Developmental appearance of Gla proteins (osteocalcin) and alkaline phosphatase in tooth germs and bones of the rat. *Bone Miner.* **2,** 361–373.
113. Mark, M. P., Prince, C. W., Gay, S., et al. (1987) A comparative immunocytochemical study on the subcellular distri-butions of a 44 kDa bone phosphoprotein and bone gamma-carboxyglutamic acid (gla)-containing protein in osteo-blasts. *J. Bone Miner. Res.* **2,** 337–346.
114. Wetterwald, A., Hofstetter, W., Cecchini, M. G., et al. (1996) Characterization and cloning of the E11 antigen, a marker expressed by rat osteoblasts and osteocytes. *Bone* **18(2),** 125–132.

6

Biology of the Vascularized Fibular Graft

Elizabeth Joneschild, MD and James R. Urbaniak, MD

BONE GRAFTING

Introduction

In the practice of orthopedics, bone grafting is a common procedure used to enhance the regeneration of bone and lead to the restoration of skeletal integrity. Bony regeneration is needed to reconstruct a wide variety of traumatic, developmental, degenerative, and neoplastic disorders that affect the skeletal system. The source of bone for grafting has evolved over the past two centuries to include autogenous cancellous or cortical, allogenic frozen, freeze-dried, or processed cortical, corticocancellous, and cancellous grafts, and demineralized bone matrix. Recently, synthetic or engineered bone graft substitutes have also been approved for use. Although this chapter concentrates on the autogenous vascularized fibular graft, a brief review of the history and basic science of bone grafting will serve as an introduction.

History

Historically, isolated cases of clinical bone grafting were described as early as 1668, when the Dutch surgeon Job van Meerkeren inserted a portion of a dog's skull to repair a soldier's cranium *(1,2)*. Further work by a fellow Dutch scientist, Antonius De Heyde, helped define the process of osteogenesis. He concluded, after his experimental observations made on frogs, that callus forms by calcification of the blood clot around the broken bone ends *(3)*. Two centuries later, the Frenchman L. Ollier published a classic paper entitled *Triate experimental et elinique de la regeneration des os*, in which he showed that autographs can be viable. Ollier also recognized that separate living bone fragments without periosteum could live and grow in a suitable environment *(4)*.

The motivation for the clinical use of bone grafting came from the simultaneous work on bone transplantation in the late 19th century by Barth in Germany and Curtis in the United States. Working independently, both published their work on bone transplantation. Barth described, *schleichenden Ersatz*, the absorption of dead tissue of the bone graft and formation of new bone, which grew into the graft from the surrounding living bone *(5)*. Curtis noted that the haversian canals, moreover, afford easy avenues for the growth of granulation tissue, and....ossification so soon takes place in the tissues which replace the bone graft as it is absorbed *(6)*. Phemeister later termed this process *creeping substitution*. He described the penetration of newly formed bone directly into the old bone, a process that required the simultaneous removal of the necrotic trabeculae of the devascularized bone and subsequent deposition of new bone *(7,8)*.

In 1820, Philips von Walter, a German surgeon, described the first clinical autograft procedure in which he replaced surgically removed parts of a skull after trepanotomy *(9)*. In 1880, William Macewen, from Scotland, described the allographic transplant of a tibia from a child with rickets to reconstruct an infected humerus in a 4-yr-old child *(10)*. However, it was not until after the publication of F. H. Albee's

From: *Bone Regeneration and Repair: Biology and Clinical Applications*
Edited by: J. R. Lieberman and G. E. Friedlaender © Humana Press Inc., Totowa, NJ

book, *Bone graft surgery*, in the United States in 1915, that bone grafting was understood and became a commonly used surgical procedure *(11)*.

Basic Science of Bone Grafting:
Osteoconduction, Osteoinduction, and Osteogenesis

Bone grafts are used to promote healing in various situations of bone loss. The principal indications for the use of bone graft include the need to fill bony defects and to enhance new bone formation. Many types of bone grafts are used, and the choice of bone graft is often tailored to the clinicial situation. Autogenous fresh cancellous and cortical bone (vascularized vs nonvascularized) are most often used, but other graft materials include allogenic fresh frozen, freeze-dried, or processed cortical, cortico-cancellous, and cancellous grafts. Synthetic or engineered bone graft substitutes are the latest addition of materials used to enhance healing of bony segments.

Each of these grafts has various capacities to provide active bone formation, to induce bone formation by cells of the surrounding soft tissue, and to serve as a substrate for bone formation. The biological activity of any graft is multifactorial. Its activity represents the sum of its inherent biological activity, its capacity to activate surrounding host tissues, and its osteoinductive capacity, which is mediated by bioactive factors within the matrix. Finally, its ability to support the ingrowth of osteogenic host tissue by its osteoconductive framework also plays a role in its behavior *(12)*.

Autogenous bone grafting is currently considered the best graft material because it provides the three elements required for bone regeneration: osteoconduction, osteoinduction, and osteogenic cells. Osteoconduction pertains to the porous, three-dimensional architecture of cancellous bone that allows for rapid ingrowth of sprouting capillaries, perivascular tissue, and osteoprogenitor cells from the recipient host bed into the three-dimensional structure of an implant or graft *(13,14)*. The structure functions as a trellis, or scaffold, for the ingrowth of new host bone *(15)*. Osteoconduction is an ordered process following predictable spatial patterns determined by the geometry of the graft, the vascular supply from the surrounding soft tissue, and the mechanical environment of the graft *(12)*. The bone graft serves as a surface on which cells attach and differentiate. Because of the graft's three-dimensional structure, it is able to support the growth, vascularization, and remodeling of bone.

Consequently, the mechanical environment of the graft site is paramount. Bone grafts are remodeled, according to Wolff's law, in response to the same mechanical stimuli as normal bone *(16)*. In addition to noting the clear relationship between bone structure and loading, Wolff made the critical observation that living bone adapts to alterations in loads by changing its structure in accordance with mathematical laws *(17,18)*. Therefore, increased motion at the interface of grafted cortical bone and host soft tissue will hinder or possibly prevent revascularization *(8)*. The healing of a bone graft is strongly influenced by the environment into which it is placed.

Within its matrix, cancellous bone contains growth factors that promote osteoinduction, a process that supports the mitogenesis of undifferentiated perivascular mesenchymal cells *(14,19)*. This cascade leads to the formation of osteoprogenitor cells with the capacity to form new bone. The osteoinductive capacity of living graft cells is related to its production of osteoindutive factors, including bone morphogenic proteins (BMPs), TGF-β, IGF-1, IGF-II, aFGFs, interleukins, and granulocyte colony-stimulating factors. In addition, the osteoinductive capacity of a specific molecule may be potentiated by other factors that influence cellular responses, such as those that enhance cellular proliferation, migration, attachment to extracellular matrix molecules, and differentiation *(13)*. All of these factors influence the differentiation of mesenchymal cells into bone-forming cells.

Finally, all living periosteal cells and other osteoblasts transplanted with the graft are osteogenic. These cells, if handled properly, can survive to produce new bone *(20)*. Cancellous bone, with its large surface area covered with quiescent lining cells or active osteoblasts, has the potential for more graft-originated new bone formation than does cortical bone *(12)*. Osteogenesis of graft origin occurs independently of the host bed, except that diffusion from the host is required for the cells to remain viable. If fresh cancellous autograft is placed in a densely fibrotic, irradiated bed, the graft will survive independently and produce new bone. Successful incorporation and bone formation is independent of the host bed. However, the health of the host bed is critical in the process of osteoinduction, because new osteoprogenitor cells are recruited by induction of residual mesenchymal cells in marrow reticulum, endosteum, periosteum, and connective tissue of the host *(12)*.

Types of Grafts

Autogenous cancellous bone is considered the best and most effective graft material because it is highly osteogenic and well revascularized, allowing for rapid integration into the recipient site. Graft incorporation occurs by creeping substitution and follows a series of stages starting with hemorrhage and inflammation, and proceeding to vascular infiltration, resorption, and bony production. Bony remodeling occurs as the final step. This process occurs in the months following surgery and is usually complete 1 yr after surgery. Autogenous cancellous graft does not provide structural support, but because autogenous cancellous bone stimulates early new bone formation, it often contributes to the early stabilization of a fracture site. This early bone formation is often critical to callus formation in the first few weeks following surgery. Although clinically successful, the harvest of autogenous cancellous graft from the iliac crest is not without significant morbidity *(21)*.

Nonvascularized cortical autographs are advantageous because they provide structural support which is often needed in large bony defects. Although cortical grafts undergo a similar process of incorporation as cancellous grafts, the density of the cortex and its lack of porosity slow this process. For cortical bone to incorporate, it must undergo a period of resorption that increases its porosity, allowing vascular invasion and subsequent osseous integration into Volkmann's and Haversian canals *(22)*. Although cortical grafts are often chosen for their structural support in large defects, they rapidly lose their structural strength as integration develops. In fact, the process of graft incorporation results in a 30% reduction in strength over 6–18 mo *(23)*. This significant weakness can persist for months to years, depending on the size of the graft, and may account for the incidence of fractures in these grafts, which is reported to range from 16 to 50% *(12,24–26)*. It is theorized that these large grafts likely sustain fatigue microdamage as cyclic loading occurs over time. Fractures occur because the necrotic bone cannot repair itself in response to damage *(27)*.

The difference between the integration of cancellous graft and nonvascularized cortical graft can also be observed radiographically. Cancellous grafts initially appear more radiodense because of the deposition of new bone on the graft infrastructure. In contrast, cortical bone becomes radiolucent as the cortical destruction progresses and revascularization occurs.

Allogenic demineralized bone prepared as fresh-frozen, freeze-dried, and demineralized bone matrix has been used extensively for skeletal reconstruction. Its most common use has been for bulk replacements in skeletal loss. Allogenic material is revascularized quickly and has moderate osteoinductive properties. However, it does not provide any structural support. In addition, allografts generate an intense immune response that interferes with graft incorporation. Studies have demonstrated that immune mismatch significantly affects incorporation *(23)*. Freezing does decrease the rejection of mismatched grafts, but frozen allograft does not incorporate as well as fresh autograph. Allograft incorporation is also much slower than that of autografts, prompting the need for rigid internal fixation to provide optimal support during allograft incorporation. In order to sustain osteogenesis, adequate cross-linking of collagen within the demineralized bone matrix must occur, and osteoinductive proteins must be present. The source and processing of demineralized bone matrix may also have a direct effect on its osteoinductive capacity. For example, storage of a bone at room temperature for more than 24 h before processing causes the recovered demineralized bone matrix to be biologically inactive *(28)*.

An additional major concern regarding the use of allographs remains the potential for disease transmission. Allograft bone readily transmits retrovirus infection despite routine processing and removal of bone marrow. Sterilization, either by ethylene oxide or gamma irradiation, is detrimental because it substantially reduces the osteoinductivity of the graft. The major defense against such disease transmission is careful screening of donors, but the potential for error always exists.

Synthetic or engineered bone graft substitutes are materials than can enhance fracture healing without the concern of disease transmission or availability. Synthetic graft substitutes consist of an osteoconductive matrix to which osteoinductive proteins and/or osteoprogenitor cells may be added. Currently, several clinically available osteoconductive substances use calcium–phosphate ceramics. These materials are available as either porous or nonporous dense implants, or porous granules. Although hydroxyapatite implants do not resorb, they cause little inflammation and are well tolerated in metaphyseal sites. However, because they are brittle and have limited potential for remodeling, the use of hydroxyapatite implants as graft substitutes for diaphyseal defects of long bone is less successful *(23)*.

The second type of synthetically engineered bone graft substitutes are the composite osteoconductive grafts that combine porous hydroxyapatite–tricalcium phosphate ceramics with type I collagen. These grafts have been shown to be as effective as autogenous cancellous bone grafts in the treatment of long bone fractures *(23)*. Additionally, use of synthetic graft avoids the morbidity associated with autogenous bone graft harvest.

The use of osteoinductive agents has not yet been tested and reported in a human series. Demineralized bone matrix prepared by acid extraction of the allograft bone is osteoconductive and slightly osteoinductive. The results of the use of demineralized bone matrix in clinical trials has been excellent. Importantly, it is also the only available source of an osteoinductive material other than autogenous bone.

VASCULARIZED CORTICAL GRAFTING

Introduction

The reconstruction of large skeletal deficiencies presents a challenge to the orthopedic surgeon. The treatment of such large defects by the conventional bone grafting techniques already mentioned was found to be limited and inadequate. Historically, massive segmental bone loss from trauma, infection, tumor resection, or congenital pseudarthrosis required amputations. However, advances in vascular and microvascular surgery over the past several decades have made it possible to transfer autogenous bone grafts on vascular pedicles to reconstruct a wide variety of defects. Vascularized grafts have added a new dimension to the science of bone grafting. Vascularized grafts are able to restore physiological blood flow by surgical anastomosis of a nutrient vascular pedicle at the recipient site. This procedure ensures the viability of cells within the transferred bone segment. The transferred living cells then immediately aid in the process of bone healing and remodeling. Consequently, reconstruction of a large defect may be accomplished by a mode of healing similar to that of a segmental fracture, rather than the usual, more lengthy process of graft incorporation.

History

The idea of transplanting a vascularized bone graft has existed since the late 1800s. In 1893, Curtis stated in his classic paper: "That calcified bone [was] at present the most practical material for use in the ordinary cases, while we are waiting the ideal of the future: the insertion of a piece of living bone which will exactly fill the gap and will continue to live without absorption" *(6)*. Early work by Phelps, in 1891, involved connecting a piece of bone from a dog as an interposition graft in a defect of the tibia of a boy *(29)*. After the operation, the boy and the dog were attached to each other for 2 wk. Unfortunately, the graft failed and was removed after 5 wk. Huntington, as early as 1905, recognized the advantages of utilizing bone graft with its own nutrient blood supply intact as a vascularized bone graft in the reconstruction of large tibial defects *(30,31)*. He assumed that the fibula's own nutrient blood supply would be preserved, although no actual anastamoses were performed. However, it was not until the publication of Alexis Carrel's original article in 1908 in *JAMA*, "Results of Transplantation of Blood Vessels, Organs, and Limbs," that distant transfer of vascularized bone grafts were recognized as a possibility *(32)*.

Advances in the field of microsurgery in the 1960s led to the development of the operating microscope as well as improvements in microsurgical instruments. These improvements made the prospect of vascularized bone grafts a reality. The first successful clinical microvascular transfer of bone was a rib to a mandible and was accomplished by McKee in 1970 *(33,34)*. In his technique, an anterior segment of rib was removed in continuity with 3 cm of internal mammary artery and vein for anastomosis to the faciolingual artery and the posterior facial vein. This graft was dependent on the periosteal circulation and not the primary medullary circulation. McKee's technique was used experimentally by Adelaar et al. to reconstruct radial defects in dogs by anastomosis of the anterior intercostal artery to a branch of the radial artery, but venous anastamosis was not performed *(35)*. However, when compared to conventional rib grafts, this study showed no improvement in clinical union, osteocyte survival, or bone viability. In addition, there was no uptake of tetracycline label. These results suggested that the anterior rib graft, in order to survive on its periosteal circulation alone, would require both arterial and venous anastomoses.

An alternate technique of rib transfer was described by Ostrup and Fredrickson in 1974 *(36)*. They developed a composite rib graft based on the posterior intercostal artery and vein as a vascular pedicle. It was used to reconstruct experimentally produced mandibular defects in dogs. They demonstrated osteocyte survival by oxytectracycline–DCAF labeling. Their technique preserved both medullary and periosteral blood supply. This technique was later successfully used to bridge defects in dog femora in an experimental study by Doi et al. *(37)*.

Successful clinical application of this posterior rib graft was first reported by Serafin et al., who used it to reconstruct a mandibular defect *(38)*. Buncke et al. pioneered the first orthopedic application of this technique in 1973, but it was not reported in the literature until 1977 *(39)*. In Buncke's case, he transferred a composite posterior rib graft with overlying soft tissue and skin to a 6-cm tibial defect by anastamoses of the intercostal vessels to the anterior tibial vessels in the leg. The paper also included two other successful applications of this technique for segmental tibial defects. Despite its successful clinical results, the rib graft had two distinct disadvantages. The first was its curved shape and the second was the significant morbidity associated with the deep intrathoracic dissection. Consequently, other bones such as the iliac crest and the fibula were studied for free vascularized transfer *(34)*.

The clinical use of free vascularized bone grafts for treatment of long bone defects was significantly advanced with the report of the free vascularized fibula transfer by Taylor et al. in 1975 *(40)*. This transfer anastomosed the peroneal artery and its venae comitantes to leg vessels for preservation of fibular blood supply. Successful transfer of a fibula into a 12.5-cm tibial defect was achieved. Weiland et al. next reported five cases of free fibula transfer for reconstruction of upper-extremity defects. (**Figs. 1A–E**). In one case, a fibula was transferred with its proximal epiphysis, preserving the inferior geniculate artery, and some degree of longitudinal growth was achieved *(41)*. In 1979, Chen et al. described the use of the free fibular vascularized graft in congenital tibial pseudoarthroses *(42)*.

Vascular Supply to Cortical Bone

The ability to transfer a segment of bone with its blood supply was firmly based on prior studies investigating the vascular pattern of cortical bone *(43–45)*. Blood flow through cortical bone depends on an intact medullary blood supply, whereas periosteal arteries play a relatively minor role in cortical nutrition. The medullary blood supply receives major contributions from nutrient arteries, which penetrate the cortex through nutrient foramina and nutrient canals. As a nutrient artery enters a bone, it sends branches proximally and distally. These branches extend radially to supply the diaphyseal cortex and then further branch longitudinally. Epiphyseal and metaphyseal arteries supply their respective areas. The surrounding muscles provide the blood supply to the periosteum, but these play a relatively minor role in cortical blood supply. Fortunately, one predominant nutrient vessel supplies most long bones *(46,47)*. If this primary vessel is preserved, a large segment of bone can be transplanted as a living graft. With the nutrient supply maintained, osteocytes and osteoblasts in the graft can survive and

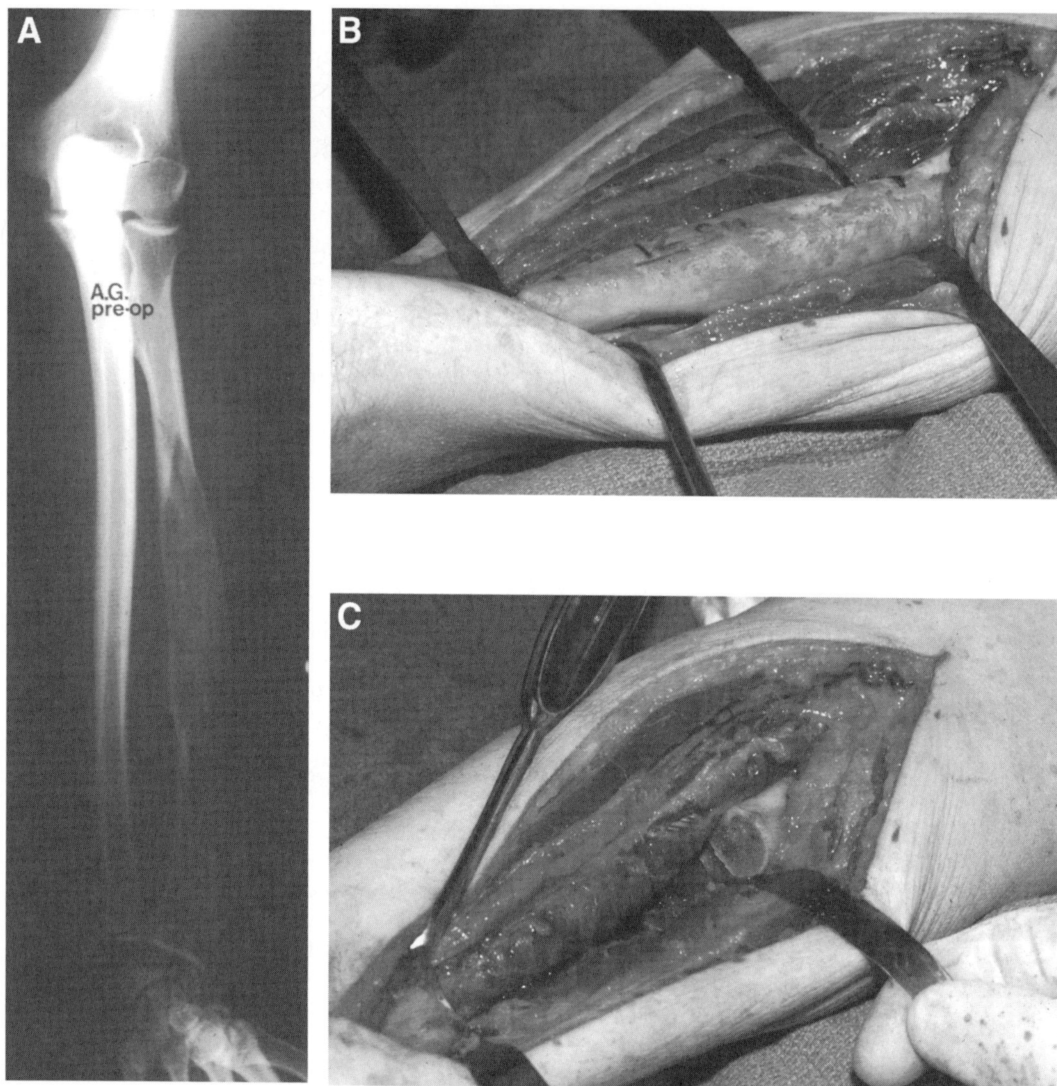

Fig. 1. (A) Preoperative radiograph showing adamantinoma. **(B)** Operative exposure of radial shaft with adamantinoma. **(C)** Excision of involved radial shaft.

the graft can heal to recipient bone rapidly, in a manner independent of the recipient bed and without the need for creeping substitution *(48)*.

Despite its relatively minor role, several studies have indicated that cells in the periosteal layer survive and are capable of osteogenesis. Thus, the lack of blood flow to the periosteum results in the formation of relatively smaller callus about the site of union and an increase in the time required for union and repair process *(49)*.

VASCULARIZED FIBULAR GRAFTING

Anatomy of the Fibula

The anatomy of the fibula makes it ideally suited for use in reconstruction of the long bones of the extremities, especially in patients with massive diaphyseal bone loss secondary to trauma or tumor

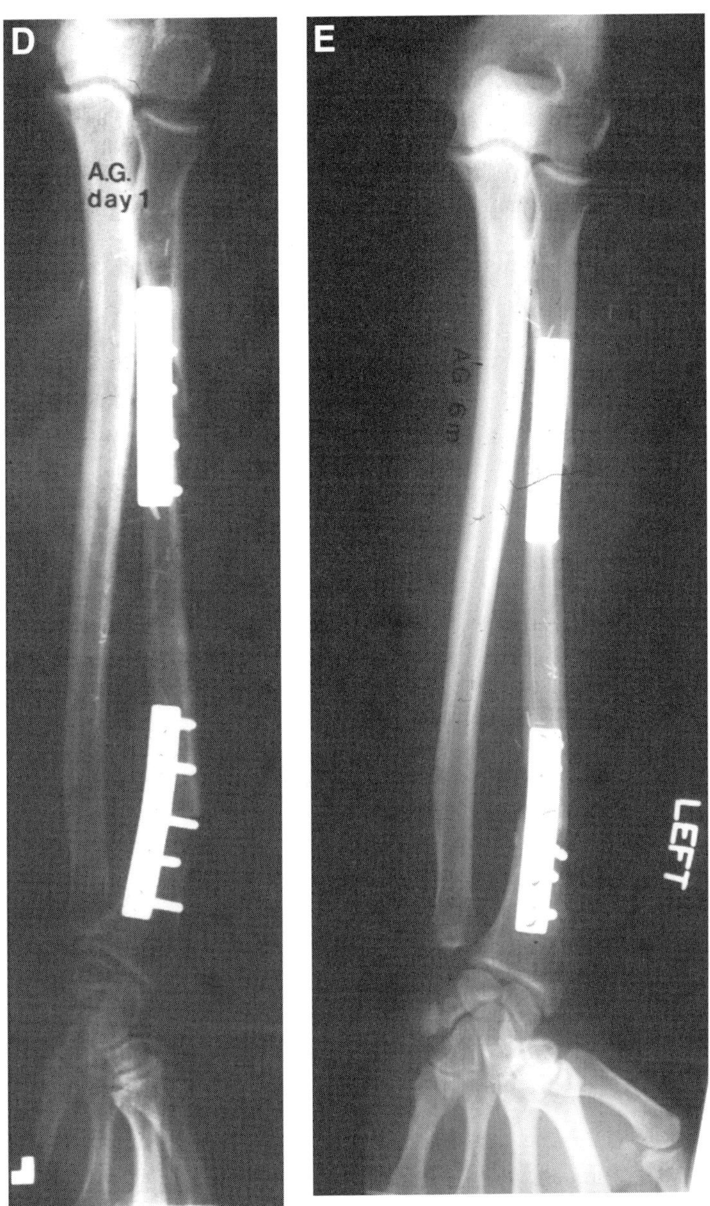

Fig. 1. (D) Postoperative d 1, vascularized fibular graft to distal radius. **(E)** Six months postoperatively with graft incorporation and hypertrophy.

resection. It is a straight cortical bone that can restore continuity in long bone defects as much as 22–26 cm *(31)*. It matches exactly the size of the radius and the ulna and fits snugly into the medullary cavity of the humerus, femur, and tibia. In addition, the high proportion of cortical bone and the triangular cross section also provide stability by resisting angular and rotational stresses *(40)*.

Blood Supply of the Fibula

The fibula also has an appropriate blood supply. It includes both endosteal and periosteal blood supplies provided by a system of vessels that are of adequate caliber for successful anastamoses *(50–52)*.

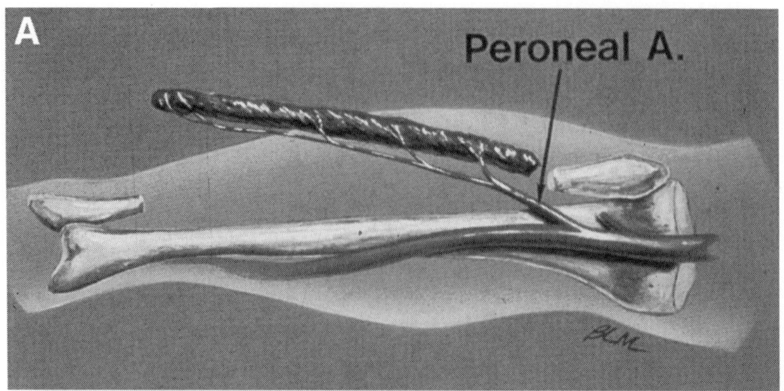

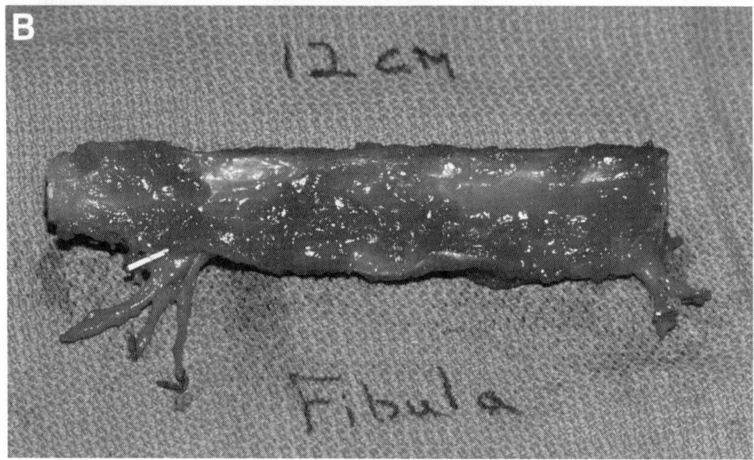

Fig. 2. (A) Schematic representation of fibula with its nutrient blood supply. **(B)** Fibula harvested with vascular pedicle (on left) with periosteum intact and minimal muscle cuff.

The nutrient artery of the fibula arises as a branch of the peroneal artery. The peroneal artery gives off several periosteal branches before giving origin to the nutrient artery, which supplies the medullary blood flow to the fibula. The nutrient artery penetrates at the mid-diaphyseal level, which may vary by 2.5 cm proximally or distally *(53)*. The length of the nutrient artery outside the fibula ranges from 5 to 115 mm, and its diameter is between 0.25 and 1.0 mm. After penetrating the bone, it divides into two branches, one passing proximally and one distally. The peroneal artery continues distally along the medial and posterior aspect of the fibular diaphysis and provides musculoperiosteal branches. By isolating the peroneal artery at its origin from the posterior tibial-peroneal trunk, it is possible to preserve the medullary and periosteal blood supplies to the fibula (**Figs. 2A,B**).

Donor Site Morbidity

Another advantage of the fibula as a free vascularized bone graft is that there is minimal donor site morbidity. We have reviewed 247 consecutive grafts in 198 patients who have had their fibula harvested for vascularized fibular transfer to the hip for avascular necrosis. At 5-yr follow-up, an abnormality was noted in 24% of lower limbs. A sensory deficit was found in 11.8% of limbs, and 2.7% of patients had some motor weakness. Pain at the ankle itself was a complaint in 11.5% of limbs; pain at other sites was reported by 8.9% of patients. Contracture of the flexor hallucis longus was present in

2% of patients due to the intramuscular plane used to protect the pedicle of the graft; this complication is avoidable with careful stretching of the toes in extension in the first few days after surgery *(54)*.

Although the morbidity of fibular graft harvest is low at our institution, the literature is inconsistent regarding the incidence of complications following fibular harvesting. In addition to flexor hallucis longus contracture, studies have reported ankle pain, ankle instability, or a combination of both pain and instability as complications following fibular graft harvest.

Previous clinical and biomechanical studies to evaluate ankle instability have concluded that a critical amount of fibula must remain intact to maintain stability. Despite quoted lengths of 6–8 cm *(55)*, this critical length was not defined until a recent paper by Pacelli *(56)*. In this study, published in 2003, the authors calculated the percentage of fibula that could be removed while still preserving ankle stability.

The study used 11 fresh paired cadaveric legs, which were tested with the foot mounted in three positions (neutral, 15° inversion, and 15° eversion) while an external and internal rotation torque and an axial load were applied across the ankle.

Finally, the ankles were tested against a varus load. Initially, the ankles were tested with an intact fibula to establish a baseline stability. Sequential fibular resections were then performed, from proximal to distal, until ankle instability was encountered.

The results of the study concluded that only 10% of the fibula was necessary to maintain ankle stability. A residual fibular length of 10% represented a mean residual fibular length of 3.9 cm. This length corresponded with a fibular osteotomy just proximal to the syndesmotic ligaments and was significantly less than the previously quoted lengths of 6–8 cm found in the literature.

Advantage of the Fibular Graft

Experimental and clinical studies have shown that autogenous bone grafts are the most favorable grafts in terms of incorporation, remodeling, and the ultimate ability to provide structural support. However, as mentioned previously, cancellous autogenous grafts merely provide the open matrix that allows for vascular invasion and the diffusion of nutrients and cells from the host tissue. Only a small percentage of the osteocytes survive transplantation, and ultimate incorporation is dependent on the process of creeping substitution. In contrast, a majority of the osteocytes and osteoblasts present in the vascularized bone graft remain viable. When the vessels are anastamosed successfully and the graft suffers only transient intraoperative ischemia, more than 90% of the osteocytes survive the transplantation procedure *(12)*. Consequently, these grafts usually heal by a repair process that does not result in resorption of a significant amount cortical bone (**Figs. 3A,B**). Because these grafts are able to maintain their structural integrity, they can be used for segmental cortical defects that are larger than 6 cm (**Figs. 4A,B**). The healing process begins immediately and can be monitored by the early hypertropy of the cortical graft.

Mechanical Properties

Since the ultimate measure of success of a bone graft depends on its ability to withstand physiological loads, mechanical testing of vascularized fibular grafts was necessary to evaluate their structural properties and validate their enhanced clinical performance. In theory, the strength and stiffness of a graft are influenced by the remodeling process associated with revascularization.

The work by Victor Goldberg has contributed the most information in the structural evaluation of vascularized grafts *(57)*. In his experimental canine model, an 8-cm proximal fibular graft was isolated on its vascular pedicle, while also preserving the flexor hallucis longus and the extensor digitorum lateralis. The vascularized fibular graft consisted of the bone, periosteum, and the muscle cuff perfused by the caudal tibial artery and vein. The vascular pedicle and the isolated fibula was removed from the surgical field and replaced into its orthotopic site. The vessels were then repaired microsurgically and the proximal fibula was stabilized by repairing the tibiofibular joint capsule. The distal fibula was

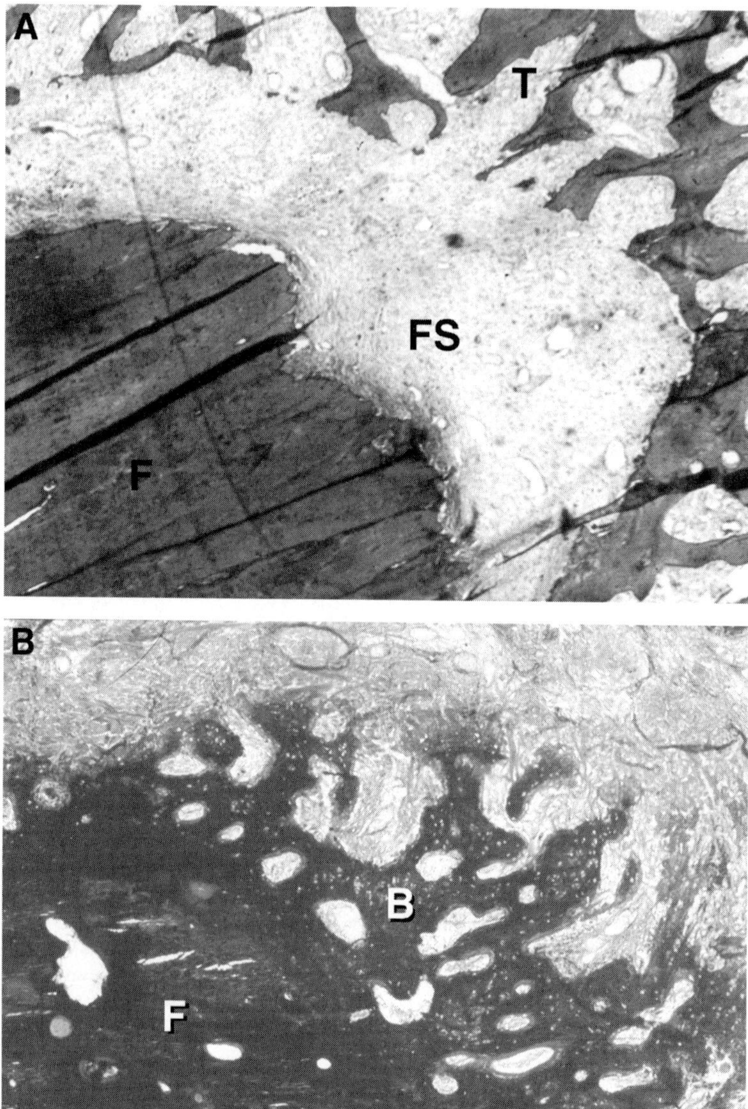

Fig. 3. (A) Histology of nonvascularized fibular graft in an animal model with fibrovascular stroma (FS) surrounding distal end of fibular graft (F). No new bone is present. Infarcted trabecula (T) can be seen. **(B)** Histology of vascularized fibular graft in an animal model with generation of new woven bone (B) from distal end of fibular graft (F).

stabilized by a transverse K-wire into the tibia. The opposite fibula was used as a nonvascularized autograft control. This nonvascularized control consisted of a subperiosteally dissected bone placed in a bed devoid of periosteum.

The bone grafts were then evaluated by multiple parameters. Blood flow to the bone graft was studied by a hydrogen washout technique. The grafts were studied by mechanical and morphometric techniques, and metabolic kinetics was used to quantify bone resorption and formation. Mechanical testing involved subjecting the grafts to torsional testing to determine torsional strength and stiffness. Morphometric analysis consisted of periodic injection of intravenous fluorochromes after surgery, so

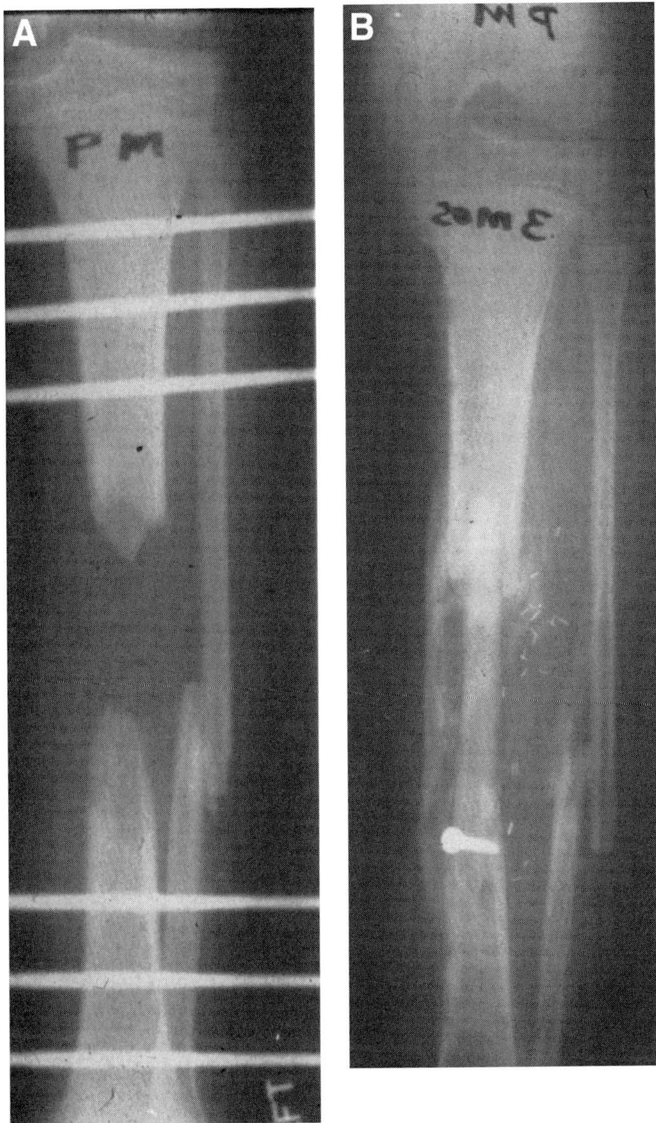

Fig. 4. (A) Traumatic 10-cm segmental midshaft tibial defect. **(B)** Bony regeneration 3 mo following vascularized fibular graft.

that bone repair could be evaluated sequentially. Cortical bone area and fractional porosity was measured from microradiographs with an image analyzer. Metabolic kinetics determined calcium levels and collagen content of the bone.

The results from the study demonstrated that the average strength and stiffness of the vascularized grafts were both significantly greater than those of nonvascularized graft 3 and 6 mo after surgery. In general, the nonvascularized grafts were extremely weak and fragile. The metabolic kinetic data demonstrated that the vascularized grafts only lost approx 30% more mineral and collagen and 25% more collagen mass than control segments, whereas nonvascularized grafts were significantly resorbed and demonstrated little new bone formation. When the vascularized grafts were compared with nonvascularized autografts, there was significantly less loss of calcium and collagen and less resorption. Furthermore, bone formation was significantly higher in these grafts than in nonvascularized autografts. Morphometric analayis demonstrated that nonvascularized grafts were significantly smaller and more porotic and had fewer osteons than vascularized grafts *(57,58)*.

The integration of these data suggests that vascularized fibular cortical autographs are similar to control segments in strength, stiffness area, total osteons, and degree of porosity. The grafts did undergo some resorption, but formation was also a significant event. They noted that the major bone loss of the vascularized autographs was in the cancellous portion of the graft. In summary, the results from Goldberg's study clearly showed the superiority of vascularized grafts over nonvascularized grafts during the period of early incorporation.

Many clinical studies have looked at the efficacy of using a free vascularized fibular graft for osteonecrosis of the femoral head, but it was not until recently that the efficacy of a vascularized fibular graft was compared directly to a nonvascularized fibular graft. In a study published in 2003, 200 patients (220 hips) with osteonecrosis of the femoral head were treated with a free vascularized fibular graft at the University of Pittsburgh Medical Center, and 99 patients (123 hips) were treated with a nonvascularized fibular graft at the Kyungpook National University Hospital in Korea *(59)*. The patient populations of the two centers were matched as closely as possible on the basis of characteristics such as the stage of the disease, extent of involvement, etiological factors, and preoperative Harris hip scores, to allow equal comparison. A retrospective case-control study of these groups was then performed to compare the postoperative Harris hip scores and the prevalence of radiographic progression and collapse of the femoral head.

The results of this study showed that the mean Harris hip score improved for 70% of the hips treated with free vascularized fibular grafting; 17 hips (34%) were rated excellent, 14 (28%) good, 9 (18%) fair, and 10 (20%) poor. In the group treated by nonvascularized fibular grafting, the mean Harris hip score improved for 36%; 5 hips (10%) were rated excellent; 9 (18%) good, 16 (32%) fair, and 20 (40%) poor. The rate of survival at 7 yr for the Stage I and II hips (precollapse) was 86% after treatment with a vascularized fibular graft, compared with 30% after nonvascularized fibular grafting. The authors reached similar conclusions to what Victor Goldberg concluded in the lab: free vascularized fibular grafting was superior to nonvascularized fibular grafting for the treatment of osteonecrosis of the femoral head *(59)*.

Blood Flow Analysis

Blood flow analysis in vascularized bone transfers has also been evaluated in a canine model by Siegert *(60)*. In this study, cortical bone blood flow was compared in undisturbed control bone, vascularized heterotopic bone transfers, and nonvascularized cortical bone grafts by means of the radionuclide-labeled microsphere technique. The results demonstrated quantitatively that significant blood flow to vascularized transfers is preserved during the early postoperative period and is augmented over the 6-wk observation period in the study.

Clinical Application

Idiopathic Osteonecrosis

Clinical application of the free vascularized fibular graft has been used in our institution (Duke University Medical Center) for the treatment of symptomatic osteonecrosis of the hip in the young patient. Osteonecrosis of the femoral head is a multifaceted process that leads to articular incongruity and subsequent osteoarthritis of the joint. Clinicians concur that the primary treatment should focus on preservation of the natural surface of the joint. However, no consensus exists on how this should be best accomplished surgically. The relative efficacy of the various treatment options (core decompression, osteotomies, electrical stimulation, and bone grafting) is difficult to evaluate because there are few prospective controlled studies in the literature. In our experience, vascularized fibular bone grafting to the femoral head has provided the most consistently successful results. The procedure allows decompression of the femoral head to halt the ischemia due to increased interosseous pressure. Necrotic bone is then removed and replaced with cancellous bone, which has both osteoinductive and conductive factors. The fibular cortical strut supplies reinforcement to the subchondral bone, and the vascular pedicle guarantees a supply of nutrients and blood to the healing femoral head. Our current results of over 1400 cases has shown an 82% success rate *(54)* (**Figs. 5A–C**).

Despite success of treating osteonecrosis of the femoral head at our institution with vascularized fibular grafting, many other treatment options have also been described. One such treatment option is core decompression. This was originally described as a diagnostic procedure, but it was later proposed to be of therapeutic benefit *(61)*. Core decompression is thought to relieve the compression caused by the interstitial edema, improve vascularity, and slow progression of necrosis within the femoral head *(62)*. The procedure gained popularity because of early promising results, low morbidity, and a lack of other alternative treatment options.

In an attempt to determine the best treatment option, we conducted a cohort statistical analysis, with multiple regression to control for covariates, to compare the results of free vascularized fibular grafting (614 femoral heads) to core decompression (98 femoral heads) carried out at another institution for the treatment of avascular necrosis of the femoral head *(63)*. The patients were stratified according to age and the stage of disease, and a survival analysis was performed with total hip arthroplasty as the end point for failure.

None of the 11 hips that had Ficat Stage I disease needed a total joint replacement after being treated with either regimen. Analysis of the hips that had Stage II disease revealed rates of survival, at 50 mo, of 65% after core decompression and 89% after vascularized fibular grafting. For Stage III hips, the rates of survival were 21% after core decompression, compared to 81% after vascularized grafting. Among the hips with either Stage II or III disease, the rate of eventual total joint arthroplasty after vascularized fibular grafting was significantly lower than that after core decompression ($p < 0.0001$).

These results clearly indicate that vascularized fibular grafting, while a more technically demanding procedure than core decompression, is justified by the associated delay in, or prevention of, articular collapse in hips with Ficat Stage II or III disease *(63)*.

Free vascularized fibular grafting has been performed at our institution for over 20 yr with great success *(64)*. The reproducibility of this technique has been well documented throughout the world literature *(65–68)*, with the most recent publication in 2001 by Soucacos *(69)*. In this study, 228 hips in 187 patients were treated with vascularized graft over an 12-yr span from 1989 to 2000. Of these 228 hips, 184 hips were assessed postoperatively with follow-up ranging from 1 to 10 yr (average 4.7 yr). Preoperatively, 21% of the hips were Steinberg *(70)* Stage II, 25% were Stage III, 42% were Stage IV, and 12% were Stage V.

Of all the hips treated, 54% remained stable postoperatively, while 38% had progression and 8% were converted to a total hip arthroplasty. Of the hips that progressed, 64% of these hips did not progress until 6–10 yr following the initial procedure. The best results were obtained in the patients with Stage II osteonecrosis, in whom 95% of the hips did not progress postoperatively. In addition, preoperative and postoperative clinical evaluation and Harris hip scores improved in all patients. Soucacos concluded that the vascularized fibular graft was an excellent procedure for the precollapse stages and a valuable alternative for those patients with more advanced stages of disease *(69)*.

The surgeons in this study used a surgical technique described and outlined by Urbaniak *(64)*. However, not all surgical techniques and approaches to the hip for free vascularized fibular grafting are similar. Judet, whose technical approach differs considerably, has recently reported his long-term results of free vascularized fibular grafting for the treatment of femoral head necrosis *(71)*. The technique, initially introduced in 1978, involves an anterior approach to the hip. Once the hip is dislocated, the necrosed bone is completely excised from the femoral head through an elevated flap of articular cartilage. A tunnel is then drilled from the femoral head to the greater trochanter for placement of the

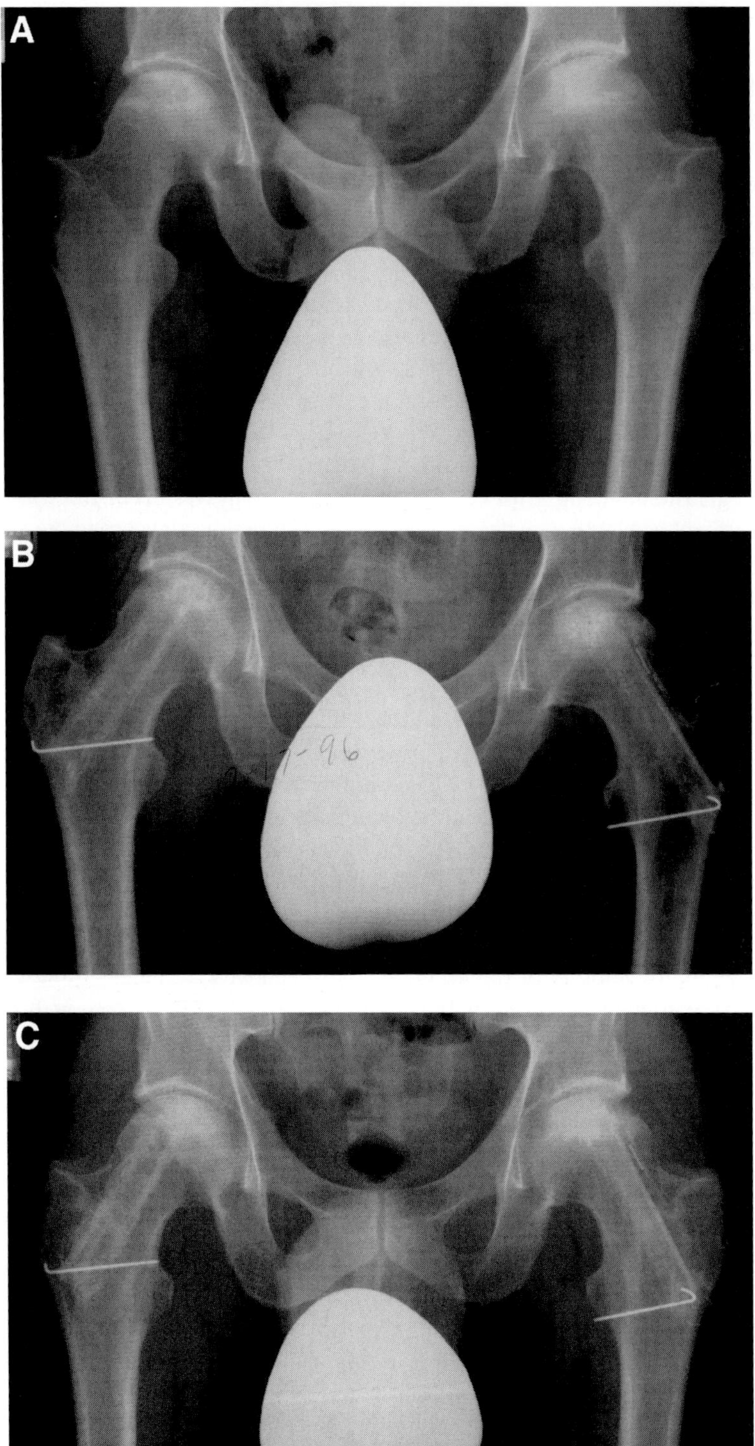

Fig. 5. (A) 34-yr-old with Stage III AVN of the right femoral head and Stage IV on the left. **(B)** Two years status post-free vascularized fibular graft (FVFG) bilaterally with well-incorporated fibular grafts, no interval bony collapse, and well-preserved joint spaces. **(C)** Five years status post-FVFG bilaterally with no interval collapse, well-incorporated grafts, and preserved joint spaces. Patient is pain-free.

vascularized fibula. The void in the femoral head is then replaced with autogenous cancellous bone from the iliac crest.

Judet's study evaluated 60 patients (68 hips) operated on from 1978 to 1985 with an average follow-up of 18 yr. The patients were assessed according to the scoring system of Merle d'Aubinge *(72)*, and 52% of the patients had good or very good results. The authors then reevaluated their data by including only the patients with early stages of osteonecrosis and ages under 40. Using these criteria, their success rate was 80%. They concluded that satisfying results could be obtained in young patients with early disease who had good sphericity of the femoral head and a healthy shell of cartilage. Although the surgical approaches differ, the key similarities are the excision of necrotic bone, replacement with autogenous cancellous graft, and use of a free vascularized fibular graft for a mechanical strut and vascular inflow. A direct comparison is difficult with different surgical techniques, but it is likely that the similarity of these three components are the critical factors responsible for successful revascularization and preservation of the femoral head.

Osteonecrosis of the femoral head has a natural history that is marked by progressive necrosis, collapse, and subsequent arthrosis of the hip. The studies quoted above *(54,67,69,71)* have all shown that the stage of disease at the time of treatment is a major factor in determining the success or failure of the surgical procedures *(73)*. All studies conclude that the earlier the intervention and the less advanced the disease, the more likely a successful long-term result will be achieved.

However, not all patients present to the clinic in the early stages of the disease. In fact, many present with various degrees of femoral head collapse. A study was conducted at our institution to review retrospectively the results in a consecutive series of 188 patients (224 hips) who had undergone free vascularized fibular grafting, between 1989 and 1999, for the treatment of osteonecrosis of the hip that had already led to collapse of the femoral head but not yet to arthrosis *(73)*. The average duration of follow-up was 4.3 yr (range 2–12 yr). The failure end point was defined as a conversion to a total hip arthroplasty. Multiple factors were analyzed to determine if they contributed to the failure. These factors included the size of the lesion, the amount of preoperative collapse of the femoral head, the etiology of the osteonecrosis, the age of the patient, and whether the lesion was bilateral. Patients were evaluated clinically with a Harris hip score both preoperatively and at the final follow-up.

The overall rate of survival was 67% for the hips that were followed for a minimum of 2 yr and 65% for those followed a minimum of 5 yr. The mean preoperative Harris hip score was 54.5, and it increased to 81 for the patients in whom the surgery was successful. While a trend toward decreased success with increasing linear collapse of the femoral head was noted, the amount of flattening or preoperative linear collapse of the femoral head was not found to be a significant predictor of survival or functional outcome. The authors postulated that the reduction of the collapsed segment intraoperatively and support of that segment with viable cancellous graft and a fibular strut diminishes the effect of the preoperative femoral head collapse, and thus reduces the influence of this variable on outcome. Although not statistically significant, there was an increased relative risk of conversion to total hip arthroplasty with increasing lesion size and amount of collapse.

The authors concluded that patients with postcollapse, predegenerative osteonecrosis of the femoral head appears to benefit from free vascularized fibular grafting, with good overall survival of the joint and significant improvement in the Harris hip score. These results differ from reports found in the literature, which caution against using femoral head-preserving procedures in patients with postcollapse osteonecrosis *(74,75)*. Although the authors remark that their success rate is higher (77–89%) in patients with early precollapse stages, this study suggests that a free vascularized fibular graft can still be a worthwhile procedure in patients with postcollapse osteonecrosis.

The evaluation of causative factors contributing to failure identify a multifactorial process with the etiology of the osteonecrosis, the amount of femoral head collapse, and the size of the femoral head lesion all interacting to affect the final outcome. In their study, larger lesions, more advanced linear collapse, and idiopathic and alcohol-related osteonecrosis increased the relative risk of failure. Finally, the authors conclude that the results of free vascularized fibular grafting in patients with postcollapse osteonecrosis are far superior to the natural history of the disease. Most important, the procedure appears to delay the need for a total hip arthroplasty in the majority of patients, and it may eliminate the need for eventual arthroplasty in many of them *(73)*.

Osteonecrosis Associated With Pregnancy

Osteonecrosis of the femoral head is a devastating disease affecting a young patient population. Although its prevalence is unknown, it has been estimated to develop in 10,000–20,000 new patients a year in the United States *(76,77)*. There are two subgroups of patients who constitute a small percentage of this total, but are unique and deserve attention. They include children, and women who develop osteonecrosis associated with pregnancy. Both of these patient groups have been successfully treated with free vascularized fibular grafting for femoral head osteonecrosis.

Pfeifer was the first to report the relatively rare association of osteonecrosis of the femoral head with pregnancy *(78)*. A study at Duke University Medical Center evaluated 13 women (17 hips) who were seen between 1992 and 1995 with an onset of hip pain during pregnancy or within the first 4 wk after delivery *(79)*. This pain persisted until a diagnosis of osteonecrosis of the femoral head was made based on magnetic resonance imaging (MRI). No patient had any other risk factors for this disease. The MRI is critical in this diagnosis because the difference between osteonecrosis and transient osteoporosis associated with pregnancy is particularly important. The two disorders have distinctly different natural histories and, thus, different treatment plans.

Although the natural history of osteonecrosis in association with pregnancy does not appear to differ from that of osteonecrosis associated with other atraumatic or idiopathic etiologies, the optimum treatment is still controversial. Because free vascularized fibular grafting has been successful at our institution in preventing articular collapse, we treated these women similarly. Eleven women (15 hips) were managed with a free vascularized fibular graft. Nine of these patients (11 hips) were evaluated with regard to relief of pain and the Harris hip score at a minimum of 2 yr postoperatively.

Of the 11 women who were managed with a free vascularized fibular graft, 9 noted marked or complete relief of their preoperative pain. Two hips in a patient who had progressive pain were treated with total hip arthroplasties. Two hips were lost to follow-up. The 9 patients (11 hips) who were available for follow-up at a minimum of 2 yr had an average improvement in the Harris hip score of 24 points. Although the numbers are small and a statistical comparison is difficult, the clinical outcomes following the use of a free vascularized fibular graft in pregnant women with osteonecrosis were similar to the outcomes in our database for the treatment of all other types of atraumatic osteonecrosis *(74)*.

Pediatric Osteonecrosis

Another use for the free vascularized fibular graft is in the pediatric population. In this age group, the condition most commonly develops after trauma, slipped capital femoral epiphysis, and steroid use. Other etiologies include Perthes disease and idiopathic causes. The natural history of osteonecrosis in pediatric patients is not as well defined as in adults because of their increased remodeling capacity.

Treatment options for pediatric and adolescent patient include observation, containment with a brace, femoral or acetabular osteotomies, fusion, core decompression, or total hip arthroplasty in patients who are skeletally mature. Free vascularized fibular grafting is an appealing alternative treatment because it maintains a viable femoral head without violating the joint capsule.

We conducted a retrospective review of all pediatric patients who had undergone a free vascularized fibular graft at our institution. The study evaluated the results of 54 hips in 50 pediatric and adolescent patients, 18 yr of age or younger, who were operated on between 1983 and 1997. All patients were followed for at least 2 yr. The range of follow-up was 2–10 yr (average, 4.3 yr). All patients were entered into the study prospectively and their function was assessed yearly with the Harris hip score, radiographic progression of the disease, and conversion to total hip arthroplasty or fusion *(80)*.

At the last follow-up, the average Harris hip scores in patients who did not undergo a total hip arthroplasty improved from a preoperative average of 55.3 points to 90.2 points. Total hip arthroplasty was performed in seven hips (seven patients), and one hip fusion was performed. Treatment of these pediatric and adolescent patients with free vascularized fibular grafting resulted in a lower rate of conversion to total hip arthroplasty or fusion (16%) when compared to conversion to total hip arthroplasty in adults (25%). The quality of life as evidenced by the increased Harris hip scores was improved significantly.

Nonunions

The majority of this chapter has focused on the use of the free vascularized fibular graft for the treatment of osteonecrosis of the femoral head. Although this is the most common use of the free vascularized fibular graft at our institution, it has also become an established procedure for the treatment of major skeletal defects and recalcitrant nonunions *(47,81)*. Nonunions in previously irradiated bone are especially difficult because ionizing radiation's detrimental effect on cortical bone *(82)* can inhibit and delay fracture healing *(83,84)*.

A recent study evaluated the functional results, rates of union, and complications associated with free vascularized fibular grafts combined with autografting for the treatment of nonunions in previously irradiated bone *(85)*. The study looked at 17 patients who had undergone 18 vascularized free fibular grafts with cancellous autografting. All patients were being treated for the nonunion of a fracture in a region of previously irradiated bone. Radiation therapy had been used on eight patients with a bony neoplasm and nine patients with a soft tissue sarcoma at an average of 111 mo prior to the pathological fracture. The fractures were initially treated with open reduction and internal fixation (ORIF) or closed reduction and casting. Two patients had bone grafting to the fracture site after their initial ORIF, and four patients had pulsed electromagnetic-field stimulation. However, despite this additional treatment, all patients had a persistent nonunion.

A total of 18 fibular graft transfers were performed for established nonunions at an average interval of 19 mo following the pathological fracture. These grafts were applied as onlay grafts and cancellous bone from the iliac crest was used at the proximal and distal junctions and at the fracture site in all patients. The average duration of follow-up after the vascularized free fibular graft was 57 mo.

Sixteen of the 18 fracture sites united after an average of 9.4 mo. Functionally, 13 patients had an excellent result; one, a good result; two; a fair result; and one, a failure of treatment. Despite the complication of infection in four patients, authors recommended that a fracture occurring within the field of therapeutic radiation should initially be treated with open reduction and stable internal fixation with a nonvascularized autograft. If there is no evidence of union at 6 mo and the internal fixation is still stable, a vascularized free fibular transfer with additional cancellous autografting should be performed.

CONCLUSION

Although many types of bone grafts exist, the vascularized fibular graft has emerged as the superior graft for large segmental bony defects, established nonunions, and osteonecrosis of the femoral head. Not only does it have an ideal vascular supply, its anatomic size and structure are perfectly suited for restoration of long bone defects and as a strut graft for the femoral head. In addition, the immediate incorporation, minimal resorption, and mechanical stability make it an ideal candidate for the bony regeneration that may be required in segmental cortical loss. Because such loss may be encountered in traumatic, developmental, degenerative, or neoplastic disorders of the skeletal system, the use of the vascularized fibular graft pervades all orthopedic subspecialties and should be considered as a valuable treatment option.

REFERENCES

1. van Meekeren, J. (1988) Heel en geneeskonstige aanmerkingen. Commelijn, Amsterdam, 1668 (quoted from deBoer).
2. DeBoer, H. H. (1988) The History of Bone Grafts. *Clin. Orthop. Rel. Res.* **226,** 292–298.
3. de Heyde, A. (1988) Anatomia Mytuli, Subjecta Centuria Observatorium. Janssonio Waesbergois, Amsterdam, 1684 (quoted from de Boer).

68. Yoo, M. C., Chung, D. W., and Hahn, C. S. (1992) Free vascularized fibular grafting for the treatment of osteonecrosis of the femoral head. *Clin. Orthop. Rel. Res.* **277,** 128–138.

69. Soucacos, P. N., Beris, A. E., Malizos, K., Koropilias, A., Zalavras, H., and Dailiana, Z. (2001) Treatment of avascular necrosis of the femoral head with vascularized fibular transplant. *Clin. Orthop. Rel. Res.* **386,** 120–130.

70. Steinberg, M. E., Hayken, G. D., and Steinberg, D. R. (1984) A new method for evaluation and staging of avascular necrosis of the femoral head, in *Bone* (Arlet, J., Ficat, R. P., and Hungerford, D. S., eds.), Williams & Wilkins, Baltimore, pp. 398–493.

71. Judet, H. and Gilbert, A. (2001) Long-term results of free vascularized fibular grafting for femoral head necrosis. *Clin. Orthop. Rel. Res.* **386,** 114–119.

72. Merle D'aubigne, R. (1970) Cotation chiffree de la fonction de la hanche. *Rev. Chir. Orthop. Reparatrice App. Mot.* **56,** 481–486.

73. Berend, K. R., Gunneson, E. E., and Urbaniak, J. R. (2003) Free vascularized fibular grafting for the treatment of post-collapse osteonecrosis of the femoral head. *J. Bone Joint Surg.* **85A,** 987–993.

74. Bozic, K. J., Zurakowski, D., and Thornhill, T. S. (1999) Survivorship analysis of hips treated with core decompression for nontraumatic osteonecrosis of the femoral head. *J. Bone Joint Surg.* **81A,** 200–209.

75. Mont, M. A., Jones, L. C., Pacheco, I., and Hungerford, D. S. (1998) Radiographic predictors of outcome of core decompression for hips with osteonecrosis stage III. *Clin. Orthop. Rel. Res.* **354,** 159–168.

76. Lavernia, C. J., Sierra. R. J., and Grieco, F. R. (1999) Osteonecrosis of the femoral head. *J. Am. Acad. Orthop. Surg.* **7,** 250–261.

77. Mont, M. A. and Hungerford, D. S. (1995) Non-traumatic avascular necrosis of the femoral head. *J. Bone Joint Surg.* **77A,** 459–474.

78. Pfeifer, W. (1957) Eine ungewohnliche Form und Genese von symmetrischen Osteonekrosen beider Femur- und Humeruskopfkappen. *Fortschr. Geb. Rontgen. Nuklearmed.* **87,** 346–349.

79. Montella, B. J., Nunley, J. A., and Urbaniak, J. R. (1999) Osteonecrosis of the femoral head associated with pregnancy. *J. Bone Joint Surg.* **81A,** 790–798.

80. Dean, G. S., Kime, R. C., Fitch, R. D., Gunneson, E., and Urbaniak, J. R. (2001) Treatment of osteonecrosis in the hip of pediatric patients by free vascularized fibular graft. *Clin. Orthop. Rel. Res.* **386,** 106–113.

81. Wood, M. B. (1990) Femoral reconstruction by vascularized bone transfer. *Microsurgery* **11,** 74–79.

82. Maeda, M., Bryant, M. H., Yamagata, M., Li, G., Earle, J. D., and Chao, E. Y. S. (1998) Effects of irradiation on cortical bone and their time-related changes. A biomechanical and histomorphological study. *J. Bone Joint Surg.* **70A,** 392–399.

83. Markbreiter, L. A., Pelker, R. R., Friedlander, G. E., Peschel, R., and Panjabi, M. M. (1989) The effect of radiation on the fracture repair process. A biomechanical evaluation of a closed fracture in a rat model. *J. Orthop. Res.* **7,** 178–183.

84. Widmann, R. F., Pelker, R. R., Friedlaender, G. E., Panjabi, M. M., and Peschel, R. E. (1993) Effects of prefracture irradiation on the biomechaniacal parameters of fracture healing. *J. Orthop. Res.* **11,** 422–428.

85. Duffy, G. P., Wood, M. B., Rock, M. G., and Sim, F. H. (2000) Vascularized free fibular transfer combined with autografting for the management of fracture nonunions associated with radiation therapy. *J. Bone Joint Surg.* **82A,** 544–554.

7

Growth Factor Regulation of Osteogenesis

Stephen B. Trippel, MD

Osteogenesis, the creation of bone, underlies all skeletal development and repair. It encompasses the differentiation of cells along specific developmental pathways and the production by these cells of the matrix required to construct, or to reconstruct, bone. The control of this process is, to a large extent, the responsibility of cell signaling molecules that include hormones, growth factors, and cytokines. This chapter reviews some of the factors that participate in regulating the creation of bone at the cellular level.

GROWTH HORMONE

Growth hormone, or somatotropin, is the prototypical regulator of skeletal growth and development. Growth hormone deficiency produces severe, generalized failure of osteogenesis at the growth plate and results in clinical dwarfism. The administration of recombinant human growth hormone to children with either growth hormone deficiency or idiopathic short stature can, at least partially, restore the kinetics of osteogenesis at the growth plate and hence the rate of linear bone growth. Excess growth hormone secretion during skeletal development increases longitudinal bone growth and produces clinical gigantism (1). Growth hormone insensitivity due to mutations in the growth hormone receptor are responsible for several forms of dwarfism, ranging from mild to severe (2,3).

The ability of growth hormone to influence osteogenesis at the site of bone repair is controversial. Growth hormone has been reported to stimulate the formation of bone in intact bones (4,5) and osseous defects (6), and to enhance healing in fracture models in rats (7–11) and dogs (12). Other investigators, however, have observed that growth hormone has no effect on bone formation (13,14), healing of defects (15), bone graft incorporation (16), or healing of fractures in rat (17,18) or rabbit models (15,19). The differences in the findings of these studies may be explained, in part, by differences in experimental design, growth hormone dosage, site of delivery, species of animal, and outcome measures employed.

Whether a deficiency of growth hormone results in failure of fracture healing is similarly controversial (20–22). Interestingly, growth hormone deficiency may increase the risk of fracture occurrence (23,24). Early reports of growth hormone treatment of human fractures were encouraging (25,26), but these studies were limited by small sample size and lack of a paralleled control group. Although growth hormone is now widely used to enhance skeletal growth, there currently appears to be little direct support for its clinical application to fracture repair.

INSULIN-LIKE GROWTH FACTOR I (IGF-I)

IGF-I was discovered in experiments testing the effect of growth hormone on sulfate incorporation into cartilage. These experiments found that a serum factor, later identified as IGF-I, mediated the effect of growth hormone on this tissue (27). Subsequent studies suggested the existence of a growth

From: *Bone Regeneration and Repair: Biology and Clinical Applications*
Edited by: J. R. Lieberman and G. E. Friedlaender © Humana Press Inc., Totowa, NJ

hormone–IGF axis that includes both endocrine and autocrine/paracrine elements. Growth hormone, secreted by the pituitary, stimulates IGF-I production by the liver *(28)* and other organs *(29)*. This IGF-I enters the systemic circulation, and from there, acts in an endocrine fashion on multiple tissues including the skeleton *(30,31)*. Evidence in support of this model, as it applies to skeletal growth, includes the identification of growth hormone receptors *(32)* and IGF-I receptors *(33,34)* on growth-plate chondrocytes, and the ability of anti-IGF-I antibodies to block the growth-enhancing effect of growth hormone delivered intraarterially to growing limbs *(35)*. In addition, growth hormone has been shown to stimulate the production of IGF-I mRNA *(36)*, and peptide *(37)* by growth-plate chondrocytes.

The role of IGF-I in the regulation of osteogenesis at the growth plate is further illuminated by studies in transgenic mice. Mice in which the IGF-I gene has been deleted manifest marked intrauterine and postnatal skeletal growth deficiency that is not corrected by growth hormone treatment *(38, 39)*. When mice were made transgenic for the IGF-I gene and for ablation of the cells that express growth hormone, the mice carrying both transgenes (IGF-I and absence of growth hormone) grew larger than litter mates that carried only the growth hormone ablation transgene *(40)*. The double-transgenic animals demonstrated weight and linear growth that were indistinguishable from those of their normal, nontransgenic siblings.

IGF-I is capable of at least partly substituting for growth hormone in humans as well as in mice. In recent clinical trials, patients with end-organ insensitivity to growth hormone resulting from an inactivating growth hormone receptor mutation were treated with IGF-I *(41,42)*. These children, who manifested severe failure of bone growth prior to therapy, experienced a substantial and sustained increase in skeletal growth during IGF-I therapy.

Not all of the skeletal effects of growth hormone can be attributed to IGF-I. Growth hormone elicits very rapid anabolic cellular responses that are unlikely to involve such mediators as IGF-I *(43)*. In addition, growth hormone administered systemically to hypophysectomized (and therefore growth hormone–deficient) rats has been found to be a more effective stimulus of skeletal growth than IGF-I, even when growth hormone was administered at 50-fold lower doses *(44)*.

The recent use of tissue-specific gene ablation techniques has permitted a partial separation of the effect of IGF-I produced in the liver and of that produced in other tissues. When the hepatic IGF-I gene was rendered nonfunctional, circulating levels of IGF-I fell by 80% while levels of growth hormone increased. Interestingly, postnatal (including pubertal) growth remained normal *(45)*. These data raise the possibility that osteogenesis at the growth plate may be less dependent on IGF-I acting by an endocrine route than on IGF-I acting in a paracrine/autocrine fashion. It is also possible that the high level of circulating growth hormone achieved in these animals augmented local production of IGF-I sufficiently to offset the loss of circulating IGF-I. The relative contributions of IGF-I acting via the circulation in an endocrine fashion, that of IGF-I acting in a paracrine/autocrine fashion, and of growth hormone acting independently of IGF-I may differ at different sites and different stages of development. The specific roles of these various components of the growth hormone–IGF-I axis remain to be elucidated.

EPIDERMAL GROWTH FACTOR

Unlike growth hormone and IGF-I, epidermal growth factor (EGF) was not initially viewed as being involved in formation of the skeleton. However, as has proved to be the case with many cell signaling molecules, the role of EGF is broader than its name implies. The view that EGF plays a role in the regulation of skeletal development *(46)* has been supported by the localization of EGF in the growth plate *(47)*, the detection of EGF receptors on growth-plate chondrocytes *(48,49)*, and the observation that EGF is present in the circulation at concentrations that are capable of initiating cellular responses in vitro *(50)*.

The potential role of EGF in skeletal growth has been clarified in recent studies that investigated the intereaction of EGF and IGF-I in the regulation of growth-plate chondrocytes. These studies found that

EGF increased cellular responsiveness to IGF-I with respect to both mitotic activity and proteoglycan synthesis *(51)*. This effect of EGF was associated with an increase in the number of IGF-I receptors per cell, but without a change in IGF-I receptor affinity. The effect of EGF on IGF-I receptors appeared to be a part of a general anabolic effect of EGF rather than a specific effect on the IGF-I receptor. These data suggest that EGF contributes to skeletal growth by increasing growth-plate chondrocyte sensitivity to IGF-I. These results may aid in understanding the previously enigmatic observation that the skeletal growth response to IGF-I does not match that achieved with growth hormone *(44)*. The inability of IGF-I to fully compensate for growth hormone presumably reflects a requirement by the growth plate for growth hormone stimulation of an element in the growth hormone–IGF-I axis other than IGF-I itself. In conjunction with the observation that growth hormone regulates EGF *(49)*, these data suggest that the IGF-I receptor is such an element.

FIBROBLAST GROWTH FACTOR

The fibroblast growth factors (FGFs) comprise a large family of polypeptides that regulate cell functions as diverse as mitogenesis, differentiation, receptor modulation, protease production, and cell maintenance *(1)*. Several lines of evidence indicate that these factors play an important role in bone formation. FGF-2 (basic FGF) has been immunolocalized to the proliferative and maturation (but not hypertrophic) zones of the growth plate of the fetal rat *(52)* and to the resting, proliferative, and perichondrial cells of the human fetus *(53)*. Indeed, during fetal development, the highest levels of FGF-2 transcripts were reported to be in the long bones *(54)*.

Growth-plate chondrocytes possess high-affinity receptors for FGF-2 *(55,56)* and, in a variety of models, FGF-2 is a potent mitogen for growth-plate chondrocytes *(57–61)*. In contrast to its reproducible effect on chondrocyte mitogenic activity, the role of FGF-2 on matrix synthesis is less clear. FGF-2 has been found to stimulate *(62)*, exert no effect on *(61,63)*, or inhibit *(61,63,64)* indices of matrix synthetic activity by growth-plate chondrocytes. FGF-2 also influences many of the cellular activities associated with chondrocyte differentiation. For example, FGF-2 effects on growth-plate chondrocytes in culture include a reduction in alkaline phosphatase *(61,65)*, calcium deposition, and calcium content *(65)*.

In a fetal rat metatarsal organ culture model of skeletal growth, the effect of FGF was biphasic *(66)*. Matrix production was stimulated by low concentrations (10 ng/mL), but inhibited by high concentrations (1000 ng/mL), of FGF-2. In this model, as in others, FGF-2 stimulated ^{3}H-thymidine incorporation, an index of DNA synthesis. However, the site of incorporation was principally in the perichondrium, and labeling was decreased in the proliferative and epiphysial chondrocytes. FGF-2 also caused a marked decrease in the number of hypertrophic chondrocytes. Taken together, these data suggest that the role of FGF-2 in osteogenesis at the growth plate is to promote an immature chondrocyte phenotype by augmenting chondrocyte proliferation and inhibiting chondrocyte differentiation *(55,65)*. The data also emphasize the complexity imposed on this role by temporal, spatial, and dosage relationships.

FGF family members also participate in regulating osteogenesis during fracture repair. FGF-2 has been shown to be widely distributed around the fracture site in a rat fibular fracture model *(67)*. FGF-2 was particularly prominent in the soft callus and periosteum. Application of a single dose of FGF-2 in a fibrin gel in this model augmented callus formation, increased the biomechanical strength of fracture repair, and restored the impaired fracture healing associated with diabetes *(67)*. Similarly, FGF-2 in a hyaluronan gel increased callus formation and biomechanical strength when injected into rabbit fibular osteotomies *(68)*. In a subperiosteal osteogenesis model, injection of FGF-2 stimulated extensive intramembranous bone formation adjacent to the parietal bone *(68)*. Injection of FGF-1 (acidic FGF) into closed rat femoral fractures resulted in a marked increase in the size of the cartilaginous callus, but also inhibited type II procollagen and proteoglycan core protein gene expression. The net result was a decrease in the mechanical strength at the fracture site *(69)*.

The effect of exogenous FGF on osteogenesis in vivo is complex. Local delivery of FGF-2 by direct infusion into the rabbit growth plate increased maximal vascular invasion and accelerated local ossification *(70)*. Systemic intravenous delivery of 0.1 mg/kg/d of FGF-2 for 7 d to growing rats increased longitudinal growth rate, cartilage cell production rate, bone formation rate, and several histomorphometric measures of bone quantity *(71)*. Endocortical mineral apposition and bone formation rates were increased, but periosteal mineral apposition and periosteal bone formation rates were depressed. These effects were not matched by the higher dose of 0.3 mg/kg/d. At this dose, FGF-2 decreased longitudinal growth rate, cartilage cell production rate, endocortical bone formation rate, and produced defective calcification of the growth-plate metaphyseal junction.

A similar biphasic effect of FGF-2 was observed in a bone chamber model. When injected into the marrow cavity of rat bone implants, a low dose (15 ng) of FGF-2 stimulated bone formation, while a high dose (1900 ng) had a profoundly inhibitory effect *(72)*. In contrast, intraosseous delivery of 400 µg or 1600 µg of FGF-2 in rabbits increased bone mineral density *(73)*.

In transgenic mice that overexpress FGF-2, the radii, ulnae, humeri, femora, and tibiae were shortened by 20–30% ($p < 0.001$) compared to nontransgenic littermate controls *(74)*. Mean body weights were not significantly different. Growth plates showed significant enlargement of the reserve and proliferative zones due to chondrocyte hyperplasia and to enhanced extracellular matrix deposition. In contrast, hypertrophic chondrocytes were substantially diminished *(74)*. Taken together, these data suggest that, in vivo, FGF may act to either augment or inhibit osteogenesis, depending on the dose, mode of delivery, and other variables.

The contribution of the FGFs to osteogenesis has been further clarified by recent studies of the receptors that mediate FGF actions. There are at least four distinct FGF receptor (FGFR) genes *(75)*, and many variants due to alternative splicing *(76)*. Like the IGF-I receptor, all four FGFRs contain intracellular tyrosine kinase domains that become activated upon FGF binding to the receptor's extracellular ligand-binding domain (**Fig. 1**). Mutations in these receptors are now known to be responsible for a variety of human chondrodysplasias. Studies of these disorders have led to extraordinary advances in our understanding of how growth factor signaling pathways influence osteogenesis during skeletal growth and development.

Achondroplasia, the most common human genetic form of dwarfism, is characterized by rhizomelic (proximal greater than distal) shortening of long bones and by narrow growth plates *(77,78)*. In more than 95% of individuals with achondrodysplasia, the cause is a point mutation in the portion of the gene encoding the transmembrane domain of FGFR3 *(79–81)* (**Fig. 2**).

Thanatophoric dysplasia, a sporadic perinatal lethal disorder, is also caused by FGFR3 mutations. This severely deforming dysplasia is characterized by micromelic limb shortening, reduced vertebral body height, and disrupted cell distribution in the growth plate *(82–84)*. Death is usually from respiratory failure associated with marked shortening of the ribs and reduced thoracic cavity volume. Thanatophoric dysplasia has been divided into two types, based on clinical features. Type I (TD-1) is characterized by curved, short femora, and type 2 (TD-2) by relatively longer, straight femora. TD-1 is associated with mutations in the extracellular region of FGFR3 or by a mutation in the stop codon of the gene *(85)*. In contrast, TD-2 is associated with a specific mutation in the intracellular tyrosine kinase domain of FGFR3 *(86)* (**Fig. 3**).

Hypochondroplasia is a rare autosomal dominant disorder with skeletal deformities similar to those of achondroplasia, but in a milder form *(87,88)*. Slightly over half of individuals with hypochondroplasia were found in a recent study to have a single mutation in the proximal tyrosine kinase domain of FGFR3 *(89)*. Interestingly, in the remaining individuals with hypochondroplasia, no mutations in FGFR3 were detected, despite screening of more than 90% of the FGFR3 coding sequence and despite the absence of phenotypic differences between the individuals who had or did not have the mutation. Thus, some other gene appears to regulate similar cell functions.

Crouzon syndrome, an autosomal dominant condition, is characterized by an abnormally shaped skull, hypertelorism, and proptosis associated with craniosynostosis. The appendicular skeleton is

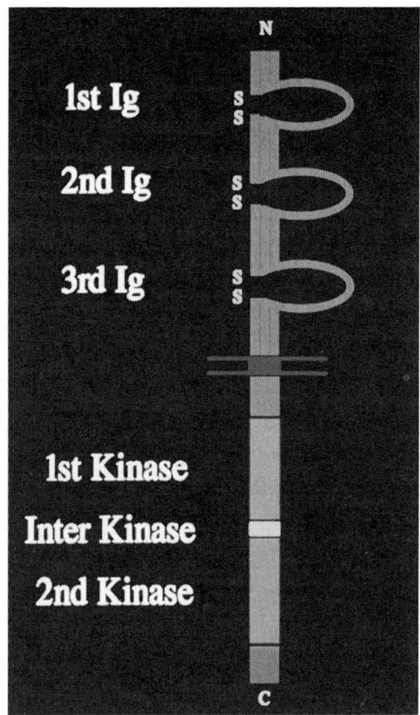

Fig. 1. Schematic illustration of a typical FGF receptor. The extracellular region contains three disulfide (S–S)-linked domains with structural homology to the immunoglobulins (Ig). The receptor traverses the cell membrane (red). The cytoplasmic region contains a bipartite kinase domain (orange). (Reproduced with permission from Trippel, S. B. (1994) Biologic regulation of bone growth, in *Bone Formation and Repair* (Brighton, C. T., Friedlaender, G., and Lane, J. M., eds.), American Academy of Orthopedic Surgeons, Rosemont, IL, pp. 39–60.) (Color illustration appears in e book.)

spared. Although it is thus quite different in its clinical picture from achondroplasia, it is in some cases similarly associated with a mutation in the transmembrane region of the FGFR3 gene. The Crouzon mutation, however, is at a slightly different location in the gene than the achondroplasia mutation *(90)*.

These genetic studies demonstrate a considerable degree of refinement in the regulation of osteogenesis by FGFR3. Subtle differences in receptor gene sequence may produce subtle, or not-so-subtle, differences in skeletal phenotype. Although the location of the mutation (near an autophosphorylation site, in the transmembrane domain, in the ligand binding region, etc.), may provide clues to the underlying mechanism of the skeletal disorder, the genotype–phenotype relationships of these receptor abnormalities are still not understood.

Of considerable interest is the demonstration in transgenic mouse models that disruption of the FGFR3 gene promotes, rather than inhibits, bone growth *(91,92)*. Mice lacking FGFR3 [FGFR3 knockout or FGFR3 (–/–)] mice developed severe, progressive bone dysplasia with expansion of proliferating and hypertrophic chondrocytes in the growth plate. Proliferating cell nuclear antigen, a marker of cell proliferation, was present in greater numbers of cells in FGFR3 (–/–) mice than in wild-type controls *(92)*. Although histological evidence of an increased height of the hypertrophic zone in the growth plate was detectable in the late embryonic period *(91)*, the FGFR3 (–/–) mice showed no obvious skeletal abnormalities during embryonic development *(92)*. By 7 wk of age, all FGFR3 (–/–) femora and 75% of humeri had become bowed. Increased femur length in FGFR3 (–/–) skeletons relative to controls was first observed at 9 wk of age, and by 4 mo or older was 6–20% that of age-matched controls *(91)*. These observations are consistent with the view that FGFR3 activation tends

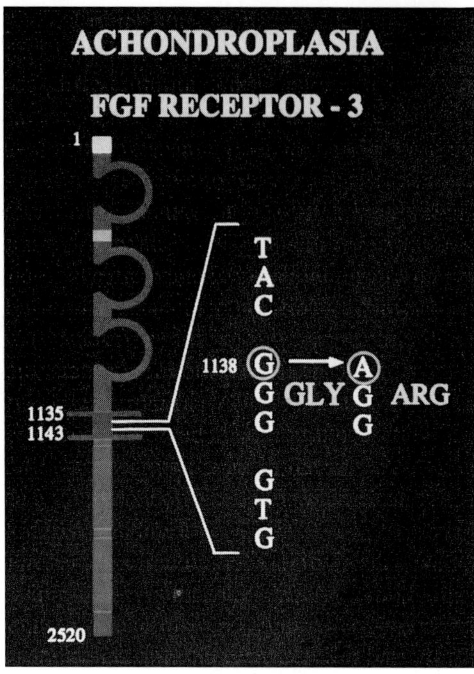

Fig. 2. Schematic illustration of the principal FGFR3 mutation associated with achondroplasia. This point mutation in the transmembrane domain of FGFR3 increases FGFR3 function. (Color illustration appears in e book.)

to suppress skeletal growth. Indeed, the achondroplasia and TD-2 mutations are associated with ligand-independent activation of FGFR3 *(93–95)*.

Thus, both activation and inhibition of FGFR3 produce disordered osteogenesis, the former characterized by deficient bone growth and the latter by bone overgrowth. Given that FGFR3 mRNA is expressed in the cartilage rudiments of all bones during endochondral ossification in the developing mouse embryo *(96)*, the observation the FGFR3 (–/–) mice show no obvious abnormalities during embryonic development suggests that alternative pathways are available for regulating the earliest phases of osteogenesis.

Other members of the FGF receptor family also participate in osteogenesis. FGFR2 mutations are, as for FGFR3, associated with a variety of craniofacial syndromes. Mutations at several sites in the FGFR-2 extracellular domain *(97,98)* have recently been linked to Crouzon syndrome (**Fig. 4**). However, 19 of the 32 Crouzon syndrome patients analyzed did not have mutations in this region and were presumed to have mutations elsewhere in the FGFR-2 gene or in other genes *(97)*. As we have seen, some of these patients have mutations in the FGFR3 gene.

Jackson–Weiss syndrome, another form of craniosynostosis, is distinguished by its foot abnormalities, including broad great toes with medial deviation and tarsal–metatarsal coalescence (Crouzon syndrome, by contrast, is characterized by an absence of digital abnormalities *[97]*). Screening of Jackson–Weiss syndrome families identified a mutation in the FGFR2 extracellular domain only 3 bp away from one of the Crouzon-associated mutations *(97)*.

The complexity in the genotype–phenotype relationships of these FGFR-based skeletal disorders is further illustrated by studies of FGFR1. Mutations in the extracellular domain of this gene cause Pfeiffer's syndrome, one of the classic autosomal dominant craniosynostosis syndromes *(99)*. Pfeiffer's

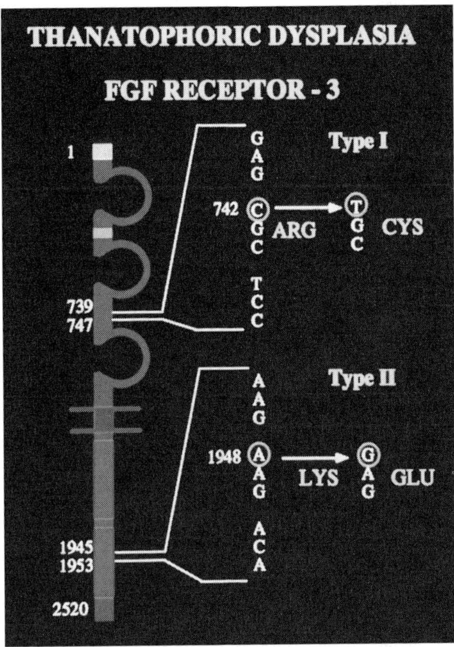

Fig. 3. Schematic illustration of the mutations associated with type I and type II thanatophoric dysplasia. These two mildly different forms of thanatophoric dysplasia are produced by mutations at two widely separated sites in FGFR3, one in the extracellular region of the receptor and the second in an intracellular tyrosine kinase domain. (Color illustration appears in e book.)

syndrome is associated with multiple digital abnormalities including broad, medially deviated great toes (as in Jackson–Weiss syndrome) and thumbs, with or without variable degrees of syndactyly or brachydactyly of other digits (unlike Jackson–Weiss syndrome) *(100)*. However, Pfeiffer's syndrome has also been shown to be caused by FGF2R mutations *(101)*, and the identical FGFR2 mutations can cause both Pfeiffer's and Crouzon's syndrome phenotypes *(102)*.

This confusing lack of correlation between genotype and phenotype is undoubtedly due in part to overlap in the clinical parameters used to identify these syndromes. Such disparities argue for a different taxonomy of skeletal anomalies, one based on genotype rather than, or in addition to, phenotype. More interestingly, however, these data demonstrate that the FGFs, acting via their receptors, regulate osteogenesis through a remarkably refined system of signaling pathways that has only begun to be understood.

Knowledge of the specific relationships between FGFR genotype and osteogenesis phenotype has recently been advanced by studies of Apert's syndrome. Apert's syndrome is a craniosynostosis associated with severe syndactyly of the hands and feet. In a recent study of 40 unrelated cases of this syndrome, missense substitutions were identified in adjacent amino acids located between the second and third immunoglobulin domains of FGFR2 *(100)* (**Fig. 5**). Both amino acid substitutions resulted from cytidine (C)-to-guanine (G) nucleic acid transversions. The C → G transversion at nucleic acid position 934 (C934G) produced a substitution from serine to tryptophan at amino acid 252. The remaining patients showed a C → G transversion at nucleic acid position 937 (C937G), resulting in a proline-to-arginine substitution at amino acid position 253. When syndactyly severity scores were correlated with mutation type, patients with the C937G mutation were found to have a higher syndactyly severity score than patients with the C934G mutation. The difference was not statistically significant for

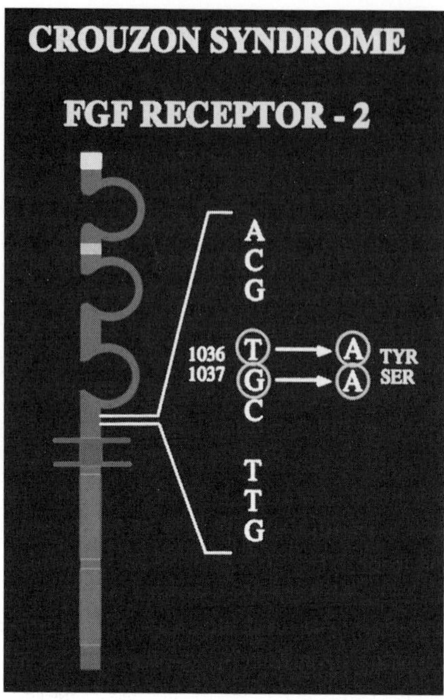

Fig. 4. Schematic illustration of two of the mutations associated with Crouzon's syndrome. The two mutations in the extracellular region of FGFR2 affect the same amino acid in the receptor and may thus be expected to produce the same clinical picture. However, Crouzon's syndrome can also be caused by mutations in the transmembrane region of FGFR3. (Color illustration appears in e book.)

the hands alone, but was statistically significant for the feet alone ($p < 0.005$) and for the hands and feet combined ($p < 0.025$). Of further interest is the fact that the C937G (Pro253Arg) mutation of FGFR-2 in Apert's syndrome corresponds precisely to the C937G (Pro252Arg) mutation of FGFR1 in some cases with Pfieffer's syndrome *(99,100)*. These observations raise the possibility that in some circumstances, the particular skeletal developmental event can be dissected down to the level of individual amino acids and their location in proteins involved in growth-factor signaling.

In contrast to the above example of a phenotypic difference associated with mutations that are extremely close to each other, some Crouzon patients with FGFR2 mutations on entirely different exons have no obvious phenotypic differences *(100)*.

The increasing number of distinct mutations that are being coupled with more carefully defined skeletal phenotypes will provide a potentially valuable resource for better understanding the role of FGF and its receptors in osteogenesis. The existence of at least 13 members of the FGF family and of multiple splice variants of the FGF receptor family yields an astronomical number of potential combinations of ligands and receptors. This permits a remarkable degree of selectivity and refinement in signaling interactions. It also creates a daunting challenge to define the specific roles of each of them.

TRANSFORMING GROWTH FACTOR-BETA (TGF-β)

The transforming growth factor-betas are members of a large superfamily of cell signaling molecules that include the bone morphogenetic proteins (BMPs), activins, inhibins, and growth and differention factors (GDFs). Of the five TGF-βs, TGF-β1, TGF-β2, and TGF-β3 are known to be important in mammalian tissues *(103–105)*. TGF-β family members have a particularly well-established participation in osteogenesis *(103,105,106)*.

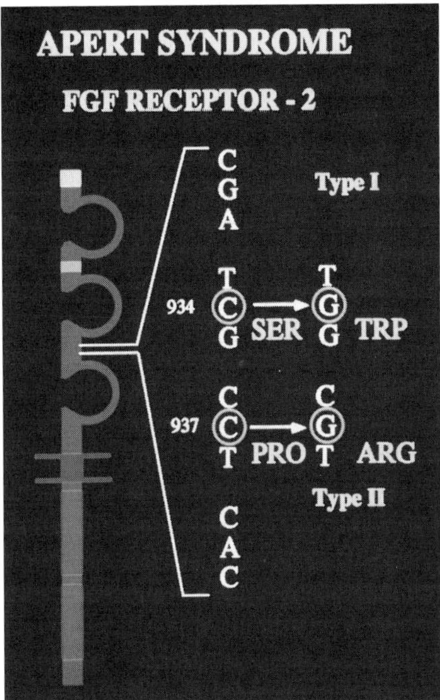

Fig. 5. Schematic illustration of two mutations that cause Apert's syndrome. Although these mutations in the extracellular region of FGFR2 are separated by only 2 bp and the affected amino acids are adjacent to each other, the mutations produce different degrees of skeletal deformity. (Color illustration appears in e book.)

In Vitro *Studies*

The actions of the TGF-βs are complex and appear to vary according to details of the experimental conditions under which they are tested. In the fetal rat calvarial osteoblast model, TGF-β has been shown to increase the production of collagen types I, II, III, V, VI, and X, osteonectin, osteopontin, fibronectin, thrombospondin, proteoglycan, and alkaline phosphatase *(104)*. TGF-β has also been reported to inhibit bone nodule formation *(107)* and mineralization *(108)* in osteoblast culture. Other reports indicate that TGF-β inhibits osteoclast formation and function *(109)*, and TGF-β has been reported to both stimulate *(110,111)* and to inhibit *(112)* type II collagen production.

In an organ culture model of fracture callus, at the relatively early time point of 7 d, TGF-β stimulated cell proliferation and inhibited expression of type II collagen and aggrecan. In contrast, at 13 d, TGF-β increased expression of type II collagen and aggrecan *(113)*. These data suggest that cell maturation may be among the factors that influence responsiveness to TGFβ.

In Vivo *Studies*

During osteogenesis by endochondral ossification, chondrocytes and osteoblasts synthesize TGF-β that accumulates in the extracellular matrix *(114)*. Indeed, bone is the largest repository of TGF-β in the body *(115)*. During fracture healing, both TGF-β mRNA and protein are present in the fracture callus *(105,116,117)*. Expression of the different TGF-β isoforms differs among the various cell types involved in fracture healing. For example, in the chick fracture model, TGF-β2 was prominently expressed in precartilaginous tissue, while TGF-β3 was present only at low levels and TGF-β1 was scarce. Later in callus formation, TGF-β1 became evident, although TGF-β2 and β3 remained relatively high *(105)*

(**Table 1**). Treatment of fractures with exogenous TGF-β has been reported to both increase *(118,119)* and to have no effect on *(31)* the quality of fracture repair. In a subperiosteal injection model, delivery of exogenous TFG-β stimulated cartilage proliferation. In this model, TGF-β2 was more effective than TGF-β1 *(114)*.

Although it is clear that TGF-β family members play a major role in osteogenesis, their mechanisms of action at the cellular and molecular biological levels remain to be elucidated. Similarly, although TGF-βs may be able to augment osteogenesis, optimization of the dose, timing, and carrier for clinical use have yet to be achieved.

PARATHYROID HORMONE (PTH) AND PARATHYROID HORMONE-RELATED PROTEIN (PTHRP)

Parathyroid hormone has long been recognized as a regulator of mineral metabolism and, in this capacity, as a stimulus of bone resorption. More recently, however, PTH has been shown to stimulate indices of osteogenesis in vitro and to enhance bone formation in vivo *(121)*.

In an in vitro rat calvarial osteoblast model, PTH increased collagen synthesis, an effect that appeared to be mediated by the production of IGF-I *(122)*. In chondrocytes, including those from the growth plate *(123–125)*, PTH stimulated both DNA and proteoglycan synthesis. It is not known whether these effects were mediated by other growth factors.

In an in vivo immature chick model, PTH deficiency increased the collagen content of tibial epiphyseal cartilage without altering the content of proteoglycan. Treatment with PTH returned collagen content toward normal *(126)*. In the rat, low dose PTH stimulated indices of bone formation when delivered in an intermittent fashion *(127)*. This anabolic effect of PTH was modulated by the growth hormone–IGF axis *(128)*. Several clinical studies have shown that PTH may be effective in the treatment of osteoporosis in humans *(129,130)*.

Recent gene therapy studies have further elucidated the role of PTH in osteogenesis. A plasmid gene encoding human PTH1-34, applied by direct gene transfer *(131)*, was tested in a rat femoral critical-sized defect model *(132)*. In contrast to controls, the group treated with human PTH 1-34 plasmids exhibited bone crossing the osteotomy gap. A similar stimulation of osteogenesis was observed when the plasmid encoding human PTH 1-34 was delivered in a collagen sponge to 8-mm defects in a canine proximal tibial bone healing model. This increase in bone was noted to originate from the existing bone surfaces *(132)*.

In contrast to PTH, which is produced in the parathyroid glands and is released into the circulation to act in a classical endocrine fashion, parathyroid hormone-related protein is produced in multiple tissues and acts in an autocrine/paracrine fashion *(133)*. PTHrP plays a central role in osteogenesis during embryonic development of the skeleton. In cultured chick growth-plate chondrocytes, PTHrP selectively inhibited type X collagen gene expression and protein synthesis without significantly changing type II collagen gene expression or protein synthesis *(134)*. In PTHrP (–/–) mice, which produce no PTHrP, chondrocyte maturation from the proliferative to the hypertrophic phase was accelerated, resulting in premature ossification *(135,136)*.

As a regulator of skeletal development, PTHrP is itself tightly regulated. Production of PTHrP in the perichondrium of embryonic bone has been shown to occur in response to a signaling polypeptide termed Indian hedgehog (IHH). The hedgehog family of proteins participates in embryonic segmentation, patterning, establishment of symmetry, and limb bud formation *(137)*. In addition to promoting PTHrP production, IHH appears to regulate early bone growth in a PTHrP-independent fashion by maintaining a high rate of division in proliferating chondrocytes *(138)*.

As is the case with growth factors, PTH and PTHrP convey information to their target cells via specific receptors. The typical PTH/PTHrP receptor is a G-protein-coupled receptor with a complement of seven transmembrane domains *(139)*. Both PTH and PTHrP bind to and activate this receptor. In growth-plate chondrocytes the PTH/PTHrP receptor is expressed predominantly in the prehypertrophic

Table 1
Representative In Vivo Studies of the Osteogenic Actions of Transform Growth Factor β

Study	Animal	Age	Transforming growth factor-β	Dose	Delivery	Site	Model	Results
Joyce et al. *(114)*	Rat	Newborn	TGFβ1,2	20–200 ng	Injection	Femur	Subperiosteal injection	Cartilage and bone formation
Lind et al. *(118)*	Rabbit	Adult	TGFβ1,2 from platelets	1–10 µg	Osmotic minipump (systemic)	Tibia	Fracture + plate	Increased callus, bending strength at 6 wk
Nielson et al. *(119)*	Rat	Young adult	TGFβ1,2 from platelets	4–40 ng	Daily injection (local)	Tibia	Fracture + intramedullary pin	Increased callus, strength at 6 wk
Critchlow et al. *(120)*	Rabbit	Adult	TGFβ2	60–600 ng	Daily injection	Tibia	Fracture + plate	Slightly increased callus, no increased strength
Beck et al. *(166)*	Rabbit	Young adult	TGFβ1	0.6–50 µg	Tricalcium phosphate carrier	Radius	Critical defect	3X increased strength and increased callus
Heckman et al. *(167)*	Dog	Adult	TGFβ1	5–50 µg	Tricalcium phosphate amylopectin carrier	Radius	Critical defect	2X increase in strength
Sun et al. *(168)*	Mouse	Adult	TGFβ1,2 from platelets	Not stated	Injection	Femur	Subperiosteal injection	Cartilage and bone formation
Beck et al. *(169)*	Rabbit	Young adult	TGFβ1	10 µg	Tricalcium phosphate amylopectin carrier	Radius	Critical defect	Increased bone and strength
Peterson et al. *(170)*	Rabbit	Adult	TGFβ1	1.5 µg	Osmotic minipump	Radius	Critical defect	Stimulated healing
Aspenberg et al. *(171)*	Rat	Adult	TGFβ1	1–1000 ng	Hydroxyapatite carrier	Tibia	Bone in growth chamber	Inhibited ingrowth

Source: Adapted from Rosier, R. N., et al. (1998) *Clin. Orthop.* **355S,** S294–S300.

stage *(140)*. From this vantage, the receptor exerts considerable control over osteogenesis in the developing skeleton.

Deletion of the PTH/PTHrP receptor gene in mice produced disproportionately short limbs with accelerated mineralization in bones formed by endochondral ossification *(141)*. In these mice, the growth plate of the proximal tibia at 18.5 d of gestation manifested irregular and shortened columns of proliferating chondrocytes. PTH/PTHrP receptor (–/–) mice also exhibited a delayed vascular invasion of the rudimentary cartilage analog, a critical step in early osteogenesis. This was associated with a dramatic decrease in trabecular bone formation in the primary spongiosa *(142)* of the developing bone. Conversely, expression by chondrocytes of constitutively active PTH/PTHrP receptors produced delayed mineralization, decelerated conversion of proliferating chondrocytes into hypertrophic chondrocytes, prolonged presence of hypertrophic chondrocytes, and delayed vascular invasion into the growth plate *(143)*. In humans, Jansen metaphyseal chondrodysplasia, a short-limbed dwarfism characterized by impaired growth-plate development, has been shown to be caused by mutation in the PTH/PTHrP receptor that results in ligand-independent constitutive receptor activation *(144)*.

Taken together, these data suggest that PTHrP and its upstream (e.g., IHH) and downstream (e.g., PTH/PTHrP receptor) network partners are importantly involved in the signaling cascade that regulates the early phases of osteogenesis in skeletal development.

BONE MORPHOGENETIC PROTEINS (BMPs)

The BMPs are, as noted previously, members of the TGF-β superfamily of cell signaling molecules. The BMPs were discovered on the basis of their ability to induce the formation of bone in bone defects and in soft tissue sites *(145,146)*. Of the many BMPs identified to date, BMP-2, -4, and -7 (also termed osteogenic protein 1) are among the most extensively studied. All three are osteogenic in multiple in vitro and in vivo systems. In vitro, BMP-2 induces the sequential expression of cartilage and bone phenotypes in osteoblast *(147,148)* and cloned limb bud *(149)* cell lines. In vivo, BMP-2 is expressed in the prechondrocytic mesenchyme of developing limb buds *(150)*, and in mesenchymal cells, chondrocytes, periosteal cells, and osteoblasts during fracture healing *(151)*. In vivo, BMP-2, BMP-4, and BMP-7/OP-1 have the remarkable capacity to initiate the full sequence of endochondral ossification from stem cell differentiation to chondrogensis to the formation of mature, marrow-containing bone following a single administration to soft tissue (ectopic) sites *(152,153)*. The BMPs also promote osteogenesis at orthotopic sites, including calvarial and long bone defects that are too large to heal spontaneously *(146,154)*.

The foregoing observations have engendered hope that the BMPs may find application in the treatment of fractures in humans. Currently, however, information about their effects on fracture healing is limited. In a rat femoral fracture model, a single injection of recombinant human BMP-2 (rhBMP-2), increased the rate of histological maturation *(155)*. In a rabbit ulnar osteotomy model, rhBMP-2 in an implantable collagen sponge accelerated the rate of healing as measured both by histological and biomechanical criteria *(156)*. Clinical trials are now in progress using rhBMP-2 in the management of open tibial shaft fractures *(157)*.

Fracture nonunion may be viewed as a failure of osteogenesis. Thus, an osteoinductive agent, such as a bone morphogenetic protein, is a logical candidate for therapy. In a recent clinical trial, OP-1/BMP-7 was compared to autologous bone grafting as a supplement to intramedullary rod fixation of tibial nonunions. Although limited by the absence of a control group, this study showed that patients treated with bone graft or OP-1/BMP-7 healed with approximately the same frequency *(158)*. The role of the BMPs in accelerating fracture healing, reducing the incidence of nonunion, or promoting the healing of established nonunion requires further investigation.

A potential application for osteogenic factors such as the BMPs is the induction of new bone at sites that are at risk for fracture. Osteoporosis, a disease characterized by insufficient bone mass, is a case in point. It has reached epidemic proportions in many parts of the world, and osteoporotic fractures, particularly of the hip, have become a major source of morbidity and mortality. A recent study tested the ability of rhBMP-2 to induce bone formation in the hip *(159)*. In an ovine (sheep) model, a single intraosseous injection of rhBMP-2 into the femoral head and neck produced dense trabecular bone along the injection track. A remarkable finding in this study was the observation that the dense new trabecular bone had completely replaced the preexisting normal trabecular bone. This resorption of normal bone in response to BMP-2 appears to be paradoxical in light of the bone-inducing actions of BMP-2 at other sites. Indeed, at sites more distant from the injection track in the sheep model, rhBMP-2 stimulated the formation of new bone onto preexisting trabeculae without evidence of prior trabecular resorption. These data indicate that intraosseous rhBMP-2 appears to function through two distinct mechanisms. One mechanism involves the initial removal of bone, followed by osteogenesis. The second appears to involve the direct formation of new bone on preexisting bone. BMP-2 has been shown to stimulate osteoclast formation and activity in vitro *(160)* and the foregoing data suggest that a similar phenomenon occurs in vivo. Whether the resorptive phase is coupled to the bone-formation phase remains to be determined.

It is possible that the osteogenic action of BMP-2 is site-specific. When delivered in contact with soft tissues, the osteogenic process includes mesenchymal cell recruitment, differentiation into chondrocytes, and subsequent endochondral ossification. At an intraosseous (trabecular) site, BMP-2 may produce direct appositional bone formation or bone resorption followed by osteogenesis. The mechanisms of BMP-induced osteogenesis will need to be considered as they are developed for therapeutic use.

In order to be useful in clinical applications, growth factors such as the BMPs, must be available for sufficient periods of time and in sufficient amounts to promote osteogenesis. One approach to achieving this goal is gene therapy. Recent studies suggest that this approach may be feasible for delivery of the BMPs. Adenoviral gene transfer was used to create rat marrow cells that produced BMP-2 *(161)*. When these cells were implanted with demineralized bone matrix into critical-sized femoral defects in syngeneic animals, 22 of 24 defects healed by 2 mo. Biomechanical parameters of healing were similar for animals treated with BMP-2-expressing cells and animals treated with BMP-2 protein. However, the cell-treated defects healed with coarser, thicker trabecular bone than did the defects treated with the BMP-2 protein. Direct application of a DNA plasmid encoding BMP-4 on a collagen sponge has also been shown to be successful in augmenting bone healing in rat critical-sized femoral defect model *(131)*.

Cbfa1

The process of osteogenesis is completed by the formation of mineralized extracellular matrix. This is the task of the osteoblast. Our understanding of osteoblast function has been substantially advanced recently by the identification of a transcription factor termed Cbfa1, which regulates osteoblast differentiation. Mice deficient in the gene encoding Cbfa1 lack osteoblasts *(162)* and mice expressing a dominant negative Cbfa1 domain become osteopenic during postnatal skeletal development *(163)*. This transcription factor binds to the promoter of, and positively regulates, a variety of genes involved in bone formation, including these encoding osteocalcin, αI (1) procollagen, bone sialoprotein, and osteopontin *(37)*. Forced expression of Cbfa1 has been shown to induce osteoblast-specific gene expression in nonosteoblastic cells *(164)*. Mutations in the Cbfa1 gene are responsible for cleidocranial dysplasia in humans *(162,165)*.

Taken together, these data suggest that Cbfa1 plays a central role in osteoblast differentiation and subsequent function, though in humans this role appears to be shared with other factors.

SUMMARY

Growth factors and other cell signaling molecules participate in all phases of osteogenesis. From the early patterning of the future skeleton to the growth and development of bone, to the remodeling

48. Halvey, O., Schindler, D., Hurwitz, S., and Pines, M. (1991) Epidermal growth factor receptor gene expression in avian epiphyseal growth-plate cartilage cells: effect of serum, parathyroid hormone and atrial natriuretic peptide. *Mol. Cell Endocrinol.* **75,** 229–235.

49. Tajima, Y., Kato, K., Kshimata, M., Hiramatsu, M., and Utsumi, N. (1994) Immunohistochemical analysis of EGF in epiphyseal growth plate from normal, hypophysectomized, and growth hormone-treated hypophysectomized rats. *Cell Tissue Res.* **278,** 279–282.

50. Murakami, Y., Nagata, H., Shizkuishi, S., et al. (1994) Histalin as a synergistic stimulator with epidermal growth factor of rabbit chondrocyte proliferation. *Biochem. Biophys. Res. Commun.* **198,** 274–280.

51. Bonassar, L. J. and Trippel, S. B. (1997) Interaction of epidermal growth factor and insulin-like growth factor-I in the regulation of growth plate chondrocytes. *Exp. Cell Res.* **234,** 1–6.

52. Gonzalez, A.-M., Buscaglia, M., Ong, M., and Baird, A. (1990) Distribution of basic fibroblast growth factor in the 18-day rat fetus: localization in the basement membranes of diverse tissues. *J. Cell Biol.* **110,** 753–765.

53. Gonzalez, A.-M., Hill, D. J., Logan, A., Maher, P. A., and Baird, A. (1996) Distribution of fibroblast growth factor (FGF)-2 and FGF receptor-1 messenger RNA expression and protein presence in the mid-trimester human fetus. *Pediatr. Res.* **39,** 375–385.

54. Hebert, J. M., Basilico, C., Goldfarb, M., Haub, O., and Martin, G. R. (1990) Isolation of cDNA's encoding four mouse FGF family members and characterization of their expression patterns during embryogenesis. *Dev. Biol.* **138,** 454–463.

55. Iwamoto, M., Shimazu, A., Nakashima, K., Suzuki, F., and Kato, Y. (1991) Reduction in basic fibroblast growth factor receptor is coupled with terminal differentiation of chondrocytes. *J. Biol. Chem.* **266,** 461–467.

56. Trippel, S. B., Whelan, M. C., Klagsbrun, M., and Doctrow, S. R. (1992) Interaction of basic fibroblast growth factor with bovine growth plate chondrocytes. *J. Orthop. Res.* **10,** 638–4646.

57. Crabb, I. D., O'Keefe, R. J., Puzas, J. E., and Rosier, R. N. (1990) Synergistic effect of transforming growth factor-β and fibroblast growth factor on DNA synthesis in chick growth plate chondrocytes. *J. Bone Miner. Res.* **5,** 1105–1112.

58. Kasper, S. and Friesen, H. G. (1986) Human pituitary tissue secretes a potent growth factor for chondrocyte proliferation. *J. Clin. Endocrinol. Metab.* **62,** 70–76.

59. Kato, Y. and Gospodarowicz, D. (1984) Growth requirements of low-density rabbit costal chondrocyte cultures maintained in serum-free medium. *J. Cell Physiol.* **120,** 354–363.

60. Rosselot, G., Vasilatos-Younken, R., and Leach, R. M. (1994) Effect of growth hormone, insulin-like growth factor I, basic fibroblast growth factor, and transforming growth factor-β on cell proliferation and proteoglycan synthesis by avian postembryonic growth plate chondrocytes. *J. Bone Miner. Res.* **9,** 431–439.

61. Trippel, S. B., Wroblewski, J., Makower, A.-M., Whelan, M. C., Schoenfeld, D., and Doctrow, S. R. (1993) Regulation of growth plate chondrocytes by insulin-like growth factor I and basic fibroblast growth factor. *J. Bone Joint Surg.* **75A,** 177–189.

62. Hill, D. J., Logan, A., McGarry, M., and DeSousa, D. (1992) Control of protein and matrix molecular synthesis in isolated ovine fetal growth plate chondrocytes by the interactions of basic fibroblast growth factor, insulin-like growth factors-I and II, and insulin and transforming growth factor-β. *J. Endocrinol.* **133,** 363–373.

63. Kato, Y. and Gospodarowicz, D. (1985) Sulfated proteoglycan synthesis by confluent cultures of rabbit costal chondrocytes grown in the presence of fibroblast growth factor. *J. Cell Biol.* **100,** 477–485.

64. Horton, W. E., Higginbotham, J. D., and Chandrasekhar, S. (1989) Transforming growth factor-beta and fibroblast growth factor act synergistically to inhibit collagen II synthesis through a mechanism involving regulatory DNA sequences. *J. Cell Physiol.* **141,** 8–15.

65. Kato, Y. and Iwamoto, M. (1990) Fibroblast growth factor is an inhibitor of chondrocyte terminal differentiation. *J. Biol. Chem.* **265,** 5903–5909.

66. Mancilla, E. E., DeLuca, F., Uyeda, J. A., Czerwiec, F. S., and Baron, J. (1998) Effects of fibroblast growth factor-2 on longitudinal bone growth. *Endocrinology* **139,** 2900–2904.

67. Kawagushi, H., Kurokawa, T., Hanada, K., et al. (1994) Stimulation of fracture repair by recombinant human basic fibroblast growth factor in normal and streptozotocin-diabetic rats. *Endocrinology* **135,** 774–781.

68. Radomsky, M. L., Thompson, A. Y., Spiro, R. C., and Poser, J. W. (1998) Potential role of fibroblast growth factor in enhancement of fracture healing. *Clin. Orthop.* **355S,** S283–S293.

69. Jingushi, S., Heydemann, A., Kana, S. K., Macey, L. R., and Bolander, M. E. (1990) Acidic fibroblast growth factor (aFGF) injection stimulates cartilage enlargement and inhibits cartilage gene expression in rat fracture healing. *J. Orthop. Res.* **8,** 364–371.

70. Baron, J., Klein, K. O., Yanovski, J. A., et al. (1994) Induction of growth plate ossification by basic fibroblast growth factor. *Endocrinology* **135,** 2790–2793.

71. Nagai, H., Tsukuda, R., and Mayahara, H. (1995) Effects of basic fibroblast growth factor on bone formation in growing rats. *Bone* **16,** 367–373.

72. Wang, J. S. and Aspenberg, P. (1993) Basic fibroblast growth factor and bone induction in rats. *Acta Orthop. Scand.* **64,** 557–561.

73. Nakamura, K., Kawaguchi, H., Aoyama, I., et al. (1997) Stimulation of bone formation by intraosseous application of recombinant basic fibroblast growth factor in normal and ovariectomized rabbits. *J. Orthop. Res.* **15**, 307–313.

74. Coffin, J. D., Florkiewicz, R. Z., Neumann, J., et al. (1995) Abnormal bone growth and selective translational regulation in basic fibroblast growth factor (FGF-2) transgenic mice. *Mol. Biol. Cell* **6**, 1861–1873.

75. Johnson, D. E. and Williams, L. T. (1993) Structural and functional diversity in the FGF receptor multigene family. *Adv. Cancer Res.* **60**, 1–41.

76. Shi, D.-L., Launay, C., Fromentoux, V., Feige, J.-J., and Boucaut, J. C. (1994) Expression of fibroblast growth factor receptor-2 splice variants is developmentally and tissue-specifically regulated in the amphibian embryo. *Dev. Biol.* **164**, 173–182.

77. Briner, J., Giedion, A., and Spycher, M. A. (1991) Variation of quantitative and qualitative changes of enchondral ossification in heterozygous achondroplasia. *Pathol. Res. Pract.* **187**, 271–278.

78. Rimoin, D. L., Hughes, G. N., Kaufman, R. L., Rosenthal, R. E., McAlister, W. H., and Silberberg, R. (1970) Endochondral ossification in achondroplastic dwarfism. *N. Engl. J. Med.* **283**, 728–735.

79. Pauli, R. M., Horton, V. K., Glinski, L. P., and Reiser, C. A. (1995) Prospective assessment of risks for cervicomedullary-junction compression in infants with achondroplasia. *Am. J. Hum. Genet.* **56**, 732–744.

80. Rousseau, F., Bonaventure, J., Legeai-Mallet, L., et al. (1994) Mutations in the gene encoding fibroblast growth factor receptor-3 in achondroplasia. *Nature* **371**, 252–254.

81. Shiang, R., Thompson, L. M., Zhu, Y.-Z., et al. (1994) Mutations in the transmembrane domain of FGFR3 cause the most common genetic form of dwarfism, achondroplasia. *Cell* **78**, 335–342.

82. Kaufman, R., Rimoin, D., McAlister, W., and Kissane, J. (1970) Thanatophoric dwarfism. *Am. J. Dis. Child* **120**, 53–57.

83. Maroteaux, P., Lamy, M., and Robert, J.-M. (1967) Le nanisme thanatophore. *Presse Med.* **49**, 2519–2524.

84. Shah, K., Astley, R., and Cameron, A. (1973) Thanatophoric dwarfism. *J. Med. Genet.* **10**, 243–252.

85. Rousseau, F., Saugier, P., LeMerrer, M., et al. (1995) Stop codon FGFR3 mutations in thanatophoric dwarfism type 1. *Nat. Genet.* **10**, 11–12.

86. Tavormina, P. L., Shiang, R., Thompson, L. M., et al. (1995) Thanatophoric dysplasia (types I and II) caused by distinct mutations in fibroblast growth factor receptor 3. *Nat. Genet.* **9**, 321–328.

87. Hall, B. D. and Spranger, J. (1979) Hypochondroplasia: clinical and radiological aspects in 39 cases. *Radiology* **133**, 95–100.

88. Walker, B. A., Murdoch, J. L., McKusick, V. A., Langer, L. O., and Beals, R. K. (1971) Hypochondroplasia. *Am. J. Dis. Child* **122**, 95–104.

89. Bellus, G. A., McIntosh, I., Smith, E. A., et al. (1995) A recurrent mutation in the tyrosine kinase domain of fibroblast growth factor receptor 3 causes hypochondroplasia. *Nat. Genet.* **10**, 357–359.

90. Meyers, G. A., Orlow, S. J., Munro, I. R., and Jabs, E. W. (1995) Fibroblast growth factor receptor 3 (FGFR3) transmembrane mutation in Crouzon syndrome with acanthosis nigricans. *Nat. Genet.* **11**, 462–464.

91. Colvin, J. S., Bohne, B. A., Harding, G. W., McEwen, D. G., and Ornitz, D. M. (1996) Skeletal overgrowth and deafness in mice lacking fibroblast growth factor receptor 3. *Nat. Genet.* **12**, 390–397.

92. Deng, C., Wynshaw-Boris, A., Zhou, F., Kuo, A., and Leder, P. (1996) Fibroblast growth factor receptor 3 is a negative regulator of bone growth. *Cell* **84**, 911–921.

93. Neilson, K. M. and Friesel, R. (1996) Ligand-independent activation of fibroblast growth factor receptors by point mutations in the extracellular, transmembrane, and kinase domains. *J. Biol. Chem.* **271**, 25049–25057.

94. Su, W.-C. S., Kitagawa, M., Xue, N., et al. (1997) Activation of Stat 1 by mutant fibroblast growth-factor receptor in thanatophoric dysplasia type II dwarfism. *Nature* **386**, 288–291.

95. Webster, M. K. and Donoghue, D. J. (1996) Constitutive activation of fibroblast growth factor receptor 3 by the transmembrane domain point mutation found in achondroplasia. *EMBO J.* **15**, 520–527.

96. Peters, K., Ornitz, D., Werner, S., and Williams, L. (1993) Unique expression pattern of the FGF receptor 3 gene during mouse organogenesis. *Dev. Biol.* **155**, 423–430.

97. Jabs, E. W., Li, X., Scott, A. F., et al. (1994) Jackson-Weiss and Crouzon syndromes are allelic with mutations in fibroblast growth factor receptor 2. *Nat. Genet.* **8**, 275–279.

98. Reardon, W., Winter, R. M., Rutland, P., Pulleyn, L. J., Jones, B. M., and Malcolm, S. (1994) Mutations in the fibroblast growth factor receptor 2 gene cause Crouzon syndrome. *Nat. Genet.* **8**, 98–103.

99. Muenke, M., Schell, U., Hehr, A., et al. (1994) A common mutation in the fibroblast growth factor receptor 1 gene in Pfeiffer syndrome. *Nat. Genet.* **8**, 269–274.

100. Wilkie, A. O. M., Slaney, S. F., Oldridge, M., et al. (1995) Apert syndrome results from localized mutations of FGFR2 and is allelic with Crouzon syndrome. *Nat. Genet.* **9**, 165–172.

101. Lajeunie, E., Ma, H. W., Bonaventure, J., Munnich, A., LeMerrer, M., and Renier, D. (1995) FGFR2 mutations in Pfeiffer syndrome. *Nat. Genet.* **9**, 108.

102. Rutland, P., Pulleyn, L. J., Reardon, W., et al. (1995) Identical mutations in the FGFR2 gene cause both Pfeiffer and Crouzon syndrome phenotypes. *Nat. Genet.* **9**, 173–176.

103. Bostrom, M. P. G. and Asnis, P. (1998) Transforming growth factor beta in fracture repair. *Clin. Orthop.* **355S,** S124–S131.

104. Massague, J. (1990) The transforming growth factor-beta family. *Ann. Rev. Cell Biol.* **6,** 597–609.

105. Rosier, R. N., O'Keefe, R. J., and Hicks, D. G. (1998) The potential role of transforming growth factor beta in frac-ture healing. *Clin. Orthop.* **355S,** S294–S300.

106. Joyce, M. E., Jingushi, S., Scully, S. P., and Bolander, M. E. (1991) Role of growth factors in fracture healing. *Prog. Clin. Biol. Res.* **365,** 391–416.

107. Harris, S. E., Bonewald, L. F., Harris, M. A., et al. (1994) Effects of transforming growth factor beta on bone nodule formation and expression of bone morphogenetic protein 2, osteocalcin, osteopontin, alkaline phosphatase, and type I collagen mRNA in long-term cultures of fetal rat calvarial osteoblasts. *J. Bone Miner. Res.* **9,** 855–863.

108. Tally-Ronsholdt, D. J., Lajiness, E., and Nagodawithana, K. (1995) Transforming growth factor-beta inhibition of mineralization by neonatal rat osteoblasts in monolayer and collagen gel culture. *In Vitro Cell Dev. Biol.* **31,** 274–282.

109. Pfeilschifter, J., Seyedin, S. M., and Mundy, G. R. (1988) Transforming growth factor beta inhibits bone resorption in fetal rat long bone cultures. *J. Clin. Invest.* **82,** 680–685.

110. Seyedin, S. M., Segarini, P. R., Rosen, D. M., Thompson, A. Y., Bentz, H., and Graycan, J. (1987) Cartilage-inducing factor-B is a unique protein structurally and functionally related to transforming growth factor-beta. *J. Biol. Chem.* **262,** 1946–1949.

111. Seyedin, S. M., Thompson, A. Y., Bentz, H., et al. (1986) Cartilage-inducing factor-A. Apparent identity to transform-ing growth factor-beta. *J. Biol. Chem.* **261,** 5693–5695.

112. Rosen, D. M., Stempien, S. A., Thompson, A. Y., and Seyedin, S. M. (1988) Transforming growth factor-beta modu-lates the expression of osteoblast and chondroblast phenotypes *in vitro. J. Cell Physiol.* **134,** 337–346.

113. Joyce, M. E., Terek, R. M., Jingushi, S., and Bolander, M. E. (1990) Role of transforming growth factor-beta in fracture repair. *Ann. N. Y. Acad. Sci.* **593,** 107–123.

114. Joyce, M. E., Roberts, A. B., Sporn, M. B., and Bolander, M. E. (1990) Transforming growth factor-beta and the initiation of chondrogenesis and osteogenesis in the rat femur. *J. Cell Biol.* **110,** 2195–2207.

115. Centrella, M., McCarthy, T. L., and Canalis, E. (1991) Transforming growth factor-beta and remodeling of bone. *J. Bone Joint Surg.* **73A,** 1418–1428.

116. Andrew, J. G., Hoyland, J., Andrew, S. M., Freemont, A. J., and Marsh, D. (1993) Demonstration of TGF-beta 1 mRNA by in situ hybridization in normal human fracture healing. *Calcif. Tissue Int.* **52,** 74–78.

117. Bolander, M. E. (1992) Regulation of fracture repair by growth factors. *Proc. Soc. Exp. Biol. Med.* **200,** 165–170.

118. Lind, M., Schumacker, B., Soballe, K., Keller, J., Melsen, F., and Bunger, C. (1993) Transforming growth factor-β enhances fracture healing in rabbit tibiae. *Acta Orthop. Scand.* **64,** 553–556.

119. Nielsen, H. M., Andreassen, T. T., Ledet, T., and Oxlund, H. (1994) Local injection of TGF-beta increases the strength of tibial fractures in the rat. *Acta Orthop. Scand.* **65,** 37–41.

120. Critchlow, M. A., Bland, Y. S., and Ashhurst, D. E. (1995) The effect of exogenous transforming growth factor-beta 2 on healing fractures in the rabbit. *Bone* **16,** 521–527.

121. Dempster, D. W., Cosman, F., Parisieu, M., Shen, V., and Lindsay, R. (1993) Anabolic actions of parathyroid hormone on bone. *Endocrine Rev.* **14,** 690–709.

122. Canalis, E., Centrella, M., Burch, W., and McCarthy, T. L. (1989) Insulin-like growth factor I mediates selective ana-bolic effects of parathyroid hormone in bone cultures. *J. Clin. Invest.* **83,** 60–65.

123. Koike, T., Iwamoto, M., Shimazu, A., Nakashima, K., Suzuki, F., and Kato, Y. (1990) Potent mitogenic effects of parathyroid hormone (PTH) on embryonic chick and rabbit chondrocytes: differential effects of age on growth, pro-teoglycan and cyclic AMP responses of chondrocytes to PTH. *J. Clin. Invest.* **85,** 626–631.

124. Pines, M. and Hurwitz, S. (1988) The effect of parathyroid hormone and atrial natriuretic peptide on cyclic nucle-otides production and proliferation of avian epiphyseal growth plate chondroprogenitor cells. *Endocrinology* **123,** 360–365.

125. Takano, T., Takigawa, M., Shirai, E., et al. (1985) Effects of synthetic analogs and fragments of bovine parathyroid hormone on adenosine 3',5'-monophosphate level, ornithine decarboxylase activity, and glycosaminoglycan synthe-sis in rabbit costal chondrocytes in culture: structure–activity relations. *Endocrinology* **116,** 2536–2542.

126. Cipera, J. D. and Cherian, A. G. (1969) Composition of epiphyseal cartilage. VI. Effect of parathyroidectomy and of a parathormone on the epiphyseal cartilage of growing chicks. *Calcif. Tissue Res.* **3,** 30–37.

127. Tam, C. S., Heersche, J. N. M., Murray, T. M., and Parsons, J. A. (1982) Parathyroid hormone stimulates the bone apposition rate independently of its resorptive action: differential effects of intermittent and continuous administration. *Endocrinology* **110,** 506–512.

128. Hock, J. M. and Fonseca, J. (1990) Anabolic effect of human synthetic parathyroid hormone-(1-34) depends on growth hormone. *Endocrinology* **127,** 1804–1810.

129. Finkelstein, J. S., Klibanski, A., Arnold, A. L., Toth, T. L., Hornstein, M. D., and Neer, R. M. (1998) Prevention of estrogen deficiency-related bone loss with human parathyroid hormone (1-34): a randomized controlled trial. *JAMA* **280,** 1067–1073.

130. Reeve, J., Meunier, P. J., Parsons, J. A., et al. (1980) Anabolic effect of human parathyroid hormone fragment on trabecular bone in involutional osteoporosis: a multicentre trial. *Br. Med. J.* **280,** 1340–1344.

131. Fang, J., Zhu, Y.-Y., Smiley, E., et al. (1996) Stimulation of new bone formation by direct transfer of osteogenic plasmid genes. *Proc. Natl. Acad. Sci. USA* **93,** 5753–5758.

132. Goldstein, S. A. and Bonadio, J. (1998) Potential role for direct gene transfer in the enhancement of fracture healing. *Clin. Orthop.* **355A,** S154–S162.

133. Broadus, A. E. and Stewart, A. F. (1994) Parathyroid hormone-related protein. Structure, processing, and physiological action, in *The Parathyroids Basic and Clinical Concepts* (Bilezikian, J. P. and Levine, M. A., eds.), Raven Press. New York, pp. 259–294.

134. O'Keefe, R. J., Loveys, L. S., Miclas, D. G., et al. (1997) Differential regulation of type-II and type-X collagen synthesis by parathyroid hormone-related protein in chick growth-plate chondrocytes. *J. Orthop. Res.* **15,** 162–174.

135. Amizuka, N., Warshawsky, H., Henderson, J. E., Goltzmman, D., and Karaplis, A. C. (1994) Parathyroid hormone-related peptide-depleted mice show abnormal epiphyseal cartilage development and altered endochondral bone formation. *J. Cell Biol.* **126,** 1611–1623.

136. Karaplis, A. C., Luz, A., Glowacki, J., et al. (1994) Lethal skeletal dysplasia from targeted disruption of the parathyroid hormone-related peptide gene. *Gene Dev.* **8,** 277–289.

137. Vortkamp, A., Lee, K., Lanske, B., Segre, G. V., Kronenberg, H. M., and Tabin, C. J. (1996) Regulation of rate of cartilage differentiation by Indian hedgehog and PTH-related protein. *Science* **273,** 613–622.

138. St-Jacques, B., Hammerschmidt, M., and McMahon, A. P. (1999) Indian hedgehog signaling regulates proliferation and differentiation of chondrocytes and is essential for bone formation. *Genes Dev.* **13,** 2072–2086.

139. Juppners, H., Abou-Samra, A.-B., Freeman, M. W., et al. (1991) A G protein-linked receptor for parathyroid hormone and parathyroid hormone-related peptide. *Science* **254,** 1024–1026.

140. Lee, K., Deeds, J. D., Chiba, S., Un-No, M., Bond, A. T., and Segre, G. V. (1994) Parathyroid hormone induces sequential c-fos expression in bone cells *in vivo*: *in situ* localization of its receptor and c-fos messenger ribonucleic acids. *Endocrinology* **134,** 441–540.

141. Lanske, B., Karaplis, A. C., Lee, K., et al. (1996) PTH/PTHrP receptor in early development and Indian hedgehog-related bone growth. *Science* **273,** 663–666.

142. Lanske, B., Amling, M., Neff, L., Guiducci, J., Baron, R., and Kronenberg, H. (1999) Ablation of the PTHrP gene or the PTH/PTHrP receptor gene leads to distinct abnormalities in bone development. *J. Clin. Invest.* **104,** 399–407.

143. Schipani, E., Lanske, B., Hunzelman, J., et al. (1997) Targeted expression of constitutively active receptors for parathyroid hormone and parathyroid hormone-related peptide delays endochondral bone formation and rescues mice that lack parathyroid hormone-related peptide. *Proc. Natl. Acad. Sci. USA* **94,** 13689–13694.

144. Schipani, E., Kruse, K., and Juppner, H. (1995) A constitutively active mutant PTH-PTHrP receptor in Jansen-type metaphyseal chondrodysplasia. *Science* **268,** 98–100.

145. Urist, M. (1965) Bone: formation by autoinduction. *Science* **150,** 893–899.

146. Wozney, J. M. and Rosen, V. (1998) Bone morphogenetic protein and bone morphogenetic protein gene family in bone formation and repair. *Clin. Orthop.* **346,** 26–37.

147. Thies, R. S., Bauduy, M., Ashton, B. A., Kurtzberg, L., Wozney, J. M., and Rosen, V. (1998) Recombinant human bone morphogenetic protein 2 induces osteoblast differentiation in W20-17 stromal cells. *Endocrinology* **130,** 1318–1324.

148. Vukicevic, S., Luyten, F. P., and Reddi, A. H. (1991) Osteogenin inhibits proliferation and stimulates differentiation in mouse osteoblast-like cells (MC3T3-E1). *Biochem. Biophys. Res. Commun.* **166,** 750–756.

149. Rosen, V., Nove, J., Song, J. J., Thies, S., Cox, K., and Wozney, J. M. (1994) Responsiveness of clonal limb bud cell lines to bone morphogenetic protein 2 reveals a sequential relationship between cartilage and bone cell phenotypes. *J. Bone Miner. Res.* **9,** 1759–1768.

150. Lyons, K. M., Pelton, R. W., and Hogan, B. L. M. (1990) Organogenesis and pattern formation in the mouse. RNA distribution patterns suggest a role for bone morphogenetic protein-2A (BMP2-A). *Development* **109,** 833–844.

151. Bostrom, M. P. G., Lane, J. M., Berberian, W. S., et al. (1995) Immunolocalization and expression of bone morphogenetic proteins 2 and 4 in fracture healing. *J. Orthop. Res.* **13,** 357–367.

152. Reddi, A. H. (1998) Role of morphogenetic proteins in skeletal tissue engineering and regeneration. *Natl. Biotech.* **16,** 247–252.

153. Rosen, V. and Wozney, J. M. (2000) Bone morphogenetic proteins and the adult skeleton, in *Skeletal Growth Factors*. Lippincott Williams Wilkins, Philadelphia, pp. 299–310.

154. Kirker-Head, C. A., Gerhart, T. N., Schelling, S. H., Henning, G. E., Wang, E., and Holtrop, M. E. (1995) Long term healing of bone using recombinant human bone morphogenetic protein 2. *Clin. Orthop.* **318,** 222–230.

155. Einhorn, T. A., Majeska, R. J., Oloumi, G., et al. (1997) Enhancement of experimental fracture healing with a local percutaneous injection of recombinant human bone morphogenetic protein-2. *Proc. V American Academy of Orthopedic Surgeons*. 64th Annual Meeting, 216 (abstr).

156. Bouxsein, M. L., Turek, T. J., Blake, C. A., et al. (1999) RHBMP-2 accelerates healing in a rabbit ulnar osteotomy model. *Trans. 45th Annual Meeting, Orthopedic Research Society* **24,** 138.

157. Riedel, G. E. and Valentin-Opran, A. (1999) Clinical evaluation of rhBMP-2/ACS in orthopaedic trauma: a progress report. *Orthopedics* **22,** 663–665.
158. Cook, S. D. (1999) Preclinical and clinical evaluation of osteogenic protein-1 (BMP-7) in bony sites. *Orthopedics* **22,** 669–671.
159. Seeherman, H. J., Kirker-Head, C. A., Li, X. J., et al. (1998) RhBMP-2 stimulates bone formation in a femoral head core decompression model in adult sheep. *Trans. Orthop. Res. Soc.* **23,** 65.
160. Kanatani, M., Sugimoto, T., Kaja, H., et al. (1995) Stimulatory effect of bone morphogenetic protein-2 on osteoblast-like cell formation and bone-resorbing activity. *J. Bone Miner. Res.* **10,** 1681–1690.
161. Lieberman, J. R., Daluiski, A., Stevenson, S., et al. (1999) The effect of regional gene therapy with bone morphogenetic protein-2-producing bone-marrow cells on the repair of segmental femoral defects in rats. *J. Bone Joint Surg.* **81A,** 905–917.
162. Otto, F., Thornell, A. P., Crompton, T., et al. (1997) Cbfa1, a candidate gene for the cleidocranial dysplasia syndrome is essential for osteoblast differentiation and bone development. *Cell* **89,** 765–771.
163. Ducy, P., Starbuck, M., Priemel, M., et al. (1999) Cbfa1-dependent genetic pathway controls bone formation beyond embryonic development. *Genes Dev.* **13,** 1025–1036.
164. Ducy, P., Zhang, R., Geoffroy, V., Ridall, A. L., and Karsenty, G. (1997) Osf2/Cbfa1: a transcriptional activator of osteoblast differentiation. *Cell* **89,** 747–754.
165. Mundlos, S., Otto, F., Mundlos, C., et al. (1997) Mutations involving the transcription factor Cbfa1 cause cleidocranial dysplasia. *Cell* **89,** 773–779.
166. Beck, L. S., Deguzman, L., Lee, W. P., Onpipattanakul, B., and Nguyen, T. (1996) Bone marrow augments the activity of transforming growth factor beta 1 in critical sized bone defects. *Trans. Orthop. Res. Soc.* **21,** 626.
167. Heckman, J. D., Aufdemorte, T. D., Athanasiou, K. A., et al. (1995) Treatment of acute ostectomy defects in the dog radius rhTGF-β1. *Trans. Orthop. Res. Soc.* **20,** 590.
168. Sun, Y., Lu, Y., Hu, Y., Ma, F., and Chen, W. (1995) Induction of osteogenesis by boving platelet transforming growth factor beta. *Chin. Med. J. (Engl.)* **108,** 914–918.
169. Beck, L. S., Deguzman, L., Lee, W. P., et al. (1995) Transforming growth factor beta 1 bound to tricalcium phosphate persists at segmental radial defects and induces bone formation. *Trans. Orthop. Res. Soc.* **20,** 593.
170. Peterson, D. R., Glancy, T. P., Bacon-Clarke, R., Plouhar, P., Malmstrom, D. F., and Nassen, P. (1997) A study of delivery timing and duration on the transforming growth factor beta 1 induced healing of critical-size long bone defects. *J. Bone Miner. Res.* **12(Suppl 1),** S304.
171. Aspenberg, P., Jeppsson, C., Wang, J. S., and Bostrom, M. (1996) Transforming growth factor beta and bone morphogenetic protein 2 for gone ingrowth: a comparison using bone chambers in rats. *Bone* **19,** 499–503.

8

Grafts and Bone Graft Substitutes

Doug Sutherland, MD and Mathias Bostrom, MD

INTRODUCTION

The value of bone transplantation is demonstrated by the frequency of its use today. Surgeons transplant bone at least 10 times more often than they do any other transplantable organ. The procedure has a rich history, dating back over 300 years to when Job van Meekeren performed the first bone graft using a canine xenograft to repair a cranial defect *(138)*. Bone grafting became critical during World War II, prompting the US Navy to establish bone banks to better treat fractures sustained in battle *(9)*. During that period a successful graft was thought to be one that could withstand the forces applied to it by the individual. Today we consider the bone graft to be a dynamic tool that should not only support normal forces, but also incorporate itself into the bed, revascularize as new bone forms, and assume the specific shape required for the healing defect. Furthermore, accelerating the normal healing process whenever possible is an obvious goal. Recombinant DNA technology might achieve this goal by allowing surgeons to apply growth factors to defects in therapeutic quantities in an effort to speed regeneration.

The ideal bone graft or bone graft substitute should provide three essential elements: (1) an osteoconductive matrix; (2) osteoinductive properties or factors; and (3) osteogenic cells. Osteoconductivity can be defined as the process of infiltration of capillaries, perivascular tissue, and osteoprogenitor cells from the host bed into the transplant *(18)*. The matrix need not be viable. However, as we will see, if the graft does not simulate the porosity and microstructure of human cancellous bone, incorporation into the bed will be delayed. Osteoinduction is the stimulation of a tissue to produce osteogenic elements *(18)*. This process is controlled primarily by growth factors such as bone morphogenetic proteins (BMPs) that are capable of inducing differentiation of mesenchymal cells into cartilage and bone-producing cells. Osteogenetic cells are mesenchymal-type cells, and they can be summoned from host or graft bone marrow *(18)*. The inclusion of osteogenic cells into grafts today remains procedurally difficult. Because few cells survive transplant, most osteogenic cells found in the graft are recruited there from the host bed by osteoinduction. This poses an obvious problem when the viability of the bed is compromised, such as a densely fibrotic defect. Thus, the creation of a bone graft or bone graft substitute that can function independently of the host bed condition is desirable.

Today, the autogenous cancellous bone graft still satisfies all three categories most completely. Hydroxyapatite and collagen serve as the osteoconductive framework, stromal cells lining the microcavities possess the necessary osteogenic potential, and the endogenous family of growth factors within the bone and adjacent hematoma fully induce both the regenerative and augmentation processes. For these reasons, the autogenous cancellous bone graft is considered the "gold standard" of bone transplantation. There are several potential complications involved with autogenous grafting, however, such as donor-site morbidity, limited availability for harvest, and increased operative blood loss. It has therefore become necessary to find suitable alternatives, particularly when a large graft is required.

From: *Bone Regeneration and Repair: Biology and Clinical Applications*
Edited by: J. R. Lieberman and G. E. Friedlaender © Humana Press Inc., Totowa, NJ

The motivation to incorporate the favorable properties of different materials into an effective bone graft compound has led to the manipulation and development of various new synthetic bone graft products. The most interesting and potentially useful substitutes are composite grafts, such as an osteoconductive graft inlayed with mesenchymal cells synthetically produced from cell culture.

Technological efforts to improve bone transplantation can be grouped into three distinct categories: osteoconductive matrices, osteoinductive factors, and osteogenic cells. All are judged in their efficacy as alternatives to autogenous and allogeneic bone grafts. Before addressing the various components of these new grafts, a review of autogenous and allogeneic bone grafts is necessary.

AUTOGENOUS BONE TRANSPLANTATION

In addition to the osteoconductive, osteoinductive, and osteogenic properties that autogenous grafting affords, they are histocompatible, do not transport disease, and retain viable osteoblasts that participate in the formation of bone. The latter is of key importance because callus formation within the first 4–8 wk after surgery is often dependent on bone formation by graft osteoblasts *(126)*. As previously mentioned, the autogenous cancellous bone graft is considered the "gold standard" of bone grafts. Autogenous cortical bone grafts contain many of the same advantages that cancellous graft provide, to a more limited extent. Less than 5% of cortical bone cells survive transplantation (nearly 100% loss of osteocytes occupying lacunae) *(111)*. As a result, cortical bone graft will retain significant islands of nonreplaced, nonviable bone throughout the life of the individual. Minced cortical graft can be used in expanding the volume of graft material, but it does not contain the same robust osteogenic potential that cancellous graft provides *(126)*. The advantage of cortical bone grafts is that their structure confers compressive strength and thus provides mechanical support. They are often used as supportive struts for this reason. The bone strut immediately enters a resorptive phase after transplant that occurs for 18 mo in canine species (presumably longer in humans), during which time nonviable bone is removed by osteoclastic tunneling. The strut loses approximately one-third of its strength before enhanced strength returns. This process of removal of necrotic bone and its replacement with new bone is known as creeping substitution *(109,110)*. In contrast, cancellous bone starts with little structural integrity and is therefore used as a means of filling small defects. Osteoclastic tunneling is not necessary when a cancellous bone graft is inserted, allowing the immediate infiltration of vessels and the initiation of osteoblast activity. The lack of structural strength rapidly changes secondary to bone augmentation and union with preexisting osteostructures. Bone regeneration starts when undifferentiated osteoprogenitor cells are recruited from the host bed and from within the graft marrow cavity. A simple scaffolding is established on which active osteoprogenitor cells can produce new bone. Bone strength increases as bone mass accumulates and the construct is remodeled along the lines of stress according to the rules defined by Wolf.

Revascularization is the defining point of contrast between cancellous and cortical bone grafting. Because mesenchymal cells (osteoblast precursors) are blood-borne, ingrowth of vasculature into the graft initiates graft incorporation. Cancellous grafts will become infiltrated with host vessels within 2 d posttransplant, which minimizes the amount of necrotic tissue and accelerates the process of creeping substitution *(109,110)*. The cortical graft is not penetrated until the d 6 posttransplant *(53)*. It is possible that this delay is attributable to the structure of cortical bone; vascular penetration follows a more extensive peripheral osteoclastic resorption into haversian canals and Volkmann's canals. The process of vascular infiltration is essential in the initiation of osteoinduction, which is mediated by numerous growth factors provided by the bone matrix itself. BMPs are the most notable of the group. BMPs are low-molecular-weight proteins that initiate endochondral bone formation, presumably by stimulating osteoblastic differentiation of mesenchymal cells and enhancing bone collagen synthesis. Transforming growth factor-beta (TGF-β) is closely related to BMPs through sequence homology, and is responsible for stimulated cell proliferation and matrix formation. TGF-β is present in the graft hematoma after release by platelets and is further synthesized by mesenchymal cells. Other growth factors present during the grafting process include fibroblast growth factors (FGFs), which are angiogenic factors important in neovascularization and wound healing. Platelet-derived growth factor (PDGF), initially isolated in blood platelets, acts as a local tissue growth regulator. Insulin-like growth factors and microglobulin-β are other examples of bone matrix-synthesizing growth factors that are important in general bone healing and graft incorporation.

Although autogenous bone grafting is effective, it is associated with several shortcomings and potential complications. Significant donor-site morbidity, with rates as high as 25%, is a major problem *(145)*. Increased postoperative pain, increased anesthesia time, and significantly increased operative blood loss are also associated with the additional harvesting procedure, primarily because of the deeply invasive techniques. Several new harvesting methods have been demonstrated recently. They provide sufficient graft material using smaller instruments that allow smaller initial incisions. Examples include using a needle biopsy kit or a simple curet to excavate cancellous bone from iliac crest. However, these novel procedures cannot remedy the limited quantity of bone available to be harvested.

Autogenous vascularized cortical bone grafts can be viewed as an attempt to accelerate the healing process of cortical bone transplantation. They provide very limited structural support, but do heal quickly at the graft–host interface if stabilized. Their incorporation differs significantly, particularly when vessels are anastomosed successfully with little intraoperative ischemia. Under these conditions, greater than 90% of osteocytes survive transplantation. Vascular infiltration by the host bed is not necessary *(37)*. Union is established without osteoclast activity and resorption as is seen in nonvascularized cortical bone grafts.

Vascularized cortical grafts are most commonly taken from fibula, although other bones have been used successfully in this process (e.g., ribs, iliac crest) *(17,41,54,79,120,129,139)*. They have been used to stabilize small fractures with compromised vasculature, such as acutely displaced femoral neck and carpal bone defects, and in radical procedures such as the re-creation of forearms following traumatic upper-extremity loss. Biomechanical studies have demonstrated that these grafts are superior to nonvascularized cortical bone grafts for 6 mo after surgery, at which time no differences in torque, bending, and tension studies can be demonstrated between them. They are clearly superior to nonvascularized cortical bone grafts when the bridging defect is greater than 12 cm *(41)*. Reported stress fractures for this distance in nonvascularized cortical bone are greater than 50%, while the rate of fracture for vascularized graft is less than 25% *(41)*. In addition, the vascularized graft has a greater ability to heal the stress-related fractures and to enhance its girth during the repair process. The obvious disadvantage is donor-site morbidity, long operative time, and greater utilization of resources.

Periosteal transplantation is another novel type of autogenous grafting *(76,82)*. These grafts take the form of pure periosteum, to be applied directly to small and large articular cartilage defects, or as periosteal flaps located at the terminal of a free vascular cortical bone graft *(76)*. The flap can be secured to cover the host–graft interface, enhancing and possibly accelerating the incorporation of the graft.

ALLOGENEIC BONE TRANSPLANTATION

The use of allogeneic bone grafts to repair skeletal defects became popular at the turn of the 20th century *(83)*. This type of grafting is not restricted by harvest availability as autogenous grafting is, nor by donor-site morbidity. Current procedures utilize allografts in the form of morselized cancellous and cortical bone chips in cavity filling, and corticocancellous and cortical struts for structural support.

Lexer, who experimented with whole and hemijoint knee allograft transplantation in the early 1900s, found that 50% of his patients did not recover well, requiring further surgery *(80)*. Infection and allograft fracture were the most common complications then, and still are today. Although stabilization and tissue testing techniques have improved allogeneic transplantation greatly, recent studies have found incidence of infection near 10–15% and incidence of fracture between 5% and 15% *(6,7,45,84,133)*. In addition, transmission of disease must be controlled. This requires stringent testing and sterilization of graft tissue prior to use. Such practices compromise the osteoinductive and osteogenic potential of allogeneic grafting significantly. Because fresh bone transplantation must be performed quickly to prevent intraoperative ischemia, very little time is allowed to test for donor disease. Hence, allogeneic bone is rarely used fresh, except for joint resurfacing (where success reflects maintenance of viable transplanted chondrocytes), and is not currently a mainstay in bone grafting. For all other needs, allograft is harvested, batch-sterilized, and preserved by deep freezing below −60°C, or freeze-drying. In an attempt to salvage the viability of allogeneic osteochondral grafts, the use of slow cryopreservation using glycerol or dimethylsulfoxide to prevent water crystallization within cells has produced some success *(96,132)*. Studies demonstrate a wide range of cartilage viability (20–70%) using these techniques, producing controversy as to their true efficacy *(96,132)*. Delipidation and deproteination using supercritical fluids and hydrogen peroxide have recently been suggested as alternatives as well. Freeze-drying fully destroys the osteoprogenitor cells and osteoinductive factors, and alters the biomechanical properties of the graft with losses of hoop and compressive strength upon rehydration. The net result of these procedures produces bone substitutes that can only provide an osteoconductive scaffold, although there is indication that they do provide decreased immunogenicity and antigenicity.

As with all allogeneic organ transplantation, the risk of disease transmission, particularly HIV and hepatitis B and C, is an important issue. The American Association of Tissue Banks (AATB) has been established to monitor hospital tissue banks to ensure compliance with comprehensive sets of standards. Regulations include donor screening, repeated infectious disease testing, sterilization of graft tissue with such substances as ethylene oxide or radiation, long-term tracking of the graft, and inspection of tissue banks. These techniques have significantly lowered the risk of disease transmission. For instance, the risk of HIV infection is now calculated at less than 1 in 10^6 from allogeneic bone transplant *(15)*. Now that infectious disease transfer appears to be suitably controlled, some feel histopathological examination should be included as part of the protocol for the collection of bone allografts. This is not currently a written protocol. Malignant tumors, osteoarthritis, Paget's disease, and avascular necrosis all represent possible allograft pathophysiology that can reasonably go unrecognized to both donor and physician without the aid of a histological examination.

Incorporation of allograft bone is markedly different than in autogenous transplant. Vascular penetration is more superficial and impeded; allograft revascularization is not as complete at 8 mo postsurgery as autogenous graft is at 1 mo *(18)*. Osteogenesis is initiated by the host bed through the process of creeping substitution, similar to the incorporation of autogenous cortical graft. There is a significant host immune response, demonstrated by an inflammatory reaction produced upon allograft transplant. This results in hyalinization of penetrating and preexisting blood vessels, prompting necrosis of allograft periosteal cells and osteocytes. The elevated quantity of necrotic bone that remains after full incorporation of the allograft due to decreased revascularization is the chief reason for the increased incidence of fracture *(6,42,130)*. This problem is most notable when using massive cortical allografts as supportive struts *(130)*. Fatigue-generated microfractures form in the necrotic bone near the fracture site, which cannot remodel itself, resulting in structural failure *(130)*.

Allografts can be used for nonstructural purposes such as reconstructing defects after curettage of a benign neoplasm or periarticular bone cyst. In addition, osteolytic cavities at time of joint arthroplasty revision can be filled with allograft. Morselized cancellous and cortical chips can be utilized in these capacities, as they provide resistance to compression due to the preservation of hydroxyapatite mineralization. The transient loss of strength, as seen in autogenous cortical graft, is not witnessed in this type of allogeneic transplantation because revascularization of morselized allograft does not require resorption. Rehydration of these chips produces an open and porous structure, without physical impediment, fully allowing the ingrowth of vasculature. Some clinicians have recommended mixing allograft with autogenous tissue or with bone marrow to enhance osteoinductivity by reintroducing osteoprogenitor cells. The efficacy of this process has yet to be tested clinically. However, when implanting into individuals with high potential for bone regeneration, such as children, this practice is theoretically unnecessary. When needed for structural roles, allografts are available in various forms including ilial bicortical and tricortical strips, cancellous cortical dowels, fibular shafts and wedges, femoral and tibial cross sections, patellae, and ribs. They can be utilized as an intercalary segment to reconstruct a diaphyseal defect of long bone, and large segments can be modeled to replace acetabular, femoral, and tibial defects during arthroplasty. Additionally, structural allografts have been used to facilitate in arthodesis about the ankle, hip, and spine. Vascularized corticocancellous allografting has also been introduced *(57,73,79)*. These grafts are used mainly in the treatment of large-scale defects, such as total knee arthroplasty and femoral reconstruction, when autogenous grafting is impossible.

Osteochondral allografts have been in use for the last 20 yr, and provide the dual purpose of replacing resected bone and providing a biological bone surface. Their use is generally confined to autogenous grafting because of the impending immune response. For that reason, these grafts are used primarily to treat large or small isolated articular cartilage defects, most often about the knee *(16,74)*. Small, cylindrical grafts are harvested arthroscopically and implanted in mosaiclike fashion into the cartilage defect. Since it has become apparent that the lifespan of conventional joint prostheses is limited, their use has become more common in the past decade. More invasive, allogeneic procedures have been suggested in response to this problem, though revascularization and host rejection make them difficult to manage. Such is the case with hemijoint and large diaphyseal reconstruction procedures. The use of an autogenous muscle tissue or a periosteal flap treaded into the medullary core of the graft, followed by anastomosing with host vasculature, has been shown to enhance graft incorporation and revascularization.

OSTEOCONDUCTIVE MATRICES

Approximately 6.2 million fractures occur in the United States each year *(106)*. Some type of bone transplantation is required in approximately 15% of all reconstructive surgical operations on the locomotive system *(69)*. The problem is therefore one of availability. Complications such as donor-site morbidity and increased intraoperative time limit autograft quantity. Tissue preparation techniques and disease transmission limit allograft use. Thus, there is a continuing search for an ideal material with adequate mechanical properties and biocompatibility that can be produced in necessary quantities. The latter functions are not as problematic; synthetic materials lack antigenicity and are rarely assaulted by an immune response, and they can be produced on demand. However, most attempts in this field lack mechanical strength and at best function as bone graft "fillers" that must be supported by other means while healing occurs.

Ceramics used in the repair of bony defects can be subdivided into three main categories based on their chemical reactivity following transplant: bioabsorbable ceramics, bioactive ceramics, and bioinert ceramics *(128)*. Bioabsorbable and bioactive substances are able to physically bond directly to the host bed, whereas bioinert substances never actually bond to the bone. Nonbiodegradable polymers form another class of osteoconductive grafts. Polymethyl methacrylate (PMMA) is such a polymer that acts similarly to a bioinert ceramic and bioglasses in that it does not incorporate into the new bone, yet it does bond adhesively to the bone surface without interfering with regeneration. Bone preparations such as decalcified bone matrix (DBM) and Pyrost, a deproteinized bone matrix, also belong to the osteoconductive group of bone grafts. These grafts hold tremendous potential because of their weak, yet present, osteoinductive capabilities.

Bioabsorbable ceramics were the first synthetic materials used in bone transplantation, and therefore have the most clinical experience. Osteogenesis follows reabsorption in this class of grafts. The chemical composition of a synthetic graft profoundly affects its rate of resorption. For example, tricalcium phosphate (TCP) will be resorbed 10–20 times faster than hydroxyapatite, another calcium phosphate ceramic.

Crystal structure also affects the total amount of resorption. Using the same example, some clinical trials have reported that TCP can be totally resorbed, or converted into hydroxyapatite, which may remain in the body indefinitely. It can be assumed that if the entirety of the ceramic is replaced, the graft has been completely replaced with bone. In the case of a hydroxyapatite graft that remains in the host bone, the intrinsic strength of the bone may be compromised at the callus site because of the weaker synthetic ceramic.

Another factor to consider in the use ceramics is porosity. The optimal osteoconductive pore sizes for ceramics appear to be between 150 and 500 μm. Cancellous bone itself has a complex trabecular pattern in which approximately 20% of the total matrix is bone and the remaining area is marrow space interconnected through pores. Synthetic ceramics, while having various sized pores, lack pore connectivity. Therefore, when they are used as graft material, the healing osteogenic process must reabsorb the bone to gain access to the interior pores. An important consideration is that with increased porosity the graft will maintain significantly less compressive strength, as in the case of TCP. The exceptions are ceramics derived from materials such as coral, which have biological pore interconnectivity mesh network. Material factors such as surface area affect the biological degradation, and in general, the larger the surface area, the greater the bioresorption. Dense ceramic blocks with small surface areas biodegrade slowly when compared to porous implants. Thus, the shape and architecture of the ceramic will also have a profound effect on resorption rates.

Plaster of Paris, a hemihydrate of calcium sulfate (CaO_4S) bioabsorbable ceramic prepared by heating gypsum, was the first substance used as a bone substitute, in the late 19th century by Dreesmann *(38)*. It has a very rapid turnover and most of it is resorbed within weeks after implantation. Several studies have shown that plaster of Paris does not inhibit osteogenesis or aggravate infection when used in infected cavities. It is very inexpensive, can be sterilized and prepared easily, and has an indefinite shelf life *(102,103)*. Interestingly, plaster of Paris can serve as a vehicle for the administration of several agents such as antimicrobials, antibiotics, or possibly osteoinductive agents. However, plaster of Paris provides no internal strength or support, and therefore can only be used to fill small bone defects such as those resulting from cyst curettage. The natural pore structure is also quite random, lacking connectivity, requiring full resorption in conjunction with ingrowth. Calcium sulfate has recently seen a resurgence in use with the recent marketing of this material in the form of tablets for use in filling osseous defects (**Fig. 1**) *(69a,86a,86b,103a,138a)*.

Calcium phosphates have received much more attention as an osteoconductive bioabsorbable ceramics (**Fig. 2**). Most calcium phosphate ceramics currently under investigation are synthetic and are composed of hydroxyapatite, TCP, or a combination of the two. Because of the wide difference in resorption rates and porosity between TCP and hydroxyapatite, a mixture of the two is clinically favorable. Most calcium phosphate ceramics are obtained by sintering calcium phosphate salts at high temperatures under the exclusion of water vapor to produce a powder that can then be molded into pellet form by high-pressure compaction. These biomaterials are commercially being produced as porous implants, nonporous dense implants, or granular particles with pores. Several injectable calcium phosphates using various crystal types are now available for restoring non-weight-bearing osseous defects and delivery of vulnerary molecules. This calcium phosphate ceramic forms in vivo and has a high carbonate substitution within the hydroxyapatite *(30)*. When injected into a bony cavity, a very firm ceramic mass forms within hours; most of its compressive strength is achieved within 24 h *(30)*. There is little control, however, over the porosity of this material. There is some demonstration that extraosseous forms can be resorbed. However, this ceramic remains stable for long periods of time because of its high density. This injectable ceramic has performed favorably in a recent clinical trial in metaphyseal fractures *(75)*.

The use of calcium phosphate bioabsorbable ceramics for applications requiring significant torsion bending or shear stress seem impracticable at present. These ceramics are brittle and have very little tensile strength. However, mechanical properties of porous calcium phosphate materials are comparable to cancellous bone once they have been incorporated and remodeled. A porous ceramic consisting solely of TCP is now available. It has a 36% porosity and contains a uniform distribution of large interconnecting pores ranging from 100 to 300 μm in size. The initial compressive strength has been shown to decrease after 4 mo *in situ* by 30–40%. Calcium phosphate ceramics must be shielded from loading forces until bony ingrowth has occurred. Rigid stabilization of surrounding and non-

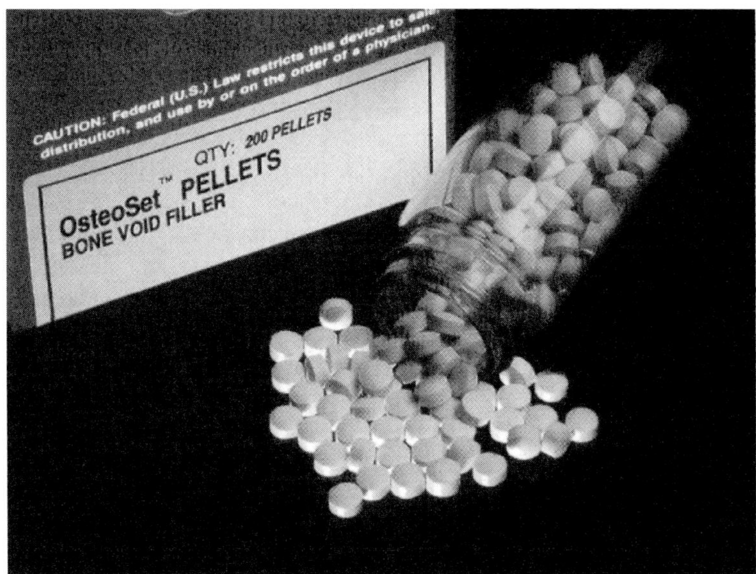

Fig. 1. Osteoset—calcium sulfate pellets.

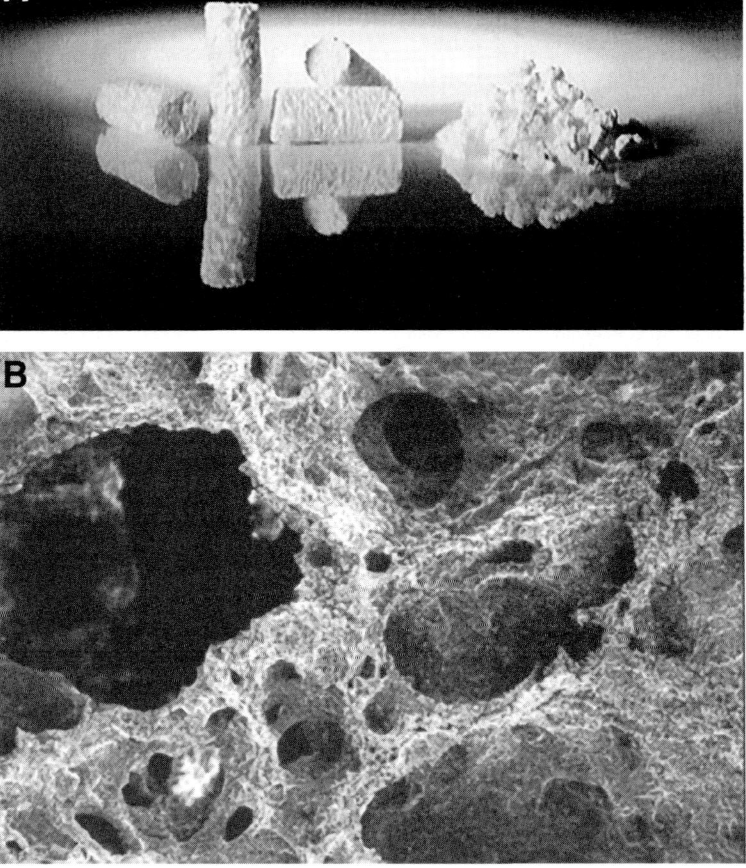

Fig. 2. (A) Vitoss—a β-tricalcium synthetic bone product. **(B)** SEM of Vitoss material.

weight bearing bone are required during this period because the ceramics can tolerate minimal bending and torque load before failing unless used in sites of relatively low mechanical stress or when forces are basically compressive. There appear to be no early adverse effects such as inflammation or foreign-body responses to these ceramics when they are in a structural block arrangement. However, small granules of material have been shown to elicit a foreign-body giant cell reaction. Radiographic findings demonstrate a continued presence of the calcium phosphate ceramic for a prolonged period of time as a result of the failure of complete remodeling. Persistent dense radiographic imagery creates difficulty in determining the degree of bony growth and incorporation into the implant. TCP, which is more biodegradable, loses more of its radiodensity and appears to be more fully incorporated into the bone. An advantageous property of calcium phosphate bioabsorbable ceramics, particularly hydroxyapatite, is that they bond to the host bone well. Because these ceramics do not contain osteoinductive properties intrinsically, the bond between host and graft provides sufficient affinity for local growth factors, which serve in the regeneration process.

Replamineform ceramics are a porous hydroxyapatite bioabsorbable ceramic derived from the calcium carbonate skeletal structure of sea coral (**Fig. 3**). A hydrothermal exchange method replaces the original calcium carbonate with a calcium phosphate replicate. In contrast to the random pore structure created in wholly synthetic porous materials, the pore structure of the coralline calcium phosphate implants is highly organized, similar to that of human cancellous bone. The porous size of this graft is determined by the genus of the coral used. *Gonipora* exhibits a microstructure similar to human cancellous bone. The hydroxyapatite ceramic derived from *Gonipora* has large pores measuring from 500 to 600 µm in diameter, with interconnections of 220–260 µm *(14)*. Hydroxyapatite derived from *Porites* has a microstructure similar to interstitial cortical bone, with a smaller pore diameter of 200–2500 µm parallel channels interconnected by 190-µm fenestrations, and a porosity of 66% *(14)*. Because these grafts lack intrinsic strength, they can be used only to fill defects up to 7–8 cm. Internal fixation is required so the material does not fail subject to cyclical loading. A recent study found that a coralline hydroxyapatite failed as a stand-alone graft in a rabbit model, and have suggested that it be used as a graft extender only. A major problem with coralline-derived hydroxyapatite is its delayed degradation. It appears only the surface of the ceramic is resorbed by osteoclasts, thereby leaving the majority of the microstructure intact. This setback may limit the coralline ceramics to anatomical regions in which bone remodeling is not critical. Interestingly, the *Gonipora* ceramic graft does not dictate the type of bone regeneration that will form *(14)*. If a medullar canal injury is grafted, trabecular bone will form. If a cortical defect is grafted, cortical bone will form. This renders the *Gonipora* hydroxyapatite graft quite versatile.

The bioactive ceramics also form a physical bond to the host bone. This phenomenon was first described in specific glasses. In general, they contain less than 60 mol% SiO_2, high NaO_2 and CaO content, and a CaO/P_2O_2 ratio similar to that found in native bone *(56)*. When exposed to an aqueous medium, these features make the glass surface highly reactive, resulting in rapid formation of hydroxyl carbonate apatite (HCA) crystals along its surface. Several combinations of different elements have been used to create different glasses. These studies have demonstrated that for a bond with host tissues to occur, a layer of HCA must form. This is true in the case of all graft materials. Changing the composition of the glasses may inhibit the formation of the HCA layer, but these changes most significantly alter the rate of HCA formation *(56)*. These materials do not reabsorb to give way to new bone growth; therefore, the value of the graft is as good as the bond between itself and the host bed. This bond can be enhanced with the addition of apatite or wollastonite crystals into the glass *(128)*. These glass-ceramic hybrids provide high mechanical strength, with bending and compressive strength values superceding native cortical bone. Bioglass without ceramic additives contains very little mechanical strength and is used to fill medium-sized bone defects. The glass–ceramic hybrids can be used to repair large defects but cannot be used in stress-bearing sites.

Bioinert ceramics do not react with living tissue and provide the highest mechanical strength of all graft material and great biocompatibility. They are often composed of metal oxides, such as alumina

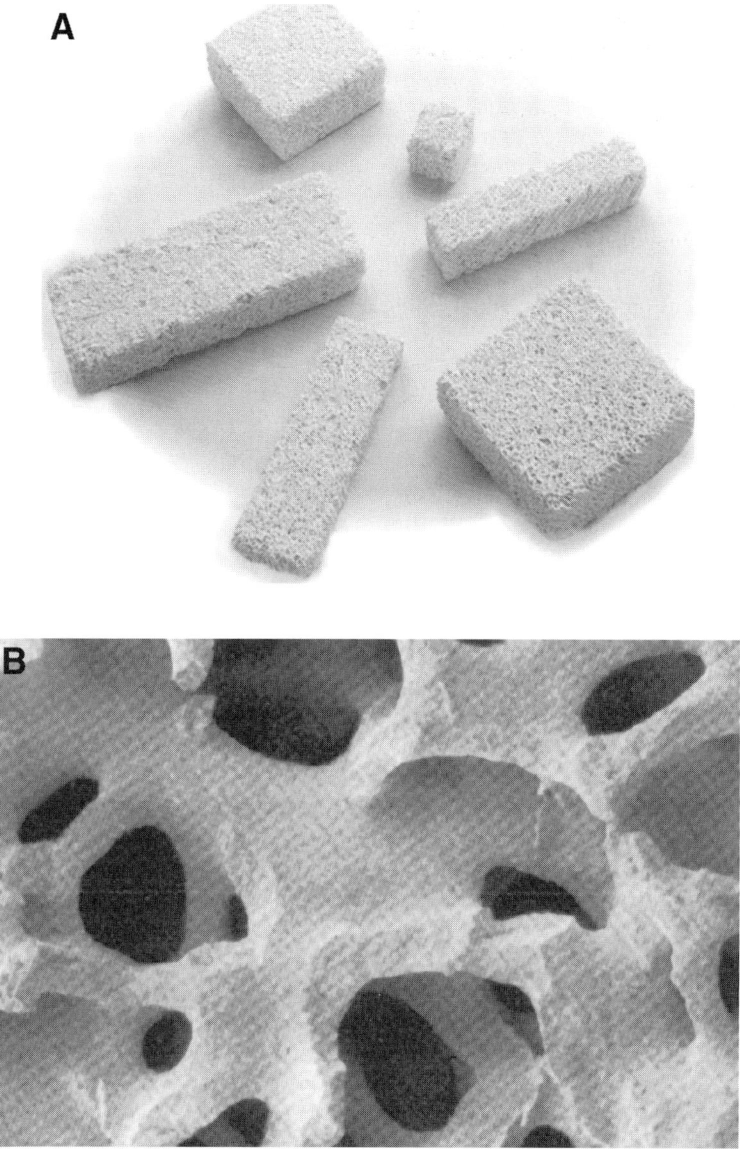

Fig. 3. (A) Interpore ProOsteon porous calcium phosphate bone graft material. **(B)** SEM of ProOsteon.

(Al_2O_3), zirconia (ZrO_2), and titania (TiO_2) *(128)*. They have been utilized predominantly for long bone defects because of their excellent compressive strength. Additionally, these ceramics are used in a few types of tumor and femoral head prostheses. Because they do not bond to bone, their application is limited to repairs that will not encounter sheering or torque forces.

Polymer cements, such as polymethylmethacrylate (PMMA), have been utilized for decades as a bone substitute to fill defects and in the reconstruction of complex fractures. PMMA initiates an osteolytic foreign-body giant cell reaction when fragmented, but is not reabsorbed and as a result new bone is not replaced at the site of the defect *(49)*. It has been speculated that polymers may even inhibit new bone growth and normal healing. Partially resorbable polymers are now being introduced, such as polyglycolide (PGA) and poly-L-lactide (PLLA), which allow for new bone growth and therefore do not degrade like PMMA *(100,105)*. The development of these polymers is advantageous. PMMA fixation to bone weakens over time as a result of degradation by foreign-body giant cell inflammatory response, which destabilizes prostheses and requires a further surgery. These novel polymers have been shaped into self-reinforcing screws, dowels, rods, and spacers, and have been utilized with some success in large bone fracture fixation. Polymers may also be important potential carriers for substances such as antibiotics and osteoinduction agents.

Demineralized bone matrix was developed as a solution to hospital power shortages that stifle cryopreservation techniques required for allogeneic bone grafts *(50)*. Demineralized bone matrix (DBM) is produced from acid extraction of bone, leaving noncollagenous proteins, bone growth factors, and collagen *(61a)*. Demineralized materials have no structural strength, but have enhanced osteoinductive capability afforded most notably by BMPs. DBM is currently prepared by bone banks as pathogen-free by virtue of donor selection and tissue processing. DBM has been utilized in clinical maneuvers to promote bone group regeneration, mainly in well-supported, stable skeletal defects. Despite the enhanced osteoinductive potential, the actual functionally accessible BMP within these demineralized grafts is exponentially lower than that used in recombinant BMP studies. The actual amount of BMP that is available from the various graft preparations has not been provided by the banks. At this time the US Food and Drug Administration requires sterilization of the DBM as prepared by bone banks, and this may in fact decrease some of the viability of the available BMP within the preparation. DBM does afford the potential of enhanced osteoinduction and to date has been used as an adjunctive to more traditional grafting materials.

DBM can also be processed from human bone by a patented technique that incorporates a permeation treatment that does not expose tissue to ethylene oxide or gamma-radiation, thus possibly protecting larger amounts of native BMP. It is processed into a gel consistency and packaged in a syringe from which it can be applied directly intraoperatively. It has no structural strength, and has been most successfully used in conjunction with internal fixation or as an adjunct of other grafting materials. Two additional forms of the gel formulation are available. One is in the form of a collagen mat retaining the noncollagenous proteins, and the other as a woven collagen mass that has the appearance of a putty.

DBM is currently available freeze-dried and processed from cortical/cancellous bone in the form of powder, crushed, chips, or as a gel (**Fig. 4**). When successful in achieving union, DBM develops bone of comparable mechanical strength of autograft. However, some commercially available DBM preparations have failed to induce bone in the Urist biological mouse muscle test. DBM is easy to mold intraoperatively, but it does not provide intrinsic strength. The clinical applications of DBM include augmentation of traditional autogenous bone grafts in repairing benign cysts, fractures, nonunions, and stable fusions.

Deproteinized bone mineral (Pyrost) represents the opposite extreme in bone tissue preparation. It is prepared from bovine bone that is put through a premaceration process and gentle combustion, followed by sintering to solidify the calcium phosphate crystal structure *(69)*. The result is a bony lattice that has maintained it original shape. The process produces a slight shrinkage of the spongy trabecular bone, but 70% porosity is maintained. Pyrost is made commercially and is available in rods of variable length and 5 × 5 mm cross sections. This process ensures zero antigenicity, minimizing the host immune response. Unlike DBM, Pyrost contains no intrinsic osteoinductive capability, as it is a simple scaffold. It is therefore favorable to inoculate Pyrost with some type of osteogenic tissue, such as bone marrow. Radiographic data indicate that Pyrost is resorbed slowly, much like synthetic hydroxyapatite, and remains in the bed up to and beyond 5 yr postimplant *(69)*. Like most osteoconductive grafts, Pyrost contains insufficient mechanical strength to be utilized in any defect placed under physiological stress. For this reason, Pyrost is generally used to fill small, metaphyseal defects, and is particularly useful in spongy bone beds. In addition, care must be taken to ensure that Pyrost does not extend out of the defect, allowing extraosseous mineralization. Pyrost is contraindicated for infected

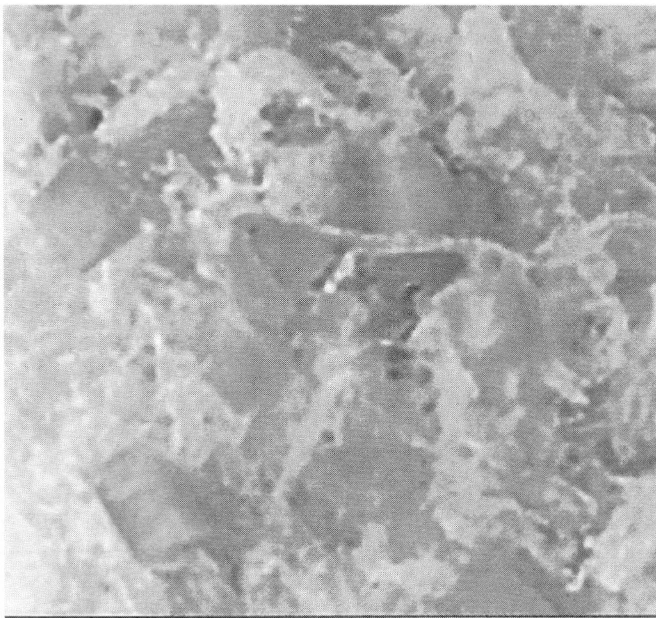

Fig. 4. Grafton Crunch demineralized bone matrix.

host beds and large resected defects *(69)*. Though it is not mandatory, covering the defect with periosteium had produced the best regenerative results and successfully contains the graft.

OSTEOGENIC CELLS

The principal downfall of purely osteoconductive bone graft materials is their inability to actually stimulate new bone growth. Alhough various preparations of demineralized bone matrix do contain variable quantities of osteoinductive growth factor, as a whole the above-mentioned graft materials solely provide a favorable meshwork for bone to grow. While many studies have shown that these biosynthetic materials speed recovery, they are not doing so by intrinsically producing or inducing bone growth. Implanting osteoblastic cells directly into a defect certainly will stimulate bone growth precisely where it is required.

The autologous cancellous graft is considered the gold standard of bone transplantation partially because bone marrow is taken up additionally and is grafted along with the osteoconductive structure. Within the marrow lies osteoprogenitor cells that differentiate into bone-forming cells upon stimulation by local growth factors. Osteoclasts are formed by monocytes of graft or host origin. Together, osteoblasts and osteoclasts produce new bone and remodel the graft and surrounding tissue. This process of creeping substitution continues until the autograft is indistinguishable from the bed. Had the original graft been denuded of ostcoprogenitor cells before grafting, it follows that the healing process would be delayed.

The osteogenic capabilities of bone marrow have been well known since the original observation in 1869 by Boujon as reported by Burwell *(21,23)*. That potential prompted surgeons to begin using it as a bone graft material 50 yr later, and then bone marrow became widely used in the 1950s. At that time the precise origin of the osteogenic cells came under scrutiny. Osteogenic cells were believed not to be transplanted with the graft. Many thought that most of the cellular contents of the autograft were removed or degenerated within days after implant. That observation led to the belief that the osteogenic cells responsible for generating the new bone were those cells within and surrounding the defect. Recent studies have found that osteoprogenitor cells are contained within both the graft and defect, as well as in the systemic circulation.

The exact origin of osteoblastic cells has yet to be elucidated. Marrow stromal fibroblasts (MSF) differentiate into functional osteoblasts *in vivo*, capable of forming bone. The MSF originates from the colony forming unit-fibroblast (CFU-F), located in the hematopoietic tissue *(77)*. However, the proportion of CFU-Fs within the marrow that demonstrate osteogenic potential is still unclear. Kuznetsov et al. have demonstrated that approx 60% of single-strain MSFs (i.e., originating from a single CFU-F) contain osteogenic potential when compared to multistrain MSF implants *(77)*. Culturing the CFU-Fs in dexamethasone and ascorbic acid enhanced the incidence of colonies with osteogenic capabilities and the extent of bone formation derived from those colonies. Why the remaining 40% of MSFs did not possess osteogenic capability, whether or not they can be induced to differentiate into osteoblasts, and the normal variance of MSF differentiation within the human population needs to be addressed.

At some specific point in their differentiation, the osteoprogenitor MSF cell can follow two routes to become bone-producing osteoblasts, determined osteogenic precursor cell (DOPC) and the inducible osteogenic precursor cell (IOPC) *(2,5,21–23,40,98,140)*. This distinction is based on eventual location and further required activation of the osteogenic cell. The DOPC is found only in marrow stroma and lining the bone cavities *(40,99)*. These cells do not require any further stimulation, as they are fully differentiated and will produce nothing but bone. They do contain stem-cell characteristics in that they are capable of self-replication and production of other DOPCs *(40,99)*. These cells are most likely responsible for regeneration of the medullary canal following injury. The IOPC is found circulating in blood and within soft tissue and connective tissue networks *(98,127)*. Although its original source is still unknown, the IOPC will not produce bone without the proper inducing agent, which is probably one of the BMPs *(98,127)*. This theory suggests that bone marrow does provide osteogenic cells when transplanted into an osseous defect, although local DOPCs and circulating IOPCs contribute as well.

Today, exactly how many osteoprogenitor cells are contained in bone marrow is unknown. This becomes important when evaluating what quantity of bone marrow is necessary to repair a defect. Fortunately, unlike the autologous bone graft, there is normally no significant donor-site morbidity or excessive blood loss with the extraction of marrow. Recent studies indicate that bone marrow contains osteoprogenitor cells of the order of 1 per 50,000 nucleated cells in children to 1 per 2,000,000 nucleated cells in the elderly *(87,88)*. This minimal quantity can be concentrated up to fivefold with new centrifugation techniques. However, the concentration of osteoprogenitor cells taken during aspiration can also vary greatly, among different patients and among different aspirations from the same patient. Clinically, Connolly et al. have demonstrated that percutaneous injections of 100–500 mL of bone marrow can successfully treat nonunions *(28,29,131)*. The marrow was injected directly into the fracture site, then set with casting or other immobilization techniques including intramedullary nailing. This study utilized 3- and 5-mL aspirates taken from the posterior iliac crest that yielded a mean of approx 3×10^9 nucleated cells. This value was obtained via a cell counter. However, a cell counter cannot specifically count osteoprogenitor cells, so osteoprogenitor concentration cannot be accurately measured using this method. Because the goal of bone marrow aspiration for bone graft is to obtain the highest concentration of osteoprogenitor cells, and not just nucleated cells (as is the case in bone marrow transplantation), it is necessary to be able to measure concentration of osteoprogenitor cells. This concentration can be determined by the prevalence of colonies displaying alkaline phosphate activity after plating prepared marrow on alpha minimum essential medium. This assay is based on the assumption that osteoprogenitor cells exhibit alkaline phosphate activity. This assay is not capable of demonstrating the ability of each individual colony to eventually differentiate into osteoblast cells, thereby leaving the possibility of displaying false-positive colonies that may exhibit alkaline phosphate activity without later osteoblastic differentiation. However, this test is accepted as one that can roughly determine the number of osteoprogenitor cells in a marrow aspirate. It is important to note that though the mean number of osteoprogenitor cells increases as the aspirate volume increases, the concentration of osteoprogenitor cells has been shown to decrease as displayed by alkaline phosphate activity. This is probably due to dilution with blood. After alkaline phosphate activity has been measured, one group has suggested that smaller aliquots be taken from different sites to enhance the concentration of osteoprogenitor cells *(89)*. Muschler et al. found that the osteoprogenitor cell count increased as the aliquot size increased, but that the concentration actually went down as the aliquot size increased *(89)*. They suggest that four 1-mL aspirates provide twice the concentration of osteoprogenitor cells as one 4-mL aspirate, and further recommend that the volume of aspirate taken from any one site not exceed 2 mL. Connolly et al. suggests taking aliquots of 2.5 mL per site, and, more important, that aspirates of this size were utilized successfully in a clinical setting *(28,29)*. More testing on the variability of quantity and function of osteoprogenitor cells as a function of age, sex, disease status, and pharmacological status is necessary.

Bone marrow by itself provides insignificant osteoconductive capabilities, and therefore is generally used in conjunction with a stabilizing agent. Multiple studies have found marrow to enhance fracture healing when transplanted with DBM, coralline hydroxyapatite and TCP, and other inorganic grafts, as well as autologous and allogeneic grafts *(21,23,87,88,117,118)*. Following bone marrow implantation, woven bone occurred initially, then progressed to early lamellar bone, and subsequently molded in a volumetric fashion. When placed in a fresh femoral defect and given in sufficient amounts, bone marrow produced a rate of union comparable to that of autogenous bone graft. Connolly et al. found that the bone formed following the percutaneous marrow graft demonstrated comparable biomechanical properties to that of a cancellous bone graft. These studies as well as those by other investigators have indicated that bone marrow can lead to structurally competent bone regeneration in an orthotopic location.

Cell culture techniques are now being used to increase the number and concentration of osteoprogenitor cells in a bone marrow graft *(13)*. Advances in cell culture technology afford researchers the opportunity to produce large quantities of nearly any cell found within the osteoblast lineage. Cells are harvested and plated on minimum essential medium that can be treated with a multitude of osteogenic factors such as various BMPs or 1,25-dihydroxyvitamin D3, or nonspecific cell culture growth factors such as dexamethasone and ascorbic acid *(78,89,91)*. Several markers are used to confirm the presence of osteogenic capability of the grown cell line. Cell counters and alkaline phosphate activity can quantitate osteoblasts grossly, while immunostaining for osteocalcin, osteonectin, and type I collagen can delineate specific osteogenic capability. Hybridization techniques for the mRNA species of the above-mentioned osteoblast-derived products have also been described to further define osteogenic capability. The primary advantage of cell culture is obviously the ability to create a higher quantity and concentration of osteogenic cells for use in bone grafting. Unfortunately, the process requires at least 4 wk for the cells to reach confluence in vitro. For that reason, autogenous intersurgical marrow aspiration and culture is not possible. Cell banking may be possible, but this method carries with it the complications found with allograft banking. ABO and HLA matching would be required, and all the factors that decrease efficacy of allografting would similarly apply.

OSTEOINDUCTIVE GROWTH FACTORS

Fracture healing, like all tissue healing processes that result from injury and acute inflammation, is orchestrated by chemical mediators. The process is both complex and redundant, with several protein factors inducing and inhibiting cell differentiation and new bone growth. Nearly all of the known mediators are beckoned from multiple locations and have multiple functions. Many are released from their respective storage vesicles immediately following injury and employed throughout the healing process, while the transcription and translation of other mediators is not initiated until after injury has occurred. All mediators utilize membrane receptors and act through cell signaling pathways that induce either cell proliferation and differentiation in the case of osteoprogenitor CFU-Fs and MSFs, or induce upregulation of expression of bone components such as collagen from differentiated osteoblasts. These local mediators, along with the microenvironment, also influence genes coding for the types of matrix that the repair cells will form.

Since the original discovery by Urist et al. in the mid-1960s, that bone fragments implanted subcutaneously or intramuscularly induced bone formation the goal to identify and understand these chemical mediators led to the discovery of numerous osteoinductive growth factors *(135.136)*. The ultimate goal of bone transplantation is to both speed and enhance the recovery from bone injury. With the advent of recombinant DNA technology that allows for the production of infinite supplies of any protein, the recently discovered osteoinductive growth factors may prove to be a large component of the solution to this goal.

The chemical mediators identified to date are (1) members of the transforming growth factor-beta (TGF-β) superfamily that include bone morphogenetic proteins (excluding BMP-1) and the BMP subclass of growth/differentiating factors 1–10, inhibins, activins, Vg-related genes, nodal-related genes, and glial-derived neurotropic factor; (2) acidic and basic fibroblast growth factors (FGF); (3) platelet-derived growth factor (PDGF); and (4) insulin-type growth factors (IGFs). The role of these factors in fracture repair without intervention has gradually been elucidated. Upon fracture, damaged cells and local macrophages release interleukin-1 and tumor necrosis factor, which produce localized vasodilation and endothelial expression of polymorphonuclear leukocyte (PMNs) adhesion molecules P-selectin and E-selectin. bFGF is released from activated macrophages, stimulating the expression of plasminogen activator and procollagenase *(26)*. PMNs adhere via selectin binding and enter the interstitial space after binding intercellular adhesion molecules. This increase in vascular permeability allows for the extravasation of platelets, red blood cells (RBCs), and coagulation cascade components along with the PMNs. The PMNs are chemotactically attracted (via N-formyl peptides, C5a, leukotriene B4, etc.) to the site of injury and proceed to endocytose microdebris and microorganisms with the local activated macrophages. The sanguineous exudate will proceed to form the hematoma surrounding the fracture under the control of PDGF, TGF-β, and bFGF *(58)*. As in the case of granulation tissue, the defect is initially characterized as hypoxic, thus stimulating macrophages to produce and release IGFs, bFGF, and chemotactic mediators to fortify cells, promote proliferation of more fibroblasts, and promote angiogenesis. As the acute inflammation progresses into a blastema, fewer PMNs and more monocytes are found at the site of injury, and macrophages develop into epitheloid cells, which may coalesce into multinucleated giant cells. Fibroblasts now begin to produce several collagen isotypes that possess the ability to selectively bind, and therefore localize, osteogenic growth factors *(112)*. This key step may actually strategically position TGF-β, BMPs, and bFGF to optimize osteogenic cell interaction directly where new bone formation is required. IOPCs are extravasated and can now be activated via BMPs to differentiate into osteoblasts *(98,127)*. DOPCs located in the remaining medullary canal and within the periosteum will also become active via BMPs, and proliferate into osteoblasts that begin laying down new bone *(40,99)*. Fibronectin and extracellular degradation products facilitate the conversion of monocytes to osteoclasts, which begin readsorbing the newly formed bone *(58)*. The process of laying down new bone and remodeling continues until the regenerated bone is indistinguishable from the tissue originally damaged.

Bone Morphogenetic Proteins

BMPs are becoming widely known as the key player in this process. Because of their role in osteoblast differentiation, growth, and activation, a multitude of preclinical research has focused on the relevance of recombinant versions of these growth factors. This drive is further fueled by the recent evidence that abnormalities in the genes of BMPs and/or their respective receptors is linked directly to bone regeneration failure *(116,123,141)*. These conclusions indicate that BMPs and their receptors are required for normal bone regeneration. Investigation initially tested BMP extracts on rodents, but has grown to utilize recombinant forms with a variety of carriers to treat segmental long bone defects in larger animals and now humans *(11a,19a,31–34,45a,47a,48,52a,52b,69b,81a,134,137a, 144)*. Additionally, BMPs are being used in marrow cell culture to induce osteoblast differentiation and proliferation *(78)*.

The BMPs were discovered and described by cDNA library searches (BMPs 1–9) and low-stringency hybridization and consensus polymerase chain reaction (BMPs 10–13) *(123)*. As noted above, BMP-1 is not a member of the TGF-β superfamily. It is a proteinase and a member of the tolloid-like protein class *(123)*. The remaining BMPs are related to the TGF-β superfamily through primary amino acid sequence homology. Their tertiary structure reveals a mature domain that undergoes cleavage and is glycosylated intracellularly, and finally expressed in its active dimer form. Variance in protein assembly (homo- vs heterodimeric assembly) and glycosylation influence the activity and effects of BMPs *(123)*. Active BMPs bind types I and II serine/threonine kinase receptors on osteoprogenitor cell membranes, with highest binding affinity achieved with type II receptor through a heterotetrameric complex *(123)*. Recent studies have pointed to BMP-2, -4, and -7 as the primary BMPs involved in the formation of this complex *(123)*. Upon formation of this complex, type II BMP receptors activate type I receptors, which are solely able to initiate an intracellular signaling reaction. This complex signaling cascade results in activated proteins, known as SMADs, which bind to DNA, leading to transcriptional activation of the osteoblast-specific factor-2/core-binding factor-1 gene, either directly or indirectly *(123)*. The resulting protein is a potent transcription activator of osteocalcin, which is the defining molecular marker of differentiated osteoblasts *(123)*. Osteoblast-specific factor-2 can also bind the promoters of α-1 collagen, bone sialoprotein, and osteopontin, thereby inducing transcription of these genes. Though further research is required to more fully understand the role of osteoblast-specific factor-2 in fracture healing, it is clearly important as a promoter of osteoblast differentiation and new bone formation, and its generation has been related directly to BMP stimulation.

Clinically, BMPs extracted from cadavers have been used successfully under the guidance of Drs. Urist, Johnson, and Dawson for the treatment of established nonunions and spine fusions *(64–66)*. In their 70 patients, to date they do not report any instances of tumor genesis of any untoward events. Urist's clinical BMP preparation represents a mixture of BMPs, although it is highly concentrated (300,000X). It has resulted in over a 93% success rate in failed nonunions, and 100% success rate in spine fusions to date. Urist's product, however, is not recombinant BMP, and in addition contains a number of different BMPs, as well as osteocalcin.

Studies reported to date with rhBMPs have been related largely to animals, although clinical trials are currently underway in the United States and Europe. Kirker-Head and coworkers evaluated the long-term healing of bone using rhBMP-2 in adult sheep *(72)*. By 12 mo all the defects were structurally intact and were rigidly healed. Both woven and lamellar bone bridged the defect site and apparently the normal sequence of ossification, modeling, and remodeling events had occurred. These reports using rhBMP-2 confirm prior studies by Toriumi and coworkers *(134)*, Gerhardt and coworkers *(48)*, Yasko and coworkers *(144)*, that BMP-2 can indeed heal skeletal defects in a wide range of animals. Cook and his coworkers have utilized BMP-7 to heal large segmental defects in rabbits, dogs, and primates *(31–33)*. In their latter studies, five of six ulna and four of five tibia treated with BMP-7 in African green monkeys exhibited complete healing in 6–8 wk, with bridging of the defect and new bone formation in 4 wk. Two unhealed defects both exhibited new bone formation. In their studies, all the tibia defects and all the ulnar defects that had been treated with autogenous bone graft developed fibrous union with little new bone formation. Thus, these studies demonstrated the efficacy of rhBMP in a nonhuman primate.

In a similar fashion, attention has been directed toward utilizing rhBMPs in the treatment of spine fusions. Spine fusions play a very important role in the treatment of a number of pathological conditions, including spine trauma, congenital anomalies, degenerative diseases, and tumors. It is estimated that more than 180,000 spine fusions were performed in the United States in 1983. Boden, Sandhu, and Muschler have developed spine models that can be utilized to test recombinant factors *(10,11,90,119,*

121,122). Sandhu performed a number of studies utilizing BMP-2 in a canine spine intraspinous process model *(119).* He found that the critical element represented the dose of BMP and it was unrelated to whether or not there was decortication of the fusion model. Sandhu further showed that rhBMP-2 with polylactic acid carrier is clearly superior at both higher and lower doses to autogenous iliac crest bone for inducing transverse process arthrodesis in a canine. BMP-2 doses ranging from 57 µg to 2.3 mg resulted in 100% clinical fusion and 85% radiologically fusion by 3 mo, compared to no autogenous fusions at that time point. Radiologically delayed union occurred with lower doses, although differences were not significant. High-dose fusions were mechanically stiffer than low-dose fusions in the axial plane, and all BMP dosed fusions were stiffer than autograft fusions in all planes. These studies clearly have demonstrated the efficacy of higher doses of rhBMP-2 for inducing posteriolateral lumbar fusion in a canine model. Muschler and coworkers also studied spine fusions in dogs, and they confirm that the use of rhBMP-2 in their experimental models demonstrated 100% fusion rate for a single-level lumbar arthrodesis without adverse neurological or systemic sequelae, and felt that this confirmed the use of this agent for spine fusions *(90).* Boden and coworkers, in a similar set of experiments performed on rabbits, further confirmed the use of rhBMP-2 as an excellent agent for spine fusion *(10,11).* Cook tested the effect of rhBMP-7 or osteogenic protein-1 in mongrel dogs in the spine *(31).* The agent was an effective bone graft substitute for achieving stable posterior spine fusion in a significantly more rapid fashion than achieved by autogenous graft. There now appears to be adequate evidence that both rhBMP-2 and -7 used in pharmacological doses in various animal models are quite efficacious and superior to autogenous grafts in achieving spine fusion.

Fibroblast Growth Factors

Fibroblast growth factors (FGFs) compose another family of growth factors evolved in wound healing and repair. Specifically, they act on both stem and differentiated cells, causing changes in migration, proliferation, further differentiation, morphology, and function *(20,27,36,47,70,113,125,142).* There are nine known members of this protein family, all of which share 30–50% sequence homology. FGF-1 and FGF-2, also termed acidic FGF (aFGF) and basic FGF (bFGF), respectively (named for their differing isoelectric points), are the prototypic and most studied members of the family. The remaining members are composed of selective growth factors and oncogene products. aFGF and bFGF have been isolated from several sources, including neural, pituitary, adrenal cortex, and placental tissues, and directly stimulate the proliferation of fibroblasts, endothelial cells, astrocytes, and many others in addition to osteoprogenitor cells *(20,43,47,51,52,113).* Four distinct FGF receptors have been identified. They all exhibit dimerization and tyrosine kinase activity upon binding their respective ligand *(62,101).* The dimerized receptor/ligand complexes are clustered together and endocytosed, where individual phosphotyrosine residues act as highly selective binding sites for specified intracellular proteins that go on to modulate DNA transcription. FGFs have a high affinity for heparin and similar polyanions. In fact, most extracellular FGF is bound to a heparin sulfate proteoglycan that is thought to protect the FGF, and possibly help store FGF in a mobile reservoir *(47).* Additionally, only heparin-bound bFGF will bind its receptor, indicating the importance of heparin-type molecules in normal FGF function *(97).*

Early studies provided mixed results on the advantage of FGFs. Human recombinant forms of FGF applied exogenously to fractures demonstrated increased differentiation among mesenchymal stem cells into chondroprogenitor cells, increased callus volume and density as well as osteoclast number, and decreased mRNA expression for type II collagen *(8,63,92,93).* Recently there has been renewed interest at the clinical relevance of aFGF and bFGF, mostly centered on identifying proper carriers that may enhance the osteogenic properties of FGFs. Radomsky et al. have experimented extensively with hyaluronate gel as a potential carrier *(107,108).* This viscous gel formulation is used in direct percutaneous injection into fresh fractures, effectively releasing FGF in a slow-release manner. These studies demonstrate increased callus size, periosteal reaction, vascularity, and cellularity in fractures treated with the FGF–hyaluronate gel implant when compared to untreated controls. Single local injections and continuous slow infusions of FGF have also demonstrated increased cytokine-induced bone formation and accelerated fracture recovery when compared to nontreated controls *(39,68,71)*.

FGF has also successfully been used within a DBM preparation, a composite that allows for easy dosing studies. Interestingly, low doses (15 ng per implant) of bFGF have been found to induce the number of chondrocytes and bone formation, whereas high doses (1900 ng per implant) greatly inhibit cartilage and bone formation *(60)*. It is known that FGF can stimulate the differentiation of osteoclasts from bone marrow stem cells through FGF-stimulated expression of prostaglandins *(60)*. However, it is not known if high doses of FGF inhibit collagen and bone synthesis through increased prostaglandin expression.

Zellin et al. have tested the effects of FGF-treated autoclaved, autogenous bone reimplantation with success *(146)*. Revitalizing bone that has been processed to inhibit disease transmission and decease antigenicity would be a major advancement in the field of allogeneous bone transplantation. The concept of reimplantation is particularly attractive because it maintains original anatomy and provides favorable immediate clinical results. Zellin et al. used a rabbit model to compared the revitalization capacity of human recombinant FGF and autogenous bone marrow. They found the FGF-treated implants maintained no difference when compared to pure autogenous bone grafts upon radiographical, histological, and histomorphometric analysis. The reimplants treated with bone marrow alone were significantly less revitalized and healed more slowly.

Transforming Growth Factor-β

TGF-β is perhaps the most extensively studied growth factor in the field of bone biology. Data to support the concept that exogenous TGF-β can stimulate bone repair is substantial, but the overall osteoinductive capacity of TGF-β is weak compared to the bone morphogenetic proteins *(25,114,115)*. TGF-β may potentiate the osteoinductive activities of the BMPs, and this may be an important role for this factor *(24)*. Thus an understanding of the biological effects of TGF-β is fundamental to the field of osteoinduction. Generally, TGF-β provides a link between inflammation and healing. It is a key chemotactic mediator for fibroblasts and macrophages, blocks plasminogen inhibitor, enhances angiogenesis, stimulates collagen and other extracellular matrix component synthesis, and inhibits procollagenase and epithelial expansion. Most important, TGF-β stimulates mesenchymal cell growth and differentiation.

Three isoforms of TGF-β exist in mammalian species, although it appears all three act on similar receptors. The most concentrated isoform is TGF-β type 1, which is stored in platelet α-granules. This is a convenient location considering TGF-β type-1's critical role in soft tissue and fracture repair. Much like the BMPs, heterotetrameric complexes form between v and its serine/theronine kinase receptors (types I and II) *(46)*. This results in the transphosphorylation of a second receptor, which initiates a Smad intracellular signaling cascade. Accordingly, v plays a key role in the induction of osteoblast differentiation and activation. It appears that this action can be enhanced with genetically engineered recombinant forms of TGF-β that contain additional collagen-binding domains. Andrades et al. engineered TGF-β 1-F2, a fusion protein with the additional collagen-binding domain, which has demonstrated an increased ability over v alone to support colonies of osteoprogenitor stem cells *(86)*. In this study, the resultant cells were implanted in conjunction with inactivated DBM, and demonstrated the ability to form new bone in a rat model.

However, studies have not proved TGF-β as favorable in the treatment of fracture healing. Several studies have separately demonstrated that there was a stimulatory effect of TGF-β on bone formation following injections of TGF-β into fetal rat and mice calveria *(12,19,61,137)*. TGF-β has also been found to stimulate the recruitment and proliferation of osteoblasts in rabbit skull defects *(67)*. Additionally, many long bone defect studies have demonstrated augmentation of bone healing with the local application of TGF-β *(3,4,55,104)*. These studies all utilized critical sized defects, and demonstrated increase in both bone formation and biomechanical strength. The effects of TGF-β specifically on fracture healing have also been addressed. These results showed that TGF-β increased bone formation and strength only when injected locally around the fracture line as opposed to continuous subcutaneous administration via osmotic pump. Furthermore, a dose-dependent increase in the cross-sectional area of the callus and bone at the fracture line was noted in both subcutaneous and local injections scenarios, although only the local injection model demonstrated an increase in ultimate load strength *(81,95)*.

Critchlow and coworkers evaluated the effect of exogenous TGF-β in a rabbit fracture model *(35)*. The investigators injected the TGF-β into the developing callus of the rat tibial fractures healing under stable or unstable mechanical conditions 4 d after fracture. The fractures were examined for 4–14 d after fracture. A large amount of edema developed around the injection site. The fractures healing under stable mechanical conditions consisted almost entirely of bone. The effects of 16-μg injections of TGF-β were minimal, but the 600-μg dose led to a small increase in the size of the callus. Callus fractures healing under unstable mechanical conditions had a large area of cartilage over the fracture site, with bone on each side. The effects of TGF-β on unstable fractures were to retard and reduce bone and cartilage formation in the callus. The overall size of the callus was not affected. In conclusion, it was felt by these authors that TGF-β does not stimulate fracture healing under either stable or unstable mechanical conditions during the initial healing phase. It was further argued that agents that stimulate callus proliferation may retard bone remodeling.

Comparing the results of these studies is difficult because of the wide differences in experimental design and execution. One conclusion that continuously reappears is that TGF-β does indeed positively augment fracture healing by increasing callus size. However, this finding is generally found in conjunction with a weaker callus. Therefore, unless a more efficacious method of delivery of TGF-β is found, it may not play a significant clinical role in augmenting fracture healing.

Platelet-Derived Growth Factor

Platelet-derived growth factor (PDGF) has been shown to have stimulatory effects on bone and cartilage cells proliferation. However, its role in fracture healing is not clearly understood. PDGF is the first growth factor to emerge at the site of injury. PDGF is expressed as a dimer formed from two homologous chains (A and B), which readily give rise to three isoforms *(59)*. The active receptors of PDGF are composed of two separate subunits (α and β) that also give rise to three isoforms *(124)*. These receptors are similar to the family of growth factor receptors in that they display intrinsic tyrosine kinase activity. PDGF is released from endothelial cells, platelets, monocytes, and activated macrophages. After PDGF binds the receptor, the receptor dimerizes and undergoes autophosphorylation, which initiates both a cellular signaling cascade and an arachidonic acid degradation cascade that results in increased intracellular calcium concentrations. Both cascades modulate DNA transcription, which eventually leads to cell growth, actin reorganization migration, and differentiation. PDGF possesses potent mitogenesis capabilities (proliferation of cell populations involved in healing), macrophage activation, and angiogenesis *(44,143)*.

Andrew and coworkers evaluated the effect of PDGFs in normally healing human fracture *(1)*. Biopsy material from 16 normally healed fractures at various times after injury were obtained and evaluated for platelet-derived growth factor by chemistry and *in situ* hybridization. PDGF α-chain was found to be expressed by many cell types over a prolonged period during fracture healing. These cells include the endothelial and mesenchymal cells, and granulation tissue in the osteoblast, chondroblast sites, osteoclasts later during fracture healing. In contrast, platelet-derived growth factor-β change to gene expression was more restricted, being directed principally in osteoclasts at the stage of bone formation. Since platelet-derived growth factor was detected using immunochemistry in various cell types during the fracture repair, the authors considered it to play an important role in the regulation of this process. Another study demonstrated radiographically apparent increases in tibial callus density and volume when the osteotomy was injected with a collagen carrier containing PDGF over collagen carrier alone *(94)*. However, once again, increased callus size did not correlate to increased strength when compared to nonfractured contralateral control limbs, as was the case in the TGF-β trials. Histologically it was observed that the PDGF-treated tibiae displayed a more advanced state of osteogenic differentiation in both the endosteum and periosteum compared to control osteotomies. For this reason, PDGF was considered by the authors to contain stimulatory effects on fracture healing.

PDGF has also been implicated in osteoclast activation. The β-receptor was recently detected on the plasma membrane of osteoclasts through immunohistochemistry techniques. Zhang et al. have demonstrated that PDGF promotes adult osteoclast bone resorption directly through the PDGF-β receptor *(147)*. A dose-dependent relationship between volume of Howship's lacuna and quantity of resorption pits was established, in addition to increased activity of both acid phosphatase and tartrate-resistant acid phosphatase, two markers of bone resorption. Moreover, PDGF influences expression of interleukin-6, a key cytokine that induces osteoclast recruitment. This role is mediated through the activation of transcription factors through interaction between PDGF and the PDGF-β receptors located on osteoblasts.

Because of PDGF's increasingly important role in bone regeneration and remodeling, a group has begun experimenting with platelet-rich plasma (PRP) and platelet-poor plasma (PPP), both rich in PDGF, TGF-α, and TGF-β, as possible bone graft materials *(85)*. Aliquots of whole blood can be centrifuged a low speeds to separate plasma from RBCs, thereby producing a PRP with a three to fourfold increase in platelet concentration and an increase in growth-factor concentration. Synthetic graft materials treated with PRP and PPP extracts demonstrated a greater trabecular bone density than did bone control grafts without the extracts through the first 6 mo.

CONCLUSION

While the development of the recombinant growth factors to a viable clinical product has been frustratingly slow, these growth factors remain a potentially powerful way of stimulating bone growth and remain essential in the development of bone graft substitutes. While these recombinant growth factors are now emerging on the market, they remain quite expensive. Alternatives such as demineralized bone matrix, allografts, and osteoconductive substitutes are less expensive and thus should remain as an essential part of the clinician's armamentarium until the next generation of substitutes becomes available.

REFERENCES

1. Andrew, J. G., Hoyland, J. A., Freemont, A. J., and Marsh, D. R. (1995) Platelet-derived growth factor expressed in normally healing human fractures. *Bone* **16**, 455–460.
2. Ashton, B. A., Allen, T. D., Howlett, C. R., et al. (1980) Formation of bone and cartilage by marrow stromal cells in diffusion chambers *in vivo*. *Clin. Orthop. Rel. Res.* **151**, 294.
3. Beck, L. S., DeGuzman, L., and Lee, W. P. (1995) Transforming growth factor-beta 1 bound to tricalcium phosphate persist at segmental radial defects and induces bone formation. *Trans. Orthop. Res. Soc.* **20**, 593.
4. Beck, L. S., DeGuzman, L., and Lee, W. P. (1996) Bone marrow augments the activity of transforming growth factor-beta 1 in critical sized defects. *Trans. Orthop. Res. Soc.* **21**, 626.
5. Beresford, J. N. (1989) Osteogenic stem cells and the stromal system of bone and marrow. *Clin. Orthop. Rel. Res.* **240**, 270.
6. Berrey, B. H., Lord, C. F., Gebhardt, M. C., and Mankin, H. J. (1990) Fractures in allografts. *J. Bone Joint Surg.* **72A**, 825–833.
7. Berrey, B. H., Lord, C. F., Gebhardt, M. C., and Mankin, H. J. (1990) Fractures of allografts. Frequency, treatment, and end-results. *J. Bone Joint Surg.* **72**, 825–833.
8. Bland, Y. S., Critchlow, M. A., and Ashhurst, D. E. (1995) Exogenous fibroblast growth factors-1 and -2 do not accelerate fracture healing in the rabbit. *Acta Orthop. Scand.* **66**, 543.

9. Blitch, E. and Ricotta, P. (1996) Introduction to bone grafting. *J. Foot Ankle Surg.* **35,** 458–462.

10. Boden, S. D., Schimandle, J. H., and Hutton, W. C. (1995) Volvo award in basic science. the use of an osteoinductive growth factor for lumbar spinal fusion. Part I: Biology of spinal fusion. *Spine* **20,** 2626–2632.

11. Boden, S. D., Schimandle, J. H., and Hutton, W. C. (1995) Volvo award in basic science. the use of an osteoinductive growth factor for lumbar spinal fusion. Part II: Study of dose, carrier, and species. *Spine* **20,** 2633–2644.

11a. Boden, S. D. (2001) Clinical application of the BMPs. *J. Bone Joint Surg.* **83A(Suppl 1, Pt 2),** S161.

11b. Boden, S. D., Zdeblick, T. A., Sandhu, H. S., and Heim, S. E. (2000) The use of rhBMP-2 in interbody fusion cages. Definitive evidence of osteoinduction in humans: a preliminary report. *Spine* **25(3),** 376–381.

12. Bonucci, E. (1981) New knowledge of the origin, function and fate of osteoclasts. *Clin. Orthop. Rel. Res.* **158,** 252–269.

13. Bruder, S. P., Fink, D. J., and Caplan, A. I. (1994) Mesenchymal stem cells in bone development, bone repair, and skeletal regeneration therapy. *J. Cell Biol.* **56,** 283–294.

14. Bucholz, R. W. (1994) Development and clinical use of coral-derived hydroxyapatite bone graft substitutes, in *Bone Grafts, Derivatives, and Substitutes* (Urist, M. R., O'Connor, B. T., and Burwell, R. G., eds.), Butterworth-Heinemann, pp. 260–270.

15. Buck, B. E., Malinin, T., and Brown, M. D. (1989) Bone transplantation and human immunodeficiency virus. *Clin. Orthop.* **240,** 129–136.

16. Bugbee, W. D. and Convery, F. R. (1999) Osteochondral allograft transplantation. *Clin. Sports Med.* **18,** 67–75.

17. Buncke, H. J., Furnas, D. W., Gordon, L., and Achaner, B. (1977) Free osteocutaneous flap for the rib to the tibia. *Plast. Reconstr. Surg.* **59,** 79–91.

18. Burchardt, H. (1983) The biology of bone graft repair. *Clin. Orthop. Rel. Res.* **174,** 28–42.

19. Burchardt, H., Glowczewskie, F. P., and Ennecking, W. F. (1981) Short-term immunosupression with fresh segmental fibular allografts in dogs. *J. Bone Joint Surg.* **63A,** 411–415.

19a. den Boer, F. C., Bramer, J. A., Blokhuis, T. J., et al. (2002) Effect of recombinant human osteogenic protein-1 on the healing of a freshly closed diaphyseal fracture. *Bone* **31(1),** 158–164.

20. Burgess, W. H. and Maciag, T. (1989) The heparin-binding (fibroblast) growth factor family of proteins. *Ann. Rev. Cell Dev. Biol.* **58,** 575–606.

21. Burwell, R. G. (1985) The function of bone marrow in the incorporation of a bone graft. *Clin. Orthop.* **200,** 125–141.

22. Burwell, R. G. (1994) The Burwell theory on the importance of bone marrow in bone grafting, in *Bone Grafts, Derivatives, and Substitutes* (Urist, M. R., O'Connor, B. T., and Burwell, R. G., eds.), Butterworth-Heinemann, pp. 103–155.

23. Burwell, R. G., Friedlaender, G. E., and Mankin, H. J. (1985) Current perspectives and future directions: the 1983 invitational conference on osteochondral allografts. *Clin. Orthop.* **200,** 141–157.

24. Carringtion, J. L., Roberts, A. B., Falnders, K. C., et al. (1988) Accumulation, localization, and compartmentation of transforming growth factor-B during endochondral bone development. *J. Cell Biol.* **107,** 1969–1975.

25. Centrella, M., Horowitz, M. C., Wozney, J., and McCarthy, T. (1994) Transforming growth factor-beta gene family members and bone. *Endocrinol. Rev.* **15,** 27.

26. Clark, R. A. F. (1996) *The Molecular and Cellular Biology of Wound Repair.* Plenum Press, New York.

27. Clarke, M. S. F., Khakee, R., and McNeil, P. L. (1993) Loss of cytoplasmic basic fibroblast growth factor for physiologically wounded myofibers of normal and dystrophic muscle. *J. Cell Sci.* **106,** 121–133.

28. Connolly, J., Guse, R., Lippiello, L., and Dehne, R. (1989) Development of an osteogenic bone-marrow preparation. *J. Bone Joint Surg.* **71,** 684–691.

29. Connolly, J., Guse, R., Tiedeman, J., and Dehne, R. (1991) Autologous marrow injection as a substitute for operative grafting of tibial nonunions. *Clin. Orthop.* **266,** 259–270.

30. Constantz, B. R., Ison, I. C., Fulmer, M. T., et al. (1995) Skeletal repair by *in situ* formation of the mineral phase of bone (see comments). *Science* **267,** 1796–1799.

31. Cook, S. D., Baffes, G. C., Wolfe, M. W., Sampath, T. K., and Rueger, D. C. (1994) Recombinant human bone morphogenetic protein-7 induces healing in a canine long-bone segmental defect model. *Clin. Orthop.* **301,** 304–312.

32. Cook, S. D., Baffes, G. C., Wolfe, M. W., et al. (1994) The effect of recombinant human osteogenic protein-1 on healing of large segmental bone defects. *J. Bone Joint Surg.* **76,** 827–838.

33. Cook, S. D., Dalton, J. E., Tan, E. H., Whitecloud, T. S., and Rueger, D. C. (1994) In vivo evaluation of recombinant human osteogenic protein (rhOP-1) implants as a bone graft substitute for spinal fusions. *Spine* **19,** 1655–1663.

34. Cook, S. D., Wolfe, M. W., Salkeld, S. L., and Rueger, D. C. (1995) Effect of recombinant human osteogenic protein-1 on healing of segmental defects in non-human primates. *J. Bone Joint Surg.* **77,** 734–750.

35. Critchlow, M. A., Bland, Y. S., and Ashhurst, D. E. (1995) The effect of exogenous transforming growth factor-beta 2 on healing fractures in the rabbit. *Bone* **16,** 521–527.

36. Cuevas, P., Burgos, J., and Baird, A. (1988) Basic fibroblast growth factor (FGF) promotes cartilage repair in vivo. *Biochemical. Biophysical. Res. Commun.* **156,** 611–618.

37. Doi, K., Tominaga, S., and Shibata, T. (1977) Bone grafts with microvascular anastomoses of vascular pedicles. *J. Bone Joint Surg.* **59A,** 809–815.

38. Dreesmann, H. (1892) Ueber Knochenplombierung. *Beitr. Klin. Chir.* **9,** 804–810.

39. Dunstan, C. R., Boyce, R., Boyce, B. F., et al. (1999) Systemic administration of acidic fibroblast growth factor (FGF-1) prevents bone loss and increases new bone formation in ovariectomized rats. *J. Bone Miner. Res.* **14,** 953–959.

40. Ekelund, A., Brosjo, O., and Nilsson, O. S. (1991) Experimental induction of heterotopic bone. *Clin. Orthop. Rel. Res.* **263,** 102.

41. Enneking, W. F., Eady, J. L., and Burchardt, H. (1980) Autogenous cortical bone grafts in the reconstruction of segmental skeletal defects. *J. Bone Joint Surg.* **62A,** 1039–1058.

42. Enneking, W. F. and Mindell, E. R. (1991) Observations on massive retrieved human allografts. *J. Bone Joint Surg.* **73A,** 1123–1142.

43. Esch, F., Baird, A., Ling, N., et al. (1985) Primary structure of bovine pituitary basic fibroblast growth factor (FGF) and comparison with the amino terminal sequence of bovine brain acidic FGF. *Proc. Natl. Acad. Sci. USA* **82,** 6507–6511.

44. Fager, G., Hansson, G. K., Ottosson, P., Dahllof, B., and Bondjers, G. (1988) Human arterial smooth muscle cells in culture: effects of platelet derived growth factor and heparin on growth in vitro. *Exp. Cell Res.* **176,** 319–335.

45. Flynn, J. M., Springfield, D. S., and Mankin, H. J. (1994) Osteoarticular allografts to treat distal femoral osteonecrosis. *Clin. Orthop.* **303,** 38–43.

45a. Friedlaender, G. E., Perry, C. R., Cole, J. D., et al. (2001) Osteogenic protein-1 (bone morphogenetic protein-7) in the treatment of tibial nonunions. *J. Bone Joint Surg. American Volume* **83A(Suppl 1, Pt 2),** S151–S158.

46. Fujimori, Y., Nakamura, T., Ijiri, S., Shimizu, K., and Yamamuro, T. (1992) Heterotopic bone formation induced by bone morphogenetic protein in mice with collagen-induced arthritis. *Biochem. Biophys. Res. Commun.* **186,** 1362–1367.

47. Galzie, Z., Kinsella, A. R., and Smith, J. A. (1997) Fibroblast growth factors and their receptors. *Biochem. Cell Biol.* **75,** 669–685.

47a. Geesink, R. G., Hoefnagels, N. H., and Bulstra, S. K. (1999) Osteogenic activity of OP-1 bone morphogenetic protein (BMP-7) in a human fibular defect. *J. Bone Joint Surg.* **81(4),** 710–718.

48. Gerhart, T., Kirker-Head, C., Kriz, M., et al. (1991) Healing of large mid-femoral segmental defects in sheep using recombinant human bone morphogenetic protein (BMP-2). *Trans. Orthop. Res. Soc.* **16,** 172.

49. Glowacki, J., Jasty, M., and Goldring, S. (1986) Comparison of multinucleated cells elicited in rats by particulate bone, polyethylene, or polymethylmethacrylate. *J. Bone Miner. Res.* **1,** 327.

50. Goel, S. C. and Tuli, S. M. (1994) Use of decalbone in healing of osseous cystic defects, in *Bone Grafts, Derivatives, and Substitutes* (Urist, M. R., O'Connor, B. T., and Burwell, R. G., eds.), Butterworth-Heinemann, pp. 210–219.

51. Gospodarowicz, D. (1974) Localisation of fibroblast growth factor and its effect alone and with hydrocortisone on 3T3 cell growth. *Nature* **249,** 123–129.

52. Gospodarowicz, D., Bialecki, H., and Greenburg, G. (1978) Purification of fibroblast growth factor activity from bovine brain. *J. Biol. Chem.* **253,** 3736–3743.

52a. Govender, S., Csimma, C., Genant, H. K., et al. (2002) BMP-2 Evaluation in Surgery for Tibial Trauma (BESTT) Study Group. Recombinant human bone morphogenetic protein-2 for treatment of open tibial fractures: a prospective, controlled, randomized study of four hundred and fifty patients. *J. Bone Joint Surg.* **84A(12),** 2123–2134.

53. Hammack, B. L. and Enneking, W. F. (1960) Comparative vascularization of autogenous and homogenous bone transplants. *J. Bone Joint Surg.* **42A,** 811.

54. Han, C. S., Wood, M. B., Bishop, A. D., et al. (1992) Vascularized bone transfer. *J. Bone Joint Surg.* **74A,** 1441–1449.

55. Heckman, J. D., Aufdemorte, T. B., and Athanasiou, K. A. (1995) Treatment of acute ostectomy defects in the dog radius with TGF-B1. *Trans. Orthop. Res. Soc.* **20,** 590.

56. Hench, L. L. (1992) Bioactive bone substitutes, in *Bone Grafts and Bone Graft Substitutes* (Habal, M. B. and Reddi, A. H., eds.), Saunders, Philadelphia, pp. 263–275.

57. Hofmann, G. O., Kirschner, M. H., Wagner, F. D., Brauns, L., Gonschorek, O., and Buhren, V. (1998) Allogeneic vascularized transplantation of human femoral diaphyses and total knee joints—first clinical experiences. *Transplant. Proc.* **30,** 2754–2761.

58. Hollinger, J. O. and Wong, M. E. (1996) The integrated processes of hard tissue regeneration with special emphasis on fracture healing. *Oral Surg. Oral Med. Oral Pathol. Oral Radiol. Endodont.* **82,** 594–606.

59. Hughes, A. D., Clunn, C. F., Refson, J., and Demoliou-Mason, C. (1996) Platelet-derived growth factor: actions and mechanisms in vascular smooth muscle. *Genet. Pharm.* **27,** 1079–1089.

60. Hurley, M. M., Lee, S. K., Raisz, L. G., Bernecker, P., and Lorenzo, J. (1998) Basic fibroblast growth factor induces osteoclast formation in murine bone marrow cultures. *Bone* **22,** 309–316.

61. Ibbotson, K. J., Harrod, J., Gowen, M., et al. (1986) Human recombinant transforming growth factor alpha stimulates bone resorption and inhibits formation in vitro. *Proc. Natl. Acad. Sci. USA* **83(7),** 2228–2232.

61a. Iwata, H., Sakano, S., Itoh, T., and Bauer, T. W. (2002) Demineralized bone matrix and native bone morphogenetic protein in orthopaedic surgery. *Clin. Orthop. Rel. Res.* **395,** 99–109.

62. Jaye, M., Schlessinger, J., and Dionne, C. A. (1992) Fibroblast growth factor receptor tyrosine kinase-molecular analysis and signal transduction. *Biochem. Biophys. Acta* **1135,** 185–199.

63. Jingushi, S., Heydemann, A., Kana, S. K., Macey, L. R., and Bolander, M. E. (1990) Acidic fibroblast growth factor (aFGF) injection stimulates cartilage enlargement and inhibits cartilage gene expression in rat fracture healing. *J. Orthop. Res.* **8,** 364–371.

nanomolar concentrations, powerfully induce new bone formation both within osseous lesions and at ectopic sites, such as skeletal muscle *(12–15)*. The US Food and Drug Administration has recently approved recombinant, human bone morphogenic proteins BMP-2 and BMP-7 for restricted clinical use. Although these are potent osteogenic agents, their clinical application is complicated by delivery problems *(16)*. The main limitation is the need for delivery systems that provide a sustained, biologically appropriate concentration of the osteogenic factor at the site of the defect. Delivery needs to be sustained, because these factors have exceedingly short biological half-lives, usually of the order of minutes or hours, rather than the days or weeks needed to stimulate a complete osteogenic response. Delivery also needs to be local to avoid ectopic ossification and other unwanted side effects.

Because systemic delivery by intravenous, intramuscular, or subcutaneous routes fails to satisfy these demands, there has been much interest in developing implantable slow-release devices from which the BMP can progressively leach. Typically, such devices comprise a biocompatible matrix impregnated with very large amounts of recombinant BMP; in the clinic they are most frequently used with autologous bone grafts. The device is surgically implanted at the site of the defect and thus satisfies the need for local delivery. However, release is not uniform over time. In most cases, there is an initial rapid efflux ("dumping") of the protein, which spikes the surrounding tissue with wildly supraphysiological concentrations of growth factor. Subsequent release, although slower, provides much lower, suboptimal concentrations of protein. Another drawback is the denaturation of the growth factor at body temperature before it is released from the matrix. Moreover, the carrier, usually bovine collagen, can provoke inflammation. Clearly, such systems, although capable of increasing osteogenesis, are clumsy and inefficient *(16,17)*. Research into the genetic manipulation of bone healing is based on the hypothesis that gene transfer can do better.

GENE THERAPY APPROACHES TO ENHANCING BONE HEALING

Advances in gene transfer technology provide the opportunity to overcome the technical limitations described above *(18–20)*. The concept, shown in **Fig. 1**, is to transfer genes encoding osteogenic factors to osseous lesions. When the transgene is expressed, the lesion becomes an endogenous, local source of the factors needed for bone healing. Thus the gene transfer approach offers great potential as a delivery system that meets the requirement of sustained and local delivery of the growth factor at the appropriate concentrations. Moreover, unlike the recombinant protein, the growth factor synthesized *in situ* as a result of gene transfer undergoes authentic posttranslational processing and is presented to the surrounding tissues in a natural, cell-based manner. This may explain why gene delivery is often more biologically potent than protein delivery. A good example of this from another area of gene therapy research is provided by the work of Makarov et al. *(21)*, who have shown that the treatment of arthritic rats with cDNA encoding the interleukin-1 receptor antagonist is 10^4 times more potent than treatment with the corresponding recombinant protein. Similar gains in potency may be achieved by local delivery of osteogenic genes to sites of osseous defect. The use of gene transfer to enhance bone repair has been previously reviewed in refs. *18, 19,* and *20)*.

A GENE TRANSFER PRIMER

Because cells do not spontaneously take up and express exogenous genes, successful gene transfer requires vectors. These can be divided into those that are derived from viruses and those that are not. The properties of the most advanced viral vectors are listed in **Table 1**. With the exception of lentivirus, all of these have been used in human clinical trials.

Retroviral vectors have the ability to integrate their genetic material into the chromosomal DNA of the cells they infect. This is a major for advantage for settings where long-term transgene expression is required. However, because the insertion site is random, there is a possibility of insertional mutagenesis. Although this possibility is extremely low, the first instances of insertional mutagenesis

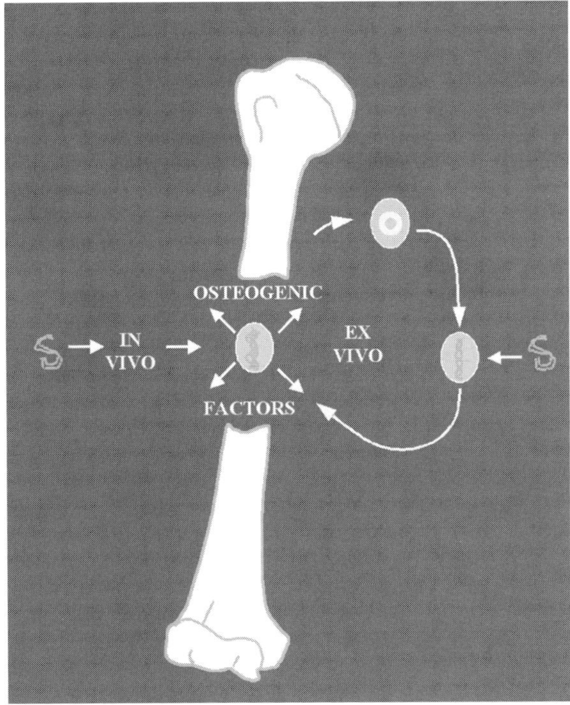

Fig. 1. Schematic representation of ex vivo and in vivo gene therapy strategies for enhancing bone healing. (From ref. *18.*)

are now emerging from human clinical trials *(23)*, and this has resurrected huge concerns about the safety of these vectors.

Because genetically enhanced bone healing should not require long-term transgene expression, use can be made of nonintegrating vectors such as adenovirus and adeno-associated virus (AAV). Both of these are DNA viruses that deliver genes episomally to the nuclei of the cells they infect. The most commonly used adenovirus vectors (so-called first-generation adenovirus vectors) have the advantage of being straightforward to construct and produce at high titers. They readily infect a wide range of dividing and nondividing cells, and usually achieve high levels of transgene expression. The big drawback of adenovirus vectors is the high antigenicity of both the virions themselves and cells infected with first-generation adenovirus. The latter problem can be eliminated by using a third-generation, so-called gutted adenovirus vector that contains no viral coding sequences, but these are difficult to manufacture. Moreover, the antigenicity of the virions is not reduced by removing viral DNA. It remains to be seen whether immune reactions limit the clinical use of adenovirus in human bone healing.

AAV is far less antigenic than adenovirus and causes no known disease in humans. Recombinant AAV vectors are of great current interest because of the perception that they are very safe. However, they are difficult to make and they do not infect all cell types well. Their carrying capacity is limited to about 4 kb, but this is probably adequate for the types of cDNAs needed to promote bone healing. As far as it is possible to tell, AAV seems to infect both dividing and nondividing cells.

Vectors derived from herpes simplex virus are difficult to manufacture, often cytotoxic, and of little immediate and obvious utility to bone healing at the present time.

Nonviral vectors (**Table 2**) can be as simple as naked, plasmid DNA. To enhance gene transfer efficiency, the DNA can be associated with carrier molecules such as various types of liposomes and synthetic or natural polymers. There is also interest in using physical techniques, such as electroporation,

Table 1
Common Viral Vectors and Their Salient Properties

Vector	Key properties	Comment
Oncoretrovirus[a] (retrovirus)	Inserts DNA into host chromosome Insertional mutagenesis a safety issue Packaging capacity ~8 kb Only transduces dividing cells Straightforward to manufacture Medium titers	Requirement for cell division usually limits use to *ex vivo* protocols Commonly derived from Moloney murine leukemia virus Human use has been associated with leukemia
Lentivirus[a] (retrovirus)	Inserts DNA into host chromosome Insertional mutagenesis a safety issue Packaging capacity ~8 kb Transduction not limited by cell division Moderately difficult to manufacture Medium titers	Commonly derived from HIV Not yet used in human clinical trials
Adeno-associated virus (AAV)	W.t. inserts DNA into host chromosome —a rare event with recombinant AAV vectors Packaging capacity ~4 kb Not all cell types are readily transduced Manufacture very difficult	Generally considered to be the safest of the viral vectors In clinical trials
Adenovirus	Noninsertional First- and second-generation vectors, packaging capacity ~8 kb Both virus and cells transduced by early-generation vectors are highly antigenic High infectivity *In vivo* use associated with inflammation Transduction not limited by cell division Straightforward to manufacture at high titer	Ease of production, high infectivity, and wide tropism ensure common experimental use, especially for *in vivo* gene delivery Human use has been associated with one death
Herpes simplex virus	Noninsertional Very large packaging potential Often cytotoxic High infectivity Transduction not limited by cell division Very difficult to manufacture High titers possible	Major clinical application may be in the CNS, where it has a natural tropism and latency

[a]Both oncoretrovirus and lentivirus are members of the *Retroviridae* family.

to improve gene transfer efficiency. Nonviral vectors are usually cheaper and safer than viral vectors, but far less efficient. Gene transfer with nonviral vectors is known as *transfection*. Gene transfer with viral vectors is known as *transduction*.

Regardless of the vector, genes may be transferred to sites in the body by ex vivo or in vivo strategies (**Fig. 1**). Other things being equal, in vivo methods are simpler, cheaper, and more expeditious, because they involve no extracorporal manipulation of the target cells. However, they raise greater safety concerns. Ex vivo methods do not involve the direct introduction of vectors into the body, and allow the target cells to be isolated, manipulated, tested, and optimized before reimplantation. Under conditions where soft tissue support for osteogenesis is compromised, ex vivo protocols allow the introduction of genetically modified osteoprogenitor cells to enhance repair.

More detailed reviews of gene therapy in an orthopedic context are to be found in refs. *24–28.*

Table 2
Common Types of Nonviral Vectors

Naked DNA
DNA combined with cationic and anionic liposomes (many different formulations)
DNA–protein complexes (many different formulations)
DNA–polymer complexes (many different synthetic and natural polymers)
Electroporation
Ballistic projection ("gene gun")

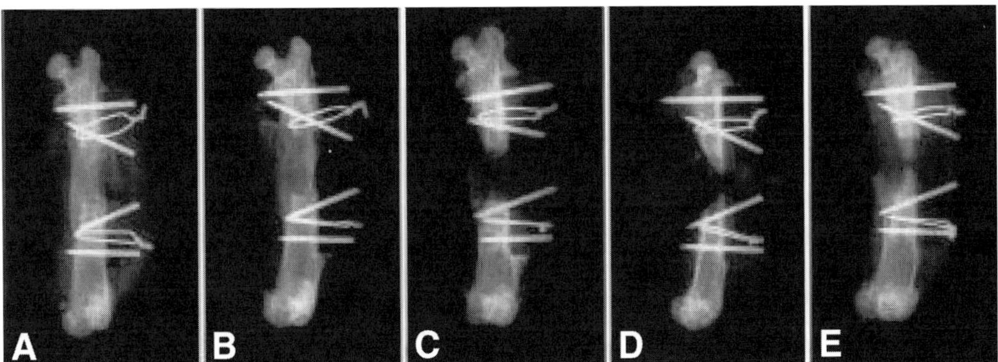

Fig. 2. Healing of rat segmental bone critical-sized defect by *ex vivo* BMP-2 gene transfer. Animals were sacrificed 2 mo postoperatively and were treated in one of the following ways: **(A)** BMP-2 producing bone marrow cells created via adenoviral gene transfer; **(B)** 20 µg of rhBMP-2; **(C)** β-galactosidase-producing bone marrow cells (cells infected with an adenovirus containing lacZ gene); **(D)** noninfected rat bone marrow cells; or **(E)** guanidine-extracted demineralized bone matrix alone. Dense trabecular bone formed within the defects that had been treated with the BMP-2-producing cells, and the bone remodeled to form a new cortex. The defects that had been treated with rhBMP-2 healed but were filled with lacelike trabecular bone. Minimal bone repair was noted in the other three groups. (From ref. *30* with permission.)

EX VIVO GENE TRANSFER

Nearly all investigators in this area have used the ex vivo approach pioneered by Lieberman and colleagues *(29,30)*. Using a rat critical-sized-defect model, Lieberman's group employed a recombinant, first-generation adenovirus to transfer a human BMP-2 cDNA to osteogenic stromal cells recovered from bone marrow. This population of cells probably includes mesenchymal stem cells (MSCs). Under the transcriptional regulation of the human cytomegalovirus early promoter, the transduced cells expressed high levels of human BMP-2. These cells were seeded onto a collagenous matrix and surgically implanted into critical-sized defects. Under conditions where control defects failed to heal, defects receiving the genetically modified cells reproducibly achieved osseous union *(29,30)* (**Fig. 2**).

BMP-2 gene therapy produced a better response than recombinant BMP-2 protein in healing osseous defects in rats. Although both approaches led to osseous union, the recombinant protein generated atypical new bone filled with lacey, delicate trabeculae, which formed a shell around the defect. The gene transfer method, in contrast, led to new bone with an authentic three-dimensional trabecular structure, remodeling to form a neocortex *(30)*.

Subsequent investigators have confirmed the success of the ex vivo approach using cells derived from skin, muscle, fat, and peripheral blood using, in addition to BMP-2, other osteogenic proteins such as BMP-4 and BMP-7 *(31–36)*. In common with marrow-derived osteoprogenitors, cells derived from muscle, fat, and, according to Krebsbach et al. *(37)*, even skin fibroblasts, have the ability to differentiate into bone under the influence of appropriate biological cues. Thus, when genetically modified, they aid osteogenesis not only as a local source of osteogenic factors, but also as an additional source of osteoprogenitor cells that enhance repair through both paracrine and autocine processes. The notion that mature fibroblasts can transdifferentiate into osteoblasts is unfamiliar, but the utility of fibroblasts is supported by the recent work of Gugala et al. *(38)*. These investigators compared the osteogenic properties of human MSCs, human skin fibroblasts, and the human fetal lung cell line MRC-5. Cells were transduced with adenovirus carrying BMP-2 cDNA and injected intramuscularly into immunodeficient mice. There was no statistically significant difference in the amount of bone formed by the three different types of human cells.

Among tissues other than skin that may contain osteoprogenitor cells, fat could be the most convenient for eventual human application. Most individuals are more than happy to donate adipose tissue, which is readily biopsied; adipose-derived stem cells are straightforward to culture, can be easily expanded, and transduced. Moreover, their abundance and proliferative properties do not appear to decline with the age of the donor. According to a recent paper by Dragoo et al. *(34)*, fat provides a richer source of osteoprogenitor cells than bone marrow, and, when genetically modified to express BMP-2, they are more efficient osteoprogenitors. These cells are also able to heal large segmental defects in rats (Lieberman et al., unpublished).

Despite the above successes, the use of first-generation adenovirus vectors remains a concern because the cells it transduces express viral proteins and thus become antigenic. Several strategies are being employed to obviate this concern. One is to make adenoviral transduction of MSCs more efficient. The major cell-surface receptor for the most commonly used recombinant adenovirus vector, serotype 5, is the Coxsackie and adenovirus receptor (CAR). It is poorly expressed on MSCs, thus requiring very high multiplicities of infection; even then, only about 20% of the cells are transduced *(39)*. Tsuda et al. *(39)* have used modified adenovirus whose coat carries the tripeptide sequence RGD, which enhances interaction with cell-surface integrins and thus engenders greater uptake. Cells transduced with the modified virus produce greater amounts of BMP-2 and are more osteogenic in vivo. It should thus be possible to reduce the antigenic load by administering fewer modified MSCs. A similar, alternative approach uses serotype 35 adenovirus, that enters cells in a CAR-independent fashion *(40)*.

Although the above strategies may reduce the antigenic burden, they will not eliminate it. For this reason there is interest in using vectors that express no foreign, antigenic proteins. Abe et al. *(41)* have successfully used a "gutted" adenovirus for this purpose. Recombinant retrovirus is also successful in animal models *(42)*, although, as discussed above, there are renewed concerns about the safety of such vectors. AAV is another candidate vector that has shown success when delivered in vivo *(43,44)* (see next section). Avoiding viral vectors altogether, Park et al. *(45)* used liposomes to transfect MSCs and heal mandibular defects in rats, by an ex vivo strategy. Healing with liposome gene delivery was slower than healing with adenoviral vectors, but was otherwise indistinguishable. Given the resistance of MSCs to transfection, this result is quite remarkable.

A major drawback of ex vivo gene delivery is the need to culture autologous cells from each patient. There is thus interest in using allogeneic cells so that a universal donor could be established. This endeavor is encouraged by the possibility that MSCs can be successfully allografted. However, in a rat segmental-defect model, allogeneic MSCs transduced with BMP-2 healed the defect only if the immunosuppressant FK 506 was administered *(46)*. Although the need for FK 506 was only transient, its clinical use in bone healing may raise difficult safety concerns. Transient immunosuppression has also been used experimentally for the in vivo delivery of osteogenic genes *(47,48)* (see next section).

IN VIVO GENE DELIVERY

Two in vivo strategies have emerged. One involves the implantation of plasmid DNA incorporated into a collagen sponge (gene activated matrix, GAM). The other involves the direct injection of vector.

GAM technologies were developed by Bonadio and Goldstein *(49,50)*, and have the advantage of using plasmid DNA. The GAM is stable upon storage, and is surgically inserted directly into the osseous lesion. Cells from the area of the lesion migrate into the matrix, where they encounter, take up, and express the DNA. GAMs containing plasmids encoding PTH 1-34 and BMP-4 healed 5-mm femoral defects in rats that would not otherwise heal *(49)*. When used in a critical sized tibial defect in dogs, a GAM containing PTH 1-34 cDNA resulted in 6 wk of transgene expression. Although impressive amounts of new bone were deposited in response, they were insufficient to heal the defect *(50)*. Human clinical trials are pending.

One of the advantages of adenoviral vectors is their ability to infect cells *in situ*, a property compatible with in vivo gene delivery. Most investigators have avoided in vivo gene delivery for bone healing, because the intramuscular injection of adenovirus vectors containing osteogenic genes leads to very little bone formation. The problem appears to lie with the immune response to the adenovirus, because considerable bone formation occurs when immunodeficient animals are used *(51)*, or when an immunosuppressant, such as cyclophosphamide, is administered *(47,48)*.

Nevertheless, Baltzer et al. were able to heal critical-sized defects in the femurs of immunocompetent rabbits by the direct, intralesional injection of adenovirus carrying a BMP-2 cDNA (**Fig. 3**). Studies were conducted with a rabbit femoral critical-sized (1.3-cm)-defect model *(52)*. Injection of a first-generation adenovirus vector carrying the human BMP-2 cDNA into such defects produced osseous union, judged radiologically and histologically, under conditions where control defects receiving an irrelevant gene failed to heal *(53)*. Injection of similar vectors carrying marker genes showed that the greatest expression of the transgene occurred in the musculature surrounding the defect, with significant expression also occurring in the gap scar and the cut ends of the bone. Marker gene expression was observed in marrow cells and lining osteoblasts. Lung, liver, and spleen were also sampled. There was transient transgene expression in the liver, but not elsewhere *(52)*.

The direct injection of Ad.BMP-2 also heals critical-sized femoral defects in rats (Betz et al., unpublished), further supporting the notion that the intraosseous environment, unlike the intramuscular one, supports osteogenesis in response to adenoviral delivery of an osteogenic transgene to immunocompetent animals. The critical difference may involve the degree to which the immune system and inflammatory responses are activated. The key question of whether redosing of the same osteogenic adenovirus will continue to promote bone formation has not yet been addressed.

The immune response to adenovirus may be further blunted by delivering the virus in conjunction with a collagenous matrix in a modified GAM strategy. Both Franceschi et al. *(35)* and Sonobe et al. *(54)* have used this tactic successfully to form bone intramuscularly and subdermally in immunocompetent rodents. The adenoviral burden may be also be reduced by using more effective serotypes of adenovirus *(40)*, or administering the virus at times when its receptor is maximally expressed. The CAR used by the type 5 adenovirus is induced upon fracture and, in mice, its expression peaks at d 5 *(55)*. The tactic of transient immunosuppression also works experimentally *(47,48)*, but its clinical applicability is questionable.

As an alternative to adenovirus, recombinant AAV vectors carrying BMP-2 *(43)* or BMP-4 *(44)* elicit bone formation after direct injection. Transcutaneous electroporation of plasmid carrying BMP-2 cDNA also stimulates bone formation in muscle *(56)*.

WHICH GENES?

Most experiments have focused on the use of transgenes expressing BMPs. Of this group, BMP-2 and BMP-7 cDNAs have advantages for overcoming the regulatory barriers to clinical application,

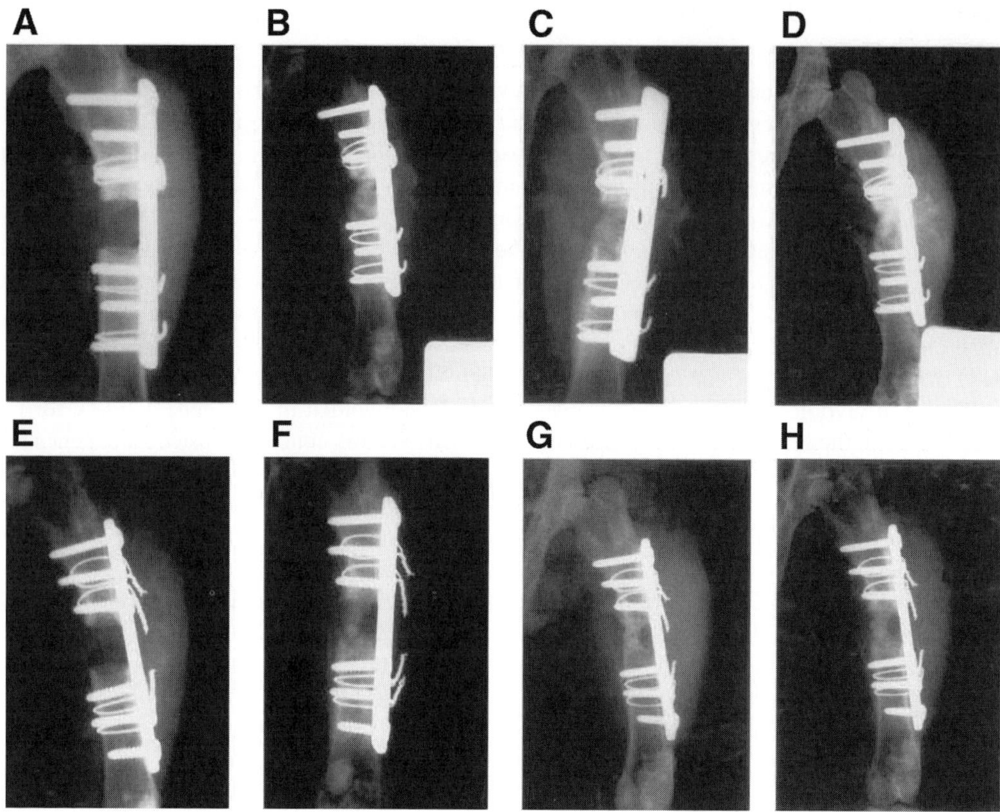

Fig. 3. Healing of rabbit segmental bone critical-sized defect by *in vivo* BMP-2 gene transfer. Defects were treated with Ad.BMP-2 (panels **A–D**) or Ad.luciferase (panels **E–H**) and radiographed at the time of surgery (panels **A** and **E**) and after 5 wk (panels **B** and **F**), 7 wk (panels **C** and **G**) and 12 wk (panels **D** and **H**). Defects treated with Ad.BMP-2 undergo osseous union, as judged radiologically, whereas those treated with Ad.luciferase do not. (From ref. *53* with permission.)

because the corresponding recombinant proteins have been widely tested in humans and shown to be safe and somewhat effective. Recent research by Helms's group, however, suggests that BMP-6 and BMP-9 cDNA are more effective osteogenic agents when delivered by adenovirus vectors *(57)*.

Growth factors are not the only class of gene product capable of eliciting bone formation. Boden's group has identified a transcription factor, LMP-1, that promotes osteogenesis at tiny concentrations *(36,58)*. Because LMP-1 acts intracellularly, gene transfer is a particularly pertinent delivery system for this protein, although advances in peptide delivery are also providing new avenues. The remarkable potency of LMP-1 is at least partially explained by its ability to induce expression of multiple, different BMPs and other osteogenic factors, thus providing a rich osteogenic environment within the osseous lesion *(59)*.

The value of combining factors has been demonstrated in a rat calvarial defect model, where healing was greater when BMP-4 and VEGF transgenes were coexpressed than when either was expressed alone *(60)*.

The types of gene products of potential use in the gene treatment of osseous lesions are listed in **Table 3**.

Table 3
Classes of Gene Products of Potential Use for Bone Healing

Class	Examples	Comment
Growth factors	BMP-2,-4,-7,-9 IGF-1 TGF-β_{1-3} PDGF	Perform well in animal models.
Transcription factors	LMP-1, Cbfa-1	Intracellular site of action compatible with gene transfer. LMP-1, very potent.
Angiogenic factors	VEGF; FGF	May act synergistically with other factors.
Antiinflammatories	sTNFR sIL-1R IL-1Ra	Of potential use under conditions of excessive bone resorption, e.g., aseptic loosening.
Osteoclast blockers	Osteoprotegerin	Good results in models of aseptic loosening.

APPLICATIONS

Gene transfer has numerous applications under circumstances where it is necessary to form bone. Long bone fractures, non- and delayed unions, as well as segmental defects, are obvious examples that have attracted the most experimental attention.

Spine fusion is another area of considerable interest, and progress has been made in the use of an abbreviated ex vivo procedure in which adenovirus carrying LMP-1 cDNA is used to transduce buffy coat cells from peripheral blood intraoperatively *(36)*. The cells are applied to a collagenous matrix and implanted. This procedure is effective in rabbits, and is now being evaluated in nonhuman primates. Successful spine fusion in a rabbit model has also been achieved with the used of MSCs expressing a BMP-2 transgene *(61)*. Percutaneous injection of adenovirus carrying cDNA for BMP-2 or BMP-9 induces spine fusion in athymic, but not immunocompetent, rodents *(33,62,63)*.

There are also many applications in the cranial and maxillofacial areas. There are numerous experimental examples of healing cranial lesions in rodents using gene transfer. Chang et al. *(64)* have recently described the repair of large maxillary defects in pigs using BMP-2 gene transfer.

The need to form bone sometimes arises under circumstances where it is necessary not only to form new bone via osteoblasts, but also to prevent bone loss via osteoclasts. Aseptic loosening provides one such example. An appropriate strategy in these conditions is to express genes whose products inhibit the activities of the cytokines that promote bone loss *(65–68)*. Discussion of this aspect is beyond the scope of this chapter, but overlaps with gene treatment of inflammatory diseases such as rheumatoid arthritis, reviewed in refs. *69–72*.

CONCLUSION

Collectively, the preclinical data provide strong experimental support for the proposition that gene transfer provides a powerful method for healing osseous defects that will not otherwise heal. However, although the application of gene therapy to clinical problems associated with bone healing has a persuasive logic and accumulating experimental support, there is a pressing need for translational studies that convert preclinical concepts and findings into clinically useful modalities. Many fundamental questions still need to be answered, including which gene or gene combinations to use, whether to use in vivo or ex vivo delivery, and which vectors to employ. There has been little work in large animal models, and safety issues remain to be addressed. The latter is of particular importance as, for the majority of prospective patients, the procedure will be elective and the condition not life-threatening.

Given the numerous different clinical circumstances under which it is necessary to promote bone formation, there will probably be no single preferred method. Not all patients will require gene therapy, and not all gene therapies will be the same. Depending on circumstances, different vectors, genes, and strategies will be indicated.

One advantage of bone healing as a target for gene therapists is the existence of a robust, natural repair process, and the observation that, at least in animal models, healing is very responsive to moderate levels of gene expression for a limited period of time. Thus clinical success may be achieved with existing gene therapy technologies. This is not the case for most other areas of gene therapy.

ACKNOWLEDGMENTS

The authors' work in this area has been supported by the Orthopaedic Trauma Association (CHE), National Institutes of Health grant number AR 050243 (CHE).

REFERENCES

1. Fernandez, D. L., Ring, D., and Jupiter, J. B. (2001) Surgical management of delayed union and nonunion of distal radius fractures. *J. Hand Surg.* **26(2)**, 201–209.
2. Rommens, P. M., Coosemans, W., and Broos, P. L. (1989) The difficult healing of segmental fractures of the tibial shaft. *Arch. Orthop. Trauma Surg.* **108(4)**, 238–242.
3. Buchler, U. and Nagy, L. (1995) The issue of vascularity in fractures and non-union of the scaphoid. *J. Hand Surg. [Br.]* **20(6)**, 726–735.
4. Einhorn, T. A. and Lane, J. M. (1998) Fracture Healing Enhancement. *Clin. Orthop. Rel. Res.* **335S**, 365.
5. Arrington, E. D., et al. (1996) Complications of iliac crest bone graft harvesting. *Clin. Orthop.* **329**, 300–309.
6. Colterjohn, N. R. and Bednar, D. A. (1997) Procurement of bone graft from the iliac crest. An operative approach with decreased morbidity. *J. Bone Joint Surg.* **79(5)**, 756–759.
7. Kwong, L. M., Jasty, M., and Harris, W. H. (1993) High failure rate of bulk femoral head allografts in total hip acetabular reconstructions at 10 years. *J. Arthroplasty* **8(4)**, 341–346.
8. Bonnarens, F. and Einhorn, T. A. (1984) Production of a standard closed fracture in laboratory animal bone. *J. Orthop. Res.* **2(1)**, 97–101.
9. Einhorn, T. A. (1998) The cell and molecular biology of fracture healing. *Clin. Orthop.* **355(Suppl)**, S7–S21.
10. Reddi, A. H. (2001) Bone morphogenetic proteins: from basic science to clinical applications. *J. Bone Joint Surg.* **83A(Suppl 1, pt 1)**, S1–S6.
11. Li, R. H. and Wozney, J. M. (2001) Delivering on the promise of bone morphogenetic proteins. *Trends Biotechnol.* **19(7)**, 255–265.
12. Lieberman, J. R., Daluiski, A., and Einhorn, T. A. (2002) The role of growth factors in the repair of bone. Biology and clinical applications. *J. Bone Joint Surg.* **84A(6)**, 1032–1044.
13. Valentin-Opran, A., et al. (2002) Clinical evaluation of recombinant human bone morphogenetic protein-2. *Clin. Orthop.* **395**, 110–120.
14. Salkeld, S. L., et al. (2001) The effect of osteogenic protein-1 on the healing of segmental bone defects treated with autograft or allograft bone. *J. Bone Joint Surg.* **83A(6)**, 803–816.
15. Bouxsein, M. L., et al. (2001) Recombinant human bone morphogenetic protein-2 accelerates healing in a rabbit ulnar osteotomy model. *J. Bone Joint Surg.* **83A(8)**, 1219–1230.
16. Uludag, H., et al. (2001) Delivery systems for BMPs: factors contributing to protein retention at an application site. *J. Bone Joint Surg.* **83A(Suppl 1, pt 2)**, S128–S135.
17. Talwar, R., et al. (2001) Effects of carrier release kinetics on bone morphogenetic protein-2-induced periodontal regeneration in vivo. *J. Clin. Periodontol.* **28(4)**, 340–347.
18. Niyibizi, C., et al. (1998) Potential role for gene therapy in the enhancement of fracture healing. *Clin. Orthop.* **355 (Suppl)**, S148–S153.
19. Lieberman, J., Ghivizzani, S., and Evans, C. (2002) Gene transfer approaches to the healing of bone and cartilage. *Mol. Ther.* **6(2)**, 141.
20. Lieberman, J. R. (2000) Orthopaedic gene therapy. Fracture healing and other nongenetic problems of bone. *Clin. Orthop.* **379(Suppl)**, S156–S158.
21. Makarov, S. S., et al. (1996) Suppression of experimental arthritis by gene transfer of interleukin 1 receptor antagonist cDNA. *Proc. Natl. Acad. Sci. USA* **93(1)**, 402–406.
22. Baltzer, A. W. A. and Lieberman, J. (2004) Regional gene therapy to enhance bone repair. *Gene Ther.* **11**, 344–350.
23. Marshall, E. (2003) Gene therapy. Second child in French trial is found to have leukemia. *Science* **299(5605)**, 320.

24. Evans, C. H., et al. (2004) Orthopaedic gene therapy. *Clin. Orthop. Rel. Res.* (in press).
25. Evans, C. H. and Robbins, P. D. (1999) Genetically augmented tissue engineering of the musculoskeletal system. *Clin. Orthop.* **367(Suppl)**, S410–S418.
26. Evans, C. H. and Robbins, P. D. (1995) Possible orthopaedic applications of gene therapy. *J. Bone Joint Surg.* **77A(7)**, 1103–1114.
27. Evans, C. H., et al. (2004) Gene therapy for the treatment of musculoskeletal diseases. *J. Am. Acad. Orthop. Surg.* (in press).
28. Oligino, T. J., et al. (2000) Vector systems for gene transfer to joints. *Clin. Orthop.* **379(Suppl)**, S17–S30.
29. Lieberman, J. R., et al. (1998) Regional gene therapy with a BMP-2-producing murine stromal cell line induces heterotopic and orthotopic bone formation in rodents. *J. Orthop. Res.* **16(3)**, 330–339.
30. Lieberman, J. R., et al. (1999) The effect of regional gene therapy with bone morphogenetic protein-2-producing bone-marrow cells on the repair of segmental femoral defects in rats. *J. Bone Joint Surg.* **81A(7)**, 905–917.
31. Lee, J. Y., et al. (2002) Enhancement of bone healing based on ex vivo gene therapy using human muscle-derived cells expressing bone morphogenetic protein 2. *Hum. Gene Ther.* **13(10)**, 1201–1211.
32. Gysin, R., et al. (2002) Ex vivo gene therapy with stromal cells transduced with a retroviral vector containing the BMP4 gene completely heals critical size calvarial defect in rats. *Gene Ther.* **9(15)**, 991–999.
33. Alden, T. D., et al. (2000) The use of bone morphogenetic protein gene therapy in craniofacial bone repair. *J. Craniofac. Surg.* **11(1)**, 24–30.
34. Dragoo, J. L., et al. (2003) Bone induction by BMP-2 transduced stem cells derived from human fat. *J. Orthop. Res.* **21(4)**, 622–629.
35. Franceschi, R. T., et al. (2000) Gene therapy for bone formation: in vitro and in vivo osteogenic activity of an adenovirus expressing BMP7. *J. Cell Biochem.* **78(3)**, 476–486.
36. Viggeswarapu, M., et al. (2001) Adenoviral delivery of LIM mineralization protein-1 induces new-bone formation in vitro and in vivo. *J. Bone Joint Surg.* **83A(3)**, 364–376.
37. Krebsbach, P. H., et al. (2000) Gene therapy-directed osteogenesis: BMP-7-transduced human fibroblasts form bone in vivo. *Hum. Gene Ther.* **11(8)**, 1201–1210.
38. Gugala, Z., et al. (2003) Osteoinduction by ex vivo adenovirus-mediated BMP2 delivery is independent of cell type. *Gene Ther.* **10(16)**, 1289–1296.
39. Tsuda, H., et al. (2003) Efficient BMP2 gene transfer and bone formation of mesenchymal stem cells by a fiber-mutant adenoviral vector. *Mol. Ther.* **7(3)**, 354–365.
40. Olmsted-Davis, E. A., et al. (2002) Use of a chimeric adenovirus vector enhances BMP2 production and bone formation. *Hum. Gene Ther.* **13(11)**, 1337–1347.
41. Abe, N., et al. (2002) Enhancement of bone repair with a helper-dependent adenoviral transfer of bone morphogenetic protein-2. *Biochem. Biophys. Res. Commun.* **297(3)**, 523–527.
42. Gysin, R., et al. (2002) Ex vivo gene therapy with stromal cells transduced with a retroviral vector containing the BMP4 gene completely heals critical size calvarial defect in rats. *Gene Ther.* **9(15)**, 991–999.
43. Chen, Y., et al. (2003) Gene therapy for new bone formation using adeno-associated viral bone morphogenetic protein-2 vectors. *Gene Ther.* **10(16)**, 1345–1353.
44. Luk, K. D., et al. (2003) Adeno-associated virus-mediated bone morphogenetic protein-4 gene therapy for in vivo bone formation. *Biochem. Biophys. Res. Commun.* **308(3)**, 636–645.
45. Park, J., et al. (2003) Bone regeneration in critical size defects by cell-mediated BMP-2 gene transfer: a comparison of adenoviral vectors and liposomes. *Gene Ther.* **10(13)**, 1089–1098.
46. Tsuchida, H., et al. (2003) Engineered allogeneic mesenchymal stem cells repair femoral segmental defect in rats. *J. Orthop. Res.* **21(1)**, 44–53.
47. Okubo, Y., et al. (2001) In vitro and in vivo studies of a bone morphogenetic protein-2 expressing adenoviral vector. *J. Bone Joint Surg.* **83A(Suppl 1, pt 2)**, S99–S104.
48. Okubo, Y., et al. (2000) Osteoinduction by bone morphogenetic protein-2 via adenoviral vector under transient immunosuppression. *Biochem. Biophys. Res. Commun.* **267(1)**, 382–387.
49. Fang, J., et al. (1996) Stimulation of new bone formation by direct transfer of osteogenic plasmid genes. *Proc. Natl. Acad. Sci. USA* **93(12)**, 5753–5758.
50. Bonadio, J., et al. (1999) Localized, direct plasmid gene delivery in vivo: prolonged therapy results in reproducible tissue regeneration. *Nat. Med.* **5(7)**, 753–759.
51. Musgrave, D. S., et al. (1999) Adenovirus-mediated direct gene therapy with bone morphogenetic protein-2 produces bone. *Bone* **24(6)**, 541–547.
52. Baltzer, A. W., et al. (1999) A gene therapy approach to accelerating bone healing. Evaluation of gene expression in a New Zealand white rabbit model. *Knee Surg. Sports Traumatol. Arthrosc.* **7(3)**, 197–202.
53. Baltzer, A. W., et al. (2000) Genetic enhancement of fracture repair: healing of an experimental segmental defect by adenoviral transfer of the BMP-2 gene. *Gene Ther.* **7(9)**, 734–739.
54. Sonobe, J., et al. (2004) Osteoinduction by bone morphogenetic protein-2 expressing adenoviral vector: application of biomaterial to mask the host immune response. *Hum. Gene Ther.* **15**, 659–668.

55. Ito, T., et al. (2003) Coxsackievirus and adenovirus receptor (CAR)-positive immature osteoblasts as targets of adenovirus-mediated gene transfer for fracture healing. *Gene Ther.* **10(18),** 1623–1628.
56. Kawai, M., et al. (2004) Ectopic bone formation by human bone morphogenetic protein-2 gene transfer to skeletal muscle using transcutaneous electroporation. *Hum. Gene Ther.* **14,** 1547–1556.
57. Li, J. Z., et al. (2004) Osteogenic potential of five different recombinant human bone morphogenetic protein (BMP) adenoviral vectors in the rat. *Gene Ther.* (in press).
58. Boden, S. D., et al. (1998) LMP-1, a LIM-domain protein, mediates BMP-6 effects on bone formation. *Endocrinology* **139(12),** 5125–5134.
59. Minamide, A., et al. (2003) Mechanism of bone formation with gene transfer of the cDNA encoding for the intracellular protein LMP-1. *J. Bone Joint Surg.* **85A(6),** 1030–1039.
60. Peng, H., et al. (2002) Synergistic enhancement of bone formation and healing by stem cell-expressed VEGF and bone morphogenetic protein-4. *J. Clin. Invest.* **110(6),** 751–759.
61. Wang, J. C., et al. (2003) Effect of regional gene therapy with bone morphogenetic protein-2-producing bone marrow cells on spinal fusion in rats. *J. Bone Joint Surg.* **85A(5),** 905–911.
62. Alden, T. D., et al. (1999) Percutaneous spinal fusion using bone morphogenetic protein-2 gene therapy. *J. Neurosurg.* **90(1 Suppl),** 109–114.
63. Helm, G. A., et al. (2000) Use of bone morphogenetic protein-9 gene therapy to induce spinal arthrodesis in the rodent. *J. Neurosurg.* **92(2 Suppl),** 191–196.
64. Chang, S. C., et al. (2003) Ex vivo gene therapy in autologous bone marrow stromal stem cells for tissue-engineered maxillofacial bone regeneration. *Gene Ther.* **10(24),** 2013–2019.
65. Carmody, E. E., et al. (2002) Viral interleukin-10 gene inhibition of inflammation, osteoclastogenesis, and bone resorption in response to titanium particles. *Arthritis Rheum.* **46(5),** 1298–1308.
66. Childs, L. M., et al. (2001) Effect of anti-tumor necrosis factor-alpha gene therapy on wear debris-induced osteolysis. *J. Bone Joint Surg.* **83A(12),** 1789–1797.
67. Sud, S., et al. (2001) Effects of cytokine gene therapy on particulate-induced inflammation in the murine air pouch. *Inflammation* **25(6),** 361–372.
68. Yang, S., et al. (2002) IL-1Ra and vIL-10 gene transfer using retroviral vectors ameliorates particle-associated inflammation in the murine air pouch model. *Inflamm. Res.* **51(7),** 342–350.
69. Robbins, P. D., Evans, C. H., and Chernajovsky, Y. (2003) Gene therapy for arthritis. *Gene Ther.* **10(10),** 902–911.
70. Gouze, E., et al. (2001) Gene therapy for rheumatoid arthritis. *Curr. Rheumatol. Rep.* **3(1),** 79–85.
71. Evans, C. H. and Robbins, P. D. (1999) Gene therapy of arthritis. *Intern. Med.* **38(3),** 233–239.
72. Evans, C. H., et al. (1999) Gene therapy for rheumatic diseases. *Arthritis Rheum.* **42(1),** 1–16.

Bone Morphogenetic Proteins and Other Growth Factors to Enhance Fracture Healing and Treatment of Nonunions

Calin S. Moucha, MD and Thomas A. Einhorn, MD

INTRODUCTION

Each year, approx 33 million people in the United States sustain an injury to their musculoskeletal system. Nearly 6.2 million of these traumatic events are fractures. Although management of these injuries has improved greatly during the past 25 yr, 5–10% of fractures go on to nonunion or delayed union *(1)*. The increasingly more aggressive acute treatment of fractures has led to an overall decrease in the incidence of nonunion and delayed union. These same treatments, however, have also increased the incidence of impaired union of some fractures, particularly those involving the tibia. Technical errors such as open reduction and internal fixation in distraction or excessive periosteal stripping may account for some of the increase in the incidence of abnormal fracture healing. The fact that limbs that were once amputated due to a high number of associated risk factors known to result in a poor outcome are now being salvaged by novel treatment modalities may have also contributed to the increased incidence of nonunions and delayed unions *(2)*.

Fracture healing is a well-orchestrated series of biological events that involves the coordinated participation of several cell types. Unlike other tissues that heal by the formation of a poorly organized scar, in fracture healing the original tissue, bone, is restored. Although full cellular and morphological regeneration occurs only in children, adult bone fracture healing also leads to a mechanically stable lamellar structure.

Urist made the first observation that implantation of demineralized lyophilized segments of bone matrix, either subcutaneously or intramuscularly, induces bone formation in animals *(3,4)*. Follow-up studies of these bone-inductive matrices resulted in identification of a family of compounds known as the bone morphogenetic proteins (BMPs) *(5)*. Several other growth factors have since been shown to play an important role in the development, repair, and induction of bone. These compounds (**Table 1**) are currently grouped into the transforming growth factor-β (TGF-β) superfamily (which includes the BMPs), the fibroblast growth factors (FGFs), the insulin-like growth factors (IGFs), and the platelet-derived growth factors (PDGFs).

Osteogenesis is the process of new bone formation. The process that promotes mitogenesis of undifferentiated mesenchymal cells, leading to formation of osteoprogenitor cells that have osteogenic capacity, is known as *osteoinduction. Osteoconduction* is the process by which fibrovascular tissue and osteoprogenitor cells invade a porous structure, often acting as a temporary scaffold, and replace it with newly formed bone.

Osteoinduction has been described as occurring in three major phases: chemotaxis, mitosis, and differentiation *(6)*. The aforementioned growth factors, all polypeptide molecules, provide a mechanism for stimulative and regulative effects on these phases. They elicit their actions by binding to transmembrane

From: *Bone Regeneration and Repair: Biology and Clinical Applications*
Edited by: J. R. Lieberman and G. E. Friedlaender © Humana Press Inc., Totowa, NJ

Table 1
Growth Factors and Fracture Repair

Fibroblast growth factor (FGF-1 and -2)	
Source	Responding cells
Inflammatory cells (macrophages)	Mitogenic effects on mesenchymal cells chondrocytes and osteoblasts
Mesenchymal cells	Angiogenic actions
Chondrocytes	FGF-1, primarily mitogenic to chondrocytes
Osteoblasts	FGF-2, possible involvement in chondrocyte maturation

Platelet-derived growth factor (PDGF-AA, -AB, -BB)	
Source	Responding cells
Degranulating platelets	Macrophage chemotaxis
Monocytes and macrophages	Mesenchymal cell chemotaxis
Hypertrophic chondrocytes (PDGF-A)	Mesenchymal cell proliferation
Osteoblasts (PDGF-B)	

Transforming growth factor-β (TGFβ-1 and -2)	
Source	Responding cells
Degranulating platelets	Pleiotropic factor
Bone extracellular matrix	Osteoprogenitor cell proliferation
Inflammatory cells	Stimulates undifferentiated mesenchymal cell and chondrocyte proliferation
Chondrocytes	
Osteoblasts	Stimulates extracellular matrix production

Bone morphogenetic protein (BMP-2, -3, -4, and -7)	
Source	Responding cells
Bone extracellular matrix	Activation of cortical osteoblasts
Osteoblasts and osteoprogenitors	Initiates differentiation of osteoprogenitor cells into osteoblasts
	Promotes differentiation of mesenchymal cells into chondrocytes

Source: Barnes, G. L., Kostenuik, P. J., Gerstenfield, L. C., and Einhorn, T. A. (1999) Growth factor regulation in fracture repair. *J. Bone Miner. Res.* **14,** 1805–1815, with permission of the American Society for Bone and Mineral Research.

receptors that are linked to gene sequences in the nucleus of various cells by a cascade of chemical reactions *(7,8)*. Because these cascades activate several genes at once, specific growth factors generate multiple effects, both within a single cell type as well as in different cell types *(7,9,10)*.

This chapter will first discuss new concepts in defining *nonunion* and *delayed union,* risk factors identified as contributory to their development, and the rationale for developing compounds that can enhance fracture healing. Then, we will highlight some of the available experimental models of normal and delayed bone healing. Lastly, we will review current knowledge on the role of growth factors in bone healing.

DELAYED AND IMPAIRED BONE HEALING

Despite advances in treatment protocols for various fractures, some heal slower than others do and some do not heal at all. Excellent reviews of this topic already exist *(11)*, and it is beyond the scope of this book to attempt a similar task. Because of the tremendous recent and anticipated future explosion of research on the role of growth factors in the treatment of nonunions, however, it is critical that the reader gain an understanding of some basic principles of impaired bone healing.

First and foremost, it is important to define the terms *delayed union* and *nonunion.* Traditionally, orthopedic surgeons have referred to a *delayed union* as a fracture that heals more slowly than average and a *nonunion* as a failure of bone healing *(11)*. These definitions, however, are vague, and considering the human body's different modes of achieving union of a fractured bone, a more specific set of definitions is required. Several authors have contributed to the task of providing relevant definitions. Tiedeman et al. *(12)* showed that radiographically visible new bone formation was a good predictor of bending stiffness. Richardson et al. *(3)*, validated stiffness as a good measure of fracture healing. They measured stiffness of 212 tibial fractures treated with an Orthofix fixator. In one group ($n = 117$), the decision to remove the fixator was taken on clinical grounds. In the other group ($n = 95$), the fixator was removed when the stiffness reached a level of 15 N-m per degree. Even though the latter group, on average, had a shorter time span to fixator removal, it also had a lower refracture rate (0% vs 6.8%). From a clinical standpoint, they viewed a threshold of 15 N-m per degree as a safe definition of union. Marsh *(14)*, considering the various sites of bony bridging that occur in a fracture (endosteal, periosteal, cortical, depending in part on different treatment modalities), questioned the clinical capability of a quantitative radiographic assessment. He reviewed 43 isolated, closed energy tibial shaft fractures treated conservatively by using a thermoplastic functional brace beginning at 3–5 wk after fracture. Callus index (the ratio of the maximum width of callus to the diameter of the original shaft at the same level) was used as a measure of periosteal new bone formation. No fracture failed to heal having reached a value of 7 N-m per degree. Stiffness measurements correlated more strongly than callus index with injury severity and functional outcome at 6 mo. The callus index, however, predicted delayed union in those fractures that showed no tendency to heal at the 10-wk stage. Based on this study, the author defined *union* as a process of structural reconstitution of the fractured bone by means of endosteal and/or periosteal regeneration. This was predicted with confidence when the bending stiffness reached 7 N-m per degree. *Delayed union* was defined as the cessation of the periosteal response before the fracture had been successfully bridged. A bending stiffness of less than 7 N-m by 20 wk was predictive of this process. *Nonunion* was defined as a cessation of both the periosteal and endosteal healing responses without bridging. Clear definitions of these terms are needed both for understanding studies on the effects of growth factors in enhancing fracture healing as well as for clinical estimates of fracture healing.

Many risk factors for impaired or delayed healing of bone have been identified. Boyd *(15)* defined several local factors that contributed to nonunions. These included (1) open fracture, (2) infection, (3) segmentation with impaired blood supply to the free fragment, (4) comminution, (5) insecure fixation, (6) insufficient length of immobilization, (7) improper open reduction, and (8) distraction. Since then, others have added to and refined this list. Systemic statuses of the patient such as nutritional status *(16)*, anemia *(17)*, diabetes mellitus *(18)*, and certain hormone deficiencies *(19)* have all been shown to have an effect on fracture healing. The nature of the traumatic injury, including the location of the fracture *(20)*, extent of soft tissue damage *(21)*, and associated compartment syndrome *(22)*, all are risk factors leading to impaired fracture healing. Inappropriate fracture care itself often goes unacknowledged as a cause of poor healing and is probably one of most readily modified. Unnecessary soft tissue insult, rigid fixation in a distracted fashion, and operative-field bacterial contamination due to poor sterility precautions or prolonged operative time are just a few of the many well-known factors that may ultimately lead to impaired healing. Fracture gaps of more than 2 mm have been shown by Claes et al. *(23)* to adversely affect healing. Smokers are at a risk of delayed union of bones *(24)*. Various pharmacological agents such as corticosteroids *(25)*, anticoagulants *(26)*, and nonsteroidal antiinflammatory drug *(27)* have all been shown to affect bone regeneration to some degree.

Even with avoidance of some or all of these risk factors, many fractures continue to go on to nonunion *(28)*. For this reason, novel modalities to enhance fracture healing have interested orthopedic surgeons for some time now. Mechanical stimulation has been shown to induce fracture healing *(29)*. Distraction osteogenesis has been used successfully in the treatment of fractures showing impaired healing *(30)*. Sharrard *(31)*, among others, has shown evidence that a pulsed electromagnetic field may be beneficial to the treatment of delayed unions of fractures. In addition, work of Xavier and Duarte *(32)* has led to a series of investigations on the use of ultrasound to enhance fracture healing, and these studies have shown enhancement of fracture healing in the tibia and distal radius and an improvement of healing in smokers *(33–35)*.

Successful treatment of a nonunion rarely consists of only one method, however, and in general, the simplest treatment modality with which the surgeon has the most familiarity should be chosen. For many years, the most frequently used method of treatment of nonunions has been bone grafting. Experience with autogenous bone grafting dates back to the early 1900s *(36)*. Some estimate that more than 250,000 bone grafts are performed annually in the United States *(37)*. Autogenous bone graft, allograft bone, or synthetic bone substitute, after proper reduction and fixation of fragments, generally help stimulate bone to unite. Autograft bone can be cancellous, nonvascularized cortical, or vascularized cortical. Cancellous autogenous grafts are generally obtained from the proximal tibia or the ilium. Allogeneic ¡•oducts are available as bone matrix, cancellous bone, cortical bone, or corticocancellous composites. Cortical bone grafts are used primarily for structural support, and cancellous bone grafts for osteogenesis.

Because this chapter focuses on the role of growth factors, we will direct our attention here to autogenous cancellous bone, as this is the type of graft that these compounds may in the future substitute. When structural support is not required, autogenous cancellous bone remains the optimal grafting material. Its osteogenic and osteoconductive properties, along with its source of osteoprogenitor cells, make it an ideal substance for nonstructural grafting.

Although autogenous bone is widely used and useful, there is morbidity associated with its harvesting. Younger and Chapman *(38)* retrospectively studied the medical records of 239 patients with 243 autogenous bone grafts to document donor-site morbidity. They found an 8.6% overall major complication rate and a 20.6% minor complication rate. Major complications included infection (2.5%), prolonged wound drainage (0.8%), large hematomas (3.3%), reoperation (3.8%), pain lasting more than 6 mo (2.5%), and sensory loss (1.2%). Minor complications included superficial infections, minor wound problems, temporary sensory loss, and mild or resolving pain. Kurz et al. *(39)* reviewed the literature for complications of harvesting autogenous iliac bone grafts with particular attention given to different operative approaches. In addition to the complications reported by Younger and Chapman, these authors specifically identified injuries to the lateral femoral cutaneous nerve (resulting in "meralgia paresthetica"), the superior cluneal nerves, and the ilioinguinal nerve. In addition, they described other, less common complications such as a postoperative gluteal gait, stress fractures of the ilium, ureteral injuries, dislocation of the sacroiliac joint, and abdominal hernias.

Reduction of fragments and stable fixation will always be necessary for the treatment of nonunions. Mechanical, electrical, and ultrasound enhancement of fracture healing need to have their risks and benefits better defined. The development of growth factors injectable into fracture sites may in the future augment if not replace autologous cancellous bone grafting. Considering the risks, complications, and additional operative time required for bone grafting, development of new strategies for its substitution is indicated. Bone morphogenetic proteins and the other peptide growth factors reviewed here may meet these expectations.

EXPERIMENTAL MODELS OF BONE HEALING

To study the enhancement of fracture repair by growth factors in such a way that clinically useful conclusions can be made, a model must fulfill three important criteria. First, it must mimic human physiology. Second, the bone defect must fail to heal in the absence of the enhancement modality under investigation. Third, the model must not heal by a method more basic than that being tested *(40)*.

Models used in research on the effect of growth factors have included systems of normal fracture healing, segmental bone defects, and various forms of nonunion. Many investigators have attempted to establish experimental models of fracture healing and nonunion *(41–46)*. Comprehensive reviews of these and other models of bone regeneration in tissue engineering research have been published *(40, 47)*. The main concept that should be retained from these reviews is that investigators should utilize models that simulate similar patient settings. Two types of models are currently being used for investigating enhancement of bone healing: critical-sized diaphyseal defects and fractures or osteotomies.

Critical-sized diaphyseal defect models mimic a clinical situation in which bone loss is so massive that normal fracture healing would not occur. Examples of such scenarios include high-energy open fractures or bone resection for musculoskeletal tumors. A critical-sized defect is defined as the smallest intraosseuous wound that would not heal by bone formation in the lifetime of the animal *(41,42)*. Although these defects would not heal without intervention, they are not truly models of normal or impaired bone healing, as it is the inherent size of the defect that leads to failed healing, rather than, as in the case of a delayed union or nonunion, host characteristics or the local fracture enviromnent. In addition, the size of the diaphyseal defect is highly variable throughout the phylogenetic tree. With the exception of the long-bone rat model described by Yasko et al. *(48)*, none of the other models have shown the smallest size defect that would not heal unless implanted with a bone graft or osteoinductive material. Although significant developments in fracture repair enhancement have been achieved using critical-sized diaphyseal defect models, they do not stimulate the more common clinical situation in which the cause of a nonunion is a compromised healing environment other than massive bone loss.

Other models of normal or impaired fracture healing involve creation of a fracture or osteotomy *(49)*. Methods to create delayed unions or nonunions in these models involve severe mechanical manipulation, metabolic alteration, neurological or proprioceptive inhibition, or induced necrosis at the fracture or osteotomy site. Utvag et al. *(50)* produced radiographically documented hypertrophic nonunions in rats by performing femoral diaphyseal osteotomies, inserting soft polyethylene nails, and then manipulating the fracture sites mechanically for 5 wk. Hollinger and Kleinschmidt *(46)* created an atrophic nonunion model in rats using a mid-diaphyseal fracture of the femur complicated by a 4-mm cauterization of surrounding periosteum. Nonunion models due to soft tissue interposition were devised by Santos Neto and Volpon *(51)* in mongrel dogs and by Lattermann et al. *(44)* in white rabbit tibias. As discussed previously, several systemic conditions have also been shown to adversely affect fracture healing. Macey et al. *(18)* showed a biomechanical and biochemical impairment in fracture healing in streptozotocin-diabetic rats. Although neither this nor another study *(52)* on diabetic rats resulted in nonunions, both showed impaired healing, which ultimately may be of greater clinical relevance with respect to testing the utility of a growth factor. Lastly, a novel atrophic nonunion model was created by Aro et al. *(41)* by surgically stripping the proprioceptive nerve receptors from the distal part of a rat fibula and producing standard fibular fractures adjacent to that site. This model could be of particular use in testing the efficacy of a growth factor to enhance healing in fractures accompanied by nerve or vascular injury.

Comparing the effects of different growth factors in these models is difficult because of their wide variability. Results of fracture-healing studies may conflict even when the same model is used in different species. Only by understanding the model system used in each particular study can one draw logical, clinically relevant conclusions from the results obtained.

OSTEOINDUCTIVE GROWTH FACTORS

Transforming Growth Factor-β

Transforming growth factor-β (TGF-β), a peptide first identified by its ability to cause phenotypic transforination of rat fibroblasts *(53)*, has since been shown to be a fundamental, multifunctional, regulatory protein that can either stimulate or inhibit several critical processes of cell function *(54)*. Since then, five different isomers of TGF-β have been identified (three of these are found in humans), as have several other related polypeptide growth factors. These are now all a part of the TGF-β superfamily, which also includes other ubiquitous compounds such as the bone morphogenetic proteins, activins, inhibins, and growth and differentiation factors. TGF-β-related proteins are found in all vertebrate species, the fruitfly *Drosophila,* and the nematode *Caenorhabditis elegans.*

All TGF-βs are disulfide-linked dimers comprising 12–18 kDa subunits *(55)*. Most are homodimers (TGF-β1, TGF-β2, and TGF-β3), but some are heterodimers (TGF-β1.2 and TGF-β2.3) *(56)*. TGF-βs are secreted in a latent propeptide form that requires activation by extracellular proteolytic activity.

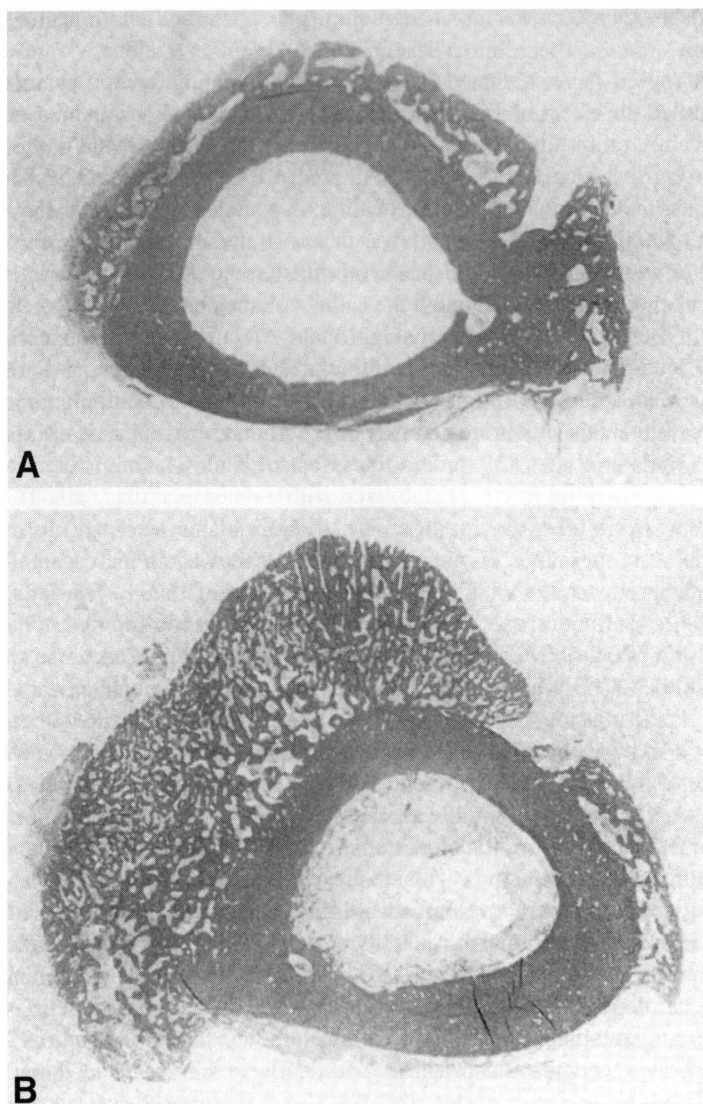

Fig. 1. Decalcified transverse section 5 mm from the osteotomy line showing diaphyseal cortical bone and callus formation in a specimen from the control group (**A**) and 10-μg/d group (**B**). A moderate callus formation is seen in the control and massive callus formation in the stimulated specimen. The callus-free area represents the AO-plate position (HE ×5). (From Lind, M., Schumacker, B., Soballe, K., et al. [1993] Transforming growth factor-β enhances fracture healing in rabbit tibiae. *Acta Orthop. Scand.* **64,** 553–556, with permission.)

signals generated by various members of the TGF-β superfamily. There are now several known members of this family of osteoinductive growth factors, which can be subdivided into several classes based on structure *(86)*. Bone morphogenetic proteins play critical roles in cell growth, differentiation, and apoptosis in a variety of cells during development, including chondrocytes and osteoblasts. Compared with TGF-β, however, bone morphogenetic proteins have been shown to have more selective and powerful effects on bone healing in animal models.

Ever since 1965, when Urist *(3,4)* first described the presence of osteoinductive proteins in bone, interest in the role of these proteins in fracture healing has led to extensive basic science and clinical research. During normal fracture repair, BMP-2, BMP-3 (osteogenin), BMP-4, and BMP-7 (osteogenic protein, OP-1) are the bone morphogenetic proteins most commonly expressed endogenously. Even in human fracture nonunions, some of these same BMPs, as well as their receptors and SMAD proteins (an indication of an activated BMP state), have been shown to be present *(87)*. In humans, BMP-2 and BMP-7 have been the most extensively studied, and they have been isolated, sequenced, and manufactured using recombinant DNA technology.

The osteoinductive effects of BMPs during bone healing have been studied extensively in vitro *(88–92)*. In vivo studies attempted to further characterize the role of BMPs in fracture healing. Using a validated rat fracture-healing model *(93)*, Bostrom et al. documented the physiological presence, localization, and chronology of BMPs in fracture healing *(94)*. Frozen undecalcified fracture calluses of rats euthanized at various times postinjury were analyzed semiquantitatively for the percentage of the different types of cells that stained positively with a monoclonal antibody against BMP-2 and BMP-4. Irnmediately after fracture, positive staining for BMP-2/4 was observed in the cambium cell layer of periosteum. As fracture healing progressed through endochondral ossification, the presence of BMP-2/4 increased dramatically, especially in the primitive mesenchymal and chondrocytic cells. While the callus cartilaginous component decreased, both the overall number of primitive cells and the number staining for the bone morphogenetic proteins decreased. When osteoblasts began laying down woven bone, they began staining more positively for the BMPs, although this decreased as lamellar bone replaced primitive woven bone. The areas of callus undergoing intramembranous ossification showed similar staining patterns. The authors concluded that BMP-2 and –4 play a significant role in the formation of intramembranous bone as well as being involved in the differentiation of primitive mesenchymal cells into chondrocytes.

Using a variety of fracture models and analytical techniques, other in vivo studies have added to present knowledge of the role of endogenous BMPs in fracture healing. Nakase et al., using reverse-transcription polymerase chain reaction (RT-PCR) and *in situ* hybridization, investigated the temporal and spatial distribution of a gene encoding murine BMP-4 in fractured and nonfractured mouse ribs *(95)*. They concluded that the BMP-4 gene is not produced by differentiated osteoblasts, but rather by less differentiated osteoprogenitor cells. Moreover, they suggested that the BMP-4 gene is upregulated by the fracture event itself, making fracture injury a local contributing factor in callus formation. Ishidou et al. *(96)*, investigated the expression of type I receptors for BMPs during mouse embryonic development and rat fracture healing using antibodies specific to these receptors. The results of their study suggested that the expression of BMP type I receptors is upregulated during bone formation in both these bone-generating models. More recent studies have shown the role of other BMPs and BMP receptors in modulating bone formation as well *(97–99)*.

Many investigators have studied the ability of exogenously administered BMPs to promote bone regeneration in osseous locations. Most of these studies have used critical-sized defect models. Yasko et al. tested the ability of subcutaneously implanted recombinant human BMP-2 to induce endochondral bone formation in 5-mm segmental defects of rat femora *(48)*. Two groups were implanted with either 1.4 or 11 µg of lyophilized rhBMP-2 in a guanidine-hydroxychloride-extracted demineralized bone matrix carrier and the other with guanidine-hydroxychloride-extracted demineralized rat bone matrix alone. The formation and healing of bone were determined by radiographic, histological, and mechanical analyses. Both doses of rhBMP-2 induced formation of endochondral bone in the defects. Only the 11.0-µg dose of subcutaneously implanted rhBMP-2 led to radiographic, histological, and mechanical evidence of union. The authors concluded that exogenously administered rhBMP-2 leads to successful union of segmental bony defects. These findings have been confirmed by other investigators, who have shown the ability of BMP-2 to bridge critical-sized bony defects *(42,100–103)*.

BMP-7 has also been widely studied in segmental long bone defects in animals. Cook et al. evaluated the use of BMP-7 (also known as osteogenic protein-1, OP-1), implanted in combination with an allogeneic bone collagen carrier, for the restoration of a large segmental defect of the ulnar diaphysis in the rabbit *(104)*. A 1.5-cm segmental defect was created in the mid-part of the ulnar shaft of adult rabbits. These were filled with either 250 µg of naturally occurring bovine osteogenic protein (bOP) or an implant consisting of a 125 mg of demineralized guanidine-extracted insoluble rabbit bone matrix, reconstituted with 3.13, 6.25, 12.5, 25, 50, 100, 200, 300, or 400 µg of recombinant OP-1. Contralateral control extremities received either the collagen carrier only or no implant at all. Radiographs of the limbs were made weekly until the animals were euthanized at 8 or 12 wk postoperatively. Histological analysis and mechanical testing were done on the harvested specimens. The radiographic results showed that all implants in the bOP group and all in the recombinant OP-1 group (except for those containing 3.13 µg of the substance) induced complete osseous union within 8 wk. The average torsional strength and energy-absorption capacity of recombinant OP-1–implanted bones was comparable to that of intact bone. Histological evaluation of the new bone at 8 wk postoperatively revealed lamellar bone with the formation of new cortices and normal-appearing marrow elements. None of the control defects showed any bone bridging. Cook et al. *(105,106)* have demonstrated similar effects of BMP-7 in canine and nonhuman primate models.

Stevenson et al. investigated the effects of natural, partially purified BMP-3 (also known as osteogenin) on bridging segmental femoral defects in rats *(107)*. An 8-mm-wide segmental defect was created in the mid-diaphysis of bilateral rat femora and stabilized with a polyacetyl plate and threaded Kirschner wires. Defects were filled with either an osteogenin (100 µg)/hydroxyapatite/tricalcium ceramic cylinder or a hydroxyapatite/tricalcium phosphate ceramic cylinder. Animals were euthanized at 1, 2, and 4 mo after the operation, and the specimens evaluated by histomorphometric and biomechanical methods. Histomorphometry revealed that the total area of bone, the area of bone outside of the implant, and the amount of bone within the pores of the implant were all greater in the femora that had an implant with BMP-3 than in those that had an implant without BMP-3. However, the femora with BMP-3 showed no significant differences in biomechanical characteristics compared to control groups.

In an attempt to find an ideal carrier/bone morphogenetic protein combination, Sciadini et al. tested the efficacy of bovine-derived bone protein (NeOsteo, Intermedics Orthopaedics, Denver, CO) in the healing of a segmental defect in a weight-bearing long bone *(108,109)*. They performed blinded, prospective, randomized studies using a well-established canine 2.5-cm radial bone defect model *(110)*. The first study *(108)* showed that using radiographic assessments, allogeneic demineralized bone matrix (DBM) administered locally into the defects with 3.0 mg of bovine-derived bone protein (BP) was more effective in healing the critical-sized segmental defects than DBM alone. Mean values for most biomechanical parameters of DBM + BP-treated radii exceeded those of their contralateral controls, which were implanted with autogenous cancellous bone graft (ABG) at 12 and 24 wk. Histology revealed evidence of normal bone healing in all ABG and DBM + BP-treated radii, while most DBM-treated radii demonstrated nonunions. In the later study *(109)*, the investigators tested the efficacy of BP associated with a natural coral carrier (calcium carbonate) in healing similar segmental defects. The first conclusion that the authors drew from their studies was that bone protein in a natural coral carrier performed better in terms of the amount of bone formed and the strength of the healed defect than autogenous cancellous bone graft. Second, it was concluded that coralline calcium carbonate alone represents a poor bone graft substitute in the canine radial critical-sized segmental defect model.

Several groups have presented data supporting the concept of using BMPs to enhance and accelerate healing of normal fractures or non-critical-sized defects. As with the aforementioned critical-sized studies, these have mostly involved BMP-2 and BMP-7.

Einhorn et al. investigated the effects of percutaneously injecting recombinant BMP-2 into standardized, closed mid-diaphyseal femur fractures in rats 6 h after injury *(111)*. Two hundred and seventy-eight male rats were first divided into three groups of 96 animals, each receiving either no injection at all, injection of an aqueous buffer, or injection of the buffer plus 80 µg of rhBMP-2. Animals in each of these groups were then further subdivided into four groups and sacrificed at 7, 14, 21,

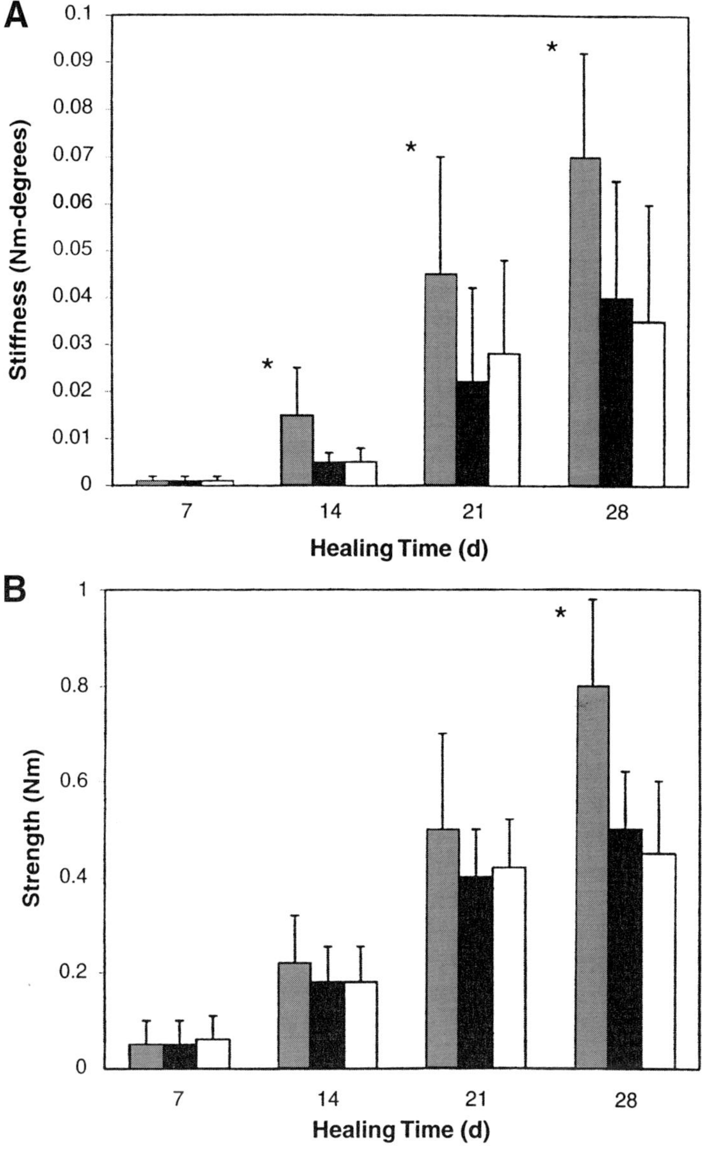

Fig. 2. Stiffness (**A**) and strength (**B**) as a function of time for the treatment groups. *Different than buffer and fracture only groups; $p < 0.01$. *Shaded bar* = BMP 2 injection, *solid bar* = buffer injection, *open bar* = fracture only. (From Bostrom, M. P. G. and Camacho, N. P. [1998] Potential role of bone morphogenetic proteins in fracture healing. *Clin. Orthop.* **355S,** 274–282, with permission.)

and 28 d after fracture. At the conclusion of the experiment, 18 femora from each subgroup were tested biomechanically and 6 were analyzed histologically. A statistically significant increase in stiffness in the rhBMP-2-treated fractures was observed by d 14 and continued at 21 and 28 d after fracture as compared to the other two groups. There was also a significant increase in strength in the rhBMP-2-treated fractures at d 28 (**Fig. 2**). A robust subperiosteal membranous bone response, greater than that seen in either of the control groups, was demonstrated histologically in the fractures treated with rhBMP-2.

Additionally, compared to controls, there was relative maturation of osteochondrogenic cells in the rhBMP-2-treated fractures. Bridging callus appeared earlier in the rhBMP-2-treated groups, and relatively increased peripheral woven bone was seen in these groups as well. The authors concluded that local percutaneous injection of rhBMP-2 into fresh fractures might accelerate the rate of normal fracture healing.

Bostrom and Camacho *(112)* and Turek et al. *(113)* studied BMP-2 combined with an absorbable collagen sponge and applied as an onlay graft in a rabbit ulnar osteotomy. Sixty rabbits were used, and limbs were randomized into three groups: 200 µg BMP-2 with collagen carrier, collagen carrier alone, and untreated controls. Radiographic and biomechanical analyses were used to evaluate healing at 2, 3, 4, and 6 wk after the osteotomy. Age-matched intact ulnae were used as controls. Radiographically, at 2 wk, the BMP-2-treated group showed slightly more mineralization in the callus compared with the collagen carrier-treated and untreated control groups. At 3 wk, the callus cross sectional area in the BMP-2-treated group was larger than that observed in the other two groups, although this was not statistically significant. In addition, bony bridging was seen in 7 of 10 specimens in the BMP-2-treated group, compared to no such bridging being evident in the other groups. At 6 wk, there was no difference among the three treatment groups in terms of overall fracture callus area of hard callus area. Biomechanically, stiffness and strength increased over time for all treatment groups. At the 2-, 4-, and 6-wk time points, there were no statistically significant differences among the treatment groups. At the 3-wk time point, energy and strength values in limbs treated with BMP-2 were greater than those treated with collagen or the controls. There was a trend of increasing stiffness in the BMP-treated control groups. The mean values of strength, energy to failure, and stiffness for the BMP-2-treated limbs at 3 wk after osteotomy were not significantly different from the values for the intact controls. The mean fracture scores as assessed by the Lane modification of White's classification were also evaluated *(48,114)*. At 2 wk, loading-to-failure scores indicated that most failures occurred through the soft callus. At 3 wk, the integrity of all fractures was improved, especially in those treated with BMP-2 as compared to control and collagen-treated ones. At 4 wk, the load-to-failure scores in the limbs treated with BMP-2 were also greater than those of the controls and were equivalent to the scores of all groups at wk 6. The investigators concluded that local application of BMP-2 on a collagen carrier could accelerate bony repair radiographically and biomechanically.

Three groups of investigators have tested the efficacy of BMP-7 in animal non-critical-sized defect models. Cook *(115)* and Poplich et al. *(116)* created bilateral 3-mm non-critical-sized defects in the mid-ulna of 35 adult male dogs. The animals were divided into three groups. One group served as a control. The second group received 0.35 mg of BMP-7 in an acetate buffer in one defect and a control solution in the contralateral defect. The third group received 0.35 mg of BMP-7 in a carboxylmethyl-cellulose carrier in one defect and carboxylmethylcellulose collagen alone in the contralateral defect. Animals in the first group were euthanized at 4, 8, 12, or 16 wk, and those in the second and third groups at 4, 8, or 12 wk only. Serial radiographic examinations revealed new bone formation as early as 2 wk in the BMP-7-treated groups, compared to similar findings at 4 wk in the untreated and carrier-treated controls. By 4 wk, the BMP-treated osteotomies had bony bridging evident, as compared to the control limbs in which bony bridging was not complete until 12 wk. Biomechanical testing revealed that it took between 4 and 8 wk for the BMP-7-treated bones to approach strengths of previously tested intact ulnae, compared to the 16-wk requirement of untreated controls to achieve similar biomechanical properties. Histological findings correlated with radiographic and mechanical testing results. In the BMP-7-treated defects, maturing bone was well incorporated with the host bone at early time periods. At later periods, dense bone filled and bridged the defects. These findings were not observed in control defects until 12–16 wk.

den Boer et al. *(117)* investigated the effect of rhOP-1 on fresh fracture gaps. Forty adult female goats underwent a closed tibial fracture that was stabilized with an external fixator. The fractures were then randomized into one of the following protocols: (1) no treatment, (2) injection of 1.0 mg of rhOP-1 dissolved in aqueous buffer, (3) injection with a collagen matrix, and (4) injection with 1.0 mg of rhOP-1 bound to a collagen matrix. The animals were euthanized at 2 or 4 wk. Three-dimensional computational tomography (CT) scanning and dual-energy X-ray absorptiometry were used to evaluate callus volume and bone mineral content, respectively. Biomechanical and histological analyses were also performed. At 2 wk, callus diameter, volume, and bone mineral content at the fracture site was significantly increased in both rhOP-1 groups as compared with the untreated group. Bending and torsional stiffness were higher, and bony bridging of the fracture gap was observed more often in the group treated with rhOP-1 in an aqueous buffer as compared to the uninjected group. At 2 wk follow-up, neither the biomechanical properties nor the bony bridging was improved by the addition of the collagen matrix. At 4 wk, there were no differences between the groups, except for a larger callus volume in the rhOP-1-plus-collagen matrix group compared to the control groups. The authors concluded that a single local administration of rhOP-1, regardless of its carrier, could accelerate the healing of a closed fracture in a goat model. The limitations of the studies conclusions, however, were well noted by the authors. These include the relatively rare use of external fixators in the treatment of human closed tibial fractures, the fact that fracture healing in goats is not very representative of human fracture healing (especially with regard to the rate of fracture healing), the relatively small number of animals per treatment group, and the lack of a group in which aqueous solution alone was injected into fractures.

Although the effect of exogenously administered BMP into acute bony defects has been studied for some time, it has not been until recently that investigators have focused their attention to the response of nonunions to BMPs. Whereas in acute defects the species specificity of a BMP does not appear to be important *(118)*, its effect on the ability of a BMP to enhance healing of a nonunion has been shown in a canine model *(119)*. Attempting to create an effective bone graft substitute for the treatment of a diaphyseal nonunion, Heckman et al. *(120)* created standardized nonunions in the midportion of the radial diaphysis in 30 mature mongrel dogs. The nonunion was treated with implantation of a carrier consisting of poly(DL-lactic acid) and polyglycolic acid polymer (50/50 polylactic acid/polyglycolic acid [PLG50]) containing canine purified BMP or recombinant human transforming growth factor-β (TGF-β1), or both, or the carrier without BMP or TGF-β1. Five groups, consisting of six dogs each, were treated with implantation of the carrier alone, implantation of the carrier with 15 mg of BMP, implantation of the carrier with 1.5 mg of BMP, implantation of the carrier with 15 mg of BMP and 10 ng of TGF-β1, or implantation of the carrier with 10 ng of TGF-β1. The specimens were examined radiographically and histomorphometrically at 12 wk after implantation. The radii treated with either 1.5 or 15 mg of BMP showed significantly increased periosteal and endosteal bone formation. No significant radiographic or histomorphometric evidence of healing was observed after implantation of the polylactic acid/polyglycolic acid carrier alone or in combination with 10 ng of TGF-β1. The authors concluded that species-specific BMP incorporated into a polylactic acid/polyglycolic acid carrier implanted at the site of an ununited diaphyseal fracture increases bone formation. Additionally, TGF-β1 at the dose used in the study did not have a similar effect and did not potentiate the effect of BMP. They suggested that the biodegradable implant containing BMP that was used in their study was an effective bone graft substitute. The importance of this study is paramount because it confirms the biocompatibility of polylactic acid/polyglycolic acid composites, the bioavailability of BMP and TGF-β1 released from this implant, and most important, the capability of BMP to augment bone healing in chronic nonunions.

Clinical experience with exogenously administered BMPs in bony defects or fresh fractures is somewhat limited. To date, studies on the enhancement of bony regeneration in humans have been limited to BMP-2 and BMP-7 (OP-1). Geesink et al. investigated the osteogenic potential of BMP-7 in a critical-size human bony defect model *(121)*. Twenty-four patients undergoing high tibial osteotomy for osteoarthritis of the knee were divided into four groups. The first group consisted of patients in whom the fibular osteotomy had been left untreated, while in the second group demineralized bone had been unchanged in the control group. These researchers concluded that systemic administration of growth hormone greatly accelerates ossification of bone regenerate in distraction osteogenesis. Their study is especially significant in that it used a species-specific GH, which, unlike other studies that examined the effects of allogeneic GH, should not induce anti-GH antibody formation.

Because of the variability in results of these and other investigations, no general conclusions can be made regarding the effects of growth hormone and insulin-like growth factors on fracture healing *(180)*. Certainly, these factors play an important role in the regulation of skeletal growth and development. The question of whether growth hormone or its mediators plays a role in the process of skeletal repair, and whether these compounds can be utilized in augmenting bone healing in normal or abnormal healing enviromnents, remains controversial.

CONCLUSION

Numerous animal-based studies have proven the effectiveness of growth factors in enhancing fracture healing. Bone morphogenetic proteins seem to be the most potent of these compounds. Given the variability in experimental design of many of these studies, further research using standardized models needs to be performed. Although human trials have already begun using some of these compounds, basic science research developing better delivery systems is needed. Gene therapy may be an effective strategy for enhancing delivery of many of these molecules. Considering the burden of disease produced in the population as a result of impaired fracture healing, the development of technologies to restore skeletal integrity will have a major impact on the future of musculoskeletal care.

REFERENCES

1. Praemer, A., Furner, S., and Price, O. P. (1992) *Musculoskeletal conditions in the United States.* American Academy of Orthopaedic Surgeons, Rosemont, IL, pp. 85–91.
2. LaVelle, D. G. (1998) Delayed union and non-union of fractures, in *Campbell's Operative Orthopaedics,* 9th ed. (Canale, S. T., ed.), Mosby, St. Louis, pp. 2579.
3. Urst, M. R. (1965) Bone: formation by autoinduction. *Science* **150,** 893–899.
4. Urist, M. R. and Strates, B. S. (1971) Bone morphogenetic protein. *J. Dent. Res.* **50,** 1392–1406.
5. Wozney, J. M., Rosen, V., Celeste, A. J., et al. (1988) Novel regulators of bone formation: molecular clones and activities. *Science* **242,** 1528–1534.
6. Reddi, A. H. (1984) Extracellular matrix and development, in *Extracellular Matrix Biochemistry* (Piez, K. A. and Reddi, A. H., eds.), Elsevier, New York, pp. 375–412.
7. Trippel, S. B. (1997) Growth factors as therapeutic agents. *Instr. Course Lect.* **46,** 473–476.
8. Massague, J. (1996) TGF-B signaling: receptors, transducers, and mad proteins. *Cell* **85,** 947–950.
9. Lein, P., Johnson, M., Guo, X., Rurger, D., and Higgins, D. (1995) Osteogenic protein-1 induces dendritic growth in rat sympathetic neurons. *Neuron* **15,** 597–605.
10. Shah, N. M., Groves, A. K., and Anderson, D. J. (1996) Alternative neural crest cell fates are instructively promoted by TGFB superfamily member. *Cell* **85,** 331–343.
11. Marsh, J. L., Buckwalter, J. A., and Evarts, C. M. (1994) Non-union, delayed union, malunion and avascular necrosis, in *Complications in Orthopaedic Surgery* (Epps, C. H., ed.), Lippincott, Philadelphia, pp. 183–211.
12. Tiedman, J. J., Lippiello, L., Connolly, J. F., and Strates, B. S. (1990) Quantitative roentgenographic densitometry for assessing fracture healing. *Clin. Orthop.* **253,** 279–286.
13. Richardson, J. B., Cunningham, J. L., Goodship, A. E., et al. (1994) Measuring stiffness can define healing of tibial fractures. *J. Bone Joint Surg.* **76B,** 389–394.
14. Marsh, D. (1998) Concepts of fracture union, delayed union, and non-union. *Clin. Orthop.* **355S,** S22–S30.
15. Boyd, H. B., Lipinski, S. W., and Wiley, J. H. (1961) Observations on non-union of the shaft of long bones, with statistical analysis of 842 patients. *J. Bone Joint Surg.* **43A,** 159–168.
16. Einhorn, T. A., Bonnarens, F., and Burnstein, A. H. (1986) The contributions of dietary protein and mineral to the healing of experimental fractures: a biomechanical study. *J. Bone Joint Surg.* **68A,** 1389–1395.
17. Rothman, R. H., Klemek, J. S., and Toton, J. J. (1971) The effect of iron deficiency anemia on fracture healing. *Clin. Orthop.* **77,** 276–283.
18. Macey, L. R., Kana, S. M., Jingushi S., et al. (1989) Defects of early fracture healing in experimental diabetes. *J. Bone Joint Surg.* **71A,** 722–733.

81. Lind, M
    rabbit ti
82. Nielson
    strength
83. Centrell
    family 1
84. Sakou,
85. Heldin,
    proteins
86. Hogan,
    10, 158
87. Kloen,
    human
88. Yamagu
    osteobla
89. Chen, P
    mesoden
    growth
90. Thies, F
    induces
91. Bouletre
    expressi
92. Fiedler,
    tic migr
93. Bonnare
    Res. **2,**
94. Bostron
    tic prote
95. Nakase,
    tein 4 n
96. Ishidou,
    proteins
97. Hayashi
    (BMPs)
98. Yonemc
    highly e
    J. Path
99. Onishi,
    genetic
100. Gerhart,
    binant h
101. Sciadini
    bone m
102. Kirker-I
    genetic
103. Kirker-I
    bone m
104. Cook, S
    ing of l
105. Cook, S
    healing
106. Cook, S
    of segm
107. Stevens
    formatic
108. Sciadini
    a canine
109. Sciadini
    coral ca

19. Walsh, W. R., Sherman, P., Howlett, C. R., Sonnabend, D. H., and Ehrlich, M. G. (1997) Fracture healing in the rat osteopenia model. *Clin. Orthop.* **342,** 218–227.
20. Uhthoff, H. K. and Rahn, B. A. (1981) Healing patterns of metaphyseal fractures. *Clin. Orthop.* **160,** 295–303.
21. Nicoll, E. A. (1964) Fractures of the tibial shaft: a survey of 705 cases. *J. Bone Joint Surg.* **46B,** 373–387.
22. Court-Brown, C. and McQueen, M. (1987) Compartment syndrome delays tibial union. *Acta Orthop. Scand.* **58,** 249–252.
23. Claes, L., Augat, P., Suger, G., and Wilke, H. J. (1997) Influence of size and stability of the osteotomy gap on success of fracture healing. *J. Orthop. Res.* **15,** 577–584.
24. Kwiatkowski, T. C., Hanley, E. N., and Ramp, W. K. (1996) Cigarette smoking and its orthopaedic consequences. *Am. J. Orthop.* **9,** 590–595.
25. Cruess, R. L. and Sakai, T. (1972) The effect of cortisone upon synthesis rates of some components of rat bone matrix. *Clin. Orthop.* **86,** 253–259.
26. Stinchfield, F. E., Sankaran, B., and Samilson, R. (1956) The effect of anticoagulant therapy on bone repair. *J. Bone Joint Surg.* **38A,** 270–282.
27. Allen, H. L., Wase, A., and Bear, W. T. (1980). Indomethacin and aspirin: effect of nonsteroidal anti-inflammatory agents on the rate of fracture repair in the rat. *Acta Orthop. Scand.* **51,** 595–600.
28. Einhorn, T. A. (1995) Current concepts review: enhancement of fracture-healing. *J. Bone Joint Surg.* **77A,** 940–956.
29. Meadows, T. H., Bronk, J. T., Chao, E. Y. S., and Kelly, P. J. (1990) Effect of weight-bearing on healing of cortical defects in the canine tibia. *J. Bone Joint Surg.* **72A,** 1074–1080.
30. Paley, D., Catagni, M., Argnani, F., et al. (1989) Ilizarov treatment of tibial non-unions with bone loss. *Clin. Orthop.* **241,** 146–165.
31. Sharrard, W. J. W. (1990) A double blind trial of pulsed electromagnetic field for delayed union on tibial fractures. *J. Bone Joint Surg.* **72B,** 347–355.
32. Xavier, C. A. M. and Duarte, L. R. (1983) Estimulaci ultra-sonica de callo osseo: applicaca clinica. *Rev. Brasileira Orthop.* **18,** 73–80.
33. Heckman, J. D., Ryaby, J. P., McCabe, J., et al. (1994) Acceleration of tibial fracture healing by non-invasive low intensity pulsed ultrasound. *J. Bone Joint Surg.* **76A,** 26–34.
34. Kristiansen, T. K. (1990) The effect of low power specifically programmed ultrasound on the healing time of fresh fractures using a Colles' model. *J. Orthop. Trauma* **4,** 227–228.
35. Cook, S. D., Ryaby, J. P., McCabe, J., et al. (1997) Acceleration of tibia and distal radius fracture healing in patients who smoke. *Clin. Orthop.* **337,** 198–207.
36. Chase, W. S. and Herndon, C. H. (1955) The fate of autogenous and homeogeneous bone grafts: a historical review. *J. Bone Joint Surg.* **37A,** 809–841.
37. Muschler, G. F. and Lane, J. M. (1992) Orthopaedic surgery, in *Bone Grafts and Bone Substitutes* (Habal, M. B. and Reddi, A. H., eds.), Saunders, New York, pp. 375–407.
38. Younger, E. M. and Chapman, M. W. (1989) Morbidity at bone graft donor sites. *J. Orthop. Trauma* **3,** 192–195.
39. Kurz, L. T., Garfin, S. R., and Booth, R. E. (1989) Harvesting autogenous iliac bone grafts: a review of complications and techniques. *Spine* **14,** 1324–1331.
40. Einhorn, T. A. (1999) Clinically applied models of bone regeneration in tissue engineering research. *Clin. Orthop.* **367S,** S59–S67.
41. Aro, H., Eerola, E., and Aho, A. J. (1985) Development of non-unions in the rat fibula after removal of periosteal neural mechanoreceptors. *Clin. Orthop.* **199,** 292–299.
42. Bostrom, M., Lane, J. M., Tomin, E., et al. (1996) Use of bone morphogenetic protein-2 in the rabbit ulnar non-union model. *Clin. Orthop.* **327,** 272–282.
43. Hietaniemi, K., Peltonen, J., and Paavolainen, P. (1995) An experimental model for non-union in rats. *Injury* **26,** 681–686.
44. Lattermann, C., Baltzer, A. W. A., Whalen, J. D., et al. (1998) Establishment and validation of an atrophic non-union model in a rabbit for use of gene therapy. *Trans. Int. Soc. Fracture Repair* **6,** 55.
45. Schmitz, J. P. and Hollinger, J. O. (1986) The critical size defect as an experimental model for craniomandibulofacial non-unions. *Clin. Orthop.* **205,** 299–308.
46. Hollinger, J. O. and Kleinschmidt, J. P. (1990) The critical sized defect as an experimental model to test bone repair materials. *Craniofacial Surg.* **1,** 60–68.
47. Nunamaker, D. M. (1998) Experimental models of fracture repair. *Clin. Orthop.* **355S,** S56–S65.
48. Yasko, A. W., Lane, J. M., Fellinger, E. J., et al. (1992) The healing of segmental bone defects induced by recombinant human morphogenetic protein (rhBMP-2). A radiographic, histological and biomechanical study in rats. *J. Bone Joint Surg.* **74A,** 659–670.
49. Einhorn, T. A. (1998) The cell and molecular biology of fracture healing. *Clin. Orthop.* **355S,** S7–S21.
50. Utvag, S. E., Grundes, O., and Reikeras, O. (1998) Graded exchange reaming and nailing of non-unions. *Arch. Orthop. Trauma Surg.* **5,** 1–6.

190

51. S
52. K
    fi
53. R
    e
54. S
55. K
    g
56. O
    tr
57. O
    b
58. W
59. J
    a
    T
60. R
    fa
61. R
    si
62. A
    de
63. Jo
    re
64. A
    in
65. R
    tu
66. Ta
    er
67. In
    dil
    en
68. W
    in
69. Pi
    tor
70. R
    an
71. Jo
    tia
72. No
    cri
73. M
    ing
74. Be
    bo
75. Be
    Re
76. Pet
    for
77. He
    rad
78. Be
    fac
79. Be
    pha
80. Cri
    2 o

routinely had fewer complications *(28,51,79,144)*. The majority of complications reported with the Wagner method have been related to bone healing *(80)*, which is exactly what the distraction osteogenesis method seeks to address.

Independent of the method and etiology, any extended lengthening routinely encounters a plethora of complications. As detailed in a comprehensive review *(57)*, complications were categorized according to pin tract (acute mechanical or thermal damage or late inflammation to frank infection of the underlying bone), bone (premature consolidation, delayed consolidation, nonunion, axial deviation, late bending or fracture), joint (contracture: hip and knee flexion, ankle equinus and subluxation), neurovascular (acute or delayed nerve or vessel injury, local edema, systemic hypertension and compartment syndromes), and psychological.

Despite the improved complication rate over the traditional Wagner technique, the reported results of distraction osteogenesis for limb lengthening still reveal higher complication rates than those reported by Ilizarov. Quite a disparity exists when comparing the complication rates of Wagner (45%), DeBastiani (14%), and Ilizarov (5%), and an even greater disparity (1–225%) exists when comparing all series *(52,128)*.

Clearly, part of the reason for such a discrepancy in the reported complication rates stems from differing definitions of "complication." Limb lengthening is a complex and prolonged procedure that extends well beyond the operating room. Problems are expected and discussed with patients before surgery. Just as problems are encountered and solved by any surgeon during an operation (e.g., bleeding, muscle disruption and repair, inaccurate osteotomy cuts or pin placements, etc.), similar problems are encountered by the patient and surgeon following the operation during an extended limb lengthening. Some authors call the latter (e.g., pin-site inflammation, pain, paresthesias, edema, and transient contractures) problems, obstacles, or minor complications *(52,128,165)*. Major or true complications are reserved for unexpected occurrences that significantly alter the treatment plan (additional operations or premature cessation of the lengthening), outcome (fracture, malunion), or function (permanent nerve injury, contracture) *(52,128,165)*.

Two independent reports *(52,165)* carefully compared major complication rates to the surgeon's experience and found that complications dropped significantly with experience, from 72% to 25% after the first 30 cases *(52)* and from 69% to 35% after the first year of using the Ilizarov method on a regular basis *(165)*. The incidence of minor complications or problems remained relatively constant, independent of surgeon experience and fixator type *(52,165)*.

Generally, the number of complications and failed lengthenings increases proportional to the length of the distraction *(52,83,165)*. Unilateral lengthenings (for congenital or acquired anisomelia) had twice the number of complications per segment as bilateral lengthening (for short stature) *(7)*. Femoral and tibial segment complication rates were similar, and both were higher than those for the humerus *(7)*. Dahl found that the number of complications was correlated to the severity of the preoperative problem and not the type of external fixator used *(52)*.

Surgical procedures routinely cause pain. When the procedure extends beyond the operating room, such as the case in distraction osteogenesis and prolonged external fixation, pain of varying degrees persists as well. In a prospective series of 23 patients (ages 11–20 yr), Young et al. compared two standardized tests for pain *(179)*. They found that the immediate postoperative pain was similar in magnitude to a standard orthopedic operation (osteotomy), but that pain of some degree persisted throughout the entire period of external fixation. There was a trend of decreasing pain from postoperative to mid-distraction to mid-consolidation time points studied.

A rare complication has been reported *(85)*, which may not be related to the method but deserves mention. Four years following an Ilizarov lengthening of the femur through an area of fibrous dysplasia, a teenage boy presented with an osteosarcoma at the site of distraction osteogenesis. Although spontaneous sarcomatous degeneration has been reported in regions of fibrous dysplasia, this case report questions a possible association between the highly activated biology of distraction osteogenesis in a region of dysplastic bone and subsequent malignant degeneration.

Joint subluxation and contracture seem to be two of the more serious complications, which can be minimized by special preoperative planning and therapy during the fixation period. When Suzuki et al. *(159)* compared two groups of femoral lengthenings according to preoperative hip stability (based on the CE angle), none of the 14 hips deteriorated if the CE angle was greater than 20°. Five of 12 hips subluxed or dislocated with CE angles less than 20°. The lengthenings averaged 5 cm using monolateral frames without crossing the hip joint for prophylaxis. The authors recommend a prelengthening osteotomy in hips at risk. Herzenberg et al. *(88)* studied knee motion before, during, and after 25 isolated femoral lengthenings without frame prophylaxis across the joint. Preoperative flexion averaged 127°, which decreased to a mean of 37° during distraction and returned to a mean of 122° at final follow-up after fixator removal. It took nearly twice the total time in the fixator to return to normal motion, however, and two patients had permanent loss of more than 15% of the preoperative motion. To avoid contractures, a specific program of physiotherapy (warm-up, isometrics, passive stretch, active motion, and weight-bearing ambulation) that involves at least 2–3 h/d (Ilizarov's patients underwent at least 6 h of group physiotherapy a day) has been recommended, in addition to night splinting *(48,75)*. Therapy progresses over four phases of treatment: inpatient, outpatient distraction, outpatient consolidation, and outpatient fixator removal *(48)*.

Deformity Correction

Deformities have been identified in the bone, joint, or contour of a limb. Bony deformities can be angulatory, rotatory, translatory, and/or involve shortening. Joint deformities can be related to motion (contracture or laxity) or articulation (subluxation or dislocation). Contour deformities involve the shape of the limb and can be related to soft tissue or bony deformity. The mechanical axis of the limb extends linearly from the center of rotation of the apical joints (in the leg from the hip to the ankle). The anatomical axis of an individual bone is derived from the diaphyseal alignment. Simple deformities can be resolved in one plane, while complex deformities involve more than one deformity in different planes. Complex deformities may accentuate each other, worsening the mechanical or anatomical axis, or they may compensate for each other, improving the mechanical or anatomical axis.

Treatment of limb deformities require a meticulous analysis of the clinical and radiographic features to determine the true deformity(ies). The Ilizarov method incorporates this information into the frame itself, using a strategy to acutely and/or gradually correct deformity through bone (distraction osteogenesis) and/or soft tissues. The frame construction generally utilizes four-point fixation to obtain mechanical advantage through fulcrum hinges and finely threaded inclined rods with long lever arms to gradually angulate, translate, rotate, and/or lengthen the bone segments. By accurate placement of the fulcrum hinge(s) and stable fixation, the mechanical axis can be corrected efficiently and gradually, allowing for spontaneous osteogenesis and soft tissue adaptation. The anatomical axes are not always corrected, because the correction may not be possible at the true level of deformity (due to local scarring or multiple deformities); compensating deformities may have to be created to correct the overall limb mechanical axis (which takes functional precedence).

The actual sequence for analyzing deformity and planning correction is beyond the scope of this review, but is well described in several papers by Paley *(131,134,160)*. Herzenberg et al. *(89,90)* expanded the methods for calculating rate and duration of deformity correction, as well as application of the method for torsional deformity correction using mathematical accuracy. Tetsworth *(161)* reviewed the clinical success of deformity correction in actual cases that used the most advanced techniques of Paley and Herzenberg. Comparing pre- to postoperative deformity, he found that the accuracy of correction improved with surgical experience. Mechanical axis deviation improved from an average of 48 mm to an average of 8.6 mm after correction, while the tibiofemoral angle improved from an average of 16° to 3°.

The Ilizarov frame also provides a convenient method for acute correction and stabilization of deformity, such as in a femoral derotational osteotomy for femoral anteversion *(117)*. The Ilizarov method of distraction osteogenesis has been used successfully to correct significant deformities in pathological bone from metabolic diseases such as renal osteodystrophy, hypophosphatemic rickets, and hypophosphatasia, if combined with medical management *(155)*. A single case of combined tibial lengthening and arthrodiastasis of a knee flexion contracture in a 12-yr-old girl with melorheostosis was reported to be successful *(23)*. Three patients (15–16 yr old) with osteogenesis imperfecta underwent significant lengthenings (5–9 cm) of four bones (three tibiae and one femur) *(61)*. Intramedullary rods were used during and after the lengthenings. Distraction osteogenesis can be used to correct deformity, even in certain types of pathological bone.

Nonunions

Ilizarov differentiated between types of nonunions based on clinical and radiographic findings to determine the treatment strategy that would biologically transform the interposed nonosseous tissue into bone. He described three basic types of nonunions: atrophic, normotrophic, or hypertrophic. Atrophic nonunions with interposed fat, loose fibrous tissue or even muscle were clinically mobile or "loose," and radiographically the bone ends were thin, osteopenic, and nonreactive, with a thick radiolucent space between bone ends. Hypertrophic nonunions were clinically stiff and radiographically expansive, with peripherally reactive bone formation and a thin radiolucent line between bone ends. Normotrophic nonunions were intermediate between atrophic and hypertrophic.

Hypertrophic nonunions with a vital blood supply from each bone end and a dense collagenous interface strongly resemble the biology of distraction osteogenesis and are therefore conducive to primary distraction to stimulate bone formation. Catagni et al. *(41)* used this strategy to treat 21 hypertrophic nonunions (11 tibiae, 9 femora, and 1 radius) in 19 patients (ages 18–65 yr). In addition to converting the nonunions to solid bone by primary distraction (0.25 mm twice a day), this method allowed for gradual correction of deformities (angular, axial, and translational) and spontaneous resolution of osteomyelitis. Stable union and deformity correction were achieved in all of the patients, while length discrepancies (average 3.9 cm, range 1–8 cm) were corrected in 18 of the 21 cases (86%) and the osteomyelitis resolved in five of six cases (83%). The sole complication was axial collapse of the regenerated bone in one patient following premature removal of the device. Fourteen of the 21 patients (66%) returned to their preinjury occupations. In another series from the same hospital *(42)*, 14 of 16 nonunions of the humerus were healed using the Ilizarov method. Of the two failures, one occurred at the site of postirradiated plasmacytoma and the other site had advanced disuse osteopenia. Three complications included a transient radial neuropraxia, which resolved following removal of a wire, and two refractures of the regenerated bone which healed with further treatment.

The strategy for atrophic nonunions requires either gradual compression of the site (to stimulate local inflammatory resorption of the atrophic interface tissues and neovascularity) followed by distraction (to transform the newly formed granulation tissue into distraction osteogenesis) or local compression with simultaneous distraction osteogenesis at an adjacent site in the same bone to increase local and regional blood flow. Twenty-two atrophic nonunions *(129)* were treated by this technique in patients ranging from 19 to 62 yr old. Of these patients, there were 13 with chronic osteomyelitis, 19 with shortening (2–11 cm), and 13 with deformity. Some of these nonunions had intercalary defects, which were regenerated at the distraction osteogenesis site. Infections were treated by resection of the necrotic bone, local compression, and adjacent distraction osteogenesis with or without actual bone transportation. Union was achieved in all cases, with a mean time to union of 13.6 mo. Ten of the 13 cases with osteomyelitis healed, 9 of the 10 deformities were corrected, and 18 of 19 length discrepancies were normalized. Complications included five equinus deformities, four cases of reflex sympathetic dystrophy, four patients with pain, and one voluntary amputation for neurogenic pain. The authors presented a classification scheme and treatment strategy for each type of nonunion *(129)*.

Bone Transportation

Intercalary defects from trauma, infection, tumor, or prosthetic replacement can be regenerated by transporting a segment of bone within the limb using mechanical methods, most commonly external

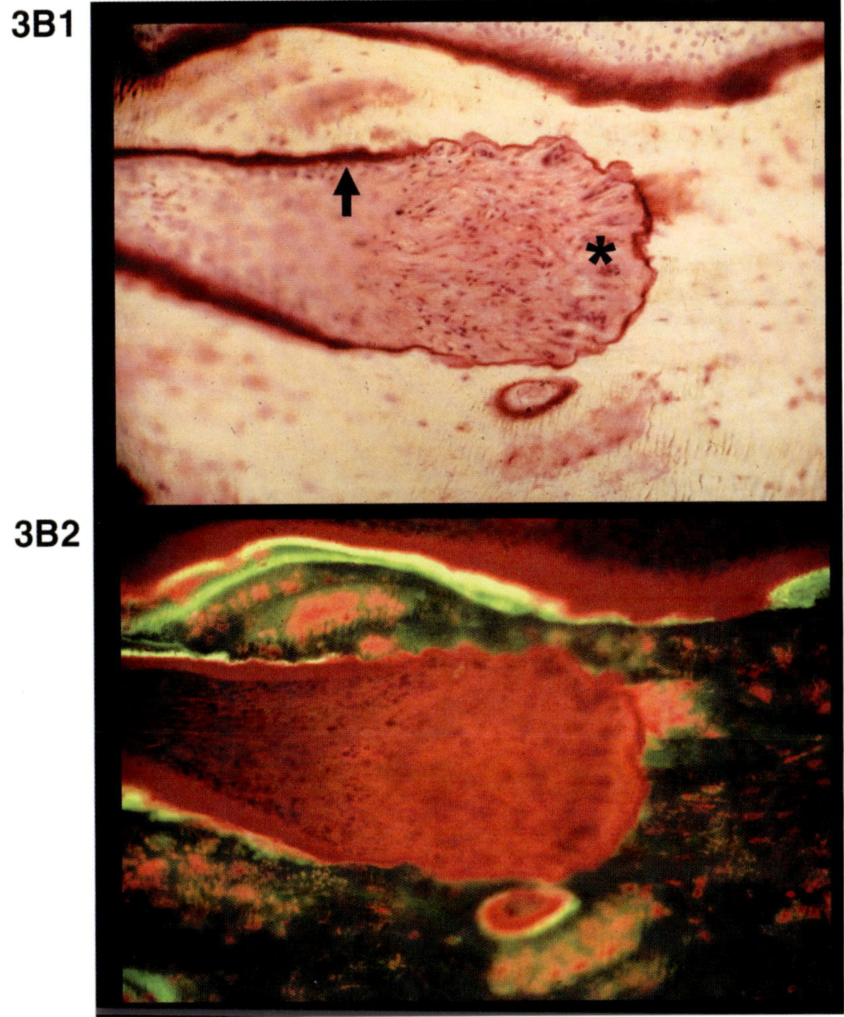

Color Plate 1, Fig. 3B. (*see* full caption and discussion in Ch. 1 on p. 10). Cutting cone in BMU.

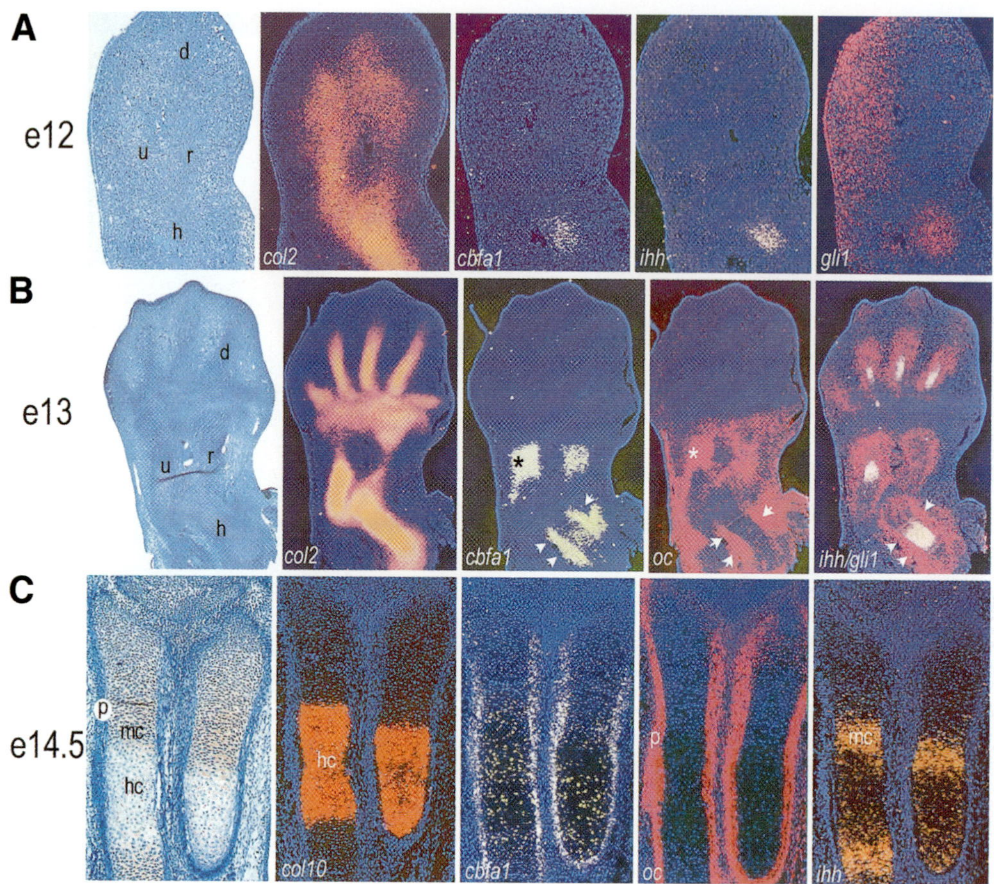

Color Plate 2, Fig. 1. (*see* full caption and discussion in Ch. 3 on p. 47.) Gene expression during mesenchymal cell condensation and cartilage development.

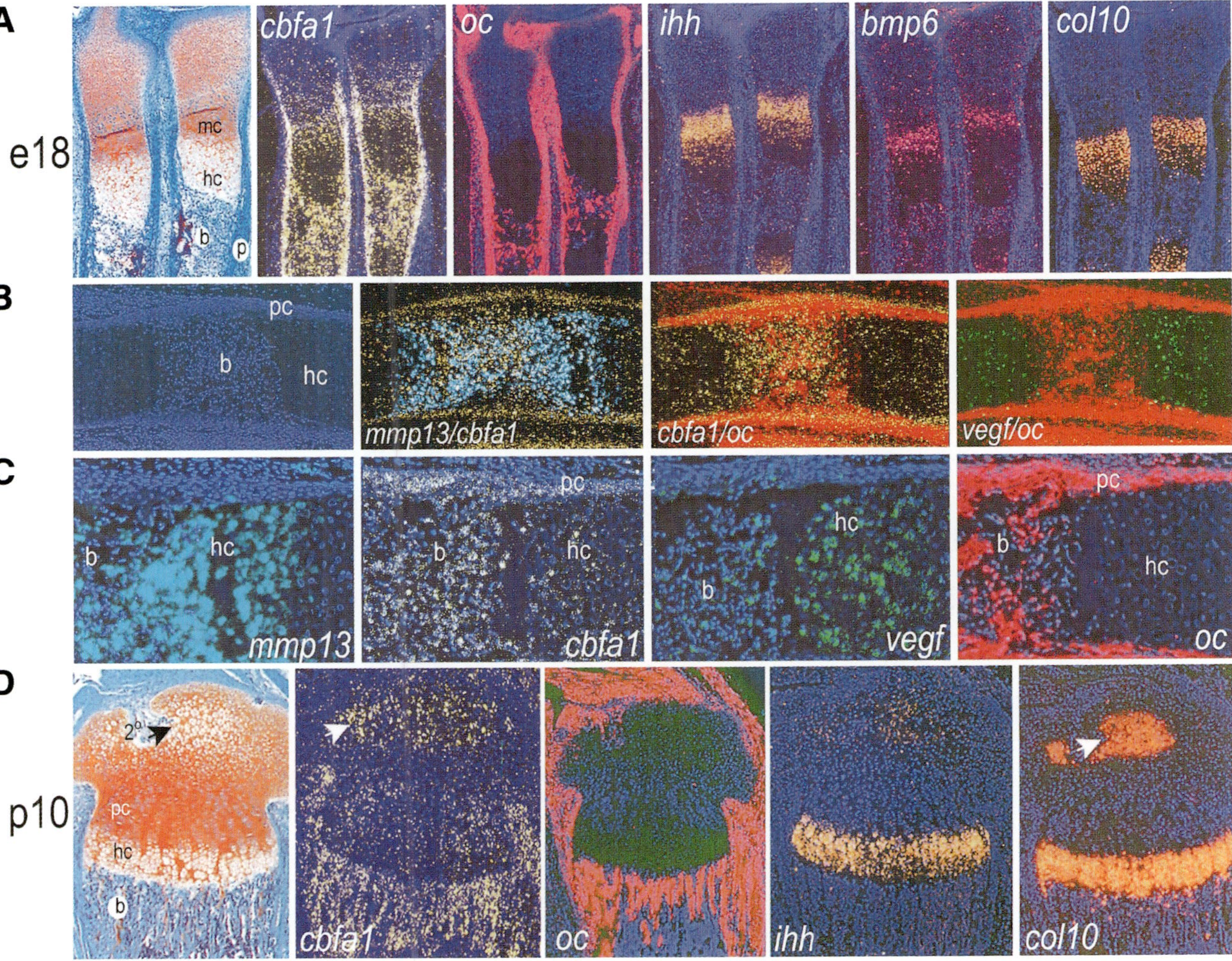

Color Plate 3, Fig. 2. (*see* full caption and discussion in p. 49.) Gene expression during cartilage maturation, vascular invasion, and ossification.

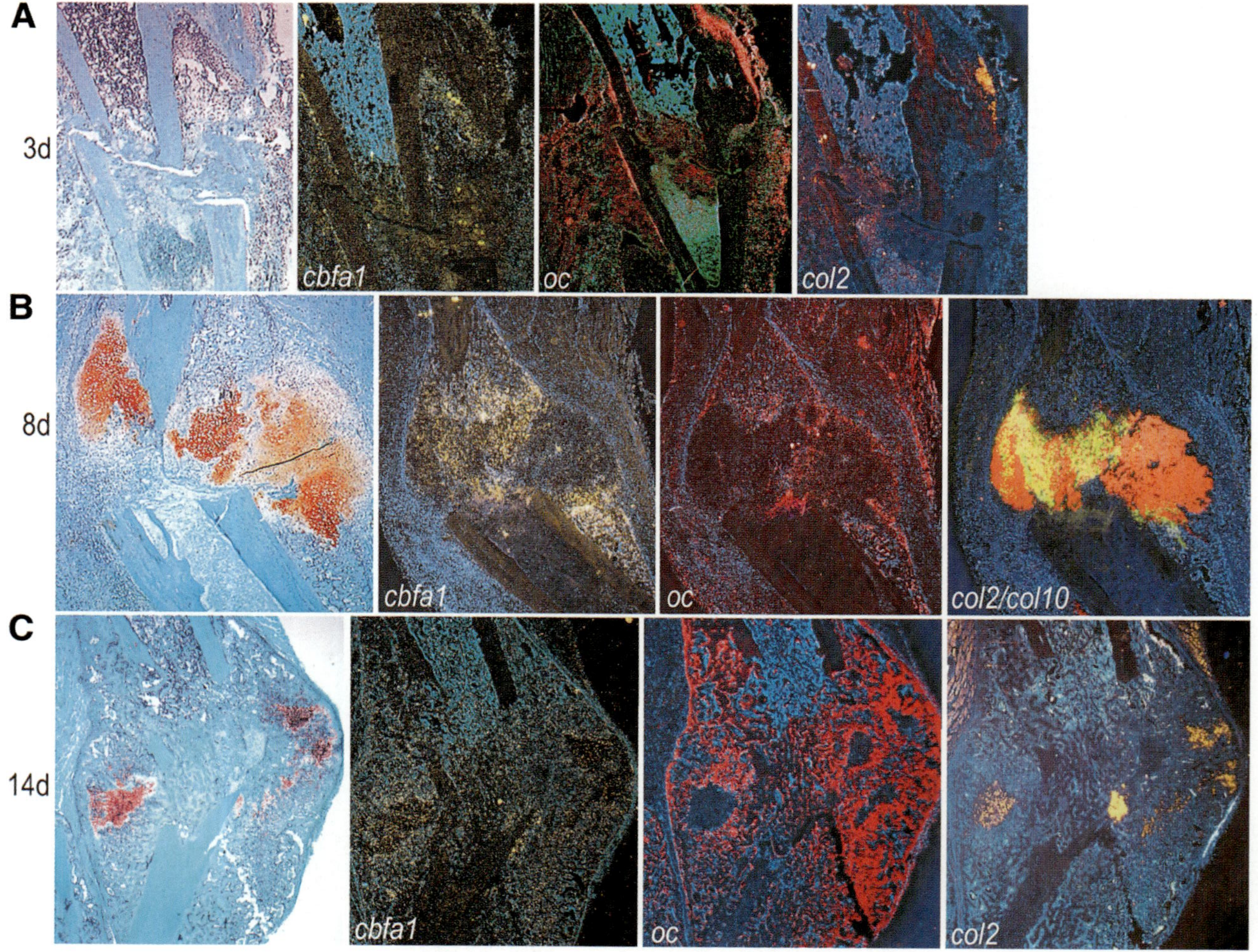

Color Plate 4, Fig. 3. (*see* full caption and discussion in Ch. 3 on p. 52.) Gene expression during the early, intermediate, and late states of nonstabilized fracture healing.

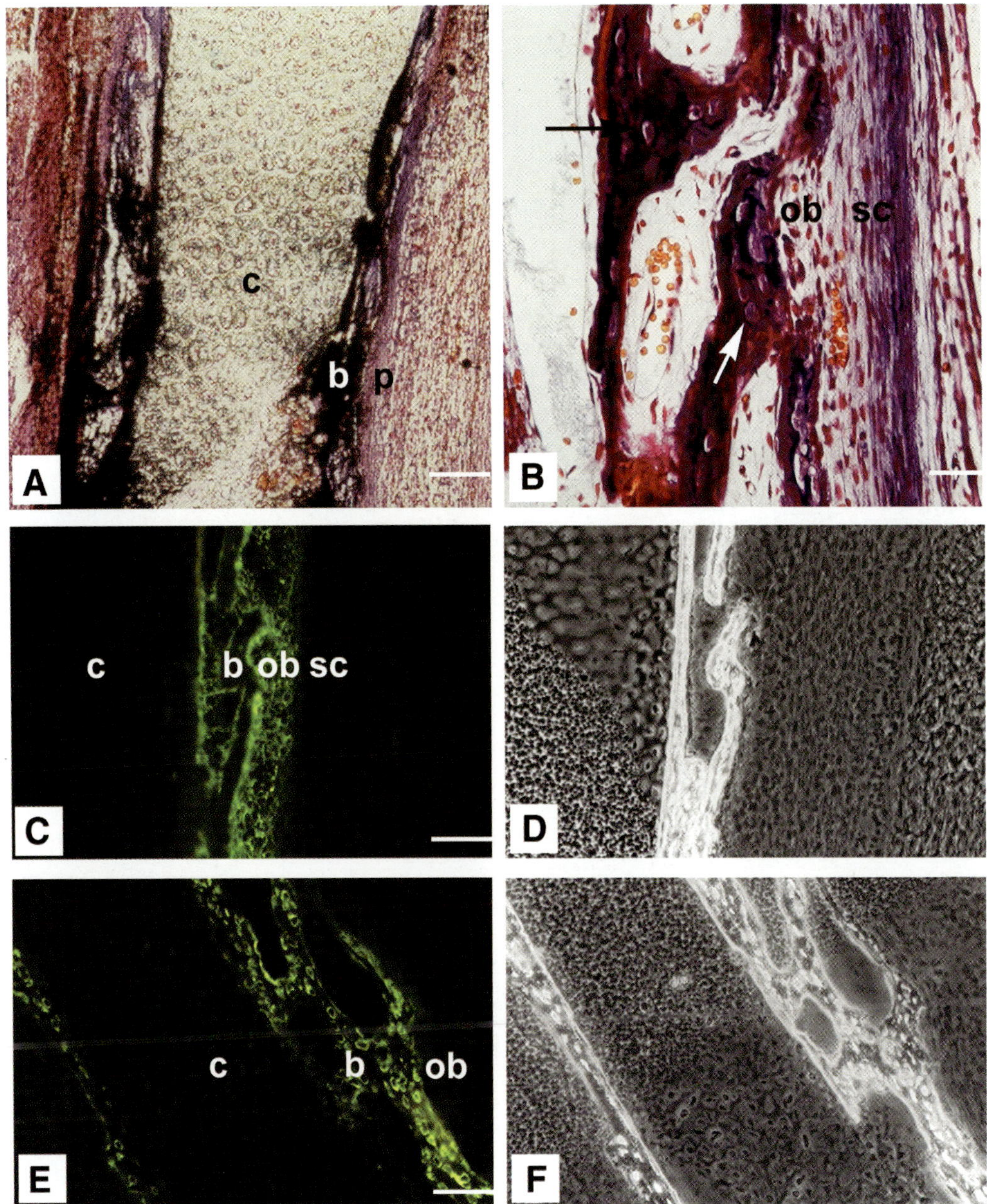

Color Plate 5, Fig. 2. (*see* full caption and discussion in Ch. 5 on on p. 69.) Mid-diaphysis of a stage 35 embryonic chick tibia.

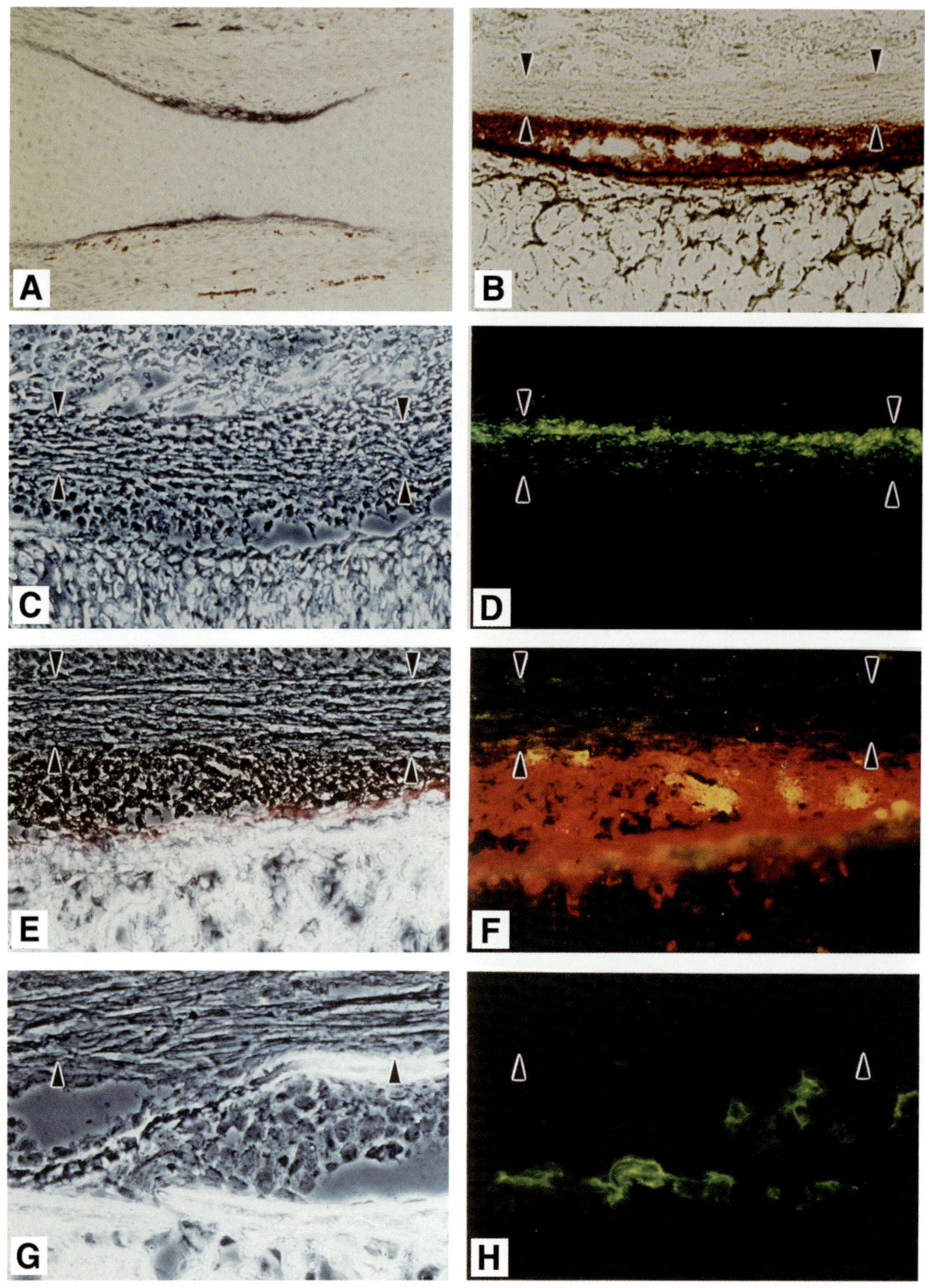

Color Plate 6, Fig. 7. (*see* full caption and discussion in Ch. 5 on p. 78.) Reactivity of antibodies SB-10 and SB-20 in longitudinal sections of developing human limbs.

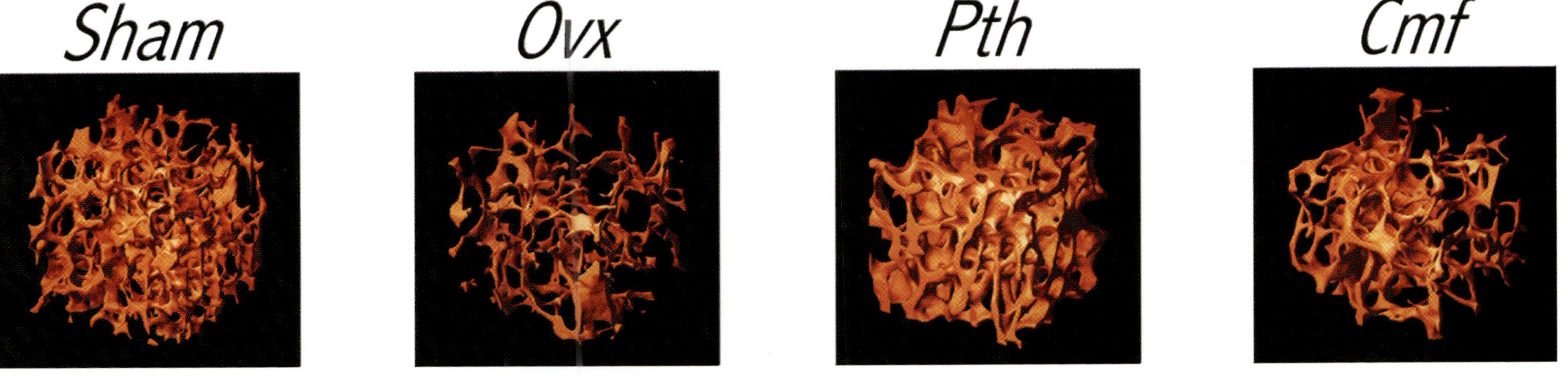

Color Plate 7, Fig. 3. (*see* full caption and discussion in Ch. 15 on p. 296.) Three-dimensional reconstructions of trabecular bone.

FRACTURE HEALING

Injury Phase

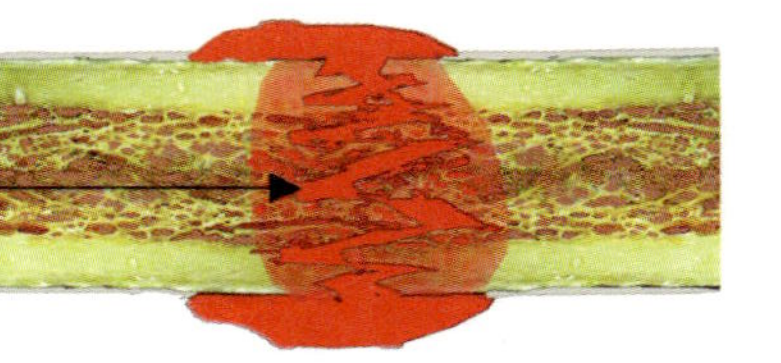

PLATELETS: Clotting, growth factors
Complement Activation
INFLAMATION: neutrophils, lymphocytes, monocytes, mast cells, macrophages
RESORPTION: osteoclasts, macrophages
FACTORS: TGF-β, PDGF, VEGF, FGF, EGF, cytokines, interleukins
ADDITIONAL CELLS: pericytes, undifferentiated cells

Proliferation Phase

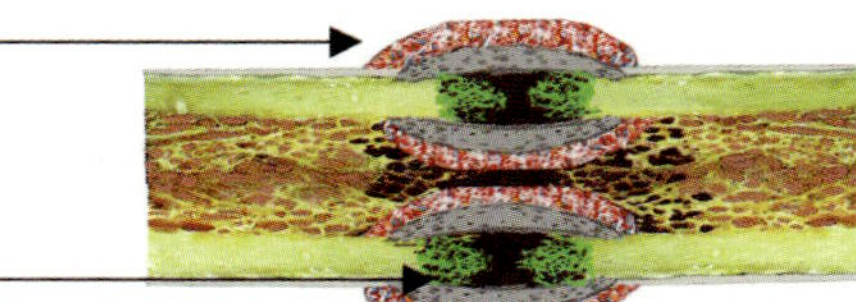

GRANULATION TISSUE: endothelial cells, Neovascularization, collagens (I, II, III, IV, IX, X), fibroblasts
FACTORS: TGF-β, PDGF, IGF, VEGF, BMPs
ADDITIONAL CELLS: macrophages, endothelial cells, chondrocytes, osteoclasts

Early Remodeling

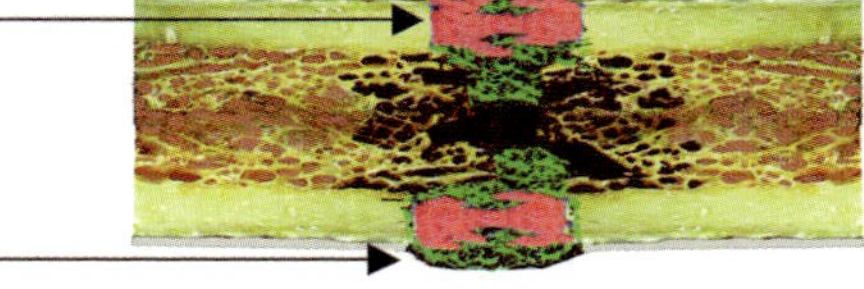

Woven bone develops into lmellar bone
FACTORS: TGF-β, CT, IGF, PTH, GH, BMPs
CELLS: osteoclasts, osteoblasts
Type I collagen predominates

Late Remodeling

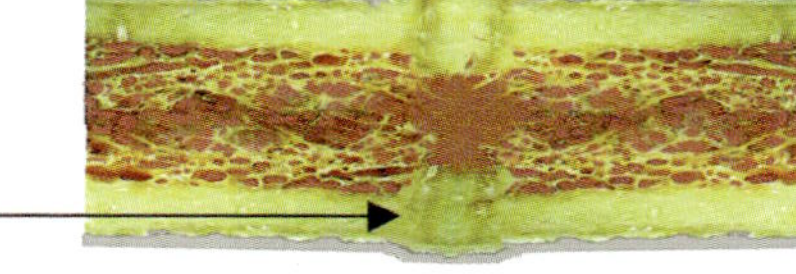

Lamellar bone, haversian bone
Bone contour restored

Color Plate 8, Fig. 1. (*see* full caption and discussion in Ch. 17 p. 338.) Four phases of fracture healing.

fixation as originally described by Ilizarov *(96)*. The bone segment must have an adequate blood supply to induce distraction osteogenesis at the trailing end and to heal when it is compressed against the docking site for transformation osteogenesis *(10,11,20)*. Granulation of soft tissue defects can also be stimulated by using such mechanical means.

The ring fixator can be used to transport bone segments in any direction using pulling wires, transverse, tensioned wires, or half-pins *(20,43,55,125)*. Monolateral fixators with half-pins can transport bone segments only in an axial direction. Partial segments of cortical bone (hemisections or less) can regenerate large defects by transportation if vascularity is maintained. Multiple segments can be transported in the same or opposite directions to accelerate regeneration of the defect. The principles of distraction osteogenesis—stable fixation, a low-energy osteotomy with gradual distraction, and bone formation by intramembranous ossification—are the same as for lengthening *(55)*. Although the classic Ilizarov method indicated that the docking site could heal by gradual and prolonged compression *(43, 96)*, Western surgeons have found that bone autograft supplementation following surgical debridement of the bone ends accelerates and facilitates healing *(46,76)*.

Pulling or oblique wires are unidirectional and cannot be used to reverse the direction of transport as is sometimes necessary to stimulate distraction osteogenesis if mineralization is delayed *(20,125)*. Since these wires often travel obliquely to the bone segment, the rate of wire distraction must be adjusted periodically to maintain a rate of 1 mm/d at the distraction osteogenesis site *(20)*. Pulling wires offer the advantage of reeling in the bone segment so as to minimize skin scarring and interference with soft tissues, which is especially useful for long transports, where nerves or vessels may cross the path of transverse wires or half-pins *(20)*. Pulling wires may be the only way to transport partial bone fragments in oblique or transverse directions *(10,43)*. Transverse wires or half-pins are used primarily for axial transport, preferably for shorter defects. They offer the advantage of transmitting more force to the soft tissues, which is helpful in closing soft tissue defects *(20,125)*. If the bone transport is axial, an intramedullary guide wire or rod may be used without interfering with bone formation and healing as long as the periosteum is maintained *(32)*.

The early results of bone transportation for the tibia were promising in a combined series of infected nonunions. These nonunions either required a segmental resection of the osteomyelitic site or presented with segmental bone loss with or without shortening *(43)*. All 28 patients healed without the addition of cancellous graft, soft tissue graft, or vascularized graft. The majority of patients did not require postoperative antibiotics, and seven patients received antibiotics for only 10 d after surgery. Limb length was equalized in 21 patients, within 1 cm in five and less than 3 cm in two patients. Functional results were good in all but one patient. Similar results were obtained in 11 adult patients (nine tibiae, one femur, and one humerus) *(70)*. Complications included two fractures of bone regenerate, one persistent leg length discrepancy, one equinus contracture, and one transient peroneal nerve palsy. Seventeen segmental skeletal defects managed by Ilizarov bone transport required an average fixation time of 9.6 mo to regenerate an average of 5.14 cm of new bone for defects ranging from 4.2% to 35% of the original bone length *(78)*.

In later series, bone transportation was compared to traditional (bone graft) and contemporary (antibiotic beads and vascularized grafts) techniques of bone defect treatment in regard to outcome, complications, and cost. In 30 sheep, a 4-cm defect was created in the mid-diaphyseal femur *(58)*. Demineralized allogeneic bone graft was compared to the Ilizarov method of bone transportation using a monolateral fixator and half-pins. The results of bone transportation were clearly superior. In a clinical series of 25 patients with infected nonunions and segmental bone loss, resection and bone transport using a monolateral device was compared to conventional methods of external fixation with bone graft and soft tissue coverage *(114)*. Each group experienced similar rates of healing, treatment time, final angulation, number of complications, and total number of surgical procedures; however, the limb length discrepancy in the bone transport group was significantly improved. In a similar series, 15 patients treated with the Papineau open bone graft technique were compared to 17 patients treated with the ring fixator and bone transportation. Treatment time was identical at 1.9 mo of fixation for each centimeter of defect reconstructed *(76)*. Each group had its unique limitations and complications. For the bone graft group, limited autograft availability, donor site morbidity, and graft fractures were problematic. For the bone transport group, lack of healing at the docking site required supplemental bone graft in seven patients, and joint contractures developed in seven patients. Another series reported results of limb salvage for 44 consecutive patients with segmental defects of the tibia. Of these patients, 21 were managed by the Ilizarov method of bone transportation and were compared to 23 patients managed by masssive cancellous grafts and tissue transfer *(46)*. Total wound healing and resolution of infection were comparable in the two groups, at about 70%. The Ilizarov group had one-half the number of complications (33% vs 60%) and saved nearly $30,000 per application, primarily as a result of 9 fewer hours in the operating room, 23 fewer inpatient days, and 5 fewer months of disability (17 vs 22 mo). In all three of these series, the authors had considerable prior experience with treatment of segmental defects and infections using conventional techniques and much less experience using the Ilizarov method.

The use of Ilizarov bone transportation does not seem to be hampered by prior microvascular muscle flap transplantation *(163)* or simultaneous use of antibiotic beads *(33)*. The treatment of cavitary osteomyelitis by fragmentary bone transport is a unique variation of bone transportation that requires advanced skills in use of the ring device with wires and an understanding of the local biology *(10,43)*.

Congenital pseudarthrosis of the tibia, a condition which has remained refractory to most conventional methods of bone grafting, is ideally treated using the bone transportation method of Ilizarov with open resection of the pseudarthrosis and intramedullary fixation. Several variations of the Ilizarov method, including closed end-to-end or side-to-side compression, primary distraction, or open resection, with or without a proximal corticotomy, have resulted in a 94% union rate after the first treatment and a 100% union rate after a second treatment *(130)*. Although the vascularized fibula technique has similar success rates for union of the pseudarthrosis, the Ilizarov method permits simultaneous lengthening. The mean treatment time in 12 patients was 5.6 mo (range 3–12 mo), with concomitant lengthening of 1.5–8 cm and correction of angular deformity *(130)*. Refractures occurred both early and late, usually through the old pseudarthrosis site if residual angular deformity persisted *(60,130)*.

Fracture Treatment

Although Ilizarov used his frame and method to treat fractures at all sites, high-energy tibia fracture treatment may benefit most from his method. In 1992, Tucker et al. *(162)* reported his results using percutaneous techniques of four ring external fixation with eight counter-opposed olive wires to manage 41 consecutive, unstable tibial diaphseal fractures. Twenty fractures of the 26 available for follow-up were open (8 grade II and 10 grade III). All fractures healed without bone grafting or chronic infections after 12–47 wk of treatment. Eleven fractures with significant bone loss were treated by simultaneus fracture site compression and adjacent distraction osteogenesis. Thirteen open fractures healed without need for skin graft. The operative time averaged 60–90 min. Of 10% of the 248 wire sites that became inflamed, one led to a ring sequestrum that required secondary curettage. Three wires broke, thought to be related to early weight bearing. The results of 25 of the 26 fractures reviewed were graded as good to excellent.

In 1992, Liljeberg and Taylor *(110)* reported on their first 43 patients with 46 complex tibial fractures. Thirty-eight patients (40 fractures) had 14-mo follow-up; 35 of these fractures (87.5%) were from high-energy trauma. Of the 15 open fractures (37.5%), 10 were type III. Nineteen of the 40 fractures (47.5%) involved the plateau, of which 14 had complete metaphyseal-diaphyseal dissociation. Thirty-nine of the 40 fractures (97.5%) healed in an average of 21 wk. The 5% deep infection and 10% bone graft rates compared favorably with other methods.

Watson *(172)* describes several advantages of small-wire external fixation for treatment of high-energy fractures of the tibial plateau: percutaneous wire placement minimizes additional devitalization of bone fragments, small wires can capture and reduce very small bone fragments, olive wires can reduce and compress condylar fractures, the circular frame can stabilize the periarticular segment for early motion and weight bearing, areas of bone loss can be stabilized like a neutralization plate, and malalignment (rotational, translational and angular) can be corrected during the consolidation process if not initially reduced. In 31 Schatzker type VI injuries treated with circular external fixation, all fractures demonstrated radiographic healing at an average of 15 wk with average motion of 106° and an average Hospital for Special Surgery knee score of 82. Twenty-seven of the 31 fractures were radiographically graded as good to excellent.

SPECIAL APPLICATIONS

Foot Reconstruction

Foot reconstruction using the methods of Ilizarov has been undertaken for a variety of conditions, including untreated, residual, or recurrent clubfeet in adults, posttraumatic deformity and degenerative joint disease, failed ankle fusions, and a variety of syndromes with deformities such as vertical talus *(53,71,126)*. In some series, deformities due to neuromuscular imbalance (e.g., cerebral palsy, spina bifida, polio), growth arrest, dyplasia, and juvenile rheumatoid arthritis as well as vascular disorders such as diabetes have been treated *(71)*.

Foot deformities can be corrected through soft tissue stretching (arthrodiastasis), osteotomies (distraction osteogenesis), or both *(53,71,126)*. Following such treatment, the foot most commonly becomes stiff. Therfore, an important prerequisite for the operation is a stiff foot with deformity, pain, nonunion, and/or shortening as secondary problems. Many of these feet are considered preamputation, and the Ilizarov method is used as a true salvage procedure.

Three types of frames, constrained (uniplanar hinges), semiconstrained (universal hinges), and nonconstrained (multiple tarsal joints form the hinges), allow for gradual correction of these complex, multiplanar deformities *(53,71,126)*. Fixation of the tibia with two rings and appropriately oriented olive wires and half-pins to provide fulcrums for leverage against the foot deformity and fixation of the foot with both olive wires and half-pins, usually placed in the calcaneus and metatarsals (less commonly in the talus and tarsals), are spanned by push or pull rods. Hinges are used to protect the internal hyaline joints from being crushed during correction.

The treatment strategy involves gradual overcorrection of the deformity at a rate of 1 mm/d at the critical structure (usually a neurovascular bundle), a holding period prior to frame removal, and frequently casting and/or bracing to prevent recurrence (rebound phenomenon) *(53,71,126)*. De la Huerta reported dramatic correction of untreated, adult clubfeet using arthrodiastasis and the nonconstrained frame *(53)*. These 12 cases in seven patients (19–42 yr old) required 5–8 mo of fixation time to achieve correction. The feet were stiff before and after treatment (especially the forefoot), and only three feet had recurrent forefoot adductus at 2–5 yr follow-up. All seven patients could wear normal shoes and walk following correction. Grant et al. *(71)* treated 17 feet (with hinges and both arthrodiastasis and distraction osteogenesis) in 23 patients (mean age of 10 yr), most of whom had neuromuscular conditions. Paley *(126)* described his results using distraction osteotomies in 25 complex foot deformities. The osteotomies were situated in the tibia (supramalleolar), calcaneus, talocalcaneal necks, midfoot (tarsals), and metatarsals, depending on the deformity. The most common complication in all three series was pin-tract inflammation, followed by wire breakage, failure of the osteotomy to open, claw toes, tarsal tunnel syndrome, direct neurovascular compromise, wire cut-out, and buckle fracture. Compared to other applications of the Ilizarov method, foot lengthening seems to be the most painful *(11)*. The majority of these patients had plantigrade feet and were satisfied following the procedure.

The Ilizarov method for ankle fusion has improved the results of conventional techniques for more difficult problems such as failed and/or infected ankle fusions, diabetic/Charcot ankle degeneration, severe deformity, and shortening of the limb *(86,98)*. Gradual compression of the ankle with or without simultaneous tibial distraction osteogenesis proximally has been used, with the former method achieving greater success *(86,98)*. Calhoun et al. *(86)* reported an 80% success rate (average follow-up of 22 mo) with this method in 18 adults, ranging from 17 to 59 yr old and two children (3 and 10 yr old). Nine of the 10 segmental defects (3–8 cm) were corrected to within 2 cm of the opposite side. Preoperative infection was resolved in 15 of 16 patients (94%). Of the 16 solid fusions (80%), 12 were well aligned. The method was also successful in patients with burn contractures *(34)*. Simple and complex equinus, cavus, rockerbottom deformities, and even toe dislocations were treated.

Arthrodiastasis

Gradual, mechanical stretching of stiff joints is not new. However, the use of a distraction hinge to protect underlying cartilage or even allow fibrocartilage to fill a narrowed joint space are unique developments associated with the Ilizarov method. The modular frame allows for both prophylactic dynamic orthoses or actual skeletal frame attachments to resist equinus contracture during extensive or rapid tibial lengthening or for postlengthening correction using the frame *(109)*.

Knee flexion contractures from a variety of etiologies can be stretched using either monolateral or ring fixators. Five knees in four patients with inflammatory arthritis, posttraumatic arthritis, and lipomyelomomeningocele underwent Ilizarov correction with a four-ring construct *(81)*. The preoperative contractures, ranging from 25° to 75°, were reduced to only 10–15° at 3 mo following fixator removal. Fourteen knees in 10 patients with a combination of etiologies including melorheostosis, arthrogryposis, congenital pterygium, tibial hemimelia, sacral agenesis, diastrophic dwarfism, amniotic band syndrome, desmoid tumor, and immobilization were treated by gradual, mechanical distraction using both monolateral and ring fixators *(87)*. The average total arc of motion improved initially, but due to a rebound phenomenon returned to the preoperative level. However, the position of this arc improved to the point that three of the patients had lasting functional improvement, two advancing from wheelchair/crawling to community ambulation and one from being bedridden to a wheelchair ambulator.

Elbow contractures in 58 patients underwent operative releases and hinged-joint distraction with lasting improvement from a mean preoperative arc of motion of 60–95° to a mean postoperative arc of 35–125° *(119)*. Arthrodiastasis of the hip using hinged distractors has also had surprising success. Nine patients with stiff hips from Perthes, slipped capital femoral epiphysis, developmental dysplasia, tuberculosis, and chondrolysis (average age of 14 yr) all had pain with an average arc of 20° preoperatively *(38)*. Following 0.5–1.0 cm of distraction for an average of 94 d, the average arc of motion improved to 65° with an average increase in the radiographic joint space of 2.8 mm and resolution of pain in six (66%). In a larger series of 80 patients with hip stiffness from avascular necrosis, hip dysplasia, osteoarthrosis, inflammatory arthritis, and chondrolysis, a monolateral fixator was used to create a 5-mm joint space while allowing flexion and extension *(3)*. Forty-two good results were found in 59 patients (71%) younger than 45 yr with osteoarthrosis, hip dyplasia, avascular necrosis, and chondrolysis. Only four of 21 patients aged 45 yr or older had a good result. Patients with inflammatory arthritis did not do well.

Special Indications

Limb salvage surgery for malignant tumors with prolonged life expectancy has led to some significant problems of leg length discrepancy, as allografts cannot grow and metal implants have limited ability to extend. Eight patients (four with osteosarcoma and four with Ewing's sarcoma) were treated with distraction osteogenesis using monolateral frames after allograft failure or progressive anisomelia. All had at least a 4-yr tumor-free interval *(39)*. In seven patients with 7–18 cm of shortening completing treatment, five had complete correction and two had residual shortening of 3 and 4 cm. Allograft pseudarthroses were healed by compression and simultaneous distraction osteogenesis at an adjacent site. Two patients required internal fixation (plates) and bone graft supplement. Prior irradiation seemed to inhibit bone formation.

The Ilizarov apparatus has been used to correct an infected spinal pseudarthrosis in a 13-yr-old thoracic-level spina bifida patient with breakdown of a kyphectomy site *(26)*. Wire fixation by open technique in the iliac wings and the thoracic laminae and spinous processes provided adequate fixation for the 13-wk fixation period. Gradual correction and compression led to healing of the spine in a well-aligned position. Six other patients with cervical spinal deformity were gradually corrected three-dimensionally using Ilizarov hinges and mechanical distractors attached to a halo-cast *(72)*. The majority of these patients had atlanto-axial problems from rheumatoid arthritis and familial cervical dysplasia. Two patients had cervico-thoracic problems associated with ankylosing spondylitis and kyphosis following laminectomy and irradiation for astrocytoma. There were no major complications in these spine cases.

Ilizarov reported success and functional improvement with stump lengthenings *(96)*. One report of a lengthening in a dyfunctionally short below-elbow stump by 100% (5 cm) supports the use of this technique for improved use of a prosthesis *(157)*.

Soft tissue distraction in chronic osteomyelitis of the humerus with shortening following sequestrectomy provided a good soft tissue envelop with undamaged host vessels for a subsequent vascularized fibula transplant *(99)*. Over 10 wk of distraction, a 10-cm gap was re-created to restore length to the humerus. A 17-cm graft was placed and then stabilized with the frame for a further 6 wk.

COSTS

When the Ilizarov method (bone transportation) was compared to conventional methods of limb salvage (sequestrectomy, antibiotic beads, free flaps, bone grafting, etc.), Cierny found approximately $30,000 savings per patient because of fewer complications, less operating room time, less inpatient time, and shorter overall disability time with the Ilizarov method *(46)*.

Major limb salvage using the Ilizarov method was also compared to the costs of amputation *(175)*. Ten patients with tibial nonunions, osteomyelitis, infected nonunions, and/or bone defects treated by the Ilizarov method were compared to six patients with similar traumatic injuries who underwent amputation (three acute and three delayed). The two groups were similar for age at treatment (average 40–41 yr) and number of operative procedures (average of four). The average inpatient hospitalization was less for the Ilizarov group at 16 d, compared to 25 d for the amputation group, but the average overall treatment time was much longer in the Ilizarov group (322 vs 175 d). The total charges (hospital and professional fees) of the limb reconstruction averaged $59,213.71 for the Ilizarov group (comparable to similar patients treated by Cierny *(46)* at $85,000), while the amputation group was much lower, at an average of $30,148.02. However, when the projected costs of prosthetic care for the remaining life expectancy was added to the amputation group, the total cost of care averaged $403,199 *(18)*.

It would seem that Ilizarov limb reconstruction methods are cost-effective for salvaging severe deformities and bone deficiencies when compared to both conventional methods and amputation.

FUTURE DIRECTIONS

Clearly, the biology and modular system of external fixation developed by Ilizarov have revolutionized orthopedic care. The biology has certain limitations, including slower osteogenesis in adults, delayed or absent growth from muscles during lengthening at the critical rate for distraction osteogenesis, rebound phenomenon in both muscles and primary collagenous tissues such as ligaments and capsules, and joint stiffness secondary to prolonged external fixation. Pin-tract inflammation and infection contribute to the discomfort, stiffness, and possibly to the poor response of muscle, as well as occasionally impinging on neurovascular structures. Research is underway in many centers to discover therapeutic innovations to accelerate bone formation, promote muscle growth, and even to avoid transcutaneous fixation. Since intramedullary rods have been compatible with distraction osteogenesis, the natural solution to avoid external fixation pins would be to develop a growing intramedullary rod. Several prototypes are in clinical trials. Perhaps one of the more exciting potential areas for future research is in the application of distraction forces at different rates to regenerate both ligaments (Aston et al. re-created a cruciate ligment by distraction histogenesis in the stifle joint of 13 dogs *[22]*) and articular cartilage (as described in the arthrodiastasis portion of this review).

Ilizarov died in 1992. He was fondly referred to as the "magician from Kurgan" by his fellow citizens. Those who had the opportunity to meet him know that he cared deeply for his patients and dedicated his life to improving care of musculoskeletal problems.

REFERENCES

1. Abbott, L. C. (1927) The operative lengthening of the tibia and fibula. *J. Bone Joint Surg.* **9A,** 128–152.
2. Aldegheri, R. (1993) Callotasis. *J. Pediatr. Orthop.* **2,** 11–15.
3. Aldegheri, R., Trivella, G., and Saleh, M. (1994) Articulated distraction of the hip. Conservative surgery for arthritis in young patients. *Clin. Orthop.* **301,** 94–101.
4. Aldegheri, R., Volino, C., Zambito, A., Tessari, G., and Trivella, G. (1993) Use of ultrasound to monitor limb lengthening by callotasis. *J. Pediatr. Orthop.* B **2,** 22–27.
5. Alho, A., Bang, G., Karaharju, E., and Armond, I. (1982) Filling of a bone defect during experimental osteotaxis distraction. *Acta Orthop. Scand.* **53,** 29–34.
6. Anderson, R. (1936) Femoral bone lengthening. *Am. J. Surg.* **31,** 479–483.
7. Aquerreta, J. D., Forriol, F., and Canadell, J. (1994) Complications of bone lengthening. *Int. Orthop.* **18,** 299–303.
8. Aronson, J. (1994) Experimental assessment of bone regenerate quality during distraction osteogenesis, in *Bone Formation and Repair*, Chapter 32. (Brighton, C. T., Friedlaender, G., and Lane, J. M., eds.), American Academy of Orthoopaedic Surgeons, Rosemont, IL. pp. 441–463.
9. Aronson, J. (1991) The biology of distraction osteogenesis, in *Operative Principles of Ilizarov. Fracture treatment, Nonunion, Osteomyelitis, Lengthening, Deformity Correction* (Bianchi Maiocchi, A. and Aronson, J., eds.), Williams and Wilkins, Baltimore, pp. 42–52.
10. Aronson, J. (1992) Cavitary osteomyelitis treated by fragmentary cortical bone transportation. *Clin. Orthop.* **280,** 153–159.
11. Aronson, J. (1994) Experimental and clinical experience with distraction osteogenesis. *Cleft Palate Craniofac. J.* **31(6),** 473–480.
12. Aronson, J. (1990) Proper wire tensioning for Ilizarov external fixation. *Techniques Orthop.* **5(4),** 27–32.
13. Aronson, J. (1994) Temporal and spatial increases in blood flow during distraction osteogenesis. *Clin. Orthop.* **301,** 124–131.
14. Aronson, J., Good, B., Stewart, C., Harrison, B., and Harp, J. H. (1990) Preliminary studies of mineralization during distraction osteogenesis. *Clin. Orthop.* **250,** 43–49.
15. Aronson, J. and Harp, J. H. (1990) Factors influencing the choice of external fixation for distraction osteogenesis, in *Instructional Course Lectures,* vol. 34. American Academy of Orthopaedic Surgeons, Park Ridge, IL, pp. 175–183.
16. Aronson, J. and Harp, J. H. (1992) Mechanical considerations in using tensioned wires in a transosseous external fixation system. *Clin. Orthop.* **280,** 23–29.
17. Aronson, J. and Harp, J. H. (1994) Mechanical forces as predictors of healing during tibial lengthening by distraction osteogenesis. *Clin. Orthop.* **301,** 73–79.
18. Aronson, J., Harrison, B., Boyd, C. M., Cannon, D. J., and Lubansky, H. J. (1988) Mechanical induction of osteogenesis: the importance of pin rigidity. *J. Pediatr. Orthop.* **8,** 396–401.
19. Aronson, J., Harrison, B. H., Stewart, C. L., and Harp, J. H. (1989) The histology of distraction osteogenesis using different external fixators. *Clin. Orthop.* **241,** 106–116.
20. Aronson, J., Johnson, E., and Harp, J. H. (1989) Local bone transportation for treatment of intercalary defects by the Ilizarov technique. Biomechanical and clinical considerations. *Clin. Orthop.* **243,** 71–79.
21. Aronson, J. and Shen, X. (1994) Experimental healing of distraction osteogenesis comparing metaphyseal with diaphyseal sites. *Clin. Orthop.* **301,** 25–30.
22. Aston, J. W., Williams, S. A., Allard, R. N., Sawamura, S., and Carollo, J. J. (1992) A new canine cruciate ligament formed through distraction histogenesis. Report of a pilot study. *Clin. Orthop.* **280,** 30–36.
23. Atar, D., Lehman, W. B., Grant, A. D., and Strongwater, A. M. (1992) The Ilizarov apparatus for treatment of melorheostosis. Case report and review of the literature. *Clin. Orthop.* **281,** 163–167.
24. Atar, D., Lehman, W. B., Grant, A. D., Strongwater, A., Frankel, V., and Golyakhovsky, V. (1991) Treatment of complex limb deformities in children with the Ilizarov technique. Original research. *Orthopaedics* **14(9),** 961–967.
25. Atar, D., Lehman, W. B., Posner, M., et al. (1991) Ilizarov technique in treatment of congenital hand anomalies. *Clin. Orthop.* **273,** 268–274.
26. Birch, J. G. (1991) Ilizarov apparatus for infected pseudarthrosis of the spine. A case report. *Bull. Hosp. Joint Dis. Orthop. Inst.* **51(1),** 93–98.

27. Blane, C. E., Herzenberg, J. E., and DiPietro, M. A. (1991) Radiographic imaging for Ilizarov limb lengthening in children. *Pediatr. Radiol.* **21**, 117–120.

28. Bonnard, C., Favard, L., Sollogoub, I., and Glorion, B. (1993) Limb lengthening in children using the Ilizarov method. *Clin. Orthop.* **293**, 83–88.

29. Bost, F. C. and Larsen, L. J. (1956) Experiences with lengthening of the femur over an intramedullary rod. *J. Bone Joint Surg.* **38A(3)**, 567–584.

30. Brockway, A. and Fowler, S. B. (1942) Experience with 105 leg lengthening operations. *Surg. Gynecol. Obstet.* **72**, 252–256.

31. Brown, R., Pedowitz, R., Rydevik, B., et al. (1993) Effects of acute graded strain on efferent conduction properties in the rabbit tibial nerve. *Clin. Orthop.* **296**, 288–294.

32. Brunner, U. H., Cordey, J., Schweiberer, L., and Perren, S. M. (1994) Force required for bone segment transport in the treatment of large bone defects using medullary nail fixation. *Clin. Orthop.* **301**, 147–155.

33. Calhoun, J. H., Anger, D. M., Ledbetter, B. R., and Mader, J. T. (1993) The Ilizarov fixator and polymethylmethacrylate-antiobiotic beads for the treatment of infected deformities. *Clin. Orthop.* **295**, 13–22.

34. Calhoun, J. H., Evans, E. B., and Herndon, D. N. (1992) Techniques for the management of burn contractures with the Ilizarov fixator. *Clin. Orthop.* **280**, 117–124.

35. Calhoun, J. H., Li, F., Ledbetter, B. R., and Gill, C. A. (1992) Biomechanics of the Ilizarov fixator for fracture fixation. *Clin. Orthop.* **280**, 15–22.

36. Canadell, J. (1993) Bone lengthening: experimental results. *J. Pediatr. Orthop. B* **2**, 8–10.

37. Canadell, J., Aquerreta, D., and Forriol, F. (1993) Prospective study of bone lengthening. *J. Pediatr. Orthop. B* **2**, 1–7.

38. Canadell, J., Gonzales, F., Barrios, R. H., and Amillo, S. (1993) Arthrodiastasis for stiff hips in young patients. *Int. Orthop.* **17**, 254–258.

39. Cara, J. A., Forriol, F., and Canadell, J. (1993) Bone lengthening after conservative oncologic surgery. *J. Pediatr. Orthop. B* **2**, 57–61.

40. Catagni, M. A., Bolano, L., and Cattaneo, R. (1991) Management of fibula hemimelia using the Ilizarov method. *Orthop. Clin. N. Am.* **22(4)**, 715–722.

41. Catagni, M. A., Guerreschi, F., Holman, J. A., and Cattaneo, R. (1994) Distraction osteogenesis in the treatment of stiff hypertrophic nonunions using the Ilizarov apparatus. *Clin. Orthop.* **301**, 159–163.

42. Catagni, M. A., Guerreschi, F., and Probe, R. A. (1991) Treatment of humeral nonunions with the Ilizarov technique. *Bull. Hosp. Joint Dis.* **51(1)**, 74–82.

43. Cattaneo, R., Catagni, M., and Johnson, E. E. (1992) The treatment of infected nonunions and segmental defects of the tibia by the methods of Ilizarov. *Clin. Orthop.* **280**, 143–152.

44. Cattaneo, R., Villa, A., Catagni, M. A., and Bell, D. (1990) Lengthening of the humerus using the Ilizarov technique. *Clin. Orthop.* **250**, 117–124.

45. Caviale, P., Herrick, W., and MacEwen, G. D. (1987) A report of leg lengthening in canines with the Ilisarov technique. *Jefferson Orthop. J.* **XIV(1)**, 43–47.

46. Cierny, G. and Zorn, K. E. (1994) Segmental tibial defects. Comparing conventional and Ilizarov methodologies. *Clin. Orthop.* **301**, 118–123.

47. Codivilla, A. (1904) On the means of lengthening, in the lower limbs, the muscles and tissues which are shortened through deformity. *Am. J. Orthop. Surg.* **2**, 353.

48. Coglianese, D. B., Herzenberg, J. E., and Goulet, J. A. (1993) Physical therapy management of patients undergoing limb lengthening by distraction osteogenesis. *J. Orthop. Sports Phys. Ther.* **17(3)**, 124–132.

49. Coleman, S. S. and Scott, S. M. (1991) The present attitude toward the biology and technology of limb lengthening. *Clin. Orthop.* **264**, 76–83.

50. Constantino, P. D., Shybut, G., Friedman, C. D., et al. (1990) Segmental mandibular resection by distraction osteogenesis. An experimental study. *Arch. Otolaryngol. Head Neck Surg.* **116**, 535–545.

51. Dahl, M. T. and Fischer, D. A. (1991) Lower extremity lengthening by Wagner's method and by callus distraction. *Orthop. Clin. N. Am.* **22(4)**, 643–649.

52. Dahl, M. T., Gulli, B., and Berg, T. (1994) Complications of limb lengthening. A learning curve. *Clin. Orthop.* **301**, 10–18.

53. De La Huerta, F. (1994) Correction of the neglected clubfoot by the Ilizarov method. *Clin. Orthop.* **301**, 89–93.

54. Delloye, C., Delefortrie, G., Coutelier, L., and Vincent, A. (1990) Bone regenerate formation in cortical bone during distraction lengthening. An experimental study. *Clin. Orthop.* **250**, 34–42.

55. De Pablos, J., Barrios, C., Alfaro, C., and Canadell, J. (1994) Large experimental segmental bone defects treated by bone transportation with monolateral external fixators. *Clin. Orthop.* **298**, 259–265.

56. De Pablos, J. and Canadell, J. (1990) Experimental physeal distraction in immature sheep. *Clin. Orthop.* **250**, 73–80.

57. Eldridge, J. C. and Bell, D. F. (1991) Problems with substantial limb lengthening. *Orthop. Clin. N. Am.* **22(4)**, 625–631.

58. Ehrnberg, A., De Pablos, J., Martinez-Lotti, G., Kreicbergs, A., and Nilsson, O. (1993) Comparison of demineralized allogenic bone matrix grafting (the Urist procedure) and the Ilizarov procedure in large diaphyseal defects in sheep. *J. Orthop. Res.* **11**, 438–447.

59. Eyres, K. S., Bell, M. J., and Kanis, J. A. (1993) New bone formation during leg lengthening evaluated by dual energy xray absorptiometry. *J. Bone Joint Surg.* **75B(1),** 96–106.
60. Fabry, G., Lammens, J., Van Melkebeek, J., and Stuyck, J. (1988) Treatment of congenital pseudarthrosis with the Ilizarov technique. Case report. *J. Pediatr. Orthop.* **8,** 67–70.
61. Fern, D., Eyres, K. S., Bell, M. J., and Saleh, M. (1993) Leg lengthening in osteogenesis imperfecta. *J. Pediatr. Orthop. B* **2,** 62–65.
62. Figuieredo, U. M., Watkins, P. E., and Goodship, A. E. (1992) The influence of micromovement in experimental leg-lengthening. *J. Bone Joint Surg.* **74B(Suppl III),** 317.
63. Fishbane, B. M. and Riley, L. H. (1976) Continuous trans-physeal traction: a simple method of bone lengthening. *Johns Hopkins Med. J.* **138,** 79–81.
64. Fischgrund, J., Paley, D., and Suter, C. (1994) Variables affecting time to bone healing during limb lengthening. *Clin. Orthop.* **301,** 31–37.
65. Frierson, M., Ibrahim, K., Boles, M., Bote, H., and Ganey, T. (1994) Distraction osteogenesis. A comparison of corticotomy techniques. *Clin. Orthop.* **301,** 19–24.
66. Ganey, T. M., Klotch, D. W., Sasse, J., Ogden, J. A., and Garcia, T. (1994) Basement membrane of blood vessels during distraction osteogenesis. *Clin. Orthop.* **301,** 132–138.
67. Gil-Albarova, J., de Pablos, J., Franzeb, M., and Canadell, J. (1992) Delayed distraction in bone lengthening. Improved healing in lambs. *Acta Orthop. Scand.* **63(6),** 604–606.
68. Goldstein, S. A., Waanders, N., Guldberg, R., et al. (1994) Stress morphology relationships during distraction osteogenesis: linkages between mechanical and architectural factors in molecular regulation, in *Bone Formation and Repair,* Chapter 29 (Brighton, C. T., Friedlaender, G., and Lane. J. M., eds.), American Academy of Orthoopaedic Surgeons, Rosemont, IL, pp. 405–419.
69. Golyakhovsky, V. and Gavriel, A. (1988) Ilizarov: "the magician from Kurgan." *Bull. Hosp. Joint Dis. Orthop. Inst.* **48(1),** 12–16.
70. Golyakhovsky, V. and Frankel, V. (1991) Ilizarov bone transport in large bone loss and in severe osteomyelitis. *Bull. Hosp. Joint Dis. Orthop. Inst.* **51(1),** 63–72.
71. Grant, A. D., Atar, D., and Lehman, W. B. (1992) The Ilizarov technique in correction of complex foot deformities. *Clin. Orthop.* **280,** 94–103.
72. Graziano, G. P., Herzenberg, J. E., and Hensinger, R. N. (1993) The halo Ilizarov cast for correction of cervical deformity. Report of six cases. *J. Bone Joint Surg.* **75A(7),** 997–1003.
73. Green, S. (1992) Editorial comment. *Clin. Orthop.* **280,** 2–6.
74. Green, S. (1991) The Ilizarov method: Rancho technique. *Orthop. Clin. N. Am.* **22(4),** 677–688.
75. Green, S. A. (1991) Postoperative management during limb lengthening. *Orthop. Clin. N. Am.* **22(4),** 723–734.
76. Green, S. A. (1994) Skeletal defects. A comparison of bone grafting and bone transport for segmental skeletal defects. *Clin. Orthop.* **301,** 111–117.
77. Green, S. A., Harris, N. L., Wall, D. M., Ishkanian, J., and Marinow, H. (1992) The Rancho mounting technique for the Ilizarov method. A preliminary report. *Clin. Orthop.* **280,** 104–116.
78. Green, S. A., Jackson, J. M., Wall, D. M., Marinow, H., and Iskanian, J. (1992) Management of segmental defects by the Ilizarov intercalary bone transport method. *Clin. Orthop.* **280,** 136–142.
79. Grill, F., Dungl, P., Steinwender, G., and Hosny, G. (1993) Congenital short femur. *J. Pediatr. Orthop. B* **2,** 35–41.
80. Guarniero, R. and Barros, T. E. P. (1990) Femoral lengthening by the Wagner method. *Clin. Orthop.* **250,** 155–159.
81. Hagglund, G., Rydholm, U., and Sunden, G. (1993) Ilizarov technique in the correction of knee flexion contracture: report of four cases. *J. Pediatr. Orthop. B* **2,** 170–172.
82. Hamanishi, C., Yoshii, T., Totani, Y., and Tanaka, S. (1994) Lengthened callus activated by axial shortening. Histological and cytomorphometrical analysis. *Clin. Orthop.* **307,** 250–254.
83. Hardy, J. M., Tadlaoui, A., Wirotius, J. M., and Saleh, M. (1991) The Sequoia circular fixator for limb lengthening. *Orthop. Clin. N. Am.* **22(4),** 663–675.
84. Harp, J. H., Aronson, J., and Hollis, M. (1994) Noninvasive determination of bone stiffness for distraction osteogenesis by quantitative computed tomography scans. *Clin. Orthop.* **301,** 42–48.
85. Harris, N. L., Eilert, R. E., Davino, N., Ruyle, S., Edwardson, M., and Wilson, V. (1994) Osteogenic sarcoma arising from bony regenerate following Ilizarov femoral lengthening through fibrous dysplasia. A case report. *J. Pediatr. Orthop.* **14,** 123–129.
86. Hawkins, B. J., Langerman, R. J., Anger, D. M., and Calhoun, J. H. (1994) The Ilizarov technique in ankle fusion. *Clin. Orthop.* **303,** 217–225.
87. Herzenberg, J. E., Davis, J. R., Paley, D., and Bhave, A. (1994) Mechanical distraction for treatment of severe knee flexion contractures. *Clin. Orthop.* **301,** 80–88.
88. Herzenberg, J. E., Scheufele, L. L., Paley, D., Bechtel, R., and Tepper, S. (1994) Knee range of motion in isolated femoral lengthening. *Clin. Orthop.* **301,** 49–54.
89. Herzenberg, J. E., Smith, J. D., and Paley, D. (1994) Correcting torsional deformities with Ilizarov's apparatus. *Clin. Orthop.* **302,** 36–41.

90. Herzenberg, J. E. and Waanders, N. A. (1991) Calculating rate and duration of distraction for deformity correction with the Ilizarov technique. *Orthop. Clin. N. Am.* **22(4),** 601–611.

91. Hope, P. G., Crawford, E. J. P., and Catterall, A. (1994) Bone growth following lengthening for congenital shortening of the lower limb. *J. Pediatr. Orthop.* **14,** 339–342.

92. Hughes, T. H., Maffulli, N., Green, V., and Fixsen, J. A. (1994) Imaging in bone lengthening. A review. *Clin. Orthop.* **308,** 50–53.

93. Ilizarov, G. A. (1990) Clinical application of the tension-stress effect for limb lengthening. *Clin. Orthop.* **250,** 8–26.

94. Ilizarov, G. A. (1989) The tension-stress effect on the genesis and growth of tissues. Part I. The influence of stability of fixation and soft-tissue preservation. *Clin. Orthop.* **238,** 249–281.

95. Ilizarov, G. A. (1989) The tension-stress effect on the genesis and growth of tissues. Part II. The influence of rate and frequency of distraction. *Clin. Orthop.* **239,** 263–285.

96. Ilizarov, G. A. (1992) *Transosseous Osteosynthesis, Theoretical and Clinical Aspects of the Regeneration and Growth of Tissue.* Springer-Verlag, Berlin.

97. Ippolito, E., Peretti, G., Bellocci, M., et al. (1994) Histology and ultrastructure of arteries, veins, and peripheral nerves during limb lengthening. *Clin. Orthop.* **308,** 54–62.

98. Johnson, E. E., Weltmer, J., Lian, G. J., and Cracchiolo, A. (1992) Ilizarov ankle arthrodesis. *Clin. Orthop.* **280,** 160–169.

99. Jupiter, J. B. and Kour, A. K. (1991) Reconstruction of the humerus by soft tissue distraction and vascularized fibula transfer. *J. Hand Surg.* **16A(5),** 940–943.

100. Kaljumae, U., Martson, A., Haviko, T., and Hanninen, O. (1995) The effect of lengthening of the femur on the extensors of the knee. *J. Bone Joint Surg.* **77A(2),** 247–250.

101. Karaharju, E. O., Aalto, K., and Kahri, A. (1993) Distraction bone healing. *Clin. Orthop.* **297,** 38–43.

102. Kawamura, B., Hosono, S., Takahashi, T., et al. (1968) Limb lengthening by means of subcutaneous osteotomy. *J. Bone Joint Surg.* **50A,** 851–878.

103. Kenwright, J. and White, S. H. (1993) A historical review of limb lengthening and bone transport. *Injury* **(Suppl 2),** S9–S19.

104. Kummer, F. J. (1992) Biomechanics of the Ilizarov external fixator. *Clin. Orthop.* **280,** 11–14.

105. Lavini, F., Renzi-Brivio, L., and De Bastiani, G. (1990) Psychologic, vascular, and physiologic aspects of lower limb lengthening in achondroplastics. *Clin. Orthop.* **250,** 138–142.

106. Lee, D. Y., Choi, I. H., Chung, C. Y., Chung, P. H., Chi, J. G., and Suh, Y. L. (1993) Effect of tibial lengthening on the gastrocnemius muscle. A histopathologic and morphometric study in rabbits. *Acta Orthop. Scand.* **64(6),** 688–692.

107. Lee, D. Y., Chung, C. Y., and Choi, I. H. (1993) Longitudinal growth of the rabbit tibia after callostasis. *J. Bone Joint Surg.* **75B,** 898–903.

108. Lee, D. Y., Han, T. R., Choi, I. H., Lee, C. K., and Chung, S. S. (1992) Changes in somatosensory-evoked potentials in limb lengthening. An experimental study on rabbits' tibias. *Clin. Orthop.* **285,** 273–279.

109. Lehman, W. B., Grant, A. D., and Atar, D. (1991) Preventing and overcoming equinus contractures during lengthening of the tibia. *Orthop. Clin. N. Am.* **22(4),** 633–641.

110. Liljeberg, R. J. and Taylor, J. C. (1992) The treatment of complex tibial fractures with Ilizarov external fixator. *J. Bone Joint Surg.* **74B(Suppl III),** 265.

111. Magnuson, P. B. (1913) Lengthening shortened bones of the leg by operation. Ivory screws with removable heads as a means of holding the two bone fragments. *Surg. Gynecol. Obstet.* **17,** 63–71.

112. Makarov, M. R., Delgado, M. R., Samchukov, M. L., Welch, R. D., and Birch, J. G. (1994) Somatosensory evoked potential evaluation of acute nerve injury associated with external fixation procedures. *Clin. Orthop.* **308,** 254–263.

113. Markel, M. D. and Chao, E. Y. S. (1993) Noninvasive monitoring techniques for quantitative description of callus mineral content and mechanical properties. *Clin. Orthop.* **293,** 37–45.

114. Marsh, J. L., Prokuski, L., and Biermann, J. S. (1994) Chronic infected tibial nonunions with bone loss. Conventional techniques versus bone transport. *Clin. Orthop.* **301,** 139–146.

115. Masada, K., Kojimoto, H., Yasui, N., and Ono, K. (1993) Progressive lengthening of forearm bones in multiple osteochondromas. *J. Pediatr. Orthop. B* **2,** 66–69.

116. Matano, T., Tamai, K., and Kurokawa, T. (1994) Adaptation of skeletal muscle in limb lengthening: a light diffraction study on the sarcomere length *in situ. J. Orthop. Res.* **12,** 193–196.

117. Moens, P., Lammens, J., Molenaers, G., and Fabry, G. (1995) Femoral derotation for increased hip anteversion. A new surgical technique with a modified Ilizarov frame. *J. Bone Joint Surg.* **77B(1),** 107–109.

118. Monticelli, G. and Spinelli, R. (1981) Distraction epiphysiolysis as a method of limb lengthening. III. Clinical applications. *Clin. Orthop.* **154,** 274–285.

119. Morrey, B. F. (1993) Distraction arthroplasty. Clinical applications. *Clin. Orthop.* **293,** 46–54.

120. Moseley, C. F. (1991) Leg lengthening: the historical perspective. *Orthop. Clin. N. Am.* **22(4),** 555–561.

121. Nele, U., Maffulli, N., and Pintore, E. (1994) Biomechanics of radiotransport circular external fixators. *Clin. Orthop.* **308,** 68–72.

122. Olney, B. W. and Jayaraman, G. (1994) Joint reaction forces during femoral lengthening. *Clin. Orthop.* **301,** 64–67.

123. Ombredanne, L. (1913) Allongement d'un femur sur un membre trop court. *Bull. Mem. Soc. Chirur.* (Paris) **39,** 1177.

124. Orbay, J. L., Frankel, V. H., Finkle, J. E., and Kummer, F. J. (1992) Canine leg lengthening by the Ilizarov technique. *Clin. Orthop.* **278,** 265–273.
125. Paley, D. (1989) Bone transport: the Ilizarov treatment for bone defects. *Techniques Orthop.* **4(3),** 80–93.
126. Paley, D. (1993) The correction of complex foot deformities using Ilizarov's distraction osteotomies. *Clin. Orthop.* **293,** 97–111.
127. Paley, D. (1988) Current techniques of limb lengthening: review article. *J. Pediatr. Orthop.* **8,** 73–92.
128. Paley, D. (1990) Problems, obstacles, and complications of limb lengthening by the Ilizarov technique. *Clin. Orthop.* **250,** 81–104.
129. Paley, D., Catagni, M., Argnani, F., Villa, A., Benedetti, G. B., and Cattaneo, R. (1989) Ilizarov treatment of tibial nonunions with bone loss. *Clin. Orthop.* **241,** 146–165.
130. Paley, D., Catagni, M., Argnani, F., Prevot, J., Bell, D., and Armstrong, P. (1992) Treatment of congenital pseudoarthrosis of the tibia using the Ilizarov technique. *Clin. Orthop.* **280,** 81–93.
131. Paley, D., Chaudray, M., Pirone, A. M., Lentz, P., and Kautz, D. (1990) Treatment of malunions and mal-nonunions of the femur and tibia by detailed preoperative planning and the Ilizarov techniques. *Orthop. Clin. N. Am.* **21(4),** 667–691.
132. Paley, D., Fleming, B., Catagni, M., Kristiansen, T., and Pope, M. (1991) Mechanical evaluation of external fixators used in limb lengthening. *Clin. Orthop.* **250,** 50–56.
133. Paley, D. and Tetsworth, K. (1992) Mechanical axis deviation of the lower limbs. Preoperative planning of multiapical frontal plane angular and bowing deformities of the femur and tibia. *Clin. Orthop.* **280,** 65–71.
134. Paley, D. and Tetsworth, K. (1992) Mechanical axis deviation of the lower limbs. Preoperative planning of uniapical angular deformities of the tibia or femur. *Clin. Orthop.* **280,** 48–64.
135. Paley, D. and Tetsworth, K. (1991) Percutaneous osteotomies: osteotome and Gigli saw techniques. *Orthop. Clin. N. Am.* **22(4),** 613–624.
136. Paterson, D. (1990) Leg-lengthening procedures. A historical review. *Clin. Orthop.* **250,** 27–33.
137. Peretti, G., Memeo, A., Paronzini, A., and Marzorati, S. (1995) Staged lengthening in the prevention of dwarfism in achondroplastic children: a preliminary report. *J. Pediatr. Orthop. B* **4,** 58–64.
138. Podolsky, A. and Chao, Y. S. (1993) Mechanical performance of Ilizarov circular external fixators in comparison with other fixators. *Clin. Orthop.* **293,** 61–70.
139. Pouliquen, J. C., Pauthier, F., Ucla, E., Kassis, B., Ceolin, J. L., and Langlais, J. (1994) Tension measurements during lengthening of the lower limbs in children and adolescents. *J. Pediatr. Orthop. B* **3,** 107–113.
140. Price, C. T. and Mann, J. W. (1991) Experience with the Orthofix device for limb lengthening. *Orthop. Clin. N. Am.* **22(4),** 651–661.
141. Price, C. T. and Cole, J. D. (1990) Limb lengthening by callotasis for children and adolescents. *Clin. Orthop.* **250,** 105–111.
142. Prokuski, L. J., Marsh, J. L., and Biermann, J. S. (1992) Comparison of conventional treatment versus bone transport for chronic infected tibial nonunions with segmental bone loss. *J. Bone Joint Surg.* **74B(Suppl III),** 264.
143. Putti, V. (1921) The operative lengthening of the femur. *JAMA* **77,** 934–935.
144. Renzi-Brivio, L., Lavini, F., and De Bastiani, G. (1990) Lengthening in the congenital short femur. *Clin. Orthop.* **250,** 112–116.
145. Ring, P. A. (1958) Experimental bone lengthening by epiphyseal distraction. *Br. J. Surg.* **46,** 169–173.
146. Saleh, M. and Burton, M. (1991) Leg lengthening: patient selection and management in achondroplasia. *Orthop. Clin. N. Am.* **22(4),** 589–599.
147. Saxby, T. and Nunley, J. A. (1992) Metatarsal lengthening by distraction osteogenesis. *Foot Ankle* **13,** 536.
148. Schenk, R. K. and Gachter, A. (1994) Histology of distraction osteogenesis, in *Bone Formation and Repair,* Chapter 27 (Brighton, C. T., Friedlaender, G., and Lane, J. M., eds.), American Academy of Orthoopaedic Surgeons, Rosemont, IL, pp. 387–394.
149. Schwartsman, V. and Schwartsman, R. (1991) Corticotomy. *Clin. Orthop.* **280,** 37–47.
150. Seitz, W. H. and Froimson, A. I. (1991) Callotasis lengthening in the upper extremity: indications, techniques, and pitfalls. *J. Hand Surg.* **16A(5),** 932–939.
151. Shearer, J. R., Roach, H. I., and Parsons, S. W. (1992) Histology of a lengthened human tibia. *J. Bone Joint Surg.* **74B,** 39–44.
152. Shen, X. C., Aronson, J., Hollis, J. M., and Lumpkin, C. K. (1995) Mechanical properties and bone mineral of regenerated bone in the rat model of tibial lengthening. *Transactions of the 41st Annual Meeting of the Orthopedic Research Society,* February 13–16, p. 569.
153. Simpson, A. H. R. W. and Kenwright, J. (1992) The response of nerve different rates of distraction. *J. Bone Joint Surg.* **74B(Suppl III),** 326.
154. Simpson, A. H. R. W., Williams, P. E., Kyberd, P., Goldspink, G., and Kenwright, J. (1995) The response of muscle to leg lengthening. *J. Bone Joint Surg.* **77B,** 630–636.
155. Stanitski, D. (1994) Treatment of deformity secondary to metabolic bone disease with the Ilizarov method. *Clin. Orthop.* **301,** 38–41.

156. Steen, H., Fjeld, T. O., Miller, J. A. A., and Ludvigsen, P. (1990) Biomechanical factors in the metaphyseal- and diaphyseal-lengthening osteotomy. *Clin. Orthop.* **259**, 282–294.

157. Stricker, S. J. (1994) Ilizarov lengthening of a posttraumatic below elbow amputation stump. A case report. *Clin. Orthop.* **306**, 124–127.

158. Strong, M., Hruska, J., Czyrny, J., Heffner, R., Brody, A., and Wong-Chung, J. (1994) Nerve palsy during femoral lengthening: MRI, electrical, and histologic findings in the central and peripheral nervous systems—a canine model. *J. Pediatr. Orthop.* **14**, 347–351.

159. Suzuki, S., Kasahara, Y., Seto, Y., Futami, T., Furukawa, K., and Nishino, Y. (1994) Dislocation and subluxation during femoral lengthening. *J. Pediatr. Orthop.* **14**, 343–346.

160. Tetsworth, K., Krome, J., and Paley, D. (1991) Lengthening and deformity correction of the upper extremity by the Ilizarov technique. *Orthop. Clin. N. Am.* **22(4)**, 689–713.

161. Tetsworth, K. D. and Paley, D. (1994) Accuracy of correction of complex lower extremity deformities by the Ilizarov method. *Clin. Orthop.* **301**, 102–110.

162. Tucker, H. L., Kendra, J. C., and Kinnebrew, T. E. (1992) Management of unstable open and closed tibial fractures using the Ilizarov method. *Clin. Orthop.* **280**, 125–135.

163. Tukiainen, E. and Asko-Seljavaara, S. (1993) Use of the Ilizarov technique after a free microvascular muscle flap transplantaion in massive trauma of the lower leg. *Clin. Orthop.* **297**, 129–134.

164. Vauhkonen, M., Peltonen, J., Karaharju, E., Aalto, K., and Alitalo, I. (1990) Collagen synthesis and mineralization in the early phase of distraction bone healing. *Bone Miner.* **10**, 171–181.

165. Velazquez, R. J., Bell, D. F., Armstrong, P. F., Babyn, P., and Tibshirani, R. (1993) Complications of use of the Ilizarov technique in the correction of limb deformities in children. *J. Bone Joint Surg.* **75A(8)**, 1148–1156.

166. Vilarrubias, J. M., Ginebreda, J., and Jimeno, E. (1990) Lengthening of the lower limbs and correction of lumbar hyperlordosis in achondroplasia. *Clin. Orthop.* **250**, 143–149.

167. Villa, A., Paley, D., Catagni, M. A., Bell, D., and Cattaneo, R. (1990) Lengthening of the forearm by the Ilizarov technique. *Clin. Orthop.* **250**, 125–137.

168. Wagner, H. (1978) Operative lengthening of the femur. *Clin. Orthop.* **136**, 125.

169. Walker, C. W., Aronson, J., Kaplan, P. A., Molpus, W. M., and Seibert, J. J. (1991) Radiologic evaluation of limb-lengthening procedures. *Am. J. Radiol.* **156**, 353–358.

170. Walsh, W. R., Hamdy, R. C., and Erlich, M. G. (1994) Biomechanical and physical properties of lengthened bone in a canine model. *Clin. Orthop.* **306**, 230–238.

171. Wasserstein, I. (1990) Twenty-five years' experience with lengthening of shortened lower extremities using cylindrical allografts. *Clin. Orthop.* **250**, 150–153.

172. Watson, J. T. (1994) High-energy fractures of the tibial plateau. *Orthop. Clin. N. Am.* **25(4)**, 723–752.

173. White, S. H. and Kenwright, J. (1991) The importance of delay in distraction of osteotomies. *Orthop. Clin. N. Am.* **22(4)**, 569–579.

174. White, S. H. and Kenwright, J. (1990) The timing of distraction of an osteotomy. *J. Bone Joint Surg.* **72B**, 356–361.

175. Williams, M. O. (1994) Long-term cost comparison of major limb salvage using the Ilizarov method versus amputation. *Clin. Orthop.* **301**, 156–158.

176. Wolfson, N., Hearn, T. C., Thomason, J. J., and Armstrong, P. F. (1990) Force and stiffness changes during Ilizarov leg lengthening. *Clin. Orthop.* **250**, 58–60.

177. Yasui, N., Kojimoto, H., Shimizu, H., and Shimomura, Y. (1991) The effect of distraction upon bone, muscle, and periosteum. *Orthop. Clin. N. Am.* **22(4)**, 563–567.

178. Yasui, N., Kojimoto, H., Sasaki, K., Kitada, A., Shimizu, H., and Shimomura, Y. (1993) Factors affecting callus distraction in limb lengthening. *Clin. Orthop.* **293**, 55–60.

179. Young, N., Bell, D. F., and Anthony, A. (1994) Pediatric pain patterns during Ilizarov treatment of limb length discrepancy and angular deformity. *J. Pediatr. Orthop.* **14**, 352–357.

180. Younger, A. S. E., Mackenzie, W. G., and Morrison, J. B. (1994) Femoral forces during limb lengthening in children. *Clin. Orthop.* **301**, 55–63.

12

Biology of Spine Fusion

Biology and Clinical Applications

K. Craig Boatright, MD and Scott D. Boden, MD

INTRODUCTION

Spinal fusion was first reported in the United States in 1911 by Hibbs and Albee for the treatment of scoliosis and tuberculosis *(1,2)*. Since that time, the indications for spinal arthrodesis have continued to increase, and, in the late 1990s, approx 250,000 spinal fusions were performed each year in the United States *(3,4)*. The rate of nonunion after spinal arthrodesis is reported to range from 5% to 35% *(5)*. Better biomechanical control of the fusion environment with instrumentation and the use of autologous bone grafting techniques have failed to eliminate the problem of pseudoarthrosis in spine surgery *(6–12)*. As spinal surgery enters the new millennium, attention is focused on the biology of spine fusion to further enhance the rate of successful arthrodesis *(13)*.

During the decade of the 1990s, spinal fusion surgery became the most common reason for autologous bone grafting *(3)*. Despite the fact that autograft bone is osteoconductive, osteoinductive, and osteogenic, it has major shortcomings as an ideal bone generator. Aside from a significant nonunion rate, bone graft harvest is associated with well-known complications in up to 25% of cases; these include infection, fracture of the ileum, abdominal herniations, increased blood loss, increased hospital stay, and increased postoperative pain *(14–16)*. In addition, autograft is available in a limited supply that can be inadequate for revisions or multilevel procedures. The clinical track record of autogenous bone graft makes it the present gold standard for spinal arthrodesis, but the motivation to find a superior alternative is obvious.

In the latter part of the 1990s, the search for a *superior* bone generator intensified. Previous goals for bone graft substitute have been to match autograft for fusion rates while avoiding the morbidity of bone graft harvest and extending the quantity of graft material available. As bone graft substitutes and growth factors become clinical realities, a new gold standard will be defined. An ideal bone-generating *combination* is now the goal. It will integrate abundant osteoconductive matrix with growth factor delivery in a localized environment and over the appropriate time course to attract and sustain osteoprogenitor cells as they differientiate into osteogenic cells. The osteoinductive growth factors will be the product of cells genetically engineered to produce these substances within a localized environment and under specific conditions controlled by the physician. Finally, the ideal substitute will provide structural support as necessary, but as the graft site matures, the graft matrix will allow transition of functional weight bearing to the new host bone.

BIOLOGY OF SPINE FUSION

Understanding the biology of spinal fusion is the primary step in advancing the systematic search for an appropriate bone graft substitute. The well-studied set of events surrounding fracture healing of long bones and healing of segmental defects in long bones differs greatly from the incorporation

From: *Bone Regeneration and Repair: Biology and Clinical Applications*
Edited by: J. R. Lieberman and G. E. Friedlaender © Humana Press Inc., Totowa, NJ

of bone graft that occurs at the site of a spinal fusion mass. The biological environment differs even between the various types of fusions found in the spine. The compressive environment of an interbody fusion is quite different from that found in posterolateral intertransverse process fusions *(17)*, the most common type of spinal arthrodesis performed in clinical practice *(4)*. Compressive forces play a much less significant role in intertransverse fusion, because consolidation is necessary prior to any weight bearing by the newly formed posterolateral bone mass *(18)*.

In a general sense, several major steps must occur for successful spinal fusion under the present paradigm of bone grafting *(18–20)*. Initially, osteoprogenitor cells must enter the fusion area. This is accomplished during surgery via decortication, which allows these specific cells to escape from the bone marrow into the fusion environment *(21)*. There they differientiate into osteoblasts, which deposit new bone matrix on the structural component of the transplanted bone graft or osteoconductive bone graft substitute. Remodeling according to Wolff's law then occurs to result in a mature fusion mass *(18)*.

Investigation of this multfactorial process has been hampered by several factors. First, it is difficult to isolate single causal factors in human trials. Moreover, an inability to determine validly whether the end point of fusion has been reached in human subjects makes results suspect even when multivariate randomized investigational design is applied. This has necessitated the development of an appropriate animal model to mimic spinal arthrodesis in humans *(22)*.

In developing such a model, a system must be sought that has similar functional and structural processes to those under study. When considering posterolateral arthrodesis in humans, an appropriate model should allow precise replication of surgical technique by providing similar spinal anatomy and availability of autogenous bone graft. In addition, a similar nonunion rate should occur in control subjects. Finally, the biology of the two systems should be analogous on a molecular level. Use of models that have successful fusion rates of 100% or use of animals with immature skeletal biology are examples of poor analogs for posterolateral fusion in adults *(22)*.

One example of an appropriately validated animal model for posterolateral fusion is that developed by Boden et al. in New Zealand white rabbits *(23)*. Rabbit anatomy allows the procedure for intertransverse process arthrodesis to be duplicated, including the use of autologous iliac crest bone graft. A nonunion rate of 30–40% occurs spontaneously, and this fusion rate has been shown to fluctuate quantitatively under various clinical circumstances, similar to human posterolateral fusion rates under the same clinical circumstances *(24–31)*.

Once this model was appropriately validated, it was used to study lumbar intertransverse fusion. The posterolateral fusion mass resulting from decortication and autologous bone grafting was characterized both geographically and temporally as the healing process proceeded *(20)* (**Fig. 1**). Membranous bone formation was found to occur early and near the transverse processes. A later repair response seen in the area designated as the "central zone" (between the adjacent transverse processes) demonstrated enchondral bone formation (**Fig. 2**). A central "lag effect" was characterized and helps explain why nonunions occur in the central zone of a fusion. The complexity of this fusion environment was clearly demonstrated, as both membranous and enchondral healing were seen to occur within the fusion mass *(20,32,33)*.

Utilizing this model, the events surrounding the fusion mass have been further characterized at the molecular level. Molecular techniques demonstrated osteoblast-related gene expression within the fusion mass that displayed a similar central lag effect as that noted in the histological studies described earlier. Proteins such as osteocalcin and various growth factors including certain bone morphogenic proteins (BMPs) were expressed in consistent patterns as each region underwent the predicted sequence toward bone formation *(33)* (**Fig. 3**).

Further studies using this model have corroborated other empirical clinical experiences for human posterolateral fusion. For example, rabbits not undergoing decortication of the posterolateral spine elements failed to fuse, and investigation using vascular injection studies revealed that the primary blood supply to the fusion mass originated from the decorticated posterolateral bony elements *(21)*. Nicotine

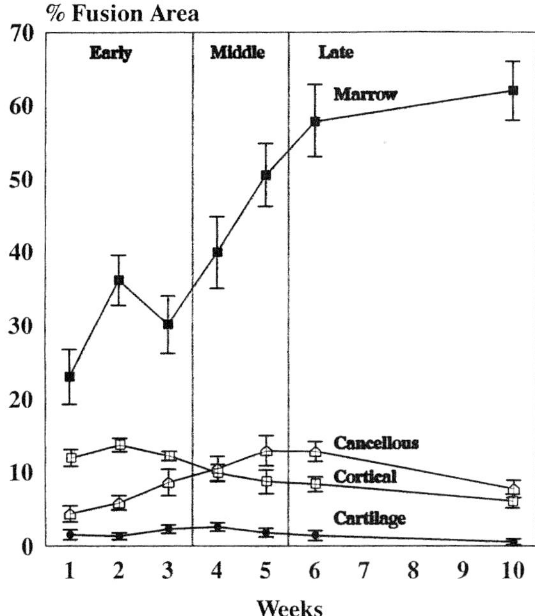

Fig. 1. Quantitative histological healing sequence of rabbit spine fusions depicted graphically. Note the continuous increase in bone marrow content of the fusion mass beginning in the early phase and continuing through the late phase of healing. During the middle phase of healing, a reversal of the cortical:cancellous bone ratio is seen, as well as a small peak in the relative percentage of cartilage corresponding to the central region endochondral ossification. (From ref. *20*, with permission.)

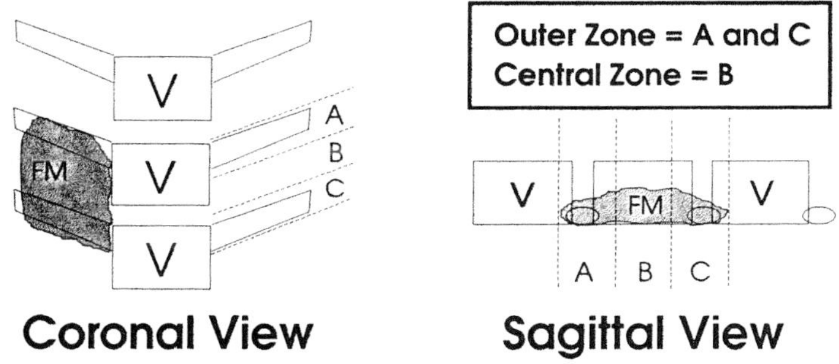

Fig. 2. Schematic diagram of lumbar spine fusion mass (FM) divided into thirds in the coronal and sagittal views and their relationship to the vertebral bodies (V). The two outer zones (A and C) are distinguished from the central zone (B). (From ref. *33*, with permission.)

exposure, nonsteroidal antiinflammatory drug exposure, and excessive postoperative handling have all been studied in this model and found to decrease the rate of spinal fusion *(24–28)*. These studies function to elucidate each particular variable and its effect on fusion while at the same time adding further support for this particular animal model as a human analog.

Spine fusion has been described empirically through outcome studies over many years. Some interventions, such as the implantation of hardware to better control the local mechanical environment have improved the success rate of spinal arthrodesis. Still, a nonunion rate of 10–15% exists *(6,7)*. Most

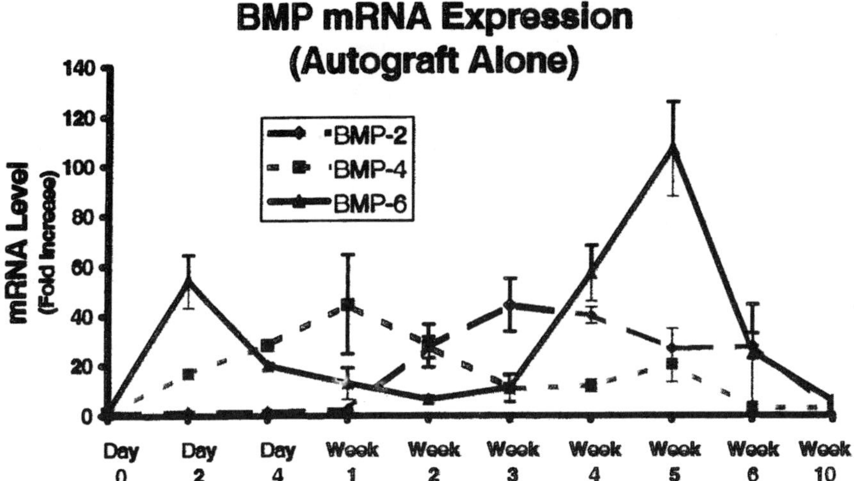

Fig. 3. BMP gene expression determined by reverse transcription polymerase chain reaction (RT-PCR) in the outer zone of the spine fusion mass at specific times after surgery. The values of mRNA levels are given as fold increases over the level present in iliac crest bone (d 0). A reproducible sequence of gene expression was seen with BMP-6 mRNA peaking earliest on d 2, followed by BMP-4 mRNA, BMP-2 mRNA, and a second peak of BMP-6 mRNA. These results suggest that different BMPs have unique temporal patterns of expression during the spine fusion healing process. (From ref. *33*, with permission.)

recently, characterization of histological and corresponding molecular events surrounding posterolateral fusion have led to exciting advances in our understanding of this process. This understanding is now being applied to the quest for a superior bone generator to replace the present gold standard of autogenous bone graft.

CLINICAL APPLICATIONS

Available bone graft substitutes can be broadly classified under three major headings: bone graft extender, bone graft enhancer, or bone graft substitute. Clinical data exist to classify many of the available materials into one of these three categories for various grafting scenarios. This discussion will concentrate on the performance results reported for spinal fusion with each substitute or combination of substitutes.

The terms *extender*, *enhancer*, and *substitute* all inherently reference the current gold standard, autogenous bone graft. One of the major shortcomings of autograft bone is the finite supply available in each patient. As the term implies, an extender is used to add to obtainable autograft in order to expand both volume and effect of the limited autograft. By definition, rates of fusion are at best equal to that of autograft alone in a successful bone graft extender/autograft combination. In contrast, a bone graft enhancer is a substance that, when used in conjunction with autograft, will increase the successful rate of fusion above that reported for autograft alone (70–90% successful fusion rate) under the specific clinical circumstance. Finally, a bone graft substitute is unique in that it can replace autogenous bone graft achieving equal rates of fusion and thereby obviate the need for autogenous bone graft harvest and avoid its concomitant morbidity. Most exciting is that the next generation of stand-alone bone graft substitutes, likely made up of an optimum combination of the substances reviewed here, can realistically be expected to have an enhanced rate of fusion when compared with autograft.

OSTEOCONDUCTIVE SUBSTANCES

The role of osteoconductive bone graft substitutes has changed considerably as more data have become available and bone grafting strategies have evolved. Reported in 1892 by Dresmann, plaster of Paris was the first substance used to fill bony defects in patients. He noted that bony voids filled with this calcium sulfate compound showed evidence of bone ingrowth *(34)*. Since that time, multiple preparations of calcium-containing compounds have been used as bone graft extenders or substitutes with varying success *(35)*.

Calcium sulfate, the primary ingredient in plaster of Paris, continues to be available for clinical use, albeit in a different form than that used by Dresmann in 1892. The work of Peltier et al. demonstrated that compounds of calcium sulfate generated very little foreign body reaction *(34)*, and Sidqui et al. went on to show that osteoclast do actively reabsorb calcium sulfate in a manner similar to physiological bone remodeling *(36)*. Despite these desirable traits, previous heterogenous compounds were unreliable and dissolved quickly, so that fibrous tissue formed instead of bone. Recently, a more crystalline form of calcium sulfate that dissolves at a more predictable rate has become available. This substance is marketed as Osteoset (Wright Bio-Orthopaedics, Arlington, TN) and has shown promising results in several animal models including a sheep posterolateral fusion model; there it performed as well as autograft *(37)*. It is marketed for clinical use as "bone void filler."

The basic science of osteoconductive materials has centered on the analysis of the porous physical structure of compounds that have demonstrated efficacy as bone graft substitutes. Porosity allows vascularization and adherence of osteogenic cells including osteoblasts and osteoclasts. The optimal pore size for bony ingrowth has been studied in detail and appears to be between 100 and 500 µm with a total porous volume of 75–80% *(38,39)*. The topographic structure of the channels appears to be most successful when it closely resembles that of natural bone. Also critical for the ultimate goal of bony union followed by physiological remodeling is the ability of a material to be reabsorbed over a time course that encourages bony replacement *(40,41)*.

Similarity between the exoskeletons of certain naturally occurring marine corals and bone was first recognized in the 1970s by Chiroff et al. *(42)*. This similarity has been exploited, as naturally occurring corals have served as templates for the generation of various implants for bone grafting. The term *coralline* was coined to classify this subset of bone graft substitutes. One of two general processes are used to prepare marine corals for implantation. The first uses the calcium carbonate exoskeleton directly after it has undergone a detergent-based process to remove the organic phase of the coral organism; this results in the product whose trade name is Biocoral (Inoteb, Saint-Gonnery, France). The second general process converts the calcium carbonate to hydroxyapatite via a hydrothermal exchange reaction known as replamineform. Products produced by this process are Prosteon and Interpore porous hydroxyapatite (Interpore Cross International, Irvine, CA) *(43)*. Multiple animal studies have demonstrated the biocompatibility as well as the bioactivity of coralline implants, as osteoblasts and vascular tissues readily migrate into their matrix *(44–46)*. Remodeling also occurs as the implants are resorbed and replaced by host bone *(40,41)*. This process is described by Wolff's law and is accomplished via osteoclastic activity similar to physiological bone remodeling *(41,43,47)*. Much of these data have been accumulated using long-bone defect models, and despite these encouraging results, these products have not proven to be stand-alone bone graft substitutes, especially in the challenging environment of posterolateral spinal fusion *(48,49)*.

Ceramic forms of calcium phosphates are formed by heating and pressurizing these nonmetallic materials. Although the biocompatibility of these substances has been excellent as a group, bioresorbability has varied among the ceramics. In fact, several ceramic substances have been abandoned because the rate of resorption is too slow. The retained implant creates a stress riser within the fusion mass and thus compromises it mechanically *(50)*. An example of poor resorbability is ceramic hydroxyapatite *(40, 41,47)*. At the other extreme of resorbability are the early calcium sulfate compounds, which dissolved so quickly that fibrous tissue ingrowth occurred instead of bone formation *(35)*. Intermediate are the tricalcium phosphate implants that dissolve within 6 wk, a time course that is still quite short for postero-lateral fusion in primates *(51)*. One strategy used to deal with these shortcomings is to integrate two substances into one composite, thus providing a more favorable timeline of dissolution *(52)*.

The use of collagen as a bone graft substitute was suggested by its role in normal bone physiology. Type I collagen functions to catalyze the events surrounding bone formation, acting in both a structural and biochemical manner *(53)*. The use of collagen as a stand-alone bone graft substitute has been unsuccessful, but it has been found to greatly potientiate the effects of other osteoinductive and conductive substances, including bone marrow *(54)* and composites of hydroxyapatite and TCP *(55)*.

Although collagen has been a successful bone graft extender in anterior spinal fusion *(56)*, its primary future role will likely be an ingredient in stand-alone bone graft substitute composites, because it appears to contribute to the ideal environment for growth factors and ceramics to form new bone *(39)*.

At this time, the use of osteoinductive implants has evolved away from use alone as bone graft substitutes. Despite this, interest and development of these substances has intensified. The goal now is to integrate the use of an ideal osteoconductive substance with a potent osteoinductive substance to create a superior bone generator.

OSTEOINDUCTIVE SUBSTANCES

Demineralized bone matrix (DBM) became available for clinical use in 1991. Since that time its use has grown, and it is estimated that in 1999 over 500,000 mL were be implanted in the United States *(3)*. The seminal work of Urist, first reported in 1965, proved the osteoinductive capacity of demineralized bone matrix *(57,58)*, which is prepared from allograft bone by decalcification of cortical bone. This process leaves the extracellular matrix that contains type I collagen and the nonstructural proteins including small amounts of growth factors. Among the growth factors are the bone morphogenic proteins, which make up approx 0.1% of the total weight of all proteins in bone and are responsible for the osteoinductive capacity of demineralized bone matrix *(3,59)*. Although DBM provides no structural integrity and is meant to be used in a stable environment, it still has variable osteoconductive potential due to the presence of collagen *(18)*.

Despite its widespread clinical use, prospective clinical data on DBM is very limited. Much of the data have been generated in small animal models, where it has been shown to have variable osteoinductive potency depending on details of preparation, composite form, and healing environment being tested *(60,61)*. Data from animal studies comprises the greatest source of information on DBM in spinal fusion, where it has been tested as a bone graft substitute and as an autogenous bone graft extender. DBM consistently performs better than mineralized allograft bone but not as well as autogenous bone in small animal models *(59,60,62–65)*.

Studies in human subjects are quite limited. Two retrospective evaluations have been reported. Sassard et al. reported a retrospective comparison of patients undergoing instrumented posterolateral fusion with local autogenous bone graft and demineralized bone matrix (Grafton, Osteotech, Eatontown, NJ) vs matched controls undergoing the same procedure using autograft alone. Radiographic comparisons were undertaken at 3-, 6-, 12-, and 24-mo intervals postoperatively. Low fusion rates were reported for both groups, 60% and 56%, respectively; these rates represented no significant difference between the two groups *(66)*. Lowery et al. reported similar findings of equivalent results when comparing Grafton DBM and autograft composite with autograft alone in posterolateral fusion *(67)*.

As the use of DBM has evolved, different variables have been studied in order to determine its optimum use. Morone and Boden *(59)*, using a validated rabbit model of lumbar posterolateral intertransverse process arthrodesis, have compared the fusion rates utilizing variable DBM gel-to-autograft ratios. Utilizing DBM as a bone graft extender, they showed that fusion rates with DBM gel: autograft ratios of 1:1 and 3:1 were both comparable to autograft alone. Further investigation in this same model compared the different composite forms of the DBM compound available for implantation.

Interestingly, this comparison between Grafton DBM gel and newer putty and matrix forms of Grafton DBM demonstrated function of both the putty and the matrix composites as bone graft enhancers, with rates of fusion of 100% when mixed in a 1:1 ratio with autograft. The matrix compound was also found to function in this rabbit model as a bone graft substitute with a stand-alone rate of fusion of 100% *(61)*.

As noted above, DBM has shown promising results as a bone graft extender and even substitute in small animal models. Care must be taken in extrapolating results form small animal models to humans because of the increased difficulty of initiating osteoinduction in primates *(18)*. In evaluating the presently available DBM compounds, the prevailing clinical attitude is that DBM compounds function as bone graft extenders, not substitutes.

Urist et al., after initially identifying demineralized bone matrix and its osteoconductive ability *(58)*, proceeded to fractionate the osteoinductive portion of the demineralized bone matrix. From this portion, they eventually reported on a series of soluble, low-molecular-weight glycoproteins that were responsible for inducing bone formation. These became known as bone morphogenic proteins and are the most widely investigated group of growth factors that result in bone formation *(68,69)*. By the early 1990s, nine specific molecules, designated BMP-1 through BMP-9 had been isolated and cloned using the molecular techniques of genetic engineering *(18)*. Through these processes, recombinant human BMPs (rhBMPs) may be generated in unlimited quantities and standard potencies, making the study of these compounds easier. As a result, these cloned growth factors are being applied to many models of bone healing, with exciting results.

Bovine-derived bone protein extract, known as NeOsteo (NeOsteo; Intermedics Orthopaedics, Denver, CO) has been investigated by Boden et al. in both rabbit and nonhuman primate models. This osteoconductive mixture is the product of improved techniques of extraction and purification and results in more concentrated bovine BMPs. Investigation in the previously discussed New Zealand rabbit model demonstrated a dose-dependent response to NeOsteo, suggesting that a dosing threshold for bone formation exists when utilizing growth factors. Further investigation in adult rhesus macaque monkeys utilizing NeOsteo vs autograft again demonstrated successful arthrodesis but over a significantly longer time course of 18–24 wk. Comparison of effect of NeOsteo delivered in several different carriers was undertaken, and results showed that NeOsteo functioned successfully for fusion when delivered in autograft, DBM, natural coral, or coralline hydroxyapatite. Variables such as time to fusion, biomechanical properties, and histology of the resulting fusion masses were studied for each different combination *(70,71)*. Systematic investigation of this compound with various delivery systems in appropriate animal models demonstrated an effective methodology to investigate optimization of new growth factors.

Recombinent human BMP-2 (Genetics Institute, Cambridge, MA) has been investigated in detail. It was demonstrated in 1990 by Wang et al. that this particular glycoprotein molecule could induce ectopic bone formation in rats *(72)*. Since that time it has been effective in several spinal fusion healing models *(22,73–78)*. Utilizing a rabbit model of posterolateral spine fusion, Boden et al. were able to demonstrate that BMP-2 used in various dosages with multiple carriers was able to induce posterolateral fusion without decortication *(79)* (**Fig. 4**). This suggests that rhBMP-2 is able to attract and effect the differentiation of bone forming cells from tissues other than bone marrow *(77)*.

Although evidence suggests that fusion in larger animals such as primates is more difficult to achieve than in the smaller animals used for many fusion models *(18,70,80)*, success has been demonstrated in primates. Hecht et al. found that in a rhesus monkey anterior interbody fusion model comparing freeze-dried allograft bone dowels filled with either autograft or RhBMP-2-soaked collagen sponges, the rhBMP-2-soaked collagen sponges had superior results. Specifically, the rhBMP-2/collagen sponge fusion sites demonstrated 100% fusion at 3 mo and extensive replacement of allograft with new host bone. The autograft-filled dowel sites showed no remodeling of initial allograft dowels *(81)*. In another nonhuman primate model, Boden et al. reported successful spinal fusion without bone graft using interbody cages filled with collagen sponges soaked in rhBMP-2 *(78,80)* (**Fig. 5**). In addition, Boden et al.

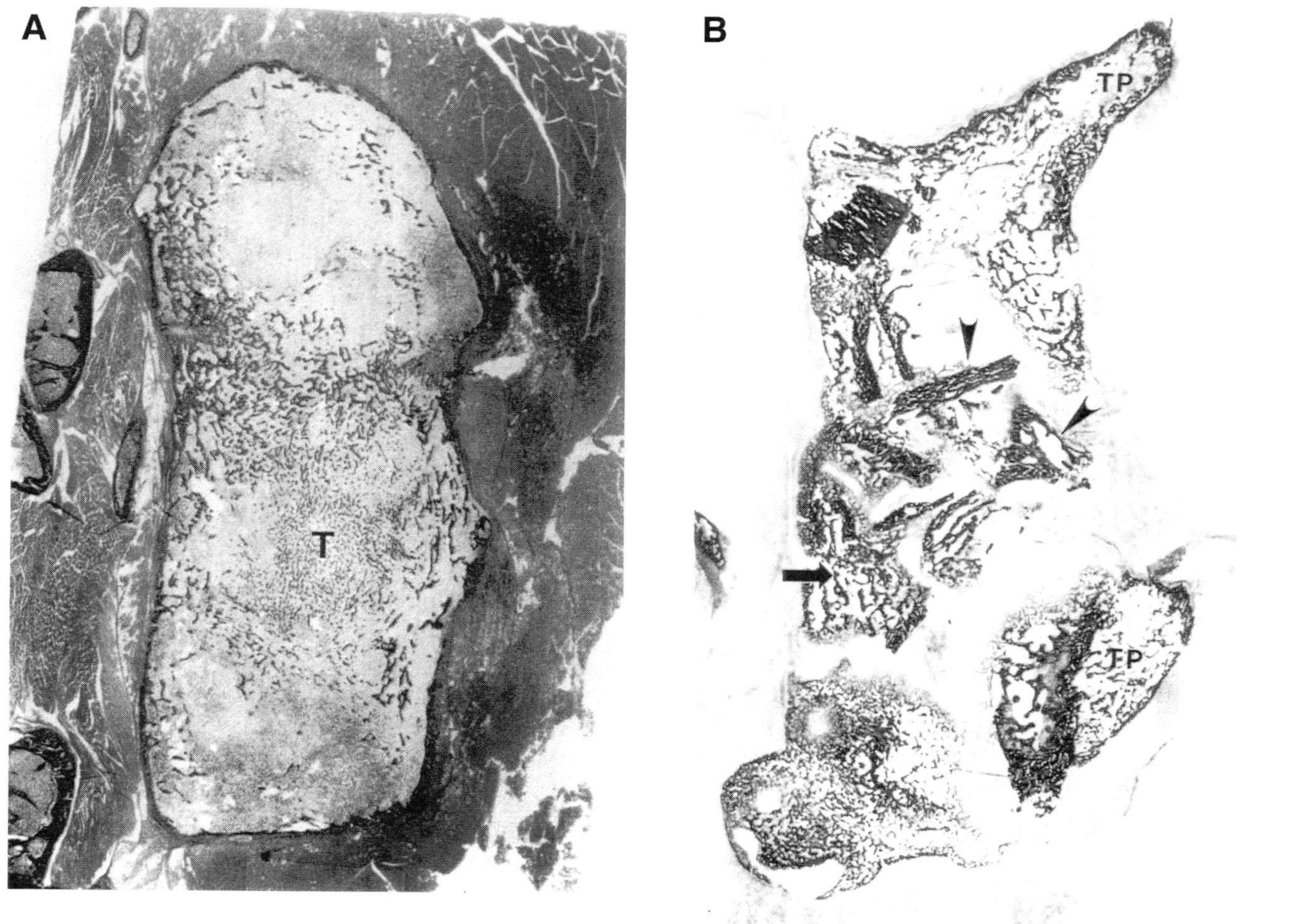

Fig. 4. (A) Coronal sections of rabbit intertranseverse process fusion masses 5-wk following arthrodesis (Goldner Trichrome, X1). **(A)** Arthrodesis with rhBMP-2 delivered in a collagen sponge carrier resulted in a mature fusion mass with a peripheral bony rim and central trabecular bone (T) and marrow. This rabbit had a solid fusion by manual palpitation. **(B)** Arthrodesis with autogenous iliac crest bone resulted in a less mature fusion mass. Remnants of unremodeled cortical bone graft can still be seen in the central zone (arrowheads) and early remodeling of the fusion mass can be seen (arrow) as well as a fibrocartilagenous gap zone in the lower third of the fusion mass. The rabbit had a nonunion based on manual palpitation. (From ref. *79*, with permission.)

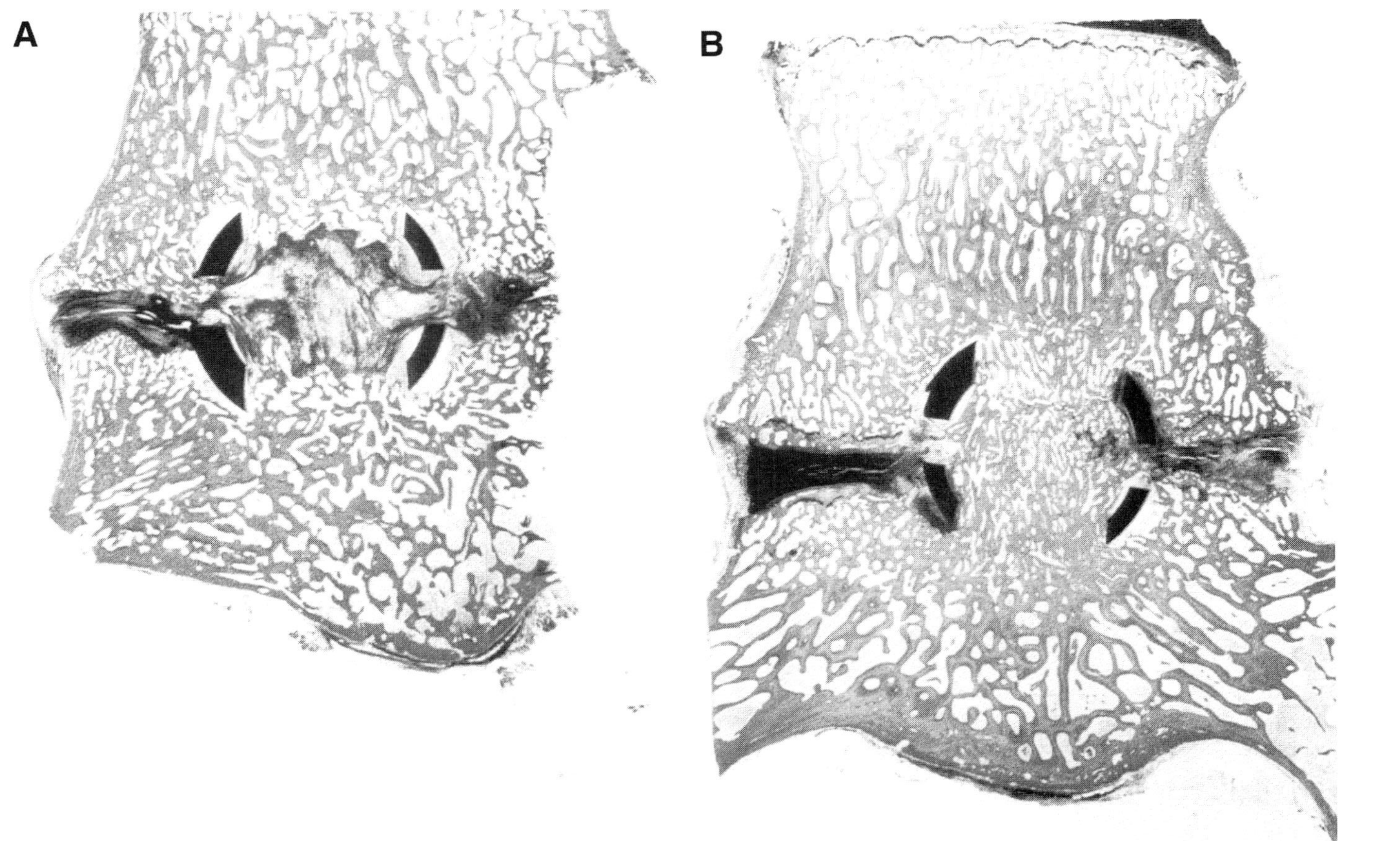

Fig. 5. Coronal section of a rhesus monkey lumbosacral spine 24 wk after interbody arthrodesis using a hollow titanium threaded fusion cage (methylene blue/basic Fuchsin, ×1). (**A**) In this monkey, the cage was filled with collagen sponge carrier without any bone growth factor to serve as a control. The cage is filled with fibrous tissue rather than new bone. (**B**) In this monkey, the cage was filled with collagen sponge carrier soaked with rhBMP-2 (1.5 mg/mL). The cage is filled with new bone that connects the two vertebral bodies through the opening in the upper and lower margin of the cage. (From ref. 78, with permission.)

have reported on a pilot study of single-level anterior lumbar interbody fusions in humans comparing rhBMP-2/collagen sponge-filled cages with iliac crest autograft-filled cages. All 11 patients randomized to the rhBMP group were fused at 6 mo postoperatively, while one of the three patients randomized to the control group receiving autograft in their cages was finally deemed a nonunion at 1 yr *(82)*. Since that time over 350 patients have received that combination of cage/BMP-2/collagen sponge, and the extremely high success rate resulted in approval of rhBMP-2 (InFuse Bone Graft, Medtronic Sofamor Danek, Memphis TN) by the US Food and Drug Administration for use inside tapered fusion cages for anterior lumbar interbody fusion. Early pilot studies using rhBMP-2 with a hydroxyapatite/ tricalcium phosphate carrier matrix have yielded encouraging results for posterolateral lumbar spine fusions.

Other bone-inducing growth factors that have been evaluated include rhBMP-7 and growth and differentiation factor-5. Cook et al. have investigated rhBMP-7, also known as osteogenic protein-1 (Stryker Biotech, Hopkinton, MA) extensively in long-bone defect models, where it has been found to be an effective bone generator in combination with collagen matrix *(83,84)*. Further work by this group utilized a canine spinal fusion model to demonstrate successful rapid posterior spinal fusion when comparing rhBMP-7 to autograft *(85)*. rhBMP-2 has also yielded a high rate of posterolateral spine fusions in the rabbit model. Early results from clinical trials in posterolateral spine have demonstrated fusion rates based only on plain radiographs (not CT scans) of 50–70%. Growth and differentiation factor 5 (GDF-5), another member of the transforming growth factor-β (TGF-β) superfamily, has also been shown to be effective in a long-bone defect model in rats and subsequently in a rabbit spinal fusion model. Spiro et al. used a rabbit posterolateral intertransverse process fusion model to compare rhGDF-5 delivered in a mineralized collagen osteoconductive bone graft matrix (Healos, Orquest, Mountainview, CA) with iliac crest autograft. The rhGDF-5/Healos combination functioned as a bone graft substitute performing as well as autograft alone *(18)*.

THE FUTURE IS HERE, BUT CHALLENGES REMAIN

As described earlier, the ideal bone generator for clinical use in spinal surgery will function to induce the migration of cells capable of becoming bone-forming cells and then activate the system of signals necessary to affect these cells to differientiate into osteoblasts. This bone generator must also supply the proper spatial environment for these bone-forming cells to function in; this requires that neovascularization occur in proximity to surface areas that provide physiologically resorbable scaffolding to act as a template for the various cells involved in bony remodeling. In this manner, the grafted material can be replaced by functional bone that can be maintained physiologically over the patient's lifetime.

As the necessary ingredients for a bone generator are better understood, it becomes clearer why no *single* substitute has been able to supplant autograft. It is also easier to explain why even autograft is not uniformly successful, because at times it fails to provide a sufficient quantity of osteoinductive substances over an appropriate time course once it has been devitalized by the grafting process. Focus has now shifted to synthesis of a composite that maximizes the potential of each ingredient.

Growth factors and an adequate supply of progenitor cells are the key to osteoinductivity. As discussed in the previous section, the glycoprotein molecules of the BMP family are effective bone-generating growth factors. The challenge now lies in delivering a potent growth factor over the appropriate time course for each specific clinical need. The time course for many spinal fusion models appears to be protracted over several months, especially in larger animal models. The normal physiological half-life of glycoprotein molecules in the cellular environment is measured in hours and days, not the weeks or months necessary for spinal fusion in primates.

In addition, it is necessary to find a "growth factor" that works early enough in the cascade of events leading to bone formation that all of the conditions for bone formation will be in place at a clinical site with appropriate physiological timing. The ideal factor will initiate bone formation by triggering the construction of the biochemical bone-forming environment, attracting and effecting differentiation of osteoprogenitor cells, and then potentiating the activity of those cells involved in physiological bone formation and remodeling.

As more physiological environments are characterized, the complexity of each has become increasingly evident. It is likely that bone generation requires a molecular milieu that is provided at specific phases of the wound-healing process. During each phase, a different milieu of permissive factors is available. These factors are substances such as transforming growth factor-β and fibroblast growth factor. It is important that these permissive and/or potentiating factors be present within the bone-forming environment for factors such as the BMPs to be maximally effective *(86,87)*.

Thus exogenous growth factors must be delivered appropriately in both a spatial and a temporal sense. Strategies for accomplishing this have included the utilization of differing doses and/or carriers with different breakdown rates, in the hope that some of the growth factor will remain and be available at the appropriate times. Pilot studies by Boden et al. have proved that it is possible for BMP to induce bone consistently in humans, but both NeOsteo and rhBMP-2 require higher dosing and take longer for osteoinduction in primates than in smaller animals *(70,82)*. These data prove that these substances can be effective in primates, but the high doses necessary and the length of time to fusion demonstrate the need to refine these systems before they will be clinically practical.

One major strategy is to develop a better delivery system for the growth factor. Multiple alternatives have been explored, which utilize the various available osteoconductive substances soaked with growth factors. These synthetic bone-graft substitute materials integrated with rhBMP have been explored in several posterolateral canine fusion models. Sandhu et al. found that rhBMP-2 in a polylactic acid carrier was superior to autogenous iliac crest bone graft for inducing transverse process arthrodesis *(73)*. Also in a canine posterolateral spine fusion model, Muschler et al. reported that rhBMP-2 in a similar biodegradable copolymer carrier of polylactic acid and glycolic acid had equivalent fusion rates and strength to autograft *(74)*.

Gene therapy is a more sophisticated delivery system for growth factors. Utilizing various molecular strategies, genes encoding for factors of the bone formation cascade are inserted into the patient's own cells that exist at the site for fusion (in vivo) or that have been removed and will be reimplanted at the site of fusion (ex vivo) *(88)*. Once these cells are in place, they will then produce a protein product from the transfected gene that leads to bone formation. In this manner, the half-life of the cell or the gene within the cell and not the actual glycoprotein is the limiting temporal factor for presence of a specific growth factor at the fusion site.

This strategy has been used in a rat posterolateral spine fusion model with excellent results. Boden et al. have reported on the use of a novel protein that was isolated via molecular methods and appears to function very early in the cascade of events leading to bone formation *(89)*. This intracellular signaling protein, named LIM mineralization protein-1 (LMP-1), has been isolated and its gene identified. This gene was then transfected into the harvested bone marrow cells of rats and reimplanted at sites for posterolateral spine fusion. Nine of nine (100%) sites implanted with cells containing the LMP-1 gene fused solidly, while 0/9 (0%) sites implanted with control cells fused *(90)* (**Fig. 6**). This study validates the feasibility of local gene therapy to induce bone formation and spinal fusion. A more recent study has demonstrated that ordinary white blood cells from venous blood can be used to deliver the LMP-1 gene with a low dose of adenovirus to achieve successful spine fusion *(91)*.

Optimizing gene therapy introduces even more challenges to the search for an ideal bone generator. Vectors for the delivery of genes into cells, the types of cells transfected, and the control of gene expression are all areas to be explored. As knowledge of each growth factor and its mechanism of action is elucidated, the most potent factor can be identified and exploited. As knowledge of spinal fusion expands on a molecular level, the search for a complete bone graft substitute will proceed in a more logical, systematic fashion and rely less on empirical trial and error.

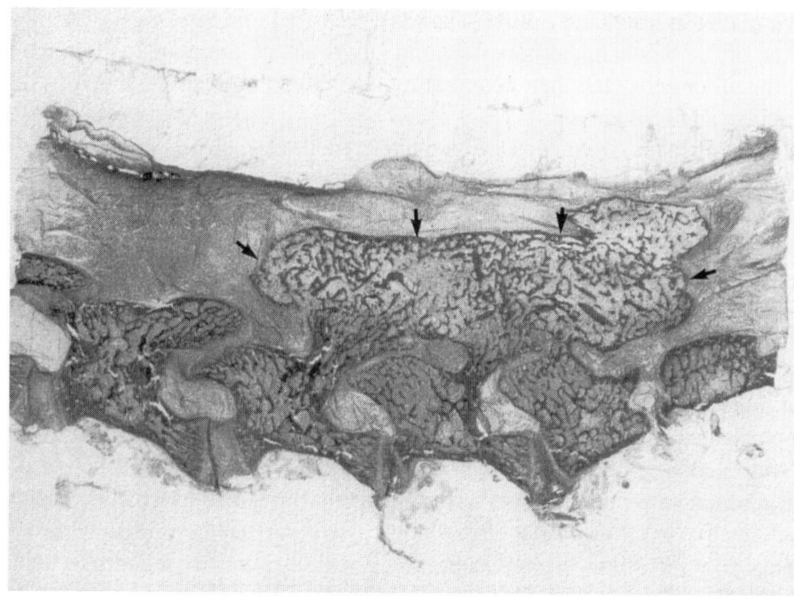

Fig. 6. Low-power sagittal view (×1) photomicrograph through a rat lumbar spine that received devitalized bone matrix with bone marrow cells transfected with active LMP-1 cDNA. A continuous solid bridge of cancellous bone (arrowheads) attached to the sinuous processes and laminae was seen. (From ref. *90*, with permission.)

REFERENCES

1. Albee, F. H. (1911) Transplantation of a portion of the tibia into the spine for pott's disease. *JAMA* **57,** 885–886.
2. Hibbs, R. A. (1911) An operation for progressive spinal deformities. A preliminary report of three cases from the service of the orthopaedic hospital. *N. Y. State J. Med.* **93,** 1013–1016.
3. Sandhu, H. S., Grewal, H. S., and Parvataneni, H. (1999) Bone grafting for spinal fusion. *Orthop. Clin. N. Am.* **30,** 685–698.
4. Katz, J. N. (1995) Lumbar spinal fusion, surgical rates, costs, and complications. *Spine* **20,** 78s–83s.
5. Steinmann, J. C. and Herkowitz, H. N. (1992) Pseudarthrosis of the spine. *Clin. Orthop.* **284,** 80–90.
6. Yuan, H. A., Garfin, S., and Dickman, C. (1994) A historical cohort study of pedicle screw fixation in thoracic, lumbar and sacral spinal fusions. *Spine* **19(Suppl 20),** 2279s–2296s.
7. Zdeblick, T. A. (1993) A prospective, randomized study of lumbar fusion: preliminary results. *Spine* **18,** 983–991.
8. Fischgrund, J. S., Mackay, M., Herkowitz, H. N., et al. (1997) Degenerative lumbar spondylolisthesis with spinal stenosis: a prospective, randomized study comparing decompressive laminectomy and arthrodesis with and without spinal instrumentation. *Spine* **22,** 2807–2812.
9. Thomsen, K., Christensen, F. B., Eiskjaer, S. P., Hansen, E. S., Fruensgaard, S., and Bunger, C. E. (1997) The effect of pedicle screw instrumentation on functional outcome and fusion rates in posterolateral lumbar spinal fusion: a prospective, Randomized clinical study. *Spine* **22,** 2813–2822.
10. Bridwell, K. H., Sedgewick, T. A., O'Brien, M. F., Lenke, L. G., and Baldus, C. (1993) The role of fusion and instrumentation in the treatment of degenerative spondylolisthesis with spinal stenosis. *J. Spinal Disord.* **6,** 461–472.
11. Mcguire, R. A. and Amundson, G. M. (1993) The use of primary internal fixation in spondylolisthesis. *Spine* **18,** 1662–1672.
12. West, J. L. III, Bradford, D. S., and Ogilvie, J. W. (1991) Results of spinal arthrodesis with pedicle screw plate fixation. *J. Bone Joint Surg.* **73A,** 1179–1184.
13. Boden, S. D. and Schimandle, J. H. (1995) Biologic enhancement of spinal fusion. *Spine* **20,** 113S–123S.
14. Banwart, J. C., Asher, M. A., and Hassanein, R. S. (1995) Iliac crest bone graft harvest donor site morbidity. A statistical evaluation. *Spine* **20,** 1055–1060.
15. Fernyhough, J. C., Schimandle, J. H., Weigel, M. C., Edwards, C. C., and Levine, A. M. (1992) Chronic donor site pain complicating bone graft harvesting from the posterior iliac crest for spinal fusion. *Spine* **17,** 1474–1480.
16. Younger, E. M. and Chapman, M. W. (1989) Morbidity at bone graft donor sites. *J. Orthop. Trauma* **3,** 192–195.

17. Mcafee, P. C., Regan, J. J., Farey, I. D., Gurr, K. R., and Warden, K. E. (1988) The biomechanical and histomorphometric properties of anterior lumb: • fusions: a canine model. *J. Spinal Disord.* **1,** 101–110.
18. Ludwig, S. C. and Boden, S. D. (1999) Osteoinductive bone graft substitutes for spinal fusion. *Orthop. Clin. N. Am.* **30,** 635–645.
19. Burchardt, H. (1987) Biology of bone transplantation. *Orthop. Clin. N. Am.* **18,** 187–196.
20. Boden, S. D., Schimandle, J. H., Hutton, W. C., and Chen, M. I. (1995) 1995 Volvo award in basic sciences. The use of an osteoinductive growth factor for lumbar spinal fusion. Part I: the biology of spinal fusion. *Spine* **20,** 2626–2632.
21. Toribatake, Y., Hutton, W. C., Boden, S. D., and Morone, M. A. (1998) Revascularization of the fusion mass in a posterolateral intertransverse process fusion. *Spine* **23,** 1149–1154.
22. Schimandle, J. H. and Boden, S. D. (1994) The use of animal models to study spinal fusion. *Spine* **19,** 1998–2006.
23. Boden, S. D., Schimandle, J. H., and Hutton, W. C. (1995) An experimental lumbar intertransverse process spinal fusion model: radiographic, histologic, and biomechanical healing characteristics. *Spine* **20,** 412–420.
24. Silcox, D. H., Daftari, T., Boden, S. D., Schimandle, J. H., Hutton, W. C., and Whitesides, T. E. (1995) The effect of nicotine on spinal fusion. *Spine* **20,** 1549–1553.
25. Riebel, G. D., Boden, S. D., Whitesides, T. E., and Hutton, W. C. (1995) The effect of nicotine on incorporation of cancellous bone graft in an animal model. *Spine* **20,** 2198–2202.
26. Keller, J., Bunger, C., Andreassen, T. T., Bak, B., and Lucht, U. (1987) Bone repair inhibited by indomethacin: effects on bone metabolism and strength of rabbit osteotomies. *Acta Orthop. Scand.* **58,** 379–383.
27. Ro, J., Sudmann, E., and Marton, P. F. (1976) Effect of indomethacin on fracture healing in rats. *Acta Orthop. Scand.* **47,** 588–599.
28. Sudmann, E., Dregelid, E., Bessesen, A., and Morland, J. (1979) Inhibition of fracture healing by indomethacin in rats. *Eur. J. Clin. Invest.* **9,** 333–339.
29. Tornkvist, H. and Lindholm, S. (1980) Effect of ibuprofen on mass and composition of fracture callus and bone. *Scand. J. Rheumatol.* **9,** 167–171.
30. Tornkvist, H., Lindholm, T. S., Netz, P., Stromberg, L., and Lindholm, T. C. (1984) Effect of ibuprofen and indomethacin on bone metabolism reflected in bone strength. *Clin. Orthop.* **187,** 255–259.
31. Tornkvist, H., Bauer, F. C. H., and Nilsson, O. S. (1985) Influence of indomethacin on experimental bone metabolism in rats. *Clin. Orthop.* **193,** 264–270.
32. Fairbank, J. C. T., Davies, J. B., Mbaot, J. C., and O'Brien, J. P. (1980) The oswestry low back pain disability questionnaire. *Physiotherapy* **66,** 271–273.
33. Morone, M. A., Boden, S. D., Martin, G., Hair, G., and Titus, L. (1998) Gene expression during autograft lumbar spine fusion and the effect of bone morphogenetic protein-2. *Clin. Orthop.* **351,** 252–265.
34. Peltier, L. (1959) The use of plaster of paris to fill large defects in bone. *Am. J. Surg.* **97,** 311–315.
35. Tay, B., Patel, V., and Bradford, D. (1999) Calcium sulfate and calcium phosphate based bone substitutes. *Orthop. Clin. N. Am.* **30,** 615–623.
36. Sidqui, M., Collin, P., and Vitte, C. (1995) Osteoblast adherence and resorption activity of isolated osteoclasts on calcium sulfate hemihydrate. *Biomaterials* **16,** 1327–1321.
37. Cunningham, B. W., Kotani, Y., McNulty, P. S., et al. (1998) Video-assisted thoracoscopic surgery versus open thoracotomy for anterior thoracic spinal fusion. A comparative radiographic, biomechanical, and histologic analysis in a sheep model. *Spine* **23,** 1333–1340.
38. White, E. and Shors E. C. (1986) Biomaterial aspects of interpore-200 porous hydroxyapatite. *Dent. Clin. N. Am.* **30,** 250–256.
39. Cornell, C. N. (1999) Osteoconductive materials as substitutes for autogenous bone grafts. *Orthop. Clin. N. Am.* **30,** 591–598.
40. Holmes, R., Mooney, V., Bucholz, R., and Tencer, A. (1984) A coralline hydroxyapatite bone graft substitute. *Clin. Orthop.* **188,** 252–262.
41. Chapman, M. W., Bucholz, R., and Cornell, C. N. (1997) Treatment of acute fractures with a collagen-calcium phosphate graft material: a randomized clinical trail. *J. Bone Joint Surg.* **18A,** 495–502.
42. Chiroff, R. T., White, E. W., Weber, J. N., and Roy, D. M. (1975) Tissue ingrowth of replamineform implants. *J. Biomed. Res. Symp.* **6,** 29–45.
43. Shors, E. C. (1999) Coralline bone graft substitutes. *Orthop. Clin. N. Am.* **30,** 599–613.
44. Begley, C. T., Doherty, M. J., and Mollan, R. A. (1995) Comparative study of the osteoinductive properties of bioceramic, coral and processed bone graft substitute. *Biomaterials* **16,** 1181–1185.
45. Doherty, M. J., Schlag, G., and Schwartz, N. (1994) Biocompatibility of xenographic bone, commercially available coral, a bioceramic and tissue sealant for human osteoblasts. *Biomaterials* **15,** 601–608.
46. Flatley, T. J., Lynch, K. L., and Benson, M. (1983) Tissue response to implants of calcium phosphate ceramic in the rabbit spine. *Clin. Orthop.* **179,** 246–252.
47. Holmes, R. E., Bucholz, R. W., and Mooney, V. (1986) Porous hydroxyapatite as a bone graft substitute in metaphyseal defects. *J. Bone Joint Surg.* **68A,** 904–911.

48. Boden, S. D., Martin, G. J., Morone, M. A., Ugbo, J. L., Titus, L., and Hutton, W. C. (1999) The use of coralline hydroxyapatite with bone marrow, autogenous bone graft, or osteoinductive bone protein extract for posterolateral lumbar spine fusion. *Spine* **24,** 320–327.

49. Boden, S. D., Schimandle, J. H., Hutton, W. C., et al. (1997) In vivo evaluation of a resorbable osteoinductive composite as a graft substitute for lumbar spinal fusion. *J. Spinal Disord.* **10,** 1–11.

50. Jarcho, M. (1981) Calcium phosphate ceramics as hard tissue prosthetics. *Clin. Orthop.* **157,** 259–278.

51. Goldberg, V. M., Stevenson, S., and Shaffer, J. W. (1991) Biology of autografts and allografts, in *Bone and Cartilage Allografts* (Friedlaender, G. E. and Goldberg, V. M., eds.), American Academy of Orthopaedic Surgeons, Park Ridge, IL, pp. 3–12.

52. Kurashina, K., Kurita, H., and Hirano, M. (1997) In vivo study of of calcium phosphate cements: implantaiton if an alpha-tricalcium phosphate/dicalcium phosphatedibasic/tetracalcium phosphate monoxide cement paste. *Biomaterials* **18,** 539–543.

53. Muschler, G. F., Lane, J. M., and Dawson, E. G. (1991) The biology of spinal fusion, in *Spinal Fusion* (Cotler, J. and Cotler, H., eds.), Springer-Verlag, New York, pp. 9–21.

54. Werntz, J., Lane, J. M., Piez, C., Seyedin, S., and Burstein, A. (1986) The repair of segmental bone defects with collagen and marrow. *Trans. Orthop. Res. Soc.* **32,** 108.

55. Johnson, K. D., Frierson, K., and Keller, T. S. (1996) Porous ceramics as bone graft substitutes in long bone defects: a biomechanical, histological, and radiographic analysis. *J. Orthop. Res.* **14,** 351–369.

56. Zerwekh, J. E., Kourosh, S., Scheinberg, R., et al. (1992) Fibrillar collagen-biphasic calcium phosphate composite as a bone graft substitute for spinal fusion. *J. Orthop. Res.* **10,** 562–572.

57. Dubuc, F. L. and Urist, M. R. (1967) The accessibility of the bone induction prinicple in surface-decalcified bone implants. *Clin. Orthop.* **55,** 217–223.

58. Urist, M. R. (1965) Bone: formation by autoinduction. *Science* **150,** 893–899.

59. Morone, M. A. and Boden, S. D. (1998) Experimental posterolateral lumbar spinal fusion with a demineralized bone matrix gel. *Spine* **23,** 159–167.

60. Cook, S. D., Dalton, J. E., Prewett, A. B., and Whitecloud, T. S. (1995) In vivo evaluation of demineralized bone matrix as a bone graft substitute for posterior spinal fusion. *Spine* **20,** 877–886.

61. Martin, G., Boden, S. D., Morone, M. A., and Titus, L. (1999) New formulations of demineralized bone matrix as a more effective graft alternative in experimental posterolateral lumbar spine arthrodesis. *Spine* **24,** 637–645

62. Turner, C. H., Akhter, M. P., Raab, D. M., Kimmel, D. B., and Recker, R. R. (1991) A noninvasive, in vivo model for studying strain adaptive bone modeling. *Bone* **12,** 73–79.

63. Feighan, J. E., Davy, D., Prewett, A. B., and Stevenson, S. (1995) Induction of bone by a demineralized bone matrix gel: a study in a rat femoral defect model. *J. Orthop. Res.* **13,** 881–891.

64. Lindholm, T. S., Ragni, P., and Lindholm, T. C. (1988) Response of bone marrow stroma cells to demineralized cortical bone matrix in experimental spinal fusion in rabbits. *Clin. Orthop.* **230,** 296–302.

65. Frenkel, S. R., Moskovitch, R., Spivak, J., Zhang, Z. H., and Prewett, A. B. (1993) Demineralized bone matrix. Enhancement of spinal fusion. *Spine* **18,** 1634–1639.

66. Sassard, W. R., Eidman, D. K., and Gray, P. M. (1994) Analysis of spine fusion utilizing demineralized bone matrix. *Orthop. Trans.* **18,** 886–887.

67. Lowery, G. L., Maxwell, K. M., Karasick, D., Block, J. E., and Russo, R. (1995) Comparison of autograft and composite grafts of demineralized bone matrix and autologous bone in posterolateral fusions: an interim report. *Innov. Tech. Biol. Med.* **16,** 1–8.

68. Urist, M. R., Dowell, T. A., Hay, P. H., and Strates, B. S. (1968) Inductive substrates for bone formation. *Clin. Orthop.* **59,** 59–96.

69. Urist, M. R., Mikulski, A., and Lietze, A. (1979) Solubilized and insolubilized bone morphogenetic protein. *Proc. Natl. Acad. Sci. USA* **76,** 1828–1832.

70. Boden, S. D., Schimandle, J. H., and Hutton, W. C. (1995) 1995 Volvo Award In Basic Sciences. The use of an osteoinductive growth factor for lumbar spinal fusion. Part II: study of dose, carrier, and species. *Spine* **20,** 2633–2644.

71. Boden, S. D., Schimandle, J. H., and Hutton, W. C. (1995) Lumbar intertransverse process spine arthrodesis using a bovine-derived osteoinductive bone protein. *J. Bone Joint Surg.* **77A,** 1404–1417.

72. Wang, E. A., Rosen, V., D'Alessandro, J. S., et al. (1990) Recombinant human bone morphogenetic protein induces bone formation. *Proc. Natl. Acad. Sci. USA* **87,** 2220–2224.

73. Sandhu, H. S., Kanim, L. E. A., Kabo, J. M., et al. (1995) Evaluation of rhbmp-2 with an OPLA carrier in a canine posterolateral (transverse process) spinal fusion model. *Spine* **20,** 2669–2682.

74. Muschler, G. F., Hyodo, A., Manning, T., Kambic, H., and Easley, K. (1994) Evaluation of human bone morphogenetic protein 2 in a canine spinal fusion model. *Clin. Orthop.* **308,** 229–240.

75. Zdeblick, T. A., Ghanayem, A. J., Rapoff, A. J., et al. (1998) Cervical interbody fusion cages: an animal model with and without bone morphogenetic protein. *Spine* **23,** 758–766.

76. Fischgrund, J. S., James, S. B., Chabot, M. C., et al. (1997) Augmentation of autograft using rhbmp-2 and different carrier media in the canine spine fusion model. *J. Spinal Disord.* **10**, 467–472.

77. Holliger, E. H., Trawick, R. H., Boden, S. D., and Hutton, W. C. (1996) Morphology of the lumbar intertransverse process fusion mass in the rabbit model: a comparison between two bone graft materials—rhbmp-2 and autograft. *J. Spinal Disord.* **9**, 125–128.

78. Boden, S. D., Martin, G. J., Horton, W. C., Truss, T. L., and Sandhu, H. S. (1998) Laparoscopic anterior spinal arthrodesis with Rhbmp-2 in a titanium interbody threaded cage. *J. Spinal Disord.* **11**, 95–101.

79. Schimandle, J. H., Boden, S. D., and Hutton, W. C. (1995) Experimental spinal fusion with recombinant human bone morphogenetic protein-2 (Rhbmp-2). *Spine* **20**, 1326–1337.

80. Boden, S. D., Moskovitz, P. A., Morone, M. A., and Toribitake, Y. (1996) Video-assisted lateral intertransverse process arthrodesis—validation of a new minimally invasive lumbar spinal fusion technique in the rabbit and nonhuman primate (rhesus) models. *Spine* **21**, 2689–2697.

81. Hecht, B., Fischgrund, J., Herkowitz, H., Penman, L., Toth, J., and Shirkhoda, A. (1999) The use of recombinent humman bone morphogenic protein 2 (Rhbmp-2) to promote spinal fusion in a nonhuman primate anterior interbody fusion model. *Spine* **24**, 629–636.

82. Boden, S. D., Zdeblick, T. A., Sandhu, H. S., and Heim, S. E. (2000) The use of Rhbmp-2 in interbody fusion cages: definitive evidence of osteoinduction in humans. *Spine* **25**, 376–381.

83. Cook, S. D., Baffes, G. C., Wolfe, M. W., Sampath, T. K., Rueger, D. C., and Whitecloud, T. S. (1994) The effect of recombinant human osteogenic protein-1 on healing of large segmental bone defects. *J. Bone Joint Surg.* **76A**, 827–838.

84. Cook, S. D., Wolfe, M. W., Salkeld, S. L., and Rueger, D. C. (1995) Effect of recombinant human osteogenic protein-1 on healing of segmental defects in non-human primates. *J. Bone Joint Surg.* **77A**, 734–750.

85. Cook, S. D., Dalton, J. E., Tan, E. H., Whitecloud, T. S., and Rueger, D. C. (1994) In vivo evaluation of recombinant human osteogenic protein (Rhop-1) implants as a bone graft substitute for spinal fusions. *Spine* **19**, 1655–1663.

86. Boden, S. D., Mccuaig, K., Hair, G., et al. (1996) Differential effects and glucocorticoid potentiation of bone morphogenetic protein action during rat osteoblast differentiation in vitro. *Endocrinology* **137**, 3401–3407.

87. Bostrom, M., Saleh, K., and Einhorn, T. (1999) Osteoinductive growth factors in preclinical fracture and long bone defects models. *Orthop. Clin. N. Am.* **30**, 647–658.

88. Scaduto, A. and Lieberman, J. (1999) Gene therapy for osteoinduction. *Orthop. Clin. N. Am.* **30**, 625–633.

89. Boden, S. D., Liu, Y., Hair, G. A., et al. (1998) LMP-1, A LIM-domain protein, mediates BMP-6 effects on bone formation. *Endocrinology* **139**, 5125–5134.

90. Boden, S. D., Titus, L., Hair, G., et al. (1998) 1998 Volvo Award In Basic Sciences: Lumbar spine fusion by local gene therapy with a Cdna encoding a novel osteoinductive protein (LMP-1). *Spine* **23**, 2486–2492.

91. Viggeswarapu, M., Boden, S. D., Liu, Y., et al. (2001) Adenoviral delivery of LIM mineralization protein-1 induces new-bone formation in vitro and in vivo. *J. Bone Joint Surg.* **83A**, 364–376.

13

Bone Allograft Transplantation

Theory and Practice

**Henry J. Mankin, MD, Francis J. Hornicek, MD, PhD,
Mark C. Gebhardt, MD, and William W. Tomford, MD**

INTRODUCTION

An amazing change has occurred in the last 30 yr in our ability to care for children and adults with bone sarcomas. Because of better imaging technology and vast improvements in our systems for treating sarcoma with chemotherapy, we no longer use amputations as our first line of care and now perform limb-sparing surgery for most of our patients. The success of metallic implants is very attractive, and many centers use these technologies *(1–14)* in the treatment of high-grade tumors involving a joint, but in addition there is a long history in orthopedics of the use of allograft implants *(15–42)*. The latter system is intriguing in many ways and may in fact outlive the metallic implants over time. The aim of this chapter is to review the history of allografting, describe the current state of knowledge, present our series of over 1000 cases and their complications, and then try to establish some rules and approaches to alloimplants of the future.

HISTORY

In the entire world of orthopedics, there has never been a more wished-for "dream" or sought after "holy grail," than osteochondral allograft implantation. When a limb is grossly diseased, a bone badly deformed, or a joint totally disabled, both the patient and physician fervently wish that they could start over with a new part, anatomically identical to the old but disease-free and completely functional. The concept of a well-accepted, low-complication, fully functional bone and cartilage alloimplant is a hope that has prevailed for centuries and remains at the present still not quite in reach. Grafts are available in appropriate sizes and shapes, the tissue is accepted with minimal problem in many cases, but still the perfect graft eludes us and remains a "dream" or perhaps may be described as the "holy grail of reconstructive orthopedics."

The dream is ancient, presumably occurring in many caretakers over the centuries but recognizably remembered as the "miracle" performed by Saints Cosmas and Damian in the sixth century AD *(43–45)*. The saints were twin physicians born in the third century AD in the town of Egea in Cilicia in Asia Minor. They were the sons of a physician and then became physicians, traveling widely in Greece, Turkey, and Rome, treating ailments and refusing payment for their services. They somehow angered Lysia, the Roman governor of Cilicia during the persecution of the Christians by the emperor Diocletian, and after a variety of attempts at killing the twins, they and their three brothers were beheaded and buried in a grave in Egea on September 27, 287 AD *(43–45)*. They were returned, however, in the fifth century to a basilica in the Roman Forum, which now bears their name, where Deacon Justinian, a faithful church retainer with a cancerous limb was so exhausted by the pain that he fell asleep during his prayers. There came to him in a dream the twin physicians, who, after amputating the limb of a

From: *Bone Regeneration and Repair: Biology and Clinical Applications*
Edited by: J. R. Lieberman and G. E. Friedlaender © Humana Press Inc., Totowa, NJ

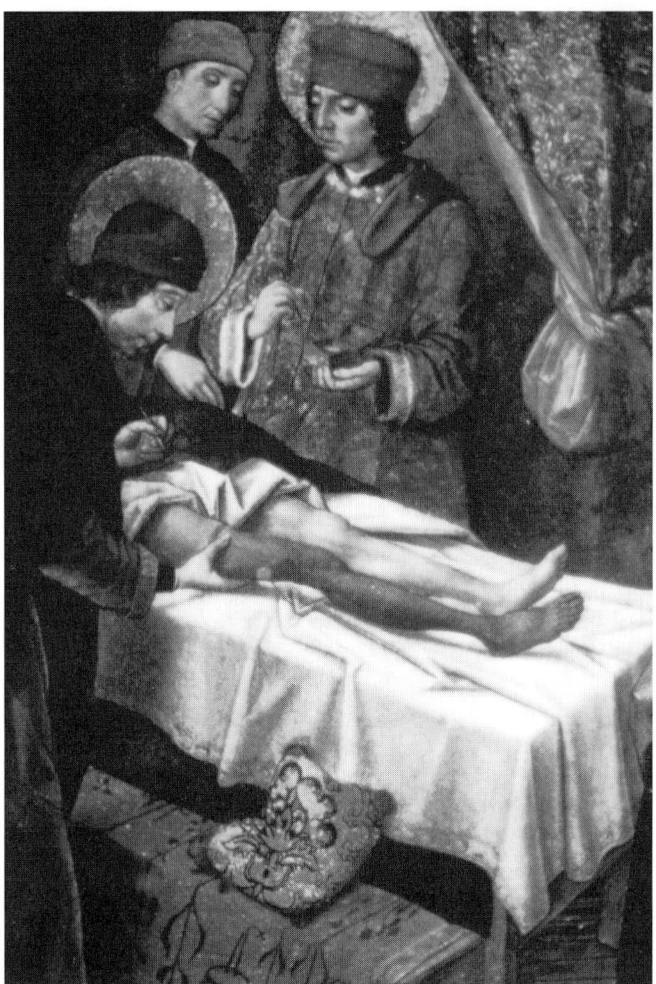

Fig. 1. Painting by Pedro de Berruguete in the 15th century hanging in the Collegiate Church in Covarrubias. Note that the saints are performing the surgical procedure on the right lower extremity and that Damian, the surgeon in the foreground, is using his left hand to suture the host–donor junction site.

Moor who had died that morning, replaced the diseased part with the obtained allograft implant. The procedure, known as the "Miracle of the Black Leg," was reportedly successful, and because of that, the twins were subsequently beatified, receiving their sainthood in approximately the year 550 AD. Of note is the fact that the occasion and drama associated with the procedure was so extraordinary that it captured the imagination of first the painter Fra Angelico and then many other artists; and literally hundreds of some of the most extraordinary paintings depicting the procedure can now be found in many of the world's museums *(43)* (**Fig. 1**).

In his exhaustive report on the history of allografting, Burwell *(46)* records several attempts by individuals over the many years that followed, but the world recognizes the first report of a successful alloimplant to be that of Macewen in 1881 *(47)*. In that procedure, Macewen transplanted segments of bone from a rachitic patient to the humerus of a 3-yr-old child who had lost a portion of the shaft as a result of osteomyelitis. The major effort, however, in the early part of the 20th century was that of Lexer, who in 1908 reported on four such procedures about the knee *(48)* and in 1925 described a reasonable success rate on 11 half joints and 23 whole joints using fresh cadaveric tissue *(49)*. Sporadic case reports and short series were presented over the next 20 or so years, but it was a Russian group under the direction of Volkov *(50)* who reported a large series of successful procedures using processed but not frozen cadaveric bone. On the basis of a sophisticated group of experimental studies, Curtiss, Chase, and Herndon *(51,52)* proposed the concept that freezing the cadaveric bony parts would reduce immunological activity and thus reduce the rejection rate. This also made it possible to develop bone banks in which bony parts obtained at surgery or autopsy (or subsequently at harvest) were stored in a freezer at −20 to −70°C and thawed prior to implantation *(16,46,53)*.

Following World War II, the US Navy became interested in preservation of allograft tissue, and in 1950 founded the Navy Tissue Bank under the direction of George Hyatt *(54)*. Subsequently, when Kenneth Sell became head, he recruited a number of Fellows to rotate through the system and perform research on graft technology. The list of graduates of Kenneth Sell's program included some very distinguished investigators, such as Andrew Bassett, Gary Friedlaender, Theodore Malinin, William Tomford, and Michael Strong, all of whom started their own banks and also performed very competent research *(34,55–69)*. Their work, along with Sell's, not only advanced the field in terms of improved success of the implant, but also added greatly to the safety of the recipient in relation to infectious bacterial and especially virus transmission *(55,64,69–72)*.

On the basis of these pioneer efforts, two major sets of experimentation started. The first of these was clinical. Frank Parrish in Houston, acting in part on the reported success of Volkov, performed a series of surgical procedures in which frozen allografts were implanted after removal of a bone tumor *(73,74)*. He carefully followed the patients and reported the complications of the procedure *(73)*. Carlos Ottolenghi in Buenos Aires started a similar series and reported on successes, and most importantly, on the causes of failures *(75)*. Stimulated by these efforts, several other groups began to look at allografting as a possibly better solution than metallic implants and further advanced the search for the "holy grail" *(18,20,21,25,27,29,31,33,34,76–81)*.

During this same period, several investigators recognized that the complications, including infection, fracture, and nonunion, that compromised the results in the clinical series were probably based on the immune response and began seeking a greater understanding of this phenomenon *(57,77,82–85)*. A group in Canada headed by Langer demonstrated that the response to allografting in animals was markedly reduced by freezing the graft, suggesting that a blocking antibody was produced by the temperature reduction *(86)*. Similar attempts to define the immune response in animal systems were reported by Burchardt *(82,83)*, Elves *(87,88)*, and Stevenson *(84,89–92)*, but it seemed that these data were really not as applicable to humans *(93)*. More recently, the studies of Friedlaender and Strong and their group showed that the clinical result was significantly improved in patients who achieved a match with MHC Class II antigens than with MHC Class I or with mismatch *(65,94)*. Simultaneously, the rules regarding bone banking were being established in a number of centers.

Methods of testing the donor for bacterial or viral diseases were established, as well as approaches to the optimal rules for freezing and thawing (most believe that slow freezing and rapid thawing is the most successful *[62,66,68,95]*) and the value and drawbacks of radiation to the graft *(71,96)*. It seemed sensible to maintain cartilage at least partially alive during the freezing and thawing process, and the use of glycerol or dimethylsulfoxide (DMSO) was proposed to achieve this important goal *(97,98)*. Establishing the Bone section of the American Association of Tissue Banks and promulgating Guidelines and Standards were major steps forward and allowed safe bone banks to spring up throughout the United States and Europe *(67,68)*.

CURRENT STATUS

It is possible to summarize the current status of our understanding of the issues surrounding allograft transplantation as follows. The response to allograft implantation appears to be species-dependent and quite variable in extent and nature in humans. As is also now well known, the operative procedure alone sometimes has a deleterious effect on the allograft response and may be the cause of some complications. In patients with high-grade bone tumors who are treated with chemotherapy *(41,99, 100)* and sometimes radiation, who have extensive resective surgery that in part damages the blood supply of the bed into which the graft is placed, it seems reasonable to blame some of the complications on these factors as well. The results of the surgery are predictably poorer in such patients as compared with others with less severe disease. *(21,22,24,25,32,34,101)*. At times the effect of the immune response is in the form of rapid dissolution of the graft (very rare in humans but known to occur in animal systems *[28,32,39,102,103]*) to a much more frequently occurring "walling off" of the segment with almost no vascular invasion *(28,102)*. It seems logical on that basis to blame at least two of the three major complications, fracture and nonunion, on this "walled off" state; and some tentative evidence has been advanced that seems to demonstrate that the high infection rate is a manifestation of the immune response as well *(28,32,69,70,101)*. As has been clearly noted by all clinical studies, then, the triad of infection, nonunion, and allograft fracture represent the major complications of the procedure and account for most of the graft failures in all the clinical series *(15,20,24,25,30,32,34,35, 38,40–42,69,71,72,77,102,104–106)*.

In similar fashion, cartilage is known to be highly antigenic and has been shown to evoke a profound cellular and humoral antibody response *(18,28,63,87,97,98)*. It is thought, however, that the cartilage matrix pore size is so small that antigen cannot pass out nor can cells or antibody enter, provided the matrix is intact *(66)*. If the cartilage is altered by surface injury, or subchondral bone fracture, it is presumed that the immune response to the matrix and cells is a major event in the development of joint disease. Cryopreservation with DMSO reduces the size of the ice crystals that form on the chondrocyte membrane and in an experimental system seems to help reduce the likelihood of such cartilage destruction. It is apparent, however, that even the most rigorous and complex of techniques for such cryopreservation have thus far been unsuccessful in maintaining cell viability in in vivo human system *(5,97,98)*. Of greater concern is the evident fact that a poor fit of the graft, which leads to surface cartilage loss or microfractures of the underlying subchondral bone, is likely to lead to an early form of osteoarthritis. In our series, about 20% of the distal femoral or proximal tibial grafts have required resurfacing procedures at a mean of over 5 yr *(30,38)*.

BONE BANKING

Most of the grafts implanted in our patients came from the Massachusetts General Hospital (MGH) Bone Bank, which was established in 1974 *(95)*. The bank uses a set of guidelines first promulgated by the American Association of Tissue Banks, which helps to establish that the parts are disease-free, appropriate in shape, and of proper size *(62,66–68)*. Prior to procurement, donors are screened by discussion with the treating physician and a careful review of the chart for evidence of occult malignancy, infection, toxic substance ingestion, drug abuse, or risk factors for AIDS *(55,62,68,96)*. All harvests are performed under sterile conditions in an operating room and almost always will follow procurement of living organs by other harvest teams. The MGH Bone Bank teams consist for the most part of physicians skilled in the harvest technique, who move rapidly to obtain the long bone and pelvic parts. Swabs from each of the individual parts are cultured separately, for bacteria, and heart blood samples are cultured and screened for hepatitis B and C and tested for HIV by seeking viral antibodies and antigens and performing polymerase chain reaction (PCR) studies *(55,62,68,69)*. The bones are stripped of soft tissue, except for ligaments and tendinous attachment sites and especially the posterior capsule and collateral ligaments of the knee joint; capsular, iliopsoas, and gluteus medius attachment sites on the proximal femur; and the rotator cuff and deltoid and pectoral insertions of the proximal humerus. The cartilage is immersed in 8% DMSO in an effort to maintain cellular viability during the freezing and thawing process *(66,97,98)*. Following the reconstruction of the cadaver with wooden dowels and plaster of Paris, the allograft parts are wrapped in gas-sterilized polyethylene bags and appropriate cloth outer wraps and labeled. Prior to freezing slowly to −80°C, all parts are X-rayed in two planes with a metal marker taped to the outer wrap. Wherever possible, a lymph node and blood are obtained and stored, and under ideal circumstances, a full autopsy is performed on the remaining parts.

Allograft parts remain in the freezer until needed, and a computerized inventory is maintained. No part is used until all the tests have been returned, supporting the sterile and virus-free status of the allograft segment. At the time of contemplated surgery, the part with best fit is selected on the basis of comparison of the radiographs of the allograft and that of the patient (the latter obtained with the same sizing device used for the donor parts in place). Following resection of the tumor, the part is brought into the operating room and thawed rapidly by immersion in warm Ringer's lactate (60°C) *(67,68)*. Additional cultures are obtained at the time of thawing the graft and are useful in retrospective analysis of infections as well as prophylactic treatment of the patient postoperatively.

THE OPERATIVE PROCEDURE

The operative procedure conforms to principles of tumor surgery as outlined in a number of protocols and texts *(107–109)*. In planning the procedure, radiographic, computerized tomographic (CT) and magnetic resonance imaging (MRI) studies should be performed to define the extent of the lesion and help the surgeon decide on the bony and soft tissue margins. The surgery demands that the tract of prior biopsy be resected and as much of the compartment(s) containing the tumor be resected with the specimen. We attempt to avoid "intralesional" margins (through the tumor mass and containing gross tumor) and strive for "wide" margins (leaving a cuff of normal tissue outside the reactive zone). Often, however, particularly for low-grade lesions or when the patient has been treated with chemotherapy and/or radiation, marginal margins (just outside the reactive zone and possibly containing microscopic tumor) are acceptable. It is rare today that we require "radical" (the entire compartment) margins *(107,110–112)*.

Once the tumor is resected and examined by the pathologist to define whether it is necessary to remove some of the tissue at the site of close margins, the specimen is sized (length and width of the articular portion) and the graft trimmed to appropriate length. If a joint is being reconstructed, the first act is to suture the capsule and ligaments. This means that for the knee it is necessary to tightly suture the medial and lateral collateral ligaments and the posterior capsule and, if possible, the cruciate ligaments *(32)*. Proximal tibial grafts require restoration of the patellar tendon attachment site on the tibial tubercle *(32)*. For the shoulder, the rotator cuff and capsule are sutured in order to fix the humeral head in close approximation to the host glenoid and then the pectoralis and deltoid are sutured to the soft tissue covering the graft *(17)*. For the hip, the capsule is sutured as is the abductors and iliopsoas *(25,42)*. To complete the grafting procedure, the host–donor junction site is approximated as tightly as possible using plates and screws rather than rods, mainly to avoid exposing distant parts to the possibility of tumor contamination *(32,35)*. We sometimes add autograft and/or synthetic materials thought to act favorably on bone healing. In recent years we have added plastic procedures to our routine, with gastrocnemius, soleus, rectus abdominis, or latissimus flaps as needed to reduce tension on the wound and cover the graft with viable muscle. The wound is closed in layers with a drain in place and immobilized appropriately, at first with a firm device that is not restrictive and subsequently often with a cast or rigid brace. The patient is maintained on intravenous antibiotics for the period of at least 1 wk, and then placed on oral tetracycline for a 3-mo period. For lower-extremity surgery in individuals over the age of 18, the patient receives coumadin for approx 4–6 wk depending on the extent of the surgery, and then is placed on aspirin as anticoagulation. The patients are seen regularly in the office setting, first at 2 wk intervals, then monthly, then every 2 mo and finally every 6–12 mo. We generally do not allow full weight bearing until the host-donor sites are healed for intercalary grafts and until in addition the joint is stable for osteoarticular or alloprosthetic grafts.

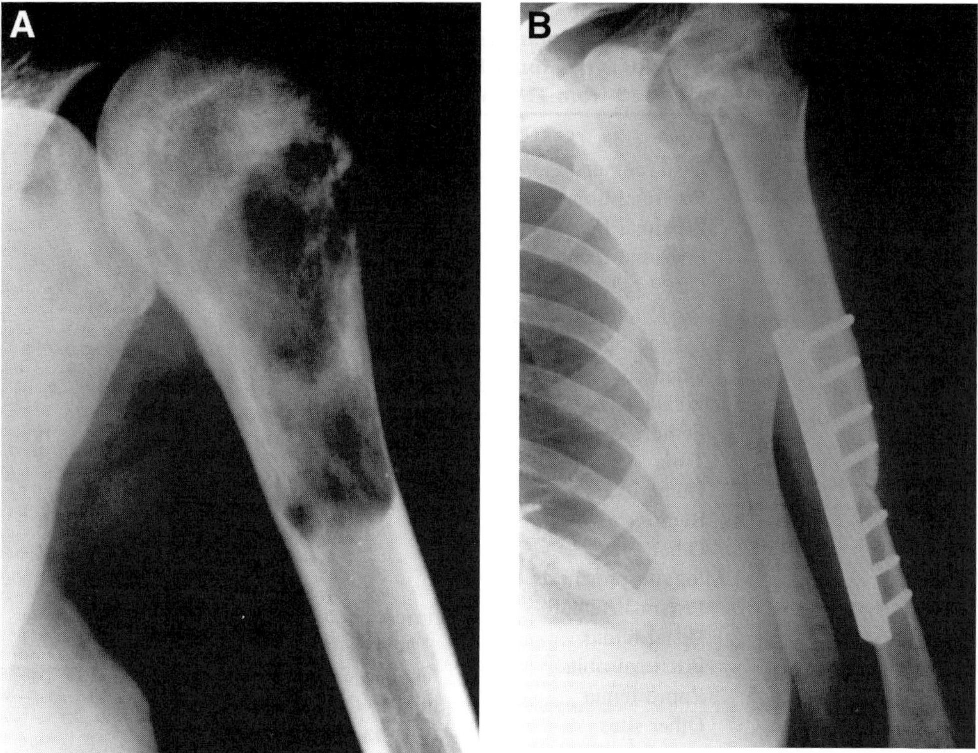

Fig. 2. *(Left)* Radiograph of the left humerus of a 17-yr-old female with a fibrosarcoma of bone. *(Right)* X-ray of the shoulder and arm of the same patient 21 yr after resection and osteoarticular allograft replacement.

as noted above, 139 patients were lost to follow-up at a mean time of 9 ± 5 yr. The scoring system utilized to evaluate their results was one originally proposed by us some years ago and remains our standard method of study *(29)*. The system is based on analysis of functional capacity. Patients were scored as "excellent" (no evident disease [NED]), return to virtually full function of the part without pain or significant disability, but could do noncontact sports); "good" (NED, modest to moderate limitation of function, no pain or major disability, limited sports activities); "fair" (NED, major limitations with a brace or support such as crutches, walker, or cane required, some tolerable pain, no sports—about half did not return to prior work activity); and "failure" (dead as a direct consequence of a local recurrence or amputation of the part or removal of the graft for recurrence or complication). The scoring system has been compared several times with that derived by analysis using the Enneking MSTS system *(107)*, and the one utilized in this study is a bit harsher but certainly easier to apply for house officers and fellows. The advantage to this system is that it is dependent on function; and one is able to compare not only the various anatomical regions, but the results of implantation of intercalary with osteoarticular segments or with grafts used in an arthrodesis or as part of an allograft-plus-prosthesis system.

The results for the 936 subjects who were followed for 2 yr or more are shown in **Table 3**. As can be noted, 71% of the 502 patients with osteoarticular grafts were characterized as excellent or good, while 29% were graded as fair or failure. The 232 patients with intercalary grafts fared considerably better, with 82% currently graded acceptable. The 124 patients with allograft–prosthesis showed a 74% acceptable score, while those 78 with an allograft–arthrodesis did relatively poorly, with only 56% presenting a good or excellent score. The overall score for the series of 936 patients showed a

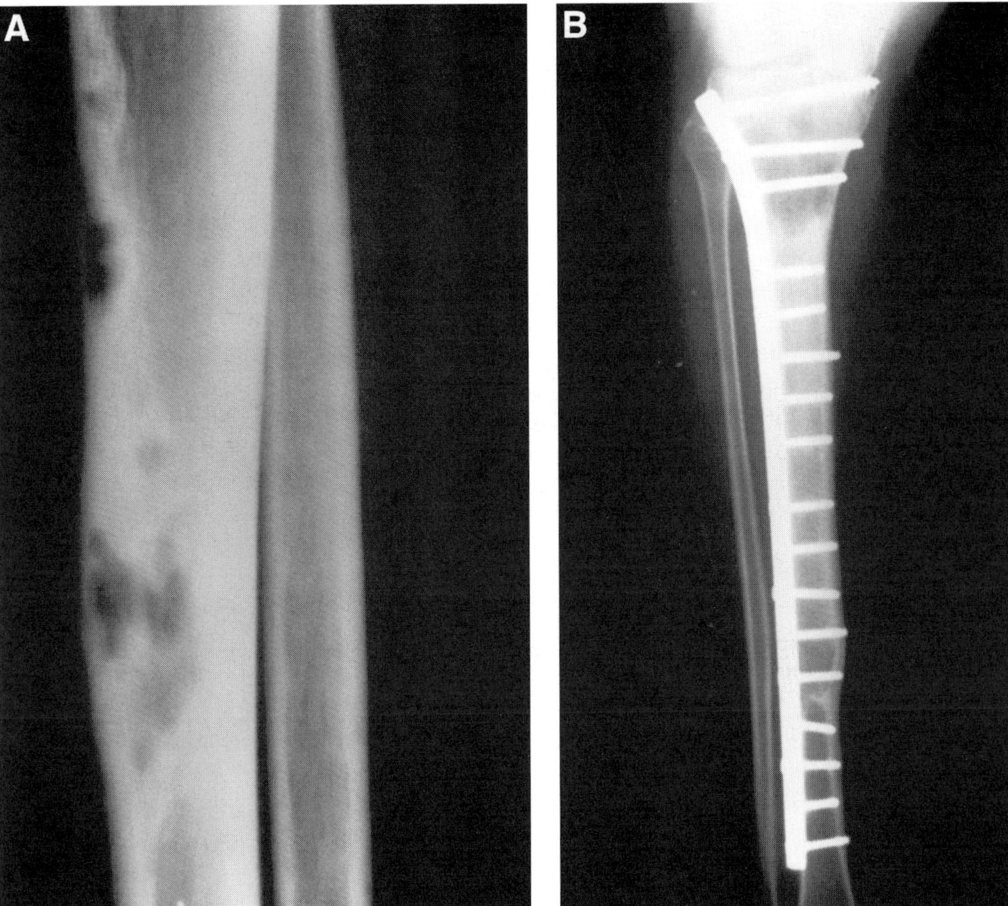

Fig. 3. *(Left)* Radiograph of the shaft of the right tibia of a 17-yr-old female with an adamantinoma. *(Right)* The same site 14 yr after intercalary allograft replacement.

figure for "acceptable" (excellent or good) at 73%, but if the 43 tumor failures are deleted from the series of 893 patients climbs to 76% (22% excellent and 54% good).

As described earlier, the complications of the operative procedure are the principal cause of failure (**Table 4**). The success or failure of an operation on a patient with a malignant bone tumor must first be considered on the basis of control of the neoplastic process. As noted in **Table 4**, for the 420 patients with high-grade tumors, 22% died of disease, 33% had metastases, and 8% sustained a local recurrence. These values are not inconsistent with any system for dealing with high-grade tumors such as osteosarcoma and Ewing's sarcoma and are to a large extent independent of the allograft surgery *(1–3,5,6,17,21,23,26,76,105,110,111,115,116)*. More characteristic and specific for the allogeneic transplant procedure itself, however, are four major complications: infection, fracture, nonunion, and instability of the joint. All of these are major issues for the patients. Infection occurred in 115 of the 936 patients (12%), fracture in 179 (19%), nonunion in 168 (18%), and unstable joint in 30 (5%). Because some of the patients presented with more than one complication, the numbers listed above are not additive and, in fact, 431 of the 936 cases (46%) of the patients had none of the allograft complications—and looking at the entire series, 429 (41%) had no complications at all. It should be noted that of this latter group, 96% remain good or excellent at mean time of 9.2 ± 5 yr following their surgery.

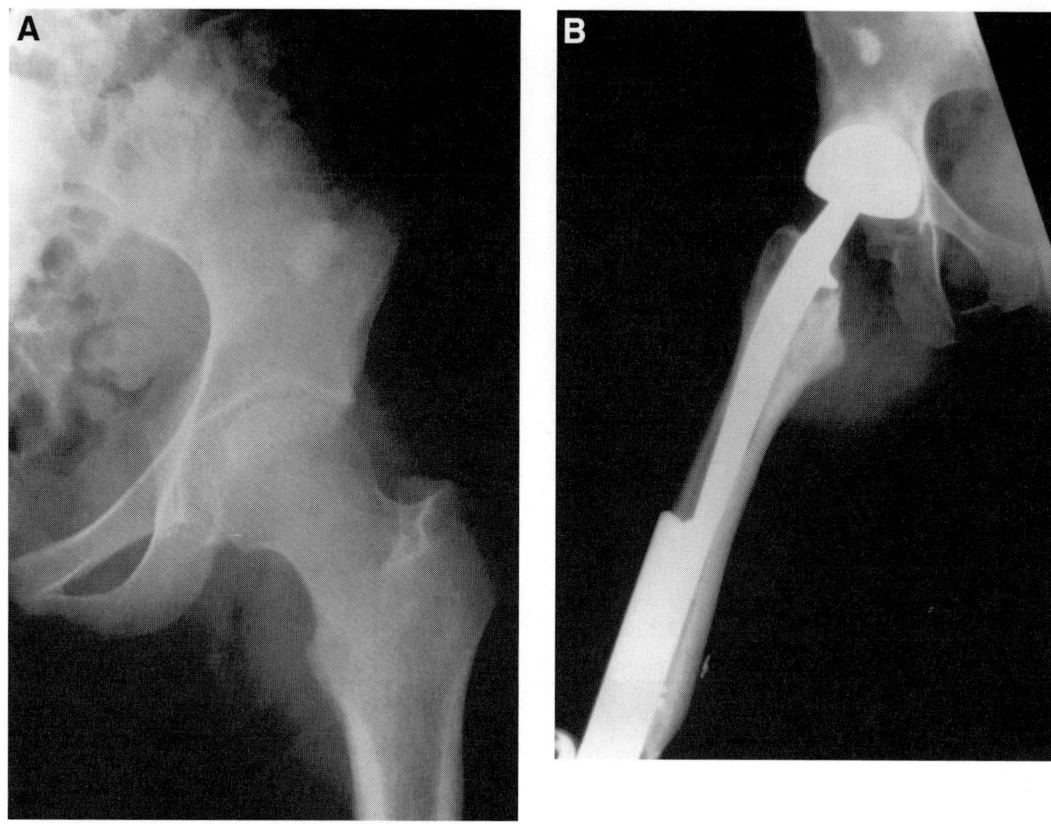

Fig. 4. *(Left)* X-ray of an osteosarcoma of the left proximal femur in a 19-yr-old female arising from the neck and trochanter. The dense structure in the pelvis is a bone island. *(Right)* Allograft–prosthesis reconstruction of the hip 13 yr after performance of the surgical procedure. The host–donor junction site is at the lower end of the photograph.

Reoperations were plentiful in this group but reflected not only the allograft complications but also the problems related to tumors. In this group of patients, 700 (75%) did not require additional surgery, but 163 had one, 46 had two, and 27 had three or more additional operative procedures. The mean time to performance of such surgery was 5.4 ± 2.2 yr. The majority of these procedures were for open reduction and internal fixation of fractures, bone graft to nonunions, drainage of infection, and in approx 17% of the patients with distal femoral, proximal femoral, or proximal tibial allografts, a total joint was implanted at a mean time of 5.8 yr.

The ultimate analytic tools for a series of cases such as this are the Kaplan–Meier Life Table Analysis *(113)* and Cox regression system *(114)*, both of which demonstrate that infection, fracture, nonunion, local recurrence, type of graft, and tumor stage had a significant impact on results. **Figure 6** is a plot depicting the outcome for the entire series. It demonstrates that most of the failures (both allograft and tumors) occur by 5 yr and that the curve declines little after that point. These data strongly suggest that once the problems of infection (almost all appear by 1 yr *[70,101]*) and fracture (most of which occur by 3 yr *[102]*), are no longer issues, the graft becomes "stable" and lasts at least through the over 20 yr of additional observation afforded by this analysis. Specifically, age of the patient (**Fig. 7**) and site of the graft (**Fig. 8**) did not provide a significant difference to the outcome of the procedure. In **Fig. 9**, it is clearly evident that the four types of graft, osteoarticular, intercalary, allograft with prosthesis and allograft with arthrodesis, have a difference in outcome, strongly suggest-

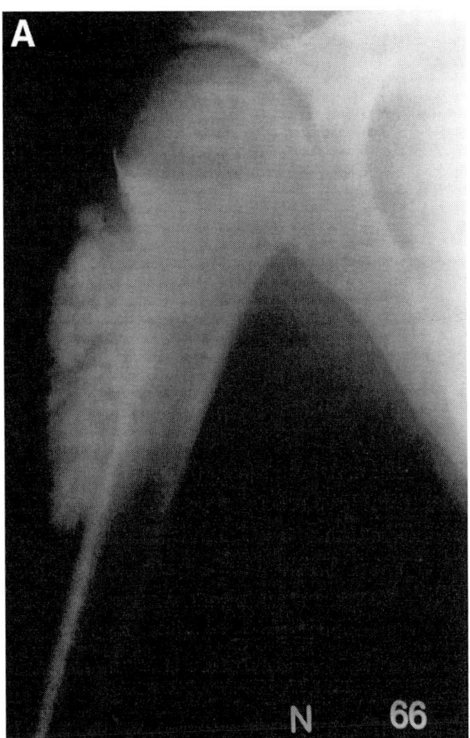
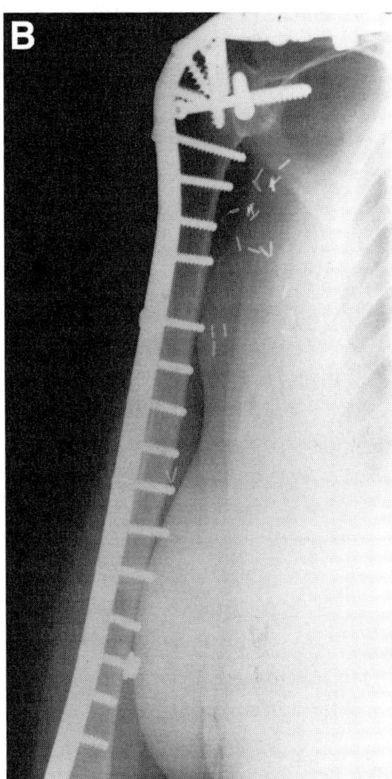

Fig. 5. *(Left)* X-ray of the proximal right humerus of an 18-yr-old female with a large osteosarcoma. The tumor was resected along with a portion of the glenoid and an allograft–arthrodesis performed. *(Right)* The graft site 11 yr following the surgery.

ing that allograft–arthrodesis is not as successful a procedure as the other three *(115)*. It also points out that the largest percentage of failures for all four types of grafts occur in the first 3 yr, and following 5 or so, the grafts become relatively stable.

Stage of the tumor (**Fig. 10**) had an effect on outcome. Patients with Stage II or III disease or more malignant diagnoses have a significantly poorer result for the allograft, presumably related to the increased frequency of recurrence, the extent of the surgery, and/or the effect of adjuvant chemotherapy and radiation on allograft incorporation. The effect of the allograft complications can be appreciated by analysis of **Fig. 11**, which clearly demonstrates the high failure rate associated with infection, and the still damaging but considerably less pernicious effect of fractures and nonunions.

The "bottom line" is best defined in terms of the results for the entire series at 5, 10, and 15 yr following the surgery. When these data are reviewed, it is noted that of 843 grafts in place longer than 5 yr, 644 (76%) are still rated as good or excellent; and of 454 in place for more than 10 yr, the percentage remains more or less the same. For the 144 patients who have had their graft in place for 15 yr or more, 105 (73%) are still successful; and for 44 which were implanted 20 yr or more ago, 29 (66%) are still rated excellent or good.

DISCUSSION

From the data presented, it is apparent that in our series as well as those from other clinical units, massive allografts are an effective method of dealing with connective tissue tumors and some benign but destructive conditions affecting the skeleton. Of some importance, however, is the clearly evident

Table 3
Allograft Transplantation Results for 936 Patients Followed for 2 yr or more, 11/71 to 1/98

Type of graft	Excellent	Good	Fair	Failure
Ostearticular (502)	87	269	22	124
	17%	54%	4%	25%
Intercalary (232)	84	107	6	35
	36%	46%	3%	15%
Allograft–prosthesis (124)	24	68	2	30
	19%	55%	2%	24%
Allograft–arthrodesis (78)	2	41	5	30
	3%	53%	6%	38%
Total series (936)	197	485	35	219
	21%	52%	4%	23%
If 43 tumor failures are deleted:				
Total series (893)	197	485	35	176
	22%	54%	4%	20%

Table 4
Allograft Transplantation Tumor
Complications in 936 Cases, 11/71 to 1/98

Tumor complications in 420 patients with high-grade tumors	
Death	91 (22%)
Metastasis	137 (33%)
Recurrence	35 (8%)
Allograft complications for all 936 procedures	
Infection	115 (12%)
Fracture	179 (19%)
Nonunion	168 (18%)
Unstable joint	30 (5%)

fact that we have not yet discovered the "holy grail," and our figures for success remain fixed at approx 72–77%. The results are significantly better for intercalary grafts and poorest for distal femoral osteoarticular grafts, presumably related to the high incidence of osteosarcoma, a disease that requires chemotherapy, in that group. Regardless of how often we review the data and how large the series gets, it remains evident that after the first 3–6 yr the grafts become stable and only exceptional events lead to failure. The three principal factors that appear to affect the end results most significantly (and account for the majority of the failures) are recurrence, infection, and fracture.

Of considerable importance in analyzing these data is that the failure rate is clearly highest in the first year and then diminishes rather sharply until at the fourth or fifth year. At that time, the system becomes stable and then remains so throughout the over 25 yr of this study. Most of the nonunions and infections occur in the first year and the bulk of the fractures are noted before the third to the fifth year, depending on the type of graft and presumably to some extent its length and the type of fixation used. Few failures occur after the sixth year, suggesting that the grafts establish an equilibrium state with the host—possibly not getting any better in terms of function over the years, but more important, not getting any worse. The exception to this rule appears to be the need for a joint resurfacing in about 17% of the patients with proximal femoral, distal femoral, or proximal tibial osteoarticular grafts at an average of 5 yr following the initial surgery. Even with this procedure, the success rate still remains above the 65% level.

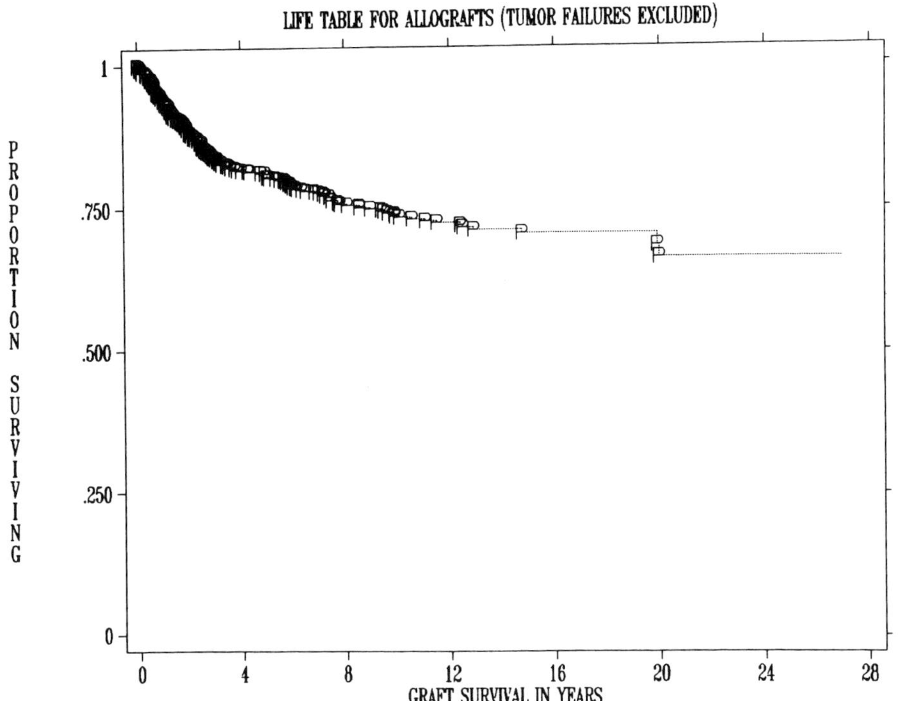

Fig. 6. The overall series as demonstrated by a life table plot (Kaplan-Meier). Note that the tumor failures are deleted in order to assess the outcome of the alloimplants themselves. As can be noted, most of the failures occur in the first 5 yr, and following this period the grafts become "stable" at an approximately 76% good or excellent status.

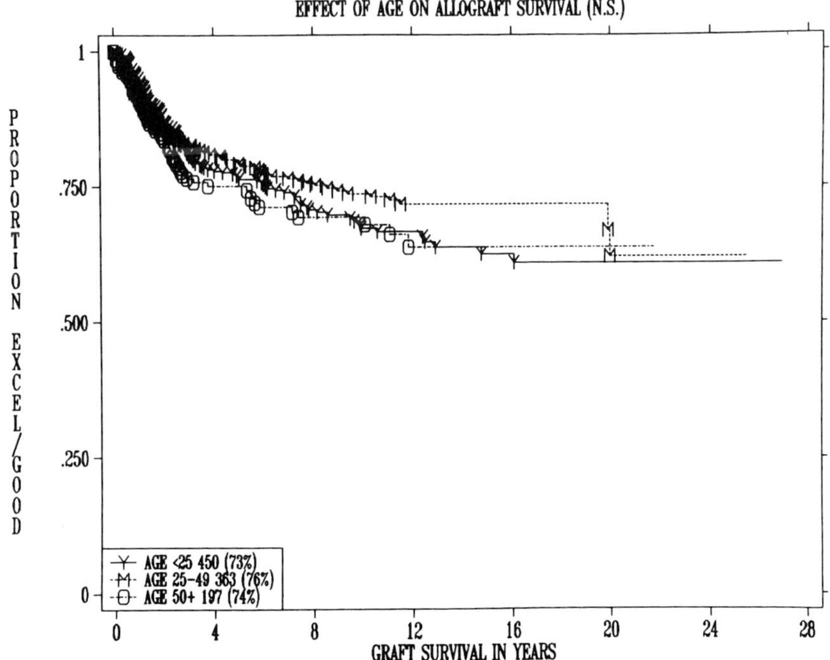

Fig. 7. Life table (Kaplan-Meier Plot) comparing the effect of age on survival. As can be noted, young, mid- and older age groups did not display a significant difference in outcome.

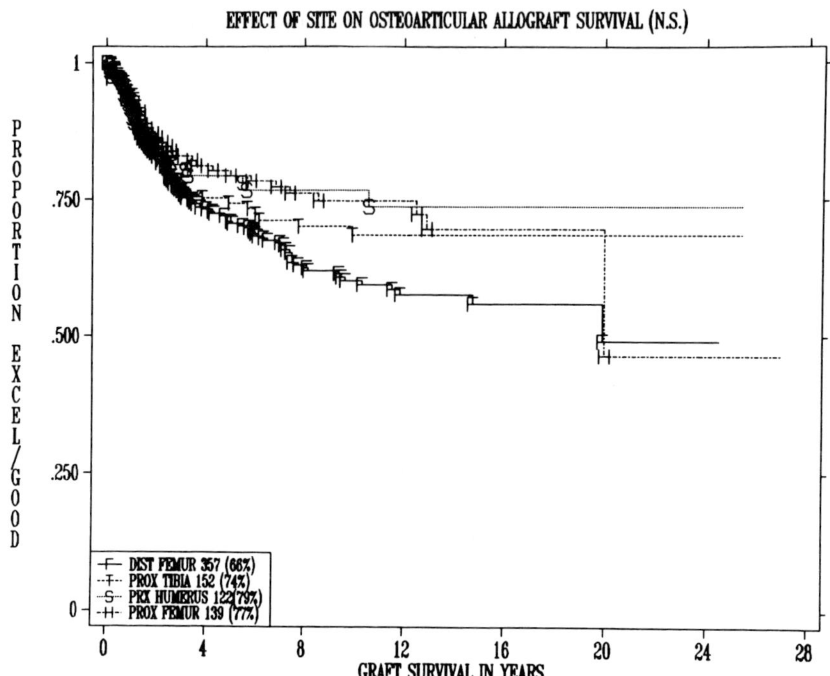

Fig. 8. Effect of site of operative procedure on survival. This plot compares the graft survival for distal femoral, proximal tibial, proximal humeral, and proximal femoral osteoarticular and allo–prosthetic grafts. As can be noted, the differences are not significant.

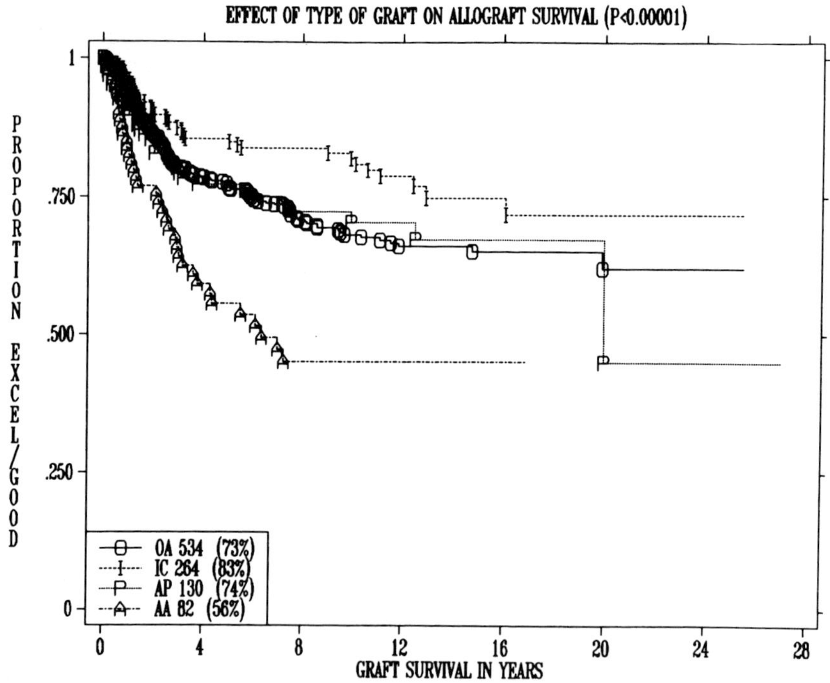

Fig. 9. Effect of type of graft on allograft survival. It is evident that the intercalary grafts have the best outcome and that the allo–arthrodeses have the poorest. This difference is highly significant.

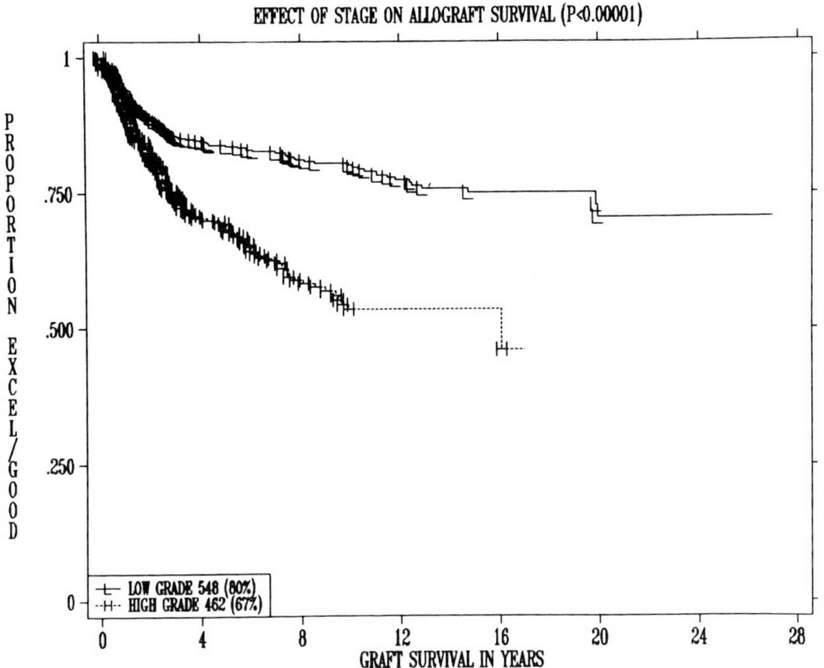

Fig. 10. This graphic demonstrates the effect of stage of the tumor on outcome. It is evident that it is highly statistically significant ($p < 0.00001$). Stage 0–1 cases have a mean survival of over 80%, while the average for those of higher grade is approximately 72%.

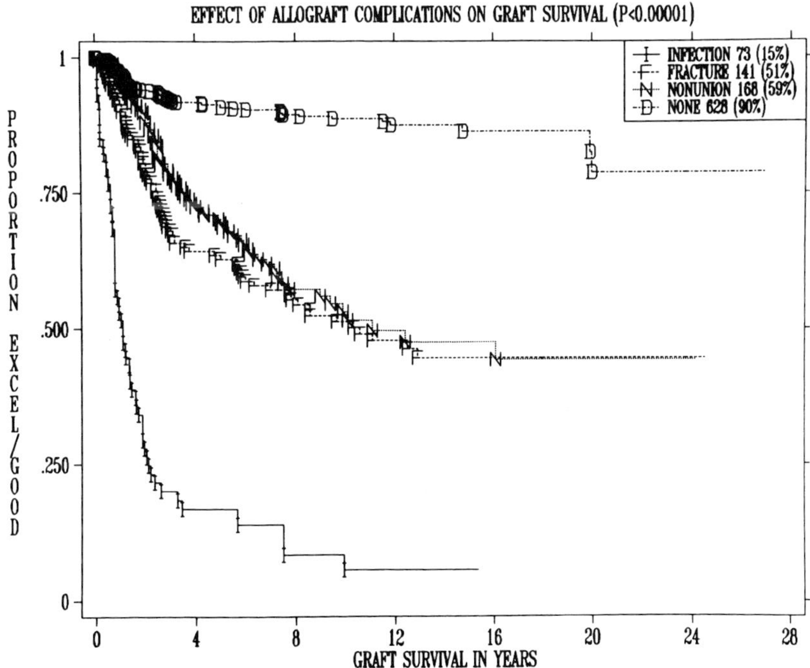

Fig. 11. Effect of allograft complications on success of the procedure. If no complications supervene, the mean score for good or excellent results is over 95%. Infection reduces the mean success rate to 16%, fracture to 65%, and nonunion to approximately the same. Note that most of the infections occur in the first year, and most of the fractures by 5 yr. These data are highly significant.

The data as presented offer for consideration two burning questions. The first of these is "What have we learned about the allograft procedure over the last 29 years?" The second question is obviously, "What can we do over the next 29 years to make things better?" In response to the first question, some axioms about the procedure can be stated as follows.

1. No matter what you do or how you do it, 80% excellent and good is the best you get right now; and that is probably because infection, fracture, and nonunion are immunologically directed *(1,28,50,70).*
2. Grafts are in fact replaced with host bone but very slowly, and it is our impression that that job is never really done *(40).* Further, the slow and unpredictable rate of replacement is at least in part the cause of fracture *(102).*
3. It is clearly essential to have access to a good bank. Sterile parts carefully obtained and studied and of the right size must be available in a timely fashion *(68).*
4. Patients must be carefully chosen for the procedure. In the early days we did the procedure for almost anyone, but it is quite clear that elderly patients, particularly with metastatic cancer, do not do as well. Furthermore, we cannot do allografts for very young children because growth will become a problem. The unanswered question is whether patients who are on chemotherapy should have grafts, and at least at this point we would say yes, but it is a tentative statement and needs more statistical support.
5. The surgeon must work rapidly, maintain as high a degree of sterility as possible, and follow the patient for a long time. We have some patients coming back yearly or at least sending X-rays at 15 yr or more after surgery.
6. It is essential to have viable muscle and good skin over the graft. We need to utilize gastrocnemius, rectus abdominus, and latissimus flaps in order to get better healing and decrease the infection rate.
7. Defects in the allograft bone probably never heal and serve as stress risers forever. If holes are necessary, they should be filled with screws or polymethylmethacrylate.
8. As tempting as it is, one should never transplant both sides of a joint. Without a competent synovial nervous system to give sensation to the joint, it is highly likely to develop into a Charcot's arthropathy.
9. If the allograft has been in place for 3 yr or more, a small percentage of fractured grafts will heal, so it is worth a brief trial of plaster before performing open reduction or other treatment.
10. Infected grafts are doomed. They should be removed and an antibiotic-impregnated polymethylmethacrylate spacer inserted and antibiotics administered. After a few months, another allograft or a metallic implant should be introduced. The salvage rate for such a procedure is reasonable *(26).*
11. Although DMSO is still used to preserve the viability of the chondrocytes, it is highly unlikely that they survive after transplant. Despite that, the cartilage holds up rather well if the size of the joint is reasonable, and thus far only a small number of our osteoarticular grafts have required metallic resurfacing *(62,66,97,98).*

The second burning question in all this is "Can these results be improved?" If one accepts the thesis that most of the complications are immunologically directed (and hence represent a form of "rejection"), the logical approach to the problem is to attempt to improve the results by either immunosuppression or better matching of the donor and host. The former is difficult to justify for two reasons. The first and most obvious is that the currently utilized immunosuppressive agents have a mortality rate of their own. Thus one finds oneself in the awkward role of advocating "life-threatening" drugs for a "limb-threatening" disease. Furthermore, treating a patient with a high-grade sarcoma with an immunosuppressive agent may cause damage to the patient's immune system and potentially will increase the growth rate or rate at which micrometastases are distributed or find a place to grow. Thus the patient is probably at increased risk from his or her tumor as a result of the treatment.

A better match is clearly advantageous for certain animal systems *(82–84,89–92)* and in theory would be of great advantage for humans, particularly in terms of the potential for successful implantation of vascularized grafts. The issue that faces us at least in theory is that a perfect match may not really be desirable, because such a graft is likely to undergo the devastating changes seen in the osteoarticular form of osteonecrosis of bone (only rarely seen in frozen cadaveric allogeneic implants). The second problem with matching is a practical one. It would seem to be very difficult to match not only for size and shape (believed to be essential to achieve good results!) and also major histocompatibility complex (MHC). In our recent study, matching in MHC Class II antigens appears to have an effect on graft survival, and indeed the reverse, i.e., pre- or postoperative transplant sensitization of Class II antigens appeared to predict a less than perfect result for the allograft procedure *(89)*. It should be possible by networking with a number of large banks to obtain such a limited match and thus run a trial to see if indeed we can obtain improved results.

Additional ways to improve the current results include further study of cryopreservation of the cartilage. Currently the best obtainable with DMSO treatment of the grafts at the time of harvest is far less than the 50% viability for ex vivo intact cartilage segments *(66,97,98)*. This is in rather sharp contrast with the almost 100% viability achievable by freezing and thawing matrix-free cells in culture with the same concentration of DMSO *(97)*. These data support the contention that the passage of DMSO through the matrix to reach the cell is not free and will require some special techniques.

It is evident, however, that the system remains imperfect and that complicating events such as infection, fracture, and nonunion make the outcome not only unpredictable but at times may lead ultimately to failure. It should also be apparent, however, that "failure" is a relative term, particularly because of 183 patients whose grafts failed for reasons of allograft complications (rather than as a result of tumor recurrence), 82% were salvaged by subsequent surgery. The number of amputations for non-tumor-related complications for the entire series of 936 patients is only 33 (4%), and even adding in the amputations related to tumor failures brings that value to 6%.

Research continues in a number of areas as described above. With more interest and more scientists studying the problem, it is likely that some major breakthrough will occur. Reduction or at least "control" over the immune response will provide a better graft, which will be less prone to complications and late failure. With improved networking in banking, a greater number of allogeneic segments will be available for each patient and perhaps allow a simple MHC match. Improvement in surgical technique continues to make the operative procedure more predictable and safer for the patient; and standardized, meticulous control of banking procedures should reduce bacterial and virus transmission to an acceptable level. Ultimately it is hoped that a sufficiently predictable and high rate of success can be achieved to allow us to approach the "dream" of Cosmas and Damian's miracle and allow our surgeons who are pure of heart to capture the "holy grail."

REFERENCES

1. Bos, G., Sim, F. H., Pritchard, D., et al. (1987) Prosthetic replacement of the proximal humerus. *Clin. Orthop.* **224,** 178–191.
2. Bradish, C. F., Kemp, H. B., Scales, J. T., and Wilson, J. N. (1987) Distal femoral replacement by custom-made prosthesis. Clinical follow-up and survivorship analysis. *J. Bone Joint Surg.* **69B,** 276–284.
3. Eckardt, J. J., Matthews, J. G., and Eilber, F. R. (1991) Endoprosthetic reconstruction after bone tumor resections of the proximal tibia. *Orthop Clin. N. Am.* **22,** 149–160.
4. Malawar, M. M. and McHale, K. A. (1989) Limb sparing surgery for high grade malignant tumors of the proximal tibia. Surgical technique and a method of extensor mechanism reconstruction. *Clin. Orthop.* **239,** 231–248.
5. Sim, F. H., Beauchamp, C. P., and Chao, E. Y. (1987) Reconstruction of musculoskeletal defects about the knee for tumor . *Clin. Orthop.* **221,** 188–201.
6. Stauffer, R. N. (1991) Problems with using metallic implants for replacement of bony defects, in *Bone and Cartilage Allografts* (Friedlaender, G. E. and Goldberg, V. M., eds.), American Academy of Orthopaedic Surgeons, Park Ridge, IL, pp. 295–299.
7. Bickels, J., Wittig, J. C., Kollender, Y., et al. (2002) Distal femoral resection with endoprosthetic reconstruction: a long term followup study. *Clin. Orthop.* **400,** 225–233.
8. Neel, M. D. and Letson, G. D. (2001) Modular endoprostheses for children with malignant bone tumors. *Cancer Control* **8,** 344–348.
9. Kotz, R., Dominkus, M., Zettl, T., et al. (2002) Advances in bone tumour treatment in 30 year with respect to survial and limb salvage. A single institution experience. *Int. Orthop.* **26,** 197–202.
10. Plotz, W., Rechl, H., Burgkart, R., et al. (2002) Limb salvage with tumor endoprostheses for malignant tumors of the knee. *Clin. Orthop.* **405,** 207–215.
11. Dominkus, M., Krepler, P., Schwameis, E., Windhager, R., and Kotz, R. (2001) Growth prediction in extendable tumor prostheses in children. *Clin. Orthop.* **390,** 212–220.
12. Mittermayer, F., Krepler, P., Dominkus, M., et al. (2001) Long-term followup of uncemented tumor endoprostheses for the lower extremity. *Clin. Orthop.* **388,** 167–177.

13. Eckardt, J. J., Kabo, J. M., Kelley, C. M., et al. (2000) Expandable endoprosthesis reconstruction in skeletally imma-ture patients with tumors. *Clin. Orthop.* **373**, 51–61.

14. Asavamongkolkul, A., Eckardt, J. J., Eilber, F. R., et al. (1999) Endoprosthetic reconstruction for malgnant upper extremity tumors. *Clin. Orthop.* **360**, 207–220.

15. Alho, A. J., Eskola, J., Ekfors, T., Manner, L., Kouri, T., and Hollmen, T. (1998) Immune responses and clinical out-come of massive human osteoarticular allografts. *Clin. Orthop.* **356**, 196–206.

16. Burwell, R. G. (1993) History of bone grafting and bone substitutes with special reference to osteogenic induction, in *Bone Grafts, Derivatives and Substitutes* (Urist, M. R., O'Connor, B. T., and Burwell, R. G., eds.), Butterworth Heine-mann, Oxford, pp. 3–102.

17. Cheng, E. Y. and Gebhardt, M. C. (1991) Allograft reconstructions of the shoulder after bone tumor resections. *Orthop. Clin. N. Am.* **22**, 37–48.

18. Czitrom, M., Langer, F., McKee, N., and Gross, A. E. (1986) Bone and cartilage allotransplantation. A review of 14 years of research and clinical studies. *Clin. Orthop.* **208**, 141–145.

19. Dean, G. S., Hollinger, E. H., and Urbaniak, J. R. (1997) Elbow allografts for reconstruction of the elbow with massive bone loss. Long term results. *Clin. Orthop.* 341-12-20.

20. Delloye, C., DeNayer, P., Allington, N., Munting, E., Coutelier, L., and Vincent, A. (1988) Massive bone allografts in large skeletal defects after tumor surgery: a clinical and microradiographic evaluation. *Arch. Orthop. Trauma Surg.* **107**, 31–41.

21. Dick, H. M., Malinin, T. I., and Mnaymneh, W. A. (1985) Massive allograft implantation following radical resection of high-grade tumors requiring adjuvant chemotherapy treatment. *Clin. Orthop.* **197**, 88–95.

22. Getty, P. I. and Peabody, T. D. (1999) Complications and functional outcomes of reconstruction with an osteoarticular allograft after intra-articular resection of the proximal humerus. *J. Bone Joint Surg.* **81A**, 1138–1146.

23. Hejna, M. J. and Gitelis, S. (1997) Allograft prosthetic composite replacements for bone tumors. *Semin. Surg. Oncol.* **13**, 18–24.

24. Hornicek, F. J., Mnymneh, W., Lackman, R. D., Exner, G. U., and Malinin, T. I. (1998) Limb salvage with osteo-articular allografts after resection of the proximal tibia bone tumors. *Clin. Orthop.* **352**, 179–186.

25. Jofe, M. H., Gebhardt, M. C., Tomford, W. W., and Mankin, H. J. (1988) Osteoarticular allografts and allografts plus prosthesis in the management of malignant tumors of the proximal femur. *J. Bone Joint Surg.* **70A**, 507–516.

26. Kocher, M. S., Gebhardt, M. C., and Mankin, H. J. (1998) Reconstruction of the distal aspect of the radius with use of an osteoarticular allograft after excision of a skeletal tumor. *J. Bone Joint Surg.* **80A**, 407–419.

27. Makley, J. T. (1985) The use of allografts to reconstruct intercalary defects of long bones. *Clin. Orthop.* **197**, 58–75.

28. Mankin, H. J. (1983) Complications of allograft surgery. in *Osteochondral Allografts* (Friedlaender, G. E., Mankin, H. J., and Sell, K. W., eds.), Little, Brown, Boston, pp. 259–274.

29. Mankin, H. I., Doppelt, S. H., Sullivan, T. R., and Tomford, W. W. (1982) Osteoarticular and intercalary allograft transplantation in the management of malignant tumors of bone. *Cancer* **50**, 613–630.

30. Mankin, H. J., Gebhardt, M. C., and Tomford, W. W. (1998) Long tenn replacement in the management of bone tumors: a retrospection, in *Advances in Tissue Banking,* vol 2 (Phillips, G. O., Strong, D. O., von Versen, R., and Nather, A., eds.), Wold Scientific, River Edge, NJ, pp. 209–239.

31. Mankin, H. J., Gebhardt, M. C., and Tomford, W. W. (1987) The use of frozen cadaveric allografts in the manage-ment of patients with bone tumors of the extremities. *Orthop. Clin. N. Am.* **18**, 275–289.

32. Mankin, H. J., Gebhardt, M. C., Jennings, L. C., Springfield, D. S., and Tomford, W. W. (1996) Long-term results of allograft replacement in the management of bone tumors. *Clin. Orthop.* **324**, 86–87.

33. Mnaymneh, W., Malinin, T. I., Makley, J. T., and Dick, H. M. (1986) Massive osteoarticular allografts in the recon-struction of extremities following resection of tumors not requiring chemotherapy and radiation. *Clin. Orthop.* **227**, 666–677.

34. Mnaymneh, W. and Malinin, T. (1989) Massive allografts in surgery of bone tumors. *Orthop. Clin. N. Am.* **20**, 455–467.

35. Ortiz-Cruz, E., Gebhardt, M. C., Jennings, L. C., Springfield, D. S., and Mankin, H. J. (1997) The result of transplanta-tion of intercalary allografts after resection of tumors. A long term follow-up study. *J. Bone Joint Surg.* **79A**, 97–106.

36. Mnaymneh, W. A., Temple, H. T., and Malinin, T. I. (2002) Allograft reconstruction after resection of malignant tumors of the scapula. *Clin. Orthop.* **405**, 223–229.

37. Hornicek, F. J., Gebhardt, M. C., Sorger, J. I., and Mankin, H. J. (1999) Tumor reconstruction. *Orthop. Clin. N. Am.* **30**, 673–684.

38. DeGroot, H. III and Mankin, H. J. (2000) Total knee arthroplasty in patients who have massive osteoarticular allo-grafts. *Clin. Orthop.* **373**, 62–72.

39. Temple, H. T., Scully, S. P., O'Keefe, R. J., Kattapurum, S., and Mankin, H. J. (2000) Clinical outcome in 38 patients with juxacortical osteosarcoma. *Clin. Orthop.* **373**, 208–217.

40. Yoshida, Y., Osaka, S., and Mankin, H. J. (2000) Hemipelvic allograft reconstruction after periacetabular bone tumor resection. *J. Orthop. Sci.* **5**, 198–204.

41. Hazan, E. J., Hornicek, F. J., Tomford, W., Gebhardt, M. C., and Mankin, H. J. (2001) The effect of adjuvant chemo-therapy on osteoarticular allografts. *Clin. Orthop.* **385**, 176–181.

42. Fox, E. J., Hau, M. A., Gebhardt, M. C., Hornicek, F. J., Tomford, W. W., and Mankin, H. J. (2002) Long term followup of proximal femoral allografts. *Clin. Orthop.* **397,** 106–113.
43. Julien, P., Ledennann, F., and Touwaide, A. (1987) *Cosma e Damiano.* 1993, Antea Edizioni, Milan.
44. Rinaldi, E. (1987) The first homoplastic limb transplant according to the legend of Saint Cosmas and Saint Damian. *Ital. J. Orthop. Traumatol.* **13,** 394–406.
45. Mankin, H. J. (2002) History of treatment of musculoskeletal tumours, in *The Evolution of Orthopaedic Surgery* (Klenerman, L., ed.), Royal Society of Medicine Press, London, pp. 191–210.
46. Burwell, R. G. (1969) The fate of bone grafts, in *Recent Advances in Orthopaedics* (Apley, O. A., ed.), Churchill, London, pp. 115–207.
47. Macewen, W. (1881) Observations concerning transplantation of bones: illustrated by a case of inter-human osseous transplantion, whereby over two-thirds of the shaft of the humerus was restored. *Proc. Roy. Soc. Lond.* **32,** 232–234.
48. Lexer, E. (1908) Die Verwendung der freien Knochenplastik nebst Versuchen uber Gelenkversteifung und Gelenk-transplantation. *Arch. Klin. Chir.* **86,** 939–954.
49. Lexer, E. (1925) Joint transplantation and arthroplasty. *Surg. Gynecol. Obstet.* **40,** 782–809.
50. Volkov, M. (1970) Allotransplantation of joints. *J. Bone Joint Surg.* **52B,** 49–53.
51. Curtiss, P. H., Powell, A. E., and Herndon, C. H. (1959) Immunological factors in homogeneous bone transplantation. III. The inability of homogeneous rabbit bone to induce circulating antibodies in rabbits. *J. Bone Joint Surg.* **41A,** 1482–1488.
52. Herndon, C. H. and Chase, S. W. (1954) The fate of massive autogenous and homogenous bone grafts including articular surfaces. *Surg. Gynecol. Obstet.* **98,** 273–290.
53. Friedlaender, G. E.. Mankin, I. U., and Sell, K. W. (1983) *Osteochondral Allografts,* Little, Brown, Boston.
54. Contreras, T. J., Blair, P. I., and Harlan, D. M. (1998) Briefhistory of the United Sates Navy Tissue Bank and Transplantation Programme. Captain Kenneth Sell's living legacy, in *Advances in Tissue Banking,* vol 2 (Phillips, G. O., Strong, D. O., von Versen, R., and Nather, A., eds.), Wold Scientific, River Edge, NJ, pp. 21–28.
55. Buck, R. E., Malinin, T. I., and Brown, M. D. (1989) Bone tranplantation and human immunodeficiency virus. An estimate of risk of acquired immunodeficiency syndrome (AIDS). *Clin. Orthop.* **240,** 129–136.
56. Conway, B., Tomford, W. W., Hirsch, M. S., Schooley, R. T., and Mankin, H. J. (1990) Effects of gamma irradiation on mV-l in a bone allograft model. *Trans. Orthop. Res. Soc.* **15,** 225.
57. Friedlaender, G. E. (1983) Immune responses to osteochondral allografts: current knowledge and future directions. *Clin. Orthop.* **174,** 58–68.
58. Friedlaender, G. E. (1991) Bone allografts: the biological consequences of immunological events. *J Bone Joint Surg.* **73A,** 1119–1122.
59. Friedlaender, G. E., Strong, D. M., and Mankin, H. J. (1998) Immunology of bone allografts: current knowledge, in *Advances in Tissue Banking,* vol 2 (Phillips, G. O., Strong, D. O., von Versen R., and Nather, A., eds.), Wold Scientific, River Edge, NJ, pp. 135–146,
60. Friedlaender, G. E.. Strong, D. M., and Sell, K. W. (1976) Studies on the antigencity of bone. I. Freeze-dried and deep frozen bone allografts in rabbits. *J. Bone Joint Surg.* **58A,** 854–858.
61. Friedlaender, G. E., Strong, D. M., and Sell, K. W. (1984) Studies on the antigencity of bone. I. Donor-specific anti-HLA antibodies in human recipients of freezed dried bone allografts. *J. Bone Joint Surg.* **66A,** 1-7-112.
62. Friedlaender, G. E. and Tomford, W. W. (1991) Approaches to the retrieval and banking of osteochondral allografts. in *Bone and Cartilage Allografts* (Friedlaender, G. E. and Goldberg, V. M., eds.), American Academy of Orthopaedic Surgeons, Park Ridge IL, pp. 185–192.
63. Mankin, H. I. and Friedlaender, G. E. (1983) Perspectives on bone allograft biology, in *Osteochondral Allografts* (Friedlaender, G. E., Mankin, H. J., and Sell, K. W., eds.), Little, Brown, Boston, pp. 3–8.
64. Strong, D. M., Sayers, M. H., and Conrad, E. U. III. (1991) Screening tissue donors for infectious markers. in *Bone and Cartilage Allografts* (Friedlaender, G. E. and Goldberg, V. M., eds.), American Academy of Orthopaedic Surgeons, Park Ridge IL, pp. 193–209.
65. Strong, D. M., Friedlaender, G. E., Tomford, W. W., et al. (1996) Immunological responses in human recipients of osseous and osteochondral allografts. *Clin. Orthop.* **326,** 107–114.
66. Tomford, W. W. (1983) Cryopreservation of articular cartilage in Osteochondral Allografts, (Friedlaender, G. E., Mankin, H. J., and Sell, K. W., eds.), Little, Brown, Boston, pp. 215–218.
67. Tomford, W. W., Doppelt, S. H., and Mankin, H. J. (1989) Organization, legal aspects and problems of bone banking in a large orthopaedic center, in *Bone Transplantation* (Aebi, M. and Regazzoni, P, eds.), Springer-Verlag, Berlin, pp. 145–150.
68. Tomford, W. W. and Mankin, H. J. (1999) Bone banking: update on methods and materials. *Orthop. Clin. N. Am.* **30,** 553–565.
69. Tomford, W. W., Thongphasuk, J., Mankin, H. J., and Ferraro, M. J. (1990) Frozen musculoskeletal allografts. A study of the clinical incidence and causes of infection associated with their use. *J. Bone Joint Surg.* **72A,** 1137–1143.
70. Lord, C. F., Gebhardt, M. C., Tomford, W. W., and Mankin, H. J. (1988) The incidence, nature and treatment of allograft infections. *J. Bone Joint Surg.* **70A,** 369–376.

71. Lietman, S. A., Tomford, W. W., Gebhardt, M. C., Springfield, D. S., and Mankin, H. J. (2000) Complications of irradiated allografts in orthopaedic tumor surgery. *Clin. Orthop.* **375,** 214–217.

72. Quinn, R. H., Mankin, H. J., Springfield, D. S., and Gebhardt, M. C. (2001) Management of infected bulk allografts with antibiotic-impregnated polymethylmethacrylate spacers. *Orthopedics* **24,** 971–975.

73. Parrish, F. F. (1973) Allograft replacement of part of the end of a long bone following excision of a tumor: report of twenty-one cases. *J. Bone Joint Surg.* **55A,** 1–22.

74. Parrish, F. F. (1966) Treatment of bone tumors by total excision and replacement with massive autologous and homologous grafts. *J. Bone Joint Surg.* **48A,** 968–990.

75. Ottolenghi, C. E. (1966) Massive osteoarticular bone grafts. *J. Bone Joint Surg.* **48B,** 646–659.

76. McGoveran, B. M., Davis, A. M., Gross, A. E., and Bell, R. S. (1986) Evaluation of the allograft-prosthesis composite technique for proximal femoral reconstruction afer resection of a primary bone tumour. *Cancer Surg.* **42,** 37–45.

77. Hornicek, F. J., Gebhardt, M. C., Tomford, W. W., et al. (2001) Mankin HJ: Factors affecting nonunion of the allograft-host junction. *Clin. Orthop.* **382,** 87–98.

78. Kattapuram, S. V., Rosol, M. S., Rosenthal, D. I., Palmer, W. E., and Mankin, H. J. (1999) Magnetic resonance imaging features of allografts. *Skeletal Radiol.* **28(7),** 383–389.

79. Lewandrowski, K. U., Hornicek, F. J., Pelow, F. X., Gebhardt, M. C., Mankin, H. J., and Tomford, W. W. (2002) Clinical and biomechanical design considerations for engineered cortical bone allografts, in *Tissue Engineering and Biodegradable Equivalents* (Lewandrowski, K. U., et al., eds.), Marcel Dekker, New York, pp. 179–194.

80. Lewandrowski, K. U., Tomford, W. W., and Mankin, H. J. (2002) Demineralization and perforation of cortical bone allografts: preparatory methods, in *Tissue Engineering and Biodegradable Equivalents* (Lewandrowski, K. U., et al., eds.), Marcel Dekker, New York, pp. 433–466.

81. Lee, F. Y., Storer, S., Hazan, E. J., Gebhardt, M. C., and Mankin, H. J. (2002) Repair of bone allograft fracture using bone morphogenetic protein-2. *Clin. Orthop.* **397,** 119–123.

82. Burchardt, H. (1987) Biology of bone transplantation. *Orthop. Clin. N. Am.* **18,** 187–196.

83. Burchardt, H., Jones, H., Glowczewskie, F., Rudner, C., and Enneking, W. F. (1978) Freeze-dried allogeneic segmental cortical bone grafts in dogs. *J. Bone Joint Surg.* **60A,** 1082–1090.

84. Goldberg, V. M., Powell, A., Schaffer, J. W., Zika, J., Bos, G. D., and Heiple, K. G. (1984) Bone grafting. Role of histocompatibility in transplantation. *J. Orthop. Res.* **3,** 389–404.

85. Horowitz, M. C. and Friedlaender, G. E. (1991) The immune response to bone grafts, in *Bone and Cartilage Allografts* (Friedlaender, G. E. and Goldberg, V. M., eds.), American Academy of Orthopaedic Surgeons, Park Ridge IL, pp. 85–102.

86. Langer, F., Gross, A. E., West, M., and Urovitz, E. P. (1978) The immunogenicity of allograft knee joint transplants. *Clin. Orthop.* **132,** 155–162.

87. Elves, M. D. (1976) Newer knowledge of the immunology of bone and cartilage. *Clin. Orthop.* **120,** 232–259.

88. Elves, M. D., Gray, J. C., and Thorogood, P. V. (1976) The cellular change in allografts of marrow-containing cortical bone. *J. Anat.* **122,** 253–269.

89. Stevenson, S. (1991) Experimental issues in histocompatability of bone grafts, in *Bone and Cartilage Allografts* (Friedlaender, G. E. and Goldberg, V. M., eds.), American Academy of Orthopaedic Surgeons, Park Ridge, IL, pp. 45–54.

90. Stevenson, S. (1999) Biology of bone grafts. *Orthop. Clin. N. Am.* **30,** 543–553.

91. Stevenson, S., Emery, S. E., and Goldberg, V. M. (1996) Factors affecting bone graft incorporation. *Clin. Orthop.* **324,** 66–75.

92. Stevenson, S. and Horowitz, M. C. (1992) Current concepts review: the response to bone allografts. *J. Bone Joint Surg.* **74A,** 939–950.

93. Muscolo, D. L., Ayerza, M. A., Calabrese, M. E., Redal, M. A., and Araujo, E. S. (1996) Human leukocyte antigen matching, radiographic score and histologic findings in massive frozen bone allografts. *Clin. Orthop.* **326,** 115–126.

94. Friedlaender, G. E., Strong, D. M., Tomford, W. W., and Mankin, H. J. (1999) Long term followup of patients with osteochondral allografts. A correlation between immunologic responses and clinical outcome. *Orthop. Clin. N. Am.* **30,** 583–590.

95. Doppelt, S. H., Tomford, W. W., Lucas, A., and Mankin, H. J. (1981) Operational and financial aspects of a hospital bone bank. *J. Bone Joint Surg.* **63A,** 1472–1479.

96. Loty, B., Courpied, J. P., Torneno, B., Postel, M., Forest, M., and Abelanet, R. (1990) Bone allografts sterilized by irradiation. Biological properties, procurement and results of 150 massive allografts. *Int. Orthop.* **14,** 237–242.

97. Schachar, N. S. and McGann, L. B. (1991) Cryopreservation of articular cartilage, in *Bone and Cartilage Allografts* (Friedlaender, G. E. and Goldberg, V. M., eds.), American Academy of Orthopaedic Surgeons, Park Ridge, IL, pp. 211–230.

98. Schachar, N. S., Cucheran, D. J., McGann, L. E., Novak, K. A., and Frank, C. B. (1994) Metabolic activity of bovine articular cartilage during refrigerated storage. *J. Orthop. Res.* **12,** 15–20.

99. Goorin, A. M., Abelson, H. T., and Frei, E. (1985) Osteosarcoma fifteen years later. *N. Engl. J. Med.* **313,** 1637–1643.

100. Rosen, G., Caparros, B., Huvos, A. G., et al. (1982) Preoperative chemotherapy for osteogenic sarcoma: selection of postoperative adjuvant chemotherapy based on the response of the primary tumor to preoperative chemotherapy. *Cancer* **49,** 1221–1230.

101. Hernigou, P., Delepine, G., and Goutallier, D. (1991) Infections after massive bone allografts in surgery of bone tumors of the limbs. Incidence, contributing factors, therapeutic problems. *Rev. Clin. Orthop.* **77,** 6–13.

102. Berrey, W. H. Jr., Lord, C. F., Gebhardt, M. C., and Mankin, H. J. (1990) Fractures of allograft fractures. Frequency, treatment and end esults. *J. Bone Joint Surg.* **72A,** 822–833.

103. Sorger, J. I., Hornicek, F. J., Zavatta, M., et al. (2001) Allograft fractures revisited. *Clin. Orthop.* **382,** 66–74.

104. Catanzariti, A. and Karlock, L. (1996) The application of allograft bone in foot and ankls surgery. *J. Foot Ankle Surg.* **35,** 440–451.

105. Clarke, H. D., Berry, D. J., and Sim, F. H. (1998) Salvage offailed femoral megaprostheses with allograft prosthesis composite. *Clin. Orthop.* **356,** 222–229.

106. Tan, M. H. and Mankin, H. J. (1997) Blood transfusion and bone allografts. Effect on infection and outcome. *Clin. Orthop.* **340,** 207–214.

107. Enneking, W. F. (1987) A system for evaluation of the surgical management of musculoskeletal tumors, in *Limb Salvage in Musculoskeletal Oncology* (Enneking, W. F., ed.), Churchill Livingstone, New York, pp. 145–150.

108. Enneking, W. F., Eady, J. L., and Burchardt, H. (1980) Autogenous cortical bone grafts in the reconstruction of segmental skeletal defects. *J. Bone Joint Surg.* **62A,** 1039–1058.

109. Enneking, W. F., Spanier, S. S., and Goodman, M. A. (1980) Current concepts review: the surgical staging of musculoskeletal sarcoma. *J. Bone Joint Surg.* **62A,** 1027–1030.

110. Simon, M. A., Aschliman, M. A., Thomas, N., and Mankin, H. J. (1986) Limb salvage treatment versus amputation for osteosarcoma of the distal end of the femur. *J. Bone Joint Surg.* **68A,** 1331–1337.

111. Refaat, Y., Gunnoe, J., Hornicek, F. J., and Mankin, H. J. (2002) Comparison of quality of life after amputation or limb salvage. *Clin. Orthop.* **397,** 298–305.

112. DiCaprio, M. R. and Friedlaender, G. E. (2003) Malignant bone tumors: limb sparing versus amputation. *J. Am. Acad. Orthop. Surg.* **11,** 25–37.

113. Kaplan, E. and Meier, P. (1958) Non-parametric estimation for incomplete observations. *J. Am. Statist. Assoc.* **53,** 457–481.

114. Cox, D. R. (1972) Regression models and life tables. *J. Roy. Statist. Soc.* **34,** 187–220.

115. Chao, E. Y. and Sim, F. H. (1985) Modular prosthetic system for segmental bone and joint replacement after tumor resection. *Orthopedics* **8,** 641–651.

116. Donati, D., Giacomini, S., Gozzi, E., et al. (2002) Allograft arthrodesis treatment of bone tumors: a two center study. *Clin. Orthop.* **400,** 217–224.

SPECTRUM OF BONE GRAFTS IN JOINT ARTHROPLASTY

Role of Bone Grafts

Bone deficiency can be treated many different ways during joint replacement surgery, and the means by which it is treated depends on various factors; these include factors specific to the bone loss itself (location and pattern of bone loss, severity of bone loss, quality of the surrounding bone), factors related to the patient (age, functional requirements), and factors related to the operation (the type of implant fixation being employed). Practical factors (such as the relative technical difficulty of each competing technique), the published results of each competing technique, and theoretical factors (such as the advantages of restoring optimal joint biomechanics or bone stock) also are considerations as the surgeon chooses a reconstructive method. The main methods available to cope with bone deficiency in joint arthroplasty include: (1) changing the planned position of an implant to compensate for the bone loss (2) (e.g., placing the socket higher or more medially than usual in a case of hip dysplasia); (2) changing the size of the implant to compensate for the deficiency *(3,4)* (e.g., using an extra-large socket during revision of a failed acetabular component); (3) filling the bone defect with metal (5) (e.g., a calcar replacement prosthesis for medial proximal femoral bone deficiency); and (4) use of bone graft to fill the bone deficiency. Each of these methods has merits, and each may be used in different circumstances.

Bone grafts are used in joint replacement for three essential indications: (1) to provide needed support for an implant that would otherwise be inadequately supported by native bone; (2) to fill bone deficiencies to provide better bone stock for the future; and (3) to protect deficient or weak bone from fracture, or to help treat an established fracture around an arthroplasty.

There are no known absolute contraindications to using bone grafts in arthroplasty reconstruction. When a joint reconstruction is performed after previous joint sepsis, surgeons tend to be circumspect about using grafts. Despite this logical caution, data to date have not demonstrated an extraordinarily high rate of reinfection when bone grafts are required for a subsequent reconstruction of a previously infected joint *(6–9)*.

Types of Bone Graft

Many different types of bone grafts are used in conjunction with joint arthroplasties, and grafts differ with respect to source of bone (autogenic or allogenic), the type of bone (bulk or particulate), and type of processing (fresh frozen, freeze-dried, radiated). Bulk bone grafts may be mostly cancellous (such as femoral head grafts) or mostly cortical (such as cortical strut grafts), or may be composites of cortical and cancellous graft (such as segmental long bone grafts). The most common bulk bone allografts are femoral heads retrieved during primary total hip arthroplasty, femoral condyles, or whole femora or tibiae. Particulate grafts vary according to the size of the particles of the graft and how densely the particles are packed into the bone.

The type of bone defect is the most important factor that determines whether particulate graft or bulk graft is employed. Large segmental uncontained bone defects more often require bulk grafts, while contained cavitary defects usually can be managed with particulate grafts. Particulate grafts usually are easier to use than bulk grafts, because they do not require contouring to fit a defect. Furthermore, particulate grafts are thought to have a better chance of being incorporated into host bone than structural grafts. Traditionally only bulk grafts have been used when structural prosthetic support is needed, but recently densely packed particulate grafts also have been used to provide structural support of implants. The type of bone graft used also is influenced by the surgeon's philosophy and the type of implant being used for reconstruction.

Source of Bone Grafts

Bone graft can be obtained from different sources: autograft bone harvested from a local or distant site during arthroplasty, allograft bone, or xenograft bone. Because of its a high immunological potential, xenograft is not in use in clinical practice. In a few circumstances, fresh autogenous bone graft from the local arthroplasty site is available to treat bone defects; most commonly this is possible during primary total hip arthroplasty (when the autogenous femoral head is available) or primary total knee arthroplasty (when resected autogenous distal femoral and proximal tibial bone is available). Autogenous bone can be harvested from the patient's own iliac crest, but the combination of the limited amount of bone available from this site and the potential for associated donor-site morbidity limits the use of this source to unusual circumstances where a smaller amount of graft with osteoinductive capability is needed (such as to promote periprosthetic fracture healing). From the practical viewpoint, for most revision joint reconstruction applications—which are the most common indication for bone grafting in joint arthroplasty—the size and shape of the bone defects encountered favors the use of allograft rather than autograft bone.

Influence of Bone Graft on Type of Reconstruction

The type of arthroplasty reconstruction influences the need for grafts and the preferred types of grafts. Likewise, the necessity of using bone grafts for an arthroplasty procedure influences the type of implants best suited for the reconstruction.

Uncemented porous coated components cannot be expected to gain bone ingrowth (or long-term stability) from areas in which they exclusively contact bone grafts. Rather, in these areas a fibrous interface between the graft and uncemented implant usually forms. Such an interface, alone, will not reliably provide good clinical results or long-term implant fixation. Thus, when a reconstruction requires that most of the bone surface on which an implant rests will be bone graft, uncemented porous coated implants usually are not used, and cemented constructs are preferred. The amount of porous implant surface that must be in contact with host bone rather than bone graft to gain reliable long-term biological fixation is not known with certainty, and probably varies with the potential biological activity of implant surface and the biological activity of the host bone that it contacts, in addition to other mechanical factors.

Large bulk bone grafts do not appear to be incorporated fully with time, thus they must provide mechanical support for the arthroplasty, with little biological renewal over long periods of time. Experience has shown that these grafts are at risk for fracture or collapse from fatigue failure, much as is any other osteonecrotic bone. When such grafts are used, implants can be chosen that protect the graft from long-term stress overload; this typically is accomplished by using implants that distribute forces over a larger surface area of bone and that allow stresses partially to bypass the bone grafts and be transmitted to host bone. For acetabular reconstruction, graft protection can be accomplished with antiprotrusio cages that bridge from native bone to native bone, and for hip femoral reconstruction this can be accomplished with long-stemmed implants that pass through and protect whole-segment femoral bulk grafts. For total knee arthroplasty these goals are accomplished by using stemmed implants that off-load stress to host diaphyseal cortical bone, and thus protect grafts of the distal femur or proximal tibia.

BIOLOGY OF BONE GRAFTS IN JOINT ARTHROPLASTY

Incorporation Process of Bone Graft

Several different fates of bone grafts are possible after implantation. Grafts can heal or fail to heal to bone; grafts can be incorporated into host bone not at all, very gradually, or completely; and grafts can be resorbed either partially or completely. The fate of the graft is important because it determines whether the goals of the joint reconstruction will be met and whether the reconstruction will be durable.

Cancellous autograft bone is the benchmark for osseous integration against which other grafts are measured, but it is of limited supply. The harvest of autograft bone is not without donor-site morbidity. Autograft is an osteoinductive material, eliciting preferentially the transformation of mesenchymal cells into osteoprogenitor stem cells *(10,11)*. A biological stimulus via local growth factors is sisted primarily of bulk cancellous bone, and were retrieved at 25 and 48 mo. They found that much of the graft–host junction consisted of a fibrous interface and that revascularization never extended beyond 2 mm from the initial host to graft interface. Despite the frequency of this form of reconstruction, limited information on the long-term histological fate of successful and failed grafts of this type is available.

Most published histological data concerning structural long-bone allografts deal with retrieved specimens in patients who underwent reconstruction following *en bloc* resection of tumors *(29,30)*. Enneking and Mindell *(29)* evaluated 16 massive frozen or freeze-dried allografts that had been implanted after tumor resection and had been *in situ* for 4 mo to 5 yr. Most were retrieved less than 2 yr after implantation. Union occurred slowly at the cortical-to-cortical junction by the formation of an external callus. On the external surface of the graft, a layer 1–2 mm thick, and marked by a distinct cement line from the necrotic cortex, was laid down by mesenchymal proliferation. Internal repair extended no more than 2 mm deep from the surface of the graft and no more than 3 mm into the cortical ends. Less than 10% of the entire graft was replaced by newly formed bone. Similar findings were made by Gouin et al. *(30)* in two biopsies performed during reoperation for minor complications after massive bone allograft implantation at 9 and 19 mo follow-up. Most of the grafted bone remained dead, and necrotic cortical bone predominated. It is possible the histological findings may have been modified by immunosuppression and chemotherapy accompanying the tumor surgery, which may alter the host cellular response to grafted bone. Most of the reconstructions following tumor resection were performed with intercalary allografts allowing host contact only at both ends of the graft. It is presumed that a similar biological process follows large segmental allograft implantation in association with allograft prosthesis composites, but there is as yet little histological confirmation.

Hamadouche et al. *(31)* studied a circumferential replacement of a deficient proximal femur retrieved at 10-yr follow-up. The reconstruction (an "intussusception" allograft) was performed with a whole proximal femur impacted into the distal host femur. The massive allograft had healed where it was surrounded by host bone. The healing and incorporation process, as reported in the literature up to 5 yr, was not substantially modified with longer implantation period. Newly formed bone was not observed beyond 5 mm into the graft thickness. Contrary to what has been reported by Hooten et al. *(16)* for bulk acetabular bone grafts retrieved at 25 and 48 mo, no fibrous interface was identified at the graft-to-host interface. The different stresses in the femur and the acetabulum may explain these contradictory findings. Microcracks were present in the nonremodeled area of the allograft bone, either parallel or perpendicular to the newly formed osteons. These microcracks are hypothesized to be of mechanical origin as hypothesized by Gouin et al. *(30)*. Irradiation at a standard dose does not alter the elasticity of the material but has been proven to diminish the bone capacity to absorb work, which may have led to the production of microfractures in the superior unsupported, nonremodeled portion of the graft.

CLINICAL APPLICATIONS AND RESULTS
OF BONE GRAFTS IN TOTAL HIP ARTHROPLASTY

Primary Acetabular Reconstruction

Most bone grafts performed in primary total hip arthroplasty are used for augmentation of medial bone in acetabular protrusion, bone cysts encountered with degenerative joint disease, or for augmentation of anterolateral bone deficits in developmental dysplasia of the hip. Grafts have been used successfully in these circumstances with both cemented and uncemented porous coated acetabular components. Most superior lateral grafts for developmental hip dysplasia have been bulk autogenous femoral head grafts (**Fig. 1**), whereas grafts for protrusio can be either structural or particulate, and most grafts for acetabular cysts are particulate.

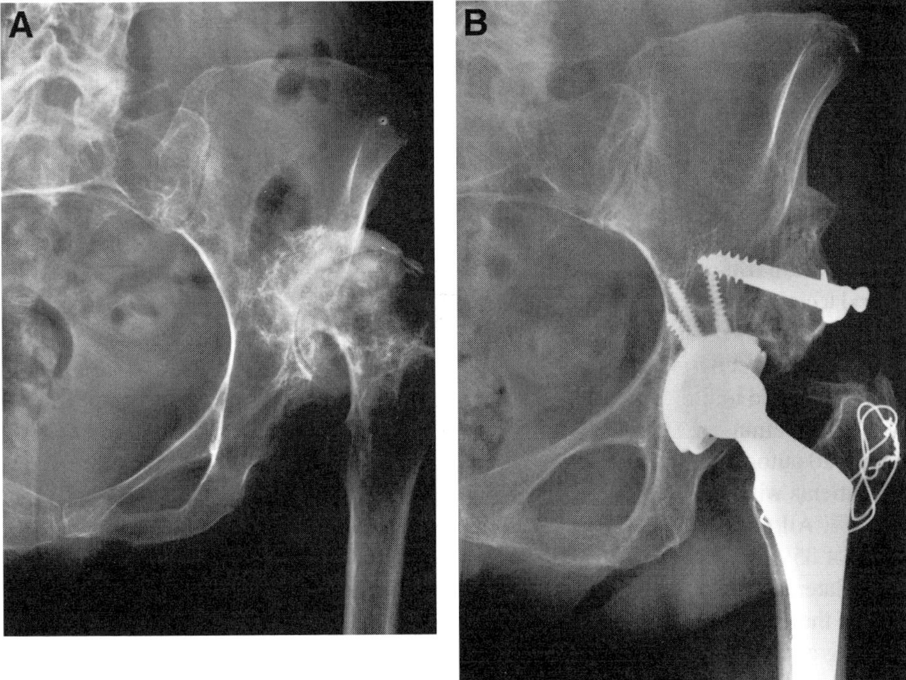

Fig. 1. (A) Preoperative radiograph of patient with hip dysplasia. **(B)** Ten years after total hip arthroplasty with autogenous femoral head graft to pelvis. The graft has healed and bone stock has been augmented.

Particulate Grafts

Particulate grafting of cavitary defects on the acetabular side can be performed in association with cemented or cementless cups. Impaction grafting of the acetabulum in association with all-polyethylene cemented cups was first described by Sloof et al. in protrusio acetabuli *(32)*. Welten et al. *(33)* reported on mean 12.3-yr follow-up of 47 hips treated with autogenous impaction allografting for acetabular protrusio in primary cemented total hip arthroplasties. All grafts were noted to have evidence of radiographic incorporation. The survival rate of the acetabular reconstructions was 94%. Bolder et al. *(34)* applied this technique to reconstructions for developmental hip dysplasia. The authors reported on 27 acetabular reconstructions at a mean 7.5 yr (range, 5–15 yr) follow-up. Index arthroplasty was performed for varying stages of hip dysplasia. Cup survival was 96.3% at 5- and 10-yr intervals.

Bulk Acetabular Autografts

Harris and coworkers have followed the long-term results of cemented sockets placed in conjunction with structural bone autografts in 55 primary total hip arthroplasties. The autografts were either screwed or bolted to the ilium. The average coverage of the acetabular component by the bulk graft was 49% (range, 15–100%). Short-term results demonstrated very encouraging results with union between the graft and the host, and notable functional improvement. A subsequent 10-yr mean follow-up report indicated a substantial increase in the failure rate and raised concern about the durability of these constructs *(35)*. By a mean 16.5 yr follow-up, the acetabular failure rate was 60% for the primary cemented total hip arthroplasties reconstructed with bulk autografts *(36)*. The authors concluded that the main parameter associated with failure was the extent of cup-to-graft coverage. When more

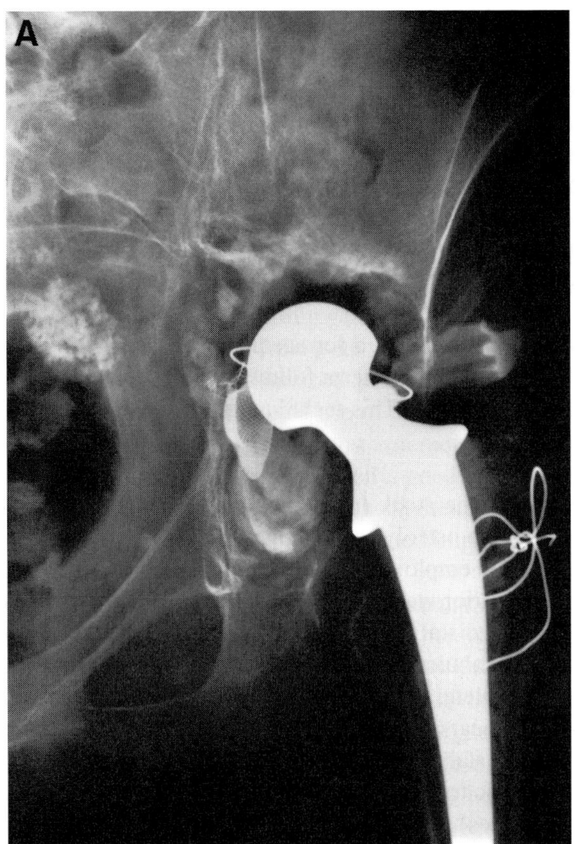

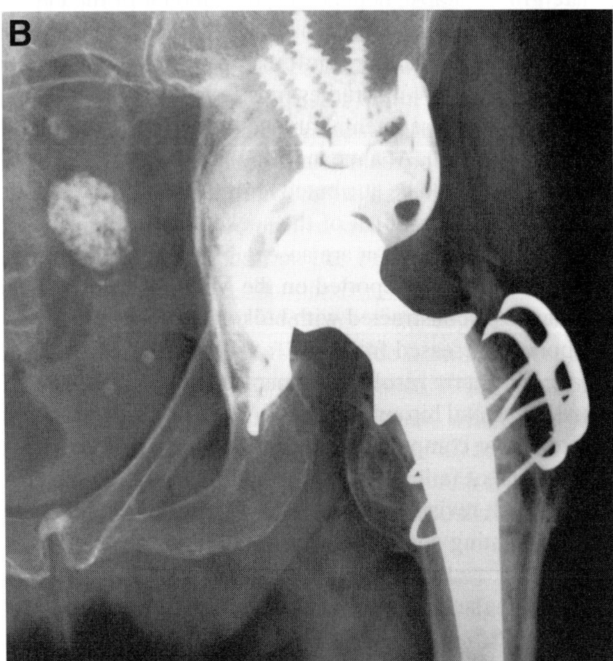

Fig. 3. (A) Preoperative radiograph of failed total hip arthroplasty with massive acetabular bone loss. **(B)** Two years after reconstruction with bulk distal femoral allograft to pelvis and antiprotrusio cage.

5 yr follow-up all grafts had healed and mild to moderate resorption was noted for 11 of the 16 hips. There were no cases of socket migration.

Steihl et al. *(72)* reported on the reconstructions of seventeen complex revisions for either cavitary and segmental bone loss or for pelvic discontinuity. Acetabular reconstructions were performed with femoral head allografts, posterior segmental acetabular allografts, or whole acetabular allografts. Anterior and posterior column plating was used to stabilize the grafts. Sockets were reconstructed with cemented cups in 10 hips and were uncemented in 7 hips. At an average follow-up of almost 7 yr, there were two allograft nonunions and a revision rate of 47%. There were two infections requiring resection arthroplasty. A higher failure rate was seen for uncemented cups placed against bulk allograft bone than was seen for the cemented cups. These reconstructions represent the more difficult scenarios faced, and the results demonstrate the limitations of minimally protected bulk allograft bone in complex revision surgery.

Because large bulk allografts may be prone to fatigue failure, using a reinforcement device to protect these grafts from overloading has been advocated. Two main types of acetabular augmentation devices have been described in the literature: rings screwed to the ileum alone (e.g., the Müller ring), and cages that span the acetabulum and are fixed to both the ileum and the ischium (e.g., the Burch–Schneider antiprotrusio cage). Several papers have reported satisfactory short-term results with the Müller acetabular reinforcement ring (MARR) *(73,74)*. However, longer-term analyses demonstrated a higher failure rate *(75)*, and it appears that unless sufficient contact is achieved with the remaining host bone, the ring cannot provide a stable and durable construct. Burch–Schneider-type antiprotrusio cages with a superior flange resting against the ileum and an inferior flange that is screwed or embedded into the ischium may yield a more durable reconstruction. Gill et al. *(76)* have reported on 37 acetabular reconstructions performed with bulk structural allografts and a cage construct. The allograft covered over 50% of the acetabular component. At an average follow-up of 7.1 yr, 97.3% of the allografts had radiographic evidence of full incorporation. Eighty-one percent of the sockets remained well fixed. This construct protected the allograft in the early postoperative period, virtually eliminating the risk of early superior migration of the cup. Saleh *(77)* reported on 20 massive structural acetabular allografts protected with a Burch–Schneider antiprotrusio cage. The defects treated were such that unprotected allograft bone would have supported greater than 50% of the acetabular component had a cage not been used. At mean follow-up of 10.5 yr, failure rate for the reconstruction was 23%.

The results of other acetabular reinforcement devices in combinations with bulk bone grafts have been described. Kerboull et al. *(78)* have published the results at a mean 10-yr follow-up of 60 reconstructions (48 type III, and 12 type IV) using bulk allograft bone and the Kerboull acetabular reinforcement device. This cruciate-shaped device is screwed to the ilium and has an inferior hook that is placed beneath the teardrop *(78)*. Apparent healing of the graft occurred in all 60 hips by 12 mo, and graft remodeling proceeded for 3–4 yr. Three failures were reported in this series, due to graft resorption and socket loosening. The survival rate at 13-yr follow-up was 92.1% with socket loosening as the end point.

From these studies, it appears that the most critical parameter related to graft failure, as with autograft in primary procedures, is the percentage of support supplied by the allograft to the reconstruction. In cases with more than 30–50% socket-to-allograft contact, the rate of failure was significantly higher, regardless of whether the graft was contained within the acetabulum or bolted to the lateral wall of the ilium. Probably the location of graft support, in addition to its magnitude, is as important in predicting success, but most retrospective studies do not allow the reader to evaluate this important parameter.

Primary Femoral Reconstruction

Grafting of the femur rarely is needed in primary total hip arthroplasty, the most common exception being patients with bone deficiencies or holes in the femur from previous trauma and previous

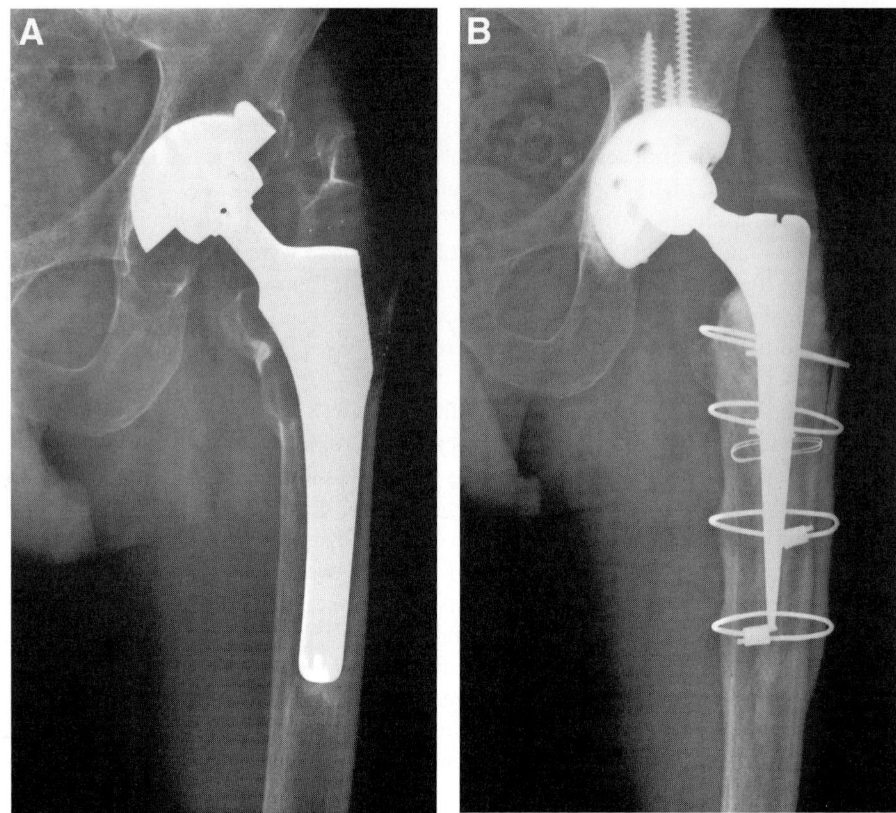

Fig. 4. (**A**) Preoperative radiograph of patient with failed total hip arthroplasty and large areas of cavitary proximal femoral bone loss. (**B**) Radiograph 1 yr after reconstruction with impacted intramedullary cancellous allograft, cemented stem, and cortical strut allograft reinforcement.

internal fixation devices. These can be addressed with particulate graft or, when necessary, cortical onlay strut allografts.

Revision Femoral Reconstruction

In revision total hip arthroplasty, the frequency and the method of bone grafting of the femur varies with the technique used for femoral reconstruction. The treatment of femoral deficiencies is guided by the type of defects. The classification of the American Academy of Orthopaedic Surgeons is a widely used grading system *(79)*. Small cavitary deficiencies of the femur usually are ignored, regardless of whether reconstruction is with conventional cemented, or uncemented implants. Large cavitary deficiencies can be treated with packed particulate bone graft in association with either cemented (**Fig. 4**) or uncemented implants (**Fig. 5**). Segmental femoral defects are treated differently depending on their location: most defects of the calcar (the medial femoral neck above the lesser trochanter) are managed by using an implant with a longer neck or a special calcar replacement implant. Segmental defects of the femoral shaft usually are bypassed by using a long femoral stem, or are reinforced with cortical strut bone allografts (**Fig. 6**). Massive proximal femoral bone loss usually is dealt with using an allograft prosthesis composite (**Fig. 7**), but occasionally is treated with a proximal femoral replacement tumor prosthesis *(80)*.

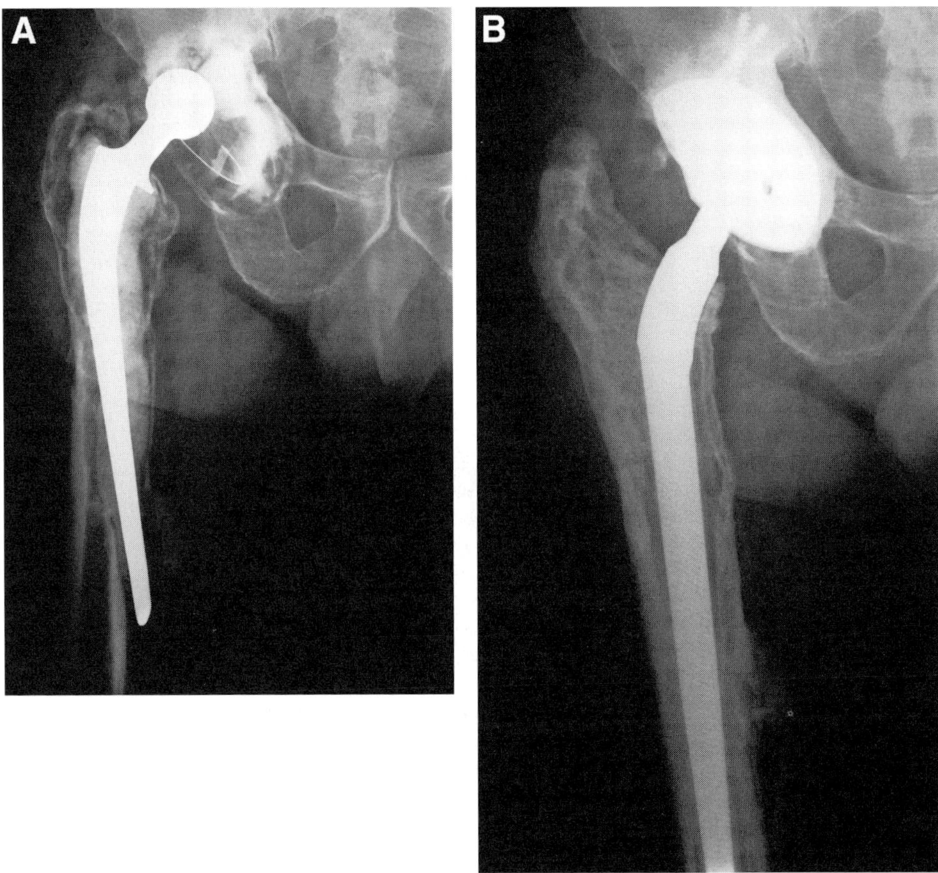

Fig. 5. (A) Radiograph of failed total hip arthroplasty with severe proximal femoral bone loss. **(B)** Four years after reconstruction with uncemented distally fixed stem (to bypass bone loss) and particulate allograft packing of bone defects.

Particulate Femoral Grafts

Most particulate bone grafting of femoral bone deficiencies is performed in association with the technique known as impaction bone grafting. The technique makes use of special instruments that allow dense packing of the particulate bone graft to create a "neomedullary canal," following which a stem is cemented into the graft *(81,82)*. If full thickness cortical defects are present, they must first be reconstituted with wire mesh or cortical bone grafts. The method relies on the densely packed cancellous graft and cement composite for early support of the implant *(83)*. Theoretically, as time goes on the graft gradually is vascularized. As discussed previously, mid-term tissue retrievals subjected to histological analysis, as well as radiographic evidence of graft remodeling (visualized as conversion of the graft from an amorphous appearance to a more trabecular appearance), support this hypothesis *(21)*. Short-term clinical results of impaction grafting reported by Gie et al. *(82)* were encouraging in 56 hips followed for a period of 1.5–4 yr. Both radiological results and histological data demonstrated bone graft incorporation and partial reconstitution of the bone stock. Other short-term studies of the method have also reported similar good results *(84–88)*, but recently several authors

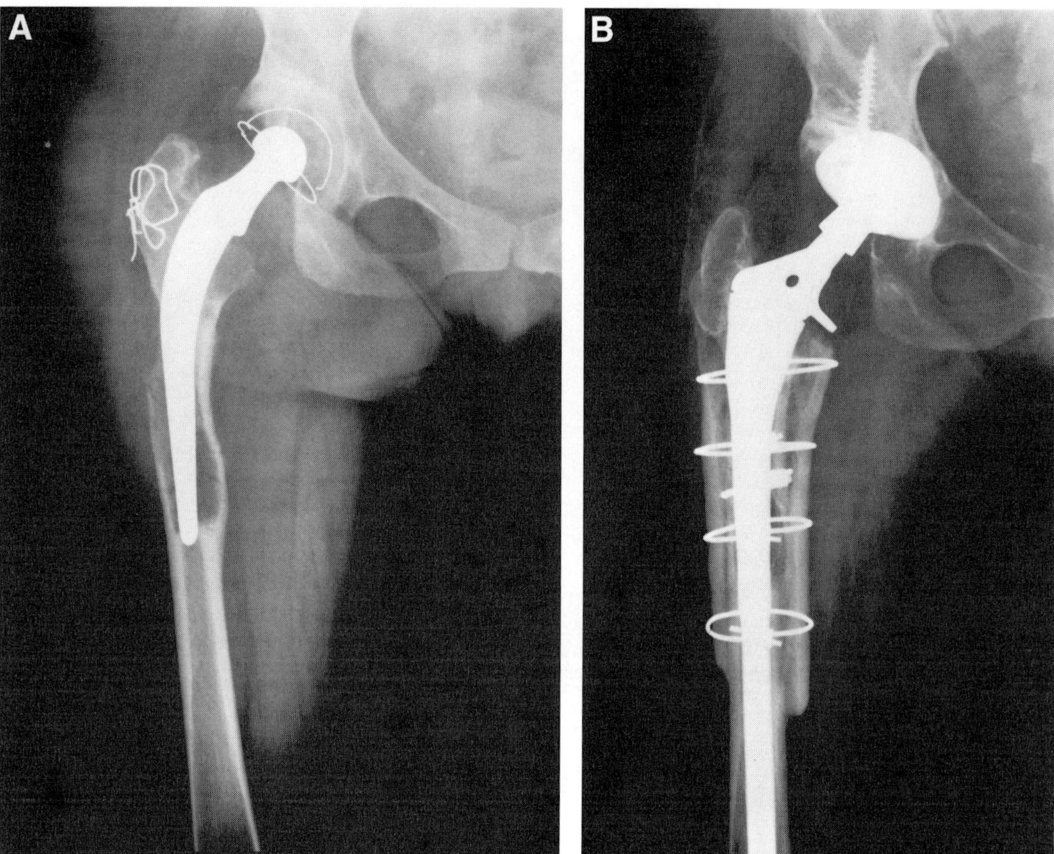

Fig. 6. (A) Radiograph of failed total hip arthroplasty with proximal femoral osteolysis and periprosthetic fracture. **(B)** Two years after reconstruction with long uncemented with cortical strut allograft reinforcement. The strut allografts have healed and are remodeling.

(89–91) have also reported early implant failures due to marked loosening and subsidence, and due to late femoral fractures near the stem tip. Recently, English et al. *(7)* have reported on the use of this technique during two-staged revisions for infection. In a series of 44 hips followed for a mean of 4.5 yr, the authors report an infection-free rate of 92.5% and a revision rate of 2%. The impaction allografting technique is appealing, especially in young patients, because it has the potential to restore bone stock. The technically demanding nature of the procedure, the potential for complications, and the unknown long-term fate of the impacted allograft highlight the need for ongoing assessment of this impaction allograft technique for femoral reconstructions *(90,92–95)*.

Cortical Strut Onlay Grafts

Cortical strut allografts usually are used to reinforce a femur with a full-thickness or near-full-thickness cortical defect or to provide structural support or augment healing of a periprosthetic femoral fracture *(26)*. Clinical and radiographic results demonstrate that cortical strut allografts heal to the femur remarkably consistently. Head et al. *(96)* have reported on 99% union rate in 265 cortical strut bone graft procedures at a mean 8.5-yr follow-up. Failures, due to stem subsidence and loosening, were observed when the graft was used as the primary source of prosthetic support. Pak et al. *(97)* found

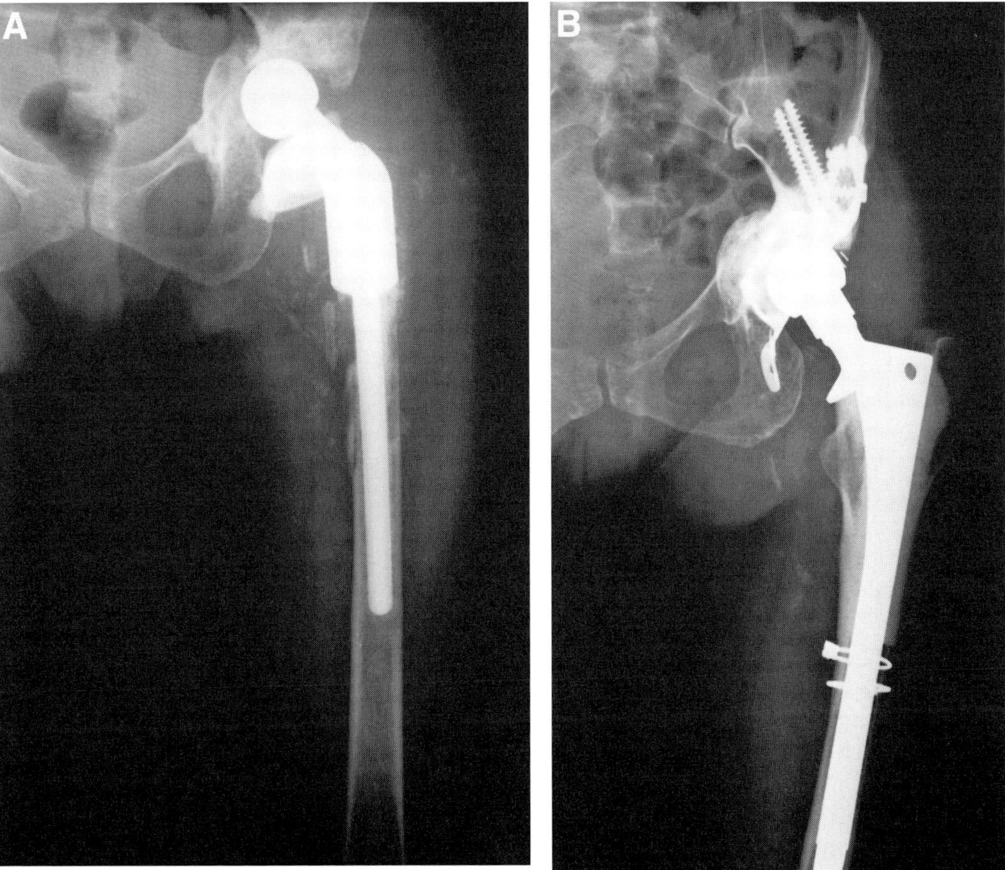

Fig. 7. (A) Radiograph of failed total hip arthroplasty with loose femoral tumor prosthesis. **(B)** Radiograph after reconstruction proximal femoral allograft prosthetic composite. The stem has been cemented into the allograft and press-fitted into the host bone.

similar results, with a 91.5% healing rate in a series of 95 strut grafts. Struts have also been used in the treatment of periprosthetic fractures associated with a stable implant that does not require revision. Haddad et al. *(98)* found 39 of 40 periprosthetic fractures treated with internal fixation using cortical strut grafts as the main source of fixation or an adjunct plate fixation healed. To promote successful healing, the cortical strut allografts should be contoured to fit the underlying bone intimately and should be fixed rigidly to the bone, usually with cerclage wires or cables. Radiographically, a typical process of strut graft union and rounding of the graft ends is followed by slow remodeling of the grafts. Presently, there is limited information on the long-term remodeling of cortical strut bone grafts.

Massive Bulk Femoral Grafts

Bulk circumferential proximal femoral allografts are used when massive proximal femoral bone loss is present. This situation usually is associated with failed total hip arthroplasty, reconstruction after infection, or resection of the proximal femur for a tumor.

Small napkin-ring segmental allografts of the proximal femur once were employed for segmental calcar bone deficiencies, but the results mostly were disappointing. Allan et al. *(99)* reported on deficiency of the proximal femur less than 5 cm in length and recommended abandoning the use of small calcar grafts due to a high rate of resorption, fragmentation, and fracture. Gross et al. *(100)* reached the same conclusion and advocated using calcar replacing prostheses or long-necked femoral implants when dealing with circumferential defects less than 5 cm in length.

Circumferential defects more than 5 cm in length have been managed by massive proximal femoral allografts. This technique has provided good short- to mid-term results in specialized units. The operative approach can be by trochanteric osteotomy, a trochanteric slide, or splitting the remaining proximal femur longitudinally. Long-stemmed femoral implants, some specially designed for this type of reconstruction, are used. Any remaining proximal femur is spilt longitudinally to preserve the native bone. The allograft is reamed and broached until a proper fit of the prosthesis is achieved. The allograft-to-host bone junction stability can be improved by a step-cut and cerclage wires to obtain rotational stability. Usually, the femoral component then is cemented into the allograft. When satisfactory rotational and axial stability of the allograft can be obtained by the geometry of the junction between the graft and the host or by the press-fit of the implant into the host femur, cement is not used in the host femur; when these criteria cannot be met, the stem can be cemented to the host femur. The residual host femur can be wrapped around the allograft and held by cerclage wires to act as a vascularized autogenous bone graft. The host trochanter is reattached to the graft with cerclage wires or a trochanteric reattachment device.

Chandler et al. *(49)* used this technique in association with a long-stemmed femoral component press-fitted in the distal host femur in 30 hips. The mean follow-up of the series was 22 mo (range, 2–46 mo). The functional outcome was notably improved, with a preoperative Harris hip score of 35 vs 78 at final examination. Union between graft and host was observed in 22 hips at a mean 7.3 mo. Complications included five dislocations, a greater trochanter escape of more than 1 cm in three hips, and one deep infection. Head et al. *(101–103)* reported on 22 procedures using proximal femoral allograft followed for an averaged of 28 mo. The authors used a cortical medial remnant of host bone as a vascularized autograft whenever possible, and autogenous bone graft routinely was packed at the host-to-allograft junction. Three methods of fixation of the prosthesis were employed: cement fixation into both the proximal femur and the distal host in 10 patients; cement fixation into only the distal host femur in three patients; and no cement in nine patients. Nonunion at the allograft–host bone junction was observed in three hips. However, only one was associated with partial resorption of the allograft and loss of fixation; in the remaining two nonunions, the implant fixation was considered stable. The functional outcome was judged as good or excellent in 16 of the 22 hips. No septic complications were identified in this series, but dislocation occurred in five patients.

The Vancouver group's latest evaluation of proximal femoral allografts was reported by Haddad et al. *(104)*, and consisted of 55 procedures in 51 patients at a mean 8.8-yr follow-up (range, 3–12.5 yr). None of the allografts were irradiated. The graft was fully cemented in 46 hips, fully uncemented in three hips, and cemented only into the allograft in six hips. Reoperation was performed for five acetabular reconstruction failures, and six failures of the proximal femoral allograft. Complications included one allograft fracture, two deep infections, and five junctional nonunions. In addition, nonunion of the greater trochanter was observed in 22 of the 55 hips, greater trochanter escape occurred in 14 hips, and instability occurred in 6 hips. Moderate to severe resorption of the allograft was seen in 11 procedures. In all seven patients with severe resorption, the host proximal femur had been discarded at the time of the reconstruction, and the prosthesis had been cemented into both the allograft and the distal host femur. Despite the complications, the clinical outcome was usually satisfactory, and overall success rate was 85%. The authors concluded that fully cementless implants should not be used in conjunction with a segmental allograft replacement. They recommended preserving any remaining femur, and cementing the prosthesis into the allograft only.

The Toronto group reported on 200 circumferential allografts longer than 5 cm at a mean 2-yr follow-up *(100,105)*. The allograft bone had been deep-frozen at −70°C and irradiated with 2.5 Mrad. A long-stemmed prosthesis cemented into the graft only was used. Complications included 11 dislocations, six infections, seven nonunions, and one loosening. Graft-to-host union usually occurred between 3 and 6 mo. Graft resorption was identified in six hips, but had not penetrated the full thickness of the cortex of the graft. Resorption measured less than 1 cm in all but one hip. Using as the definition for success an increase in the functional score of at least 20 points, a stable implant, and no further surgery related to the allograft, the success rate was 85% in 130 hips with an average of 4.8 yr follow-up. In a follow-up study of 65 hips with a mean 9-yr follow-up, using the previous definition for success, success was observed in 55 of the 65 hips (85%) *(100)*. In their most recent follow-up, at a mean of 11 yr, 48 allograft reconstructions had a 78% success rate *(106)*.

Kerboull *(107,108)* in France has proposed a different method of using femoral allografts in these challenging situations. This author has proposed using a proximal femoral structural allograft impacted into the remaining host femur. A femoral component of standard length then is cemented only into the allograft. The clinical and radiological results have been satisfactory, with one revision of 27 procedures at a mean 5-yr follow-up. The revision was performed because of resorption of the proximal allograft.

Although most of these reports identify a relatively high rate of complications, including infection, instability, nonunion, and trochanteric escape *(109)*, the majority of patients have a satisfactory clinical result. As other reconstructive methods and more sophisticated implants have become available, whole-segment proximal femoral grafts are used less frequently. Nevertheless, proximal femoral allografts still allow the successful reconstruction of difficult hip problems with massive proximal femoral bone loss and provide a good alternative to tumor prostheses (which have been reported to have a reasonably high failure rate due to loosening and which do not provide good options for abductor muscle reattachment).

CLINICAL RESULTS OF BONE GRAFTS IN TOTAL KNEE ARTHROPLASTY

Bone grafts are needed less frequently in total knee arthroplasty than in total hip arthroplasty, because bone deficiency often can be managed with metallic augmentation of the metallic arthroplasty implants. As is the case for the hip joint, loss of bone stock can be classified as either cavitary or segmental. In primary total knee arthroplasty some segmental bone deficiencies of the proximal tibia need bulk grafts, and large cysts in the femur and tibia often are treated with particulate bone graft *(110)*. The source of most bone grafts in primary knee total knee arthroplasty is the autologous bone removed routinely during the tibial and femoral bone resection. In revision total knee arthroplasty, large deficiencies of the femur or tibia can be treated with particulate or bulk grafts when they are treated with a metal implant *(110)*. Most cavitary deficiencies are filled with cement or with packed particulate bone allograft. Most segmental distal femur and proximal tibia defects are managed with wedge- or block-shaped metal component augmentation, but they also, depending on shape and size, can be managed with structural bone allografts derived from femoral heads, the distal femur, or proximal tibia. Finally, large segmental bone loss of the distal femur or proximal tibia can be treated with large segmental distal femoral or proximal tibial allografts (**Fig. 8**).

Particulate Grafts in Revision Total Knee Arthroplasty

Samuelson *(111)* reported the use of bone graft in revision knee surgery in a series of 22 patients at an average of 15 mo follow-up (range, 6 mo to 3 yr). Bone graft was of three types: finely milled, coarsely milled (5–8 mm), and blocks. Cemented stemmed components were used in all cases. Radiological graft incorporation occurred between 6 mo to 1 yr. No revisions and no infections were noted. Görlich et al. *(112)* and Ries *(48)* used autogenous bone graft harvested from the resected articular surfaces or the contralateral knee in the case of cemented bilateral knee replacements. Graft incorporation was observed in both studies between 3 and 6 mo. The Nijmegen group in the Netherlands *(113)* has reported on allograft and autogenous bone in 36 knees (23 primary and 13 revision procedures) followed for 2–5 yr. According to the defects, bone graft was either morcellized or solid corticocan-

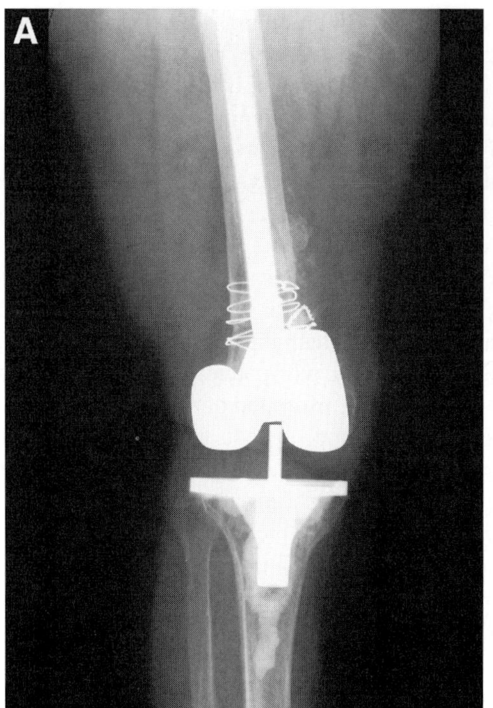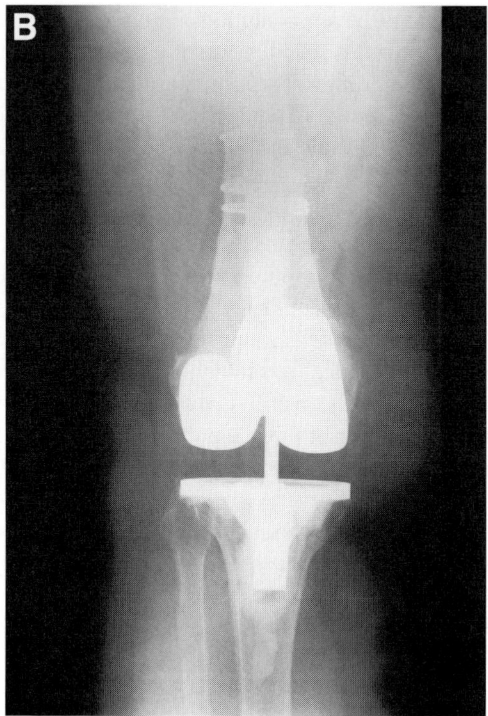

Fig. 8. (A) Radiograph of patient with nonunion of supracondylar femur fracture above total knee arthroplasty. **(B)** Radiograph after reconstruction with distal femoral allograft prosthetic composite.

cellous. There was no significant difference between allograft and autograft bone in terms of incorporation, which occurred at a mean 1 yr after the surgery. Graft resorption was noted in two of the eight solid corticocancellous allografts used on the femoral side. The same group evaluated the mechanical properties of morcellized bone graft in a cadaver model *(114)*. A unicondylar noncontained femoral defect was filled with impacted morcellized bone graft, and a stemless total knee arthroplasty was used. The authors found no collapse of the graft under load-bearing conditions. However, this study addressed the immediate postoperative situation, and therefore did not investigate the long-term stability of the construct during bone remodeling.

Benjamin et al. *(115)* has reported on 2-yr follow-up of 33 cemented knee revisions in which particulate bone allograft was used to reconstruct contained femoral and tibial defects. No failures were observed in this short-term follow-up study. Graft remodeling was noted and was believed to signal successful graft incorporation. Bradley *(116)* has reported success in 18 of 19 revisions treated with this technique.

Lonner et al. *(117)* utilized impaction grafting with a wire mesh for graft containment to treat uncontained defects in 17 cemented knee revision arthroplasties. At 18 mo mean follow-up there were no revisions, but three knees had nonprogressive tibial lucencies. The long-term durability of this construct cannot yet be predicted. Beharie and Nelson *(118)* reported on the use of impacting grafting in conjunction with a long-stemmed tibial component to treat a periprosthetic tibial fracture associated with a loose tibial component. The authors believe the this technique provides stable fixation, provides an osteoconductive substrate at the fracture site, potentially restores bone stock, and prevents cement extrusion at the fracture site.

Particulate bone grafting has also been used with cementless fixation in revision total knee arthroplasty. Whiteside *(119)* reported encouraging short-term results with cementless fixation in a series of 20 patients with a minimum of 2 yr of follow-up. Radiological evidence of graft incorporation was observed by 1 yr, and no component had migrated. Whiteside and Bicalho *(24)* subsequently reported on a larger series of 63 cementless revision procedures with at least 5 yr of follow-up in which morcellized bone allograft combined with a demineralized bone matrix was used to treat major bone defects. The overall complication rate was 22%. Radiographically, formation of trabecular pattern and presumed healing was identified in all allografts by 1 yr after surgery. Stable fixation of the stemmed implants fixed with supplemental screws was noted in 97% of the knees.

The use of particulate grafting has also been expanded to the treatment of severe patellar bone loss in revision total knee arthroplasty. Hanssen *(120)* has described a technique for impaction grafting of the patella. A pocket of tissue is created from peripatellar fibrotic tissue, fascia lata, or suprapatellar tissue and overlies the remnant of host patella. This soft tissue flap is sutured into place and either autograft or allograft bone is impacted into the pouch to reconstruct the patellar bone stock. At average mean follow-up just over 3 yr, 10–12 mm millimeters of patellar bone thickness had been restored.

Structural Grafts Revision Total Knee Arthroplasty

In the case of a major structural defect, a number of authors have advocated the use of bulk allograft bone, usually in association with a long-stemmed prosthesis to reduce load on the graft. Short- to mid-term studies have demonstrated encouraging results, with a high allograft-to-host union rate when adequate allograft fixation was obtained. Mnaymneh et al. *(121)* reported on 14 massive allografts in 10 patients followed for an average of 40 mo. Components were cemented to the allograft, but the stem was uncemented. Union of the allograft to the host bone occurred radiologically in 12 of the 14 procedures. Complications included one femoral allograft fracture and resorption, one deep infection, marked knee instability in two cases, and tibial loosening in two cases. Tsahakis et al. *(122)* reviewed 19 structural allografts (13 in the distal femur, and six in the proximal tibia) after an average 2.1 yr follow-up. The components were cemented to the allograft, and the stems were press-fitted in the medullary canal. Functional outcome was greatly improved in all patients, and the allografts healed by 1 yr. No infections and no reoperations were reported in this series. A larger series of 35 bulk allografts in 30 patients at a mean 4.2-yr follow-up (range, 2–10 yr) was reported by Engh et al. *(123)* in 1997. Allografts included two femoral heads, five distal femoral allografts, and one proximal tibial allograft. Stemmed components were used in all patients. Clinical results were judged as good or excellent in 26 of the 30 patients. Incorporation of the graft was demonstrated in 20 of the 30 patients, and in 10 it was uncertain radiographically whether the graft was incorporated. No case of graft resorption was noted. Three out of four prosthetic components (two in the femur, and one in the tibia) that were not porous coated and uncemented subsided 5–9 mm over a period of 9 yr. No complications related directly to the grafts occurred. In light of these results, the authors concluded that structural allograft in conjunction with a stemmed component inserted with cement provided excellent results for the treatment of large defects during knee reconstruction procedures. Other series, including those reported by Mow and Wiedel *(124)*, and Ghazavi et al. *(125)* on structural allografting with a stemmed knee prosthesis, also have shown a high mid-term rate of graft-to-host union.

Lindstrand et al. *(126)* using radiostereometric analysis (RSA) to evaluate tibial implants stability following revision total knee arthroplasty performed with structural autograft bone. Autogenous structural bone resected from either the intact femoral condyle or the tibial plateau was used. The tibial components were always cemented, and the graft-to-host fixation was augmented by screws. The mean migration was 0.5 mm (range, 0.2–1.5 mm) at a mean 5-yr follow-up, and no case of continuous migration was recorded. Radiologically, all but one graft had united to the host.

Clatworthy et al. *(127)* reported medium to long-term follow-up of 52 revision total knee arthroplasties treated with structural allograft and stemmed components. At a mean follow-up of 8 yr, there were 13 failures of the reconstruction, yielding a success rate of 75%. There were two nonunions of the host–allograft junction, four infections, and five instances of graft resorption resulting in implant loosening.

Bone–Tendon Grafts in Revision Total Knee Arthroplasty

Extensor mechanism disruption is an infrequent but catastrophic complication after total knee arthroplasty *(128)*. One method of reconstruction of chronic quadriceps or patellar tendon deficiencies is the use of tendon–patella–bone or tendon–bone allografts. Emerson et al. *(129,130)* reported good initial results in a series of 13 knees, but at longer follow-up, an extensor lag between 20° and 40° was found in three patients. All of the allograft bone–host interfaces healed without complication. Nazarian and Booth *(131)* have modified this technique by creating a tight-fitting trough in the native tibia, into which the distal attachment of the extensor allograft is impacted and fixed with wires or screws. In addition, the graft is tensioned in full extension. In a series of 40 patients at 3.6 yr mean follow-up, they reduced the incidence of extensor lag to 42% of the patients and reduced the magnitude of the lag that occurred to a mean of 13°. There were no failures at the graft–host junction. The limitations of this reconstruction do not appear to involve the bony interfaces, but rather the response of the allograft tendon to repetitive loading with subsequent elongation. For chronic patellar tendon disruptions, the use of an Achilles tendon allograft has been described. Crossett et al. *(132)* recently reported on the results in nine patients at 2.3 yr mean follow-up. The attachment of the Achilles tendon bone block to the tibia was fixed in a similar manner to that described above. There were two graft failures in the tendinous region and no allograft–host bone nonunions. A significant reduction in extensor lag was achieved. For extensor tendon disruption associated with massive proximal tibial bone loss, Barrack and Lyons *(133)* have the use of a composite allograft of proximal tibia–patellar tendon–patella–quadriceps tendon.

OTHER JOINTS

For primary and revision shoulder arthroplasties, segmental glenoid deficiencies can be managed with structural bone grafts or prosthetic or cement augmentation. Reconstruction of humeral deficiencies in total shoulder arthroplasty is mostly analogous to revision techniques of the femur around the hip. For primary elbow arthroplasties, some designs make standard use of autologous bone grafts to enhance humeral implant stability, but otherwise grafting is needed uncommonly. In revision elbow arthroplasty, the types of bone grafts and the techniques are analogous to the hip and knee arthroplasty, with cancellous grafts used for cavitary defects, strut grafts for long-bone reinforcement, and segmental grafts reserved for severe distal humeral or proximal ulnar segmental deficiencies.

COMPLICATIONS OF BONE GRAFTS

The main complications of using bone grafts in joint arthroplasty are graft resorption, graft collapse, or graft fracture. Graft collapse and graft fracture may occur secondary to resorption, osteolysis, or mechanical stress overload. The success of a graft will depend on the host environment into which it is implanted, as well as the loads to which it is subjected.

Infection is one of the most serious complications of joint reconstruction with associated bone grafting. There is a higher risk of infection in arthroplasties in which graft is used, but it is uncertain whether this relates specifically to the presence of the bone graft or to the selection factor of grafts being used in complex reconstructions *(109)*. A number of studies support the idea that bone grafts (autograft and allograft) can be used successfully in some cases for reconstruction after deep infection *(6,8,9,84,87,88)*.

REFERENCES

1. Goldring, S. R., Clark, C. R., and Wright, T. M. (1993) The problem in total joint arthroplasty: aseptic loosening. *J. Bone Joint Surg.* **75A**, 799–801.
2. Harris, W. H. (1993) Management of the deficient acetabulum using cementless fixation without bone grafting. *Orthop. Clin. N. Am.* **24**, 663–665.
3. Berry, D., Sutherland, C., Trousdale, R., et al. (2000) Bilobed oblong porous coated acetabular components in revision total hip arthroplasty. *Clin. Orthop.* **371**, 154–160.
4. Dearborn, J. and Harris, W. (2000) Acetabular revision arthroplasty using so-called jumbo cementless components: an average 7-year follow-up study. *J. Arthroplasty* **15**, 8–15.
5. Malkani, A., Paiso, J., and Sim, F. (2000) Proximal femoral replacement with megaprosthesis. *Instr. Course Lect.* **49**, 141–146.
6. Berry, D. J., Chandler, H. P., and Reilly, D. T. (1991) The use of bone allografts in two-stage reconstruction after failure of hip replacements due to infection. *J. Bone Joint Surg.* **73A**, 1460–1468.
7. English, H., Timperly, A. J., Dunlop, D., et al. (2002) Impaction grafting of the femur in two stage revision for infected total hip replacement. *J. Bone Joint Surg.* **84B**, 700–705.
8. Haddad, F. S., Garbuz, D. S., Masri, B. A., et al. (2000) Structural proximal femoral allografts for failed total hip replacements. *J. Bone Joint Surg.* **82B**, 830–836.
9. Haddad, F. S., Muirhead-Allwood, S. K., Manktelow, A. R., et al. (2000) Two-stage uncemented revision hip arthroplasty for infection. *J. Bone Joint Surg.* **82B**, 689–694.
10. Bauer, T. and Muschler, G. (2000) Bone graft materials. An overview of the basic science. *Clin. Orthop.* **371**, 10–27.
11. Goldberg, V. M. (2000) Selection of bone grafts for revision total hip arthroplasty. *Clin. Orthop.* **381**, 68–76.
12. Dunlop, D. G., Brewster, N. T., Madabhushi, S. P., et al. (2003) Techniques to improve the shear strength of impacted bone graft. The effect of particle size and washing of the graft. *J. Bone Joint Surg.* **85A**, 639–646.
13. Schreurs, B. W., Slooff, T. J., Gardeniers, J. W., et al. (2001) Acetabular reconstruction with bone impaction grafting and a cemented cup. 20 years' experience. *Clin. Orthop.* **393**, 202–215.
14. Schreurs, B. W., Gardeniers, J. W., and Slooff, T. J. (2001) Acetabular reconstruction with bone impaction grafting: 20 years of experience. *Instr. Course Lect.* **50**, 221–228.
15. Schreurs, B. W., Slooff, T. J., Buma, P., et al. (2001) Basic science of impaction grafting. *Instr. Course Lect.* **50**, 211–220.
16. Garbuz, D., Masri, B., and Czitrom, A. (1998) Biology of allografting. *Orthop. Clin. N. Am.* **29**, 199–204.
17. Huo, M., Friedlaender, G., and Salvati, E. (1992) Bone graft and total hip arthroplasty. A review. *J. Arthroplasty* **7**, 109–120.
18. Ullmark, G. and Obrant, K. J. (2002) Histology of impacted bone graft incorporation. *J. Arthroplasty* **17**, 150–157.
19. van der Donk, S., Buma, P., Slooff, T. J., et al. (2002) Incorporation of morselized bone grafts: a study of 24 acetabular biopsy specimens. *Clin. Orthop.* **396**, 131–141.
20. Heekin, R., Engh, C., and Vinh, T. (1995) Morcellized allograft in acetabular reconstruction. A postmortem retrieval analysis. *Clin. Orthop.* **319**, 184–190.
21. Ling, R., Timperley, A., and Linder, L. (1993) Histology of cancellous impaction grafting in the femur. A case report. *J. Bone Joint Surg.* **75B**, 693–696.
22. Mikhail, W., Weidenhielm, L., Wretenberg, P., et al. (1999) Femoral bone regeneration subsequent to impaction grafting during hip revision: histologic analysis of a human biopsy specimen. *J. Arthroplasty* **14**, 849–853.
23. Nelissen, R., Bauer, T., Weidenhielm, L., et al. (1995) Revision hip arthroplasty with the use of cement and impaction grafting. Histological analysis of four cases. *J. Bone Joint Surg.* **77A**, 412–422.
24. Whiteside, L. (1989) Cementless reconstruction of massive tibial bone loss in revision total knee arthroplasty. *Clin. Orthop.* **248**, 80–86.
25. Van Loon, C. J., Buma, P., de Waal Malefijit, M. C., et al. (2000) Morsellised bone allografting in revision total knee replacement. A case report with a 4 year histologic follow-up. *Acta Orthop. Scand.* **71**, 98–101.
26. Emerson, R. J., Malinin, T., Cuellar, A., et al. (1992) Cortical strut allografts in the reconstruction of the femur in revision total hip arthroplasty. A basic science and clinical study. *Clin. Orthop.* **285**, 35–44.
27. Hamer, A., Suvarna, S., and Stockley, I. (1997) Histologic evidence of cortical allograft bone incorporation in revision hip surgery. *J. Arthroplasty* **12**, 785–789.
28. Hooten, J. J., Engh, C., Heekin, R., et al. (1996) Structural bulk allografts in acetabular reconstruction. Analysis of two grafts retrieved at post-mortem. *J. Bone Joint Surg.* **78B**, 270–275.
29. Enneking, W. and Mindell, E. (1991) Observations on massive retrieved human allografts. *J. Bone Joint Surg.* **73A**, 1123–1142.
30. Gouin, F., Passuti, N., Verriele, V., et al. (1996) Histological features of large bone allografts. *J. Bone Joint Surg.* **78B**, 38–41.

31. Hamadouche, M., Mathieu, M., Meunier, A., et al. (2002) Histological findings in a proximal femoral allograft ten years following revision total hip arthroplasty. A case report. *J. Bone Joint Surg.* **84A**, 269–273.
32. Slooff, T. (1986) Bone grafting in total hip replacement for acetabular protrusion. *Acta Orthop. Belg.* **52**, 324–327.
33. Welten, M. L., Schreurs, B. W., Buma, P., et al. (2000) Acetabular reconstruction with impacted morcellized cancellous bone autograft and cemented primary total hip arthroplasty. *J. Arthroplasty* **15**, 819–824.
34. Bolder, S. B., Melenhorst, J., Gardeniers, J. W., et al. (2001) Cemented total hip arthroplasty with impacted morcellized bone-grafts to restore acetabular bone defects in congenital hip dysplasia. *J. Arthroplasty* **16**, 164–169.
35. Kwong, L. M., Jasty, M., and Harris, W. H. (1993) High failure rate of bulk femoral head allografts in total hip acetabular reconstructions at 10 years. *J. Arthroplasty* **8**, 341–346.
36. Shinar, A. A. and Harris, W. H. (1997) Bulk structural autogenous grafts and allografts for reconstruction of the acetabulum in total hip arthroplasty. Sixteen-year-average follow-up. *J. Bone Joint Surg.* **79A**, 159–168.
37. Lee, B., Cabanela, M., Wallrichs, S., et al. (1997) Bone-graft augmentation for acetabular deficiencies in total hip arthroplasty. Results of long-term follow-up evaluation. *J. Arthroplasty* **12**, 503–510.
38. Kobayashi, S., Saito, N., Nawata, M., et al. (2003) Total hip arthroplasty with bulk femoral head autograft for acetabular reconstruction in developmental dysplasia of the hip. *J. Bone Joint Surg.* **85A**, 615–621.
39. Inao, S. and Matsuno, T. (2000) Cemented total hip arthroplasty with autogenous acetabular bone grafting for hips with developmental dysplasia in adults. The results a minimum of ten years. *J. Bone Joint Surg.* **82B**, 375–377.
40. Ritter, M. and Trancik, T. (1985) Lateral acetabular bone graft in total hip arthroplasty. A three- to eight-year follow-up study without internal fixation. *Clin. Orthop.* **193**, 156–159.
41. Cameron, H. and MacDiarmid, A. (1987) Major acetabulum reconstruction using autograft bone in cemented hip arthroplasty. *Orthop. Rev.* **16**, 845–850.
42. Garbuz, D., Morsi, E., Mohamed, N., et al. (1996) Classification and reconstruction in revision acetabular arthroplasty with bone stock deficiency. *Clin. Orthop.* **324**, 98–107.
43. Garbuz, D. and Penner, M. (1998) Role and results of segmental allografts for acetabular segmental bone deficiency. *Orthop. Clin. N. Am.* **29**, 263–275.
44. Gross, A., Allan, D., Catre, M., et al. (1993) Bone grafts in hip replacement surgery. The pelvic side. *Orthop. Clin. N. Am.* **24**, 679–695.
45. Gross, A. and Catre, M. (1994) The use of femoral head autograft shelf reconstruction and cemented acetabular components in the dysplastic hip. *Clin. Orthop.* **298**, 60–66.
46. Gross, A. and Solomon, M. (1997) The flying buttress acetabular bone-graft. *J. Arthroplasty* **12**, 706–708.
47. Marti, R., Schuller, H., and van, S. M. (1994) Superolateral bone grafting for acetabular deficiency in primary total hip replacement and revision. *J. Bone Joint Surg.* **76B**, 728–734.
48. Ries, M. (1996) Impacted cancellous autograft for contained bone defects in total knee arthroplasty. *Am. J. Knee Surg.* **9**, 51–54.
49. DeWal, H., Chen, F., Su, E., et al. (2003) Use of structural bone graft with cementless acetabular cups in total hip arthroplasty. *J. Arthroplasty* **18**, 23–28.
50. Morsi, E., Garbuz, D., and Gross, A. (1996) Total hip arthroplasty with shelf grafts using uncemented cups. A long-term follow-up study. *J. Arthroplasty* **11**, 81–85.
51. Spangehl, M. J., Berry, D. J., Trousdale, R. T., and Cabanela, M. E. (2001) Uncemented acetabular components with bulk femoral head autografts for acetabular reconstruction in developmental dysplasia of the hip: results at 5 to 12 years. *J. Bone Joint Surg.* **83A**, 1484–1489.
52. D'Antonio, J., Capello, W., Borden, L., et al. (1989) Classification and management of acetabular abnormalities in total hip arthroplasty. *Clin. Orthop.* **243**, 126–137.
53. Schreurs, B., Slooff, T., Buma, P., et al. (1998) Acetabular reconstruction with impacted morcellized cancellous bone graft and cement. A 10- to 15-year follow-up of 60 revision arthroplasties. *J. Bone Joint Surg.* **8B**, 391–395.
54. Slooff, T., Buma, P., Schreurs, B., et al. (1996) Acetabular and femoral reconstruction with impacted graft and cement. *Clin. Orthop.* **324**, 108–115.
55. Slooff, T., Schimmel, J., and Buma, P. (1993) Cemented fixation with bone grafts. *Orthop. Clin. N. Am.* **24**, 667–677.
56. Slooff, T., Schreurs, B., Buma, P., et al. (1999) Impaction morcellized allografting and cement. *Instr. Course Lect.* **48**, 79–89.
57. Schreurs, B. W., Thien, T. M., de Waal Malefijt, M. C., et al. (2003) Acetabular revision with impacted morselized cancellous bone graft and a cemented cup in patients with rheumatoid arthritis: three to fourteen-year follow-up. *J. Bone Joint Surg.* **85A**, 647–652.
58. Azuma, T., Yasuda, H., Okagaki, K., et al. (1994) Compressed allograft chips for acetabular reconstruction in revision hip arthroplasty. *J. Bone Joint Surg.* **76B**, 40–44.
59. Kinzinger, P., Karthaus, R., and Slooff, T. (1991) Bone grafting for acetabular protrusion in hip arthroplasty. 27 cases of rheumatoid arthritis followed for 2-8 years. *Acta Orthop. Scand.* **62**, 110–112.
60. Knight, J., Fujii, K., Atwater, R., et al. (1993) Bone-grafting for acetabular deficiency during primary and revision total hip arthroplasty. A radiographic and clinical analysis. *J. Arthroplasty* **8**, 371–382.

61. Paprosky, W. G., Bradford, M. S., and Jablonsky, W. S. (1996) Acetabular reconstruction with massive acetabular allografts. *Instr. Course Lect.* **45,** 149–159.
62. Lachiewicz, P. and Poon, E. (1998) Revision of a total hip arthroplasty with a Harris-Galante porous-coated acetabular component inserted without cement. A follow-up note on the results at five to twelve years. *J. Bone Joint Surg.* **80A,** 980–984.
63. Leopold, S., Rosenberg, A., Bhatt, R., et al. (1999) Cementless acetabular revision. Evaluation at an average of 10.5 years. *Clin. Orthop.* **369,** 179–186.
64. Jasty, M. (1998) Jumbo cups and morcellized graft. *Orthop. Clin. N. Am.* **29,** 249–254.
65. Whaley, A., Berry, D. J., and Harmsen, W. S. (2001) Extra-large uncemented hemispherical acetabular components for revision total hip arthroplasty: results in 89 hips at a minimum of 5 years. *J. Bone Joint Surg.* **83A,** 1352–1357.
66. Berry, D. J. and Muller, M. E. (1992) Revision arthroplasty using an anti-protrusio cage for massive acetabular bone deficiency. *J. Bone Joint Surg.* **74B,** 711–715.
67. Gill, T. J., Sledge, J. B., and Muller, M. E. (1998) The Burch-Schneider anti-protrusio cage in revision total hip arthroplasty: indications, principles and long-term results. *J. Bone Joint Surg.* **80B,** 946–953.
68. Maloney, W., Herzwurm, P., Paprosky, W., et al. (1997) Treatment of pelvic osteolysis associated with a stable acetabular component inserted without cement as part of a total hip replacement. *J. Bone Joint Surg.* **79A,** 1628–1634.
69. Schmalzried, T., Fowble, V., and Amstutz, H. (1998) The fate of pelvic osteolysis after reoperation. No recurrence with lesional treatment. *Clin. Orthop.* **350,** 128–137.
70. Brien, W. W., Bruce, W. J., Salvati, E. A., et al. (1990) Acetabular reconstruction with a bipolar prosthesis and morcellized bone grafts. *J. Bone Joint Surg.* **72A,** 1230–1235.
71. Somers, J. F. A., Timperley, A. J., Norton, M., et al. (2002) Block allografts in revision total hip arthroplasty. *J. Arthroplasty* **17,** 562–568.
72. Stiehl, J. B., Saluja, R., and Diener, T. (2000) Reconstruction of major column defects and pelvic discontinuity in revision total hip arthroplasty. *J. Arthroplasty* **15,** 848–857.
73. Cabanela, M. (1998) Reconstruction rings and bone graft in total hip revision surgery. *Orthop. Clin. N. Am.* **29,** 255–262.
74. Haentjens, P., de Boeck, H., Handelberg, F., et al. (1993) Cemented acetabular reconstruction with the Muller support ring. A minimum five-year clinical and roentgenographic follow-up study. *Clin. Orthop.* **290,** 225–235.
75. Zehntner, M. K. and Ganz, R. (1994) Midterm results (5.5–10 years) of acetabular allograft reconstruction with the acetabular reinforcement ring during total hip revision. *J. Arthroplasty* **9,** 469–479.
76. Gill, T., Sledge, J., and Muller, M. (2000) The management of severe acetabular bone loss using structural allograft and acetabular reinforcement devices. *J. Arthroplasty* **15,** 1–7.
77. Saleh, K. J., Jaroszynski, G., Woodgate, I., et al. (2000) Revision total hip arthroplasty with the use of structural acetabular allograft and reconstruction ring. A case series with a 10-year average follow-up. *J. Arthroplasty* **15,** 951–958.
78. Kerboull, M., Hamadouche, M., and Kerboull, L. (2000) The Kerboull acetabular reinforcement device in major acetabular reconstruction. *Clin. Orthop.* **278,** 155–168.
79. Chandler, H., Clark, J., Murphy, S., et al. (1994) Reconstruction of major segmental loss of the proximal femur in revision total hip arthroplasty. *Clin. Orthop.* **298,** 67–74.
80. Gross, A. E., Blackley, H., Wong, P., et al. (2002) The role of allografts in revision arthroplasty of the hip. *Instr. Course Lect.* **51,** 103–113.
81. Gie, G., Linder, L., Ling, R., et al. (1993) Contained morcellized allograft in revision total hip arthroplasty. Surgical technique. *Orthop. Clin. N. Am.* **24,** 717–725.
82. Gie, G., Linder, L., Ling, R., et al. (1993) Impacted cancellous allografts and cement for revision total hip arthroplasty. *J. Bone Joint Surg.* **75B,** 14–21.
83. Malkani, A., Voor, M., Fee, K., et al. (1996) Femoral component revision using impacted morcellized cancellous graft. A biomechanical study of implant stability. *J. Bone Joint Surg.* **78B,** 973–978.
84. Boldt, J. G., Dilawari, P., Argarwal, S., et al. (2001) Revision total hip arthroplasty using impaction bone grafting with cemented nonpolished stems and Charnley cups. *J. Arthroplasty* **16,** 943–952.
85. Duncan, C., Masterson, E., and Masri, B. (1998) Impaction allografting with cement for the management of femoral bone loss. *Orthop. Clin. N. Am.* **29,** 297–305.
86. Kligman, M., Con, V., and Roffman, M. (2002) Cortical and cancellous morselized allograft in revision total hip replacement. *Clin. Orthop.* **401,** 139–148.
87. Lind, M., Krarup, N., Mikkelsen, S., et al. (2002) Exchange impaction allografting for femoral revision hip arthroplasty. Results in 87 cases after 3.6 years follow-up. *J. Arthroplasty* **17,** 158–164.
88. Ullmark, G., Hallin, G., and Nilsson, O. (2002) Impacted corticocancelous allografts and cement for revision of the femur component in total hip arthroplasty using Lubinus and Charnley prostheses. *J. Arthroplasty* **17,** 325–334.
89. Eldridge, J., Smith, E., Hubble, M., et al. (1997) Massive early subsidence following femoral impaction grafting. *J. Arthroplasty* **12,** 535–540.
90. Jazrawi, L., Della, V. C., Kummer, F., et al. (1999) Catastrophic failure of a cemented, collarless, polished, tapered cobalt-chromium femoral stem used with impaction bone-grafting. A report of two cases. *J. Bone Joint Surg.* **81A,** 844–847.

91. Meding, J., Ritter, M., Keating, E., et al. (1997) Impaction bone-grafting before insertion of a femoral stem with cement in revision total hip arthroplasty. A minimum two-year follow-up study. *J. Bone Joint Surg.* **79A,** 1834–1841.
92. Haddad, F. and Duncan, C. (1999) Impaction allografting of the proximal femur: fact or fad? *Orthopedics* **22,** 855–858.
93. Knight, J. and Helming, C. (2000) Collarless polished tapered impaction grafting of the femur during revision total hip arthroplasty: pitfalls of the surgical technique and follow-up in 31 cases. *J. Arthroplasty* **15,** 159–165.
94. Leopold, S., Jacobs, J., and Rosenberg, A. (2000) Cancellous allograft in revision total hip arthroplasty. A clinical review. *Clin. Orthop.* **371,** 86–97.
95. Leopold, S. and Rosenberg, A. (2000) Current status of impaction allografting for revision of a femoral component. *Instr. Course Lect.* **49,** 111–118.
96. Head, W. (2000) Results of onlay allografts. *Clin. Orthop.* **371,** 108–112.
97. Pak, J., Paprosky, W., Jablonsky, W., et al. (1993) Femoral strut allografts in cementless revision total hip arthroplasty. *Clin. Orthop.* **295,** 172–178.
98. Haddad, R., Duncan, C. P., Berry, D. J., Lewallen, D. G., Gross, A. E., and Chandler, H. P. (2002) Periprosthetic femoral fractures around well fixed implants: use of ortical onlay allografts with and without plates. *J. Bone Joint Surg.* **84A,** 945–950.
99. Allan, D., Lavoie, G., McDonald, S., et al. (1991) Proximal femoral allografts in revision hip arthroplasty. *J. Bone Joint Surg.* **73B,** 235–240.
100. Gross, A. and Hutchison, C. (1998) Proximal femoral allografts for reconstruction of bone stock in revision arthroplasty of the hip. *Orthop. Clin. N. Am.* **29,** 313–317.
101. Head, W., Berklacich, F., Malinin, T., et al. (1987) Proximal femoral allografts in revision total hip arthroplasty. *Clin. Orthop.* **225,** 22–36.
102. Head, W., Emerson, R. J., and Malinin, T. (1999) Structural bone grafting for femoral reconstruction. *Clin. Orthop.* **369,** 223–229.
103. Head, W., Malinin, T., and Berklacich, F. (1987) Freeze-dried proximal femur allografts in revision total hip arthroplasty. A preliminary report. *Clin. Orthop.* **215,** 109–121.
104. Haddad, F. (2000) Circumferential allograft replacement of the proximal femur. A critical analysis. *Clin. Orthop.* **371,** 98–107.
105. Gross, A., Hutchison, C., Alexeeff, M., et al. (1995) Proximal femoral allografts for reconstruction of bone stock in revision arthroplasty of the hip. *Clin. Orthop.* **319,** 151–158.
106. Blackley, H. R. L., Davis, A. M., Hutchison, C. R., et al. (2001) Proximal femoral allografts for reconstruction of bone stock in revision arthroplasty of the hip. A nine to fifteen year follow-up. *J. Bone Joint Surg.* **83A,** 346–354.
107. Kerboull, M. (1985) in *Arthroplastie totale de hanche* (Postel, M., Kerboull, M., Evrard, J., and Courpied, J. P., eds.), Springer-Verlag, New York, pp. 89–96.
108. Kerboull, M. (1996) in *Conférence d'enseignement de la SOFCOT Expansion Française,* Paris, pp. 1–17.
109. Martin, W. and Sutherland, C. (1993) Complications of proximal femoral allografts in revision total hip arthroplasty. *Clin. Orthop.* **295,** 161–167.
110. Stulberg, S. D. (2003) Bone loss in revision total knee arthroplasty: graft options and adjuncts. *J. Arthroplasty* **18 (3 Suppl 1),** 48–50.
111. Samuelson, K. (1988) Bone grafting and noncemented revision arthroplasty of the knee. *Clin. Orthop.* **226,** 93–101.
112. Görlich, Y., Lebek, S., and Reichel, H. (1999) Substitution of tibial bony defects with allogeneic and autogeneic cancellous bone: encouraging preliminary results in 18 knee replacements. *Arch. Orthop. Trauma Surg.* **119,** 220–222.
113. DeWaal Malefit, M., Van Kampen, A., and Slooff, T. (1995) Bone grafting in cemented knee replacement. 45 primary and secondary cases followed for 2-5 years. *Acta Orthop. Scand.* **66,** 325–328.
114. Van Loon, C., DeWall Malefit, M., Verdonschot, N., et al. (1999) Morcellized bone grafting compensates for femoral bone loss in revision total knee arthroplasty. An experimental study. *Biomaterials* **20,** 85–89.
115. Benjamin, J., Engh, G., Parsley, B., et al. (2001) Morselized bone grafting of defects in revision total knee arthroplasty. *Clin. Orthop.* **392,** 62–67.
116. Bradley, G. W. (2000) Revision total knee arthroplasty by impaction bone grafting. *Clin. Orthop.* **371,** 113–118.
117. Lonner, J. H., Lotke, P. A., Kim, J., et al. (2002) Impaction grafting and wire mesh for uncontained defects in revision knee arthroplasty. *Clin. Orthop.* **404,** 145–151.
118. Beharrie, A. W. and Nelson, C. L. (2003) Impaction bone-grafting in the treatment of a periprosthetic fracture of the tibia: a case report. *J. Bone Joint Surg.* **85A,** 703–707.
119. Whiteside, L. and Bicalho, P. (1998) Radiologic and histologic analysis of morcellized allograft in revision total knee replacement. *Clin. Orthop.* **357,** 149–156.
120. Hanssen, A. D. (2001) Bone-grafting for severe patellar bone loss during revision knee arthroplasty. *J. Bone Joint Surg.* **83A,** 171–176.
121. Mnaymneh, W., Emerson, R., Borja, F., et al. (1990) Massive allografts in salvage revisions of failed total knee arthroplasties. *Clin. Orthop.* **260,** 144–153.

122. Tsahakis, P., Beaver, W., and Brick, G. (1994) Technique and results of allograft reconstruction in revision total knee arthroplasty. *Clin. Orthop.* **303,** 86–94.
123. Engh, G., Herzwurm, P., and Parks, N. (1997) Treatment of major defects of bone with bulk allografts and stemmed components during total knee arthroplasty. *J. Bone Joint Surg.* **79A,** 1030–1039.
124. Mow, C. and Wiedel, J. (1996) Structural allografting in revision total knee arthroplasty. *J. Arthroplasty* **11,** 235–241.
125. Ghazavi, M., Stockley, I., Yee, G., et al. (1997) Reconstruction of massive bone defects with allograft in revision total knee arthroplasty. *J. Bone Joint Surg.* **79A,** 17–25.
126. Lindstrand, A., Hansson, U., Toksvig-Larsen, S., et al. (1999) Major bone transplantation in total knee arthroplasty: a 2- to 9-year radiostereometric analysis of tibial implant stability. *J. Arthroplasty* **14,** 144–148.
127. Clatworthy, M. G., Balance, J., Brick, G. W., et al. (2001) The use of structural allograft for uncontained defects in revision total knee arthroplasty. A minimum five-year review. *J. Bone Joint Surg.* **83A,** 404–411.
128. Parker, D. A., Dunbar, M. J., and Rorabeck, C. H. (2003) Extensor mechanism failure associated with total knee arthroplasty: prevention and management. *J. Am. Acad. Orthop. Surg.* **11,** 238–247.
129. Emerson, R. H. Jr., Head, W. C., and Malinin, T. I. (1994) Extensor mechanism reconstruction with an allograft after total knee arthroplasty. *Clin. Orthop.* **303,** 79–85.
130. Emerson, R. H. Jr., Head, W. C., and Malinin, T. I. (1990) Reconstruction of patellar tendon rupture after total knee arthroplasty with an extensor mechanism allograft. *Clin. Orthop.* **260,** 154–161.
131. Nazarian, D. G. and Booth, R. E. Jr. (1999) Extensor mechanism allografts in total knee arthroplasty. *Clin. Orthop.* **367,** 123–129.
132. Crossett, L. S., Sinha, R. K., Sechriest, V. F., et al. (2002) Reconstruction of a ruptured patellar tendon with achilles tendon allograft following total knee arthroplasty. *J. Bone Joint Surg.* **84A,** 1354–1361.
133. Barrack, R. L. and Lysons, T. (2000) Proximal tibia-extensor mechanism composite allograft for revision total knee arthroplasty with chronic patellar rupture. *Acta Orthop. Scand.* **71,** 419–421.

Table 1
Selected Effects of EF/EMF Stimuli on Growth Factor Release/Synthesis *in Vitro* and *in Vivo*

Authors	Year (ref.)	Method	Growth factor(s)	Experimental system
Fitzsimmons et al.	1992 *(19)*	CCEF	IGF-II	Osteoblast cultures
Ryaby et al.	1994 *(24)*	CMF	IGF-I/II	Rat fracture callus
Fitzsimmons et al.	1995 *(22)*	CMF	IGF-II	Osteoblast cultures
Zhuang et al.	1997 *(35)*	CCEF	TGF-β1	Osteoblast cultures
Bodamayli et al.	1998 *(36)*	PEMF	BMP-2, 4	Osteoblast cultures
Lohmann et al.	2000 *(32)*	PEMF	TGF-β	Osteoblast cultures
Guerkov et al.	2001 *(33)*	PEMF	TGF-β	Human nonunion cultures
Aaron et al.	2002 *(28)*	PEMF	TGF-β1	DBM bone induction model

to the recommended clinical use for treatment of nonunions. Increases in TGF-β1, collagen, and osteocalcin synthesis were noted with PEMF stimulation. This study was followed by the first assessment of the effect of PEMF on an osteocytic cell line, MLO-Y4 *(33)*. In these osteocytic cells, PEMF also showed upregulation of alkaline phosphatase and TGF-β1, with a decrease in connexin 43 protein. The most significant study from Boyan's group is the first use of human nonunion cells to assess the effects of EF/EMF *(34)*. Cells from both hypertrophic and atrophic nonunion tissues were assessed using the identical exposure conditions stated above. PEMF stimulated an increase in TGF-β1 in both hypertrophic cells at d 2 and in atrophic cells at d 4. The conclusion from these studies is that stimulation of growth factors is an important signaling event in PEMF interaction.

Is this upregulation of growth factor production a common denominator in the tissue-level mechanisms underlying all electrical and electromagnetic stimulation technologies? Recent work by the groups of Brighton *(35)* and Stevens *(36)* has supported this mechanism as a common underlying concept. The Brighton group, using CCEF, showed an increase in both TGF-β1 mRNA and protein in osteoblast cultures after CCEF exposure. Using specific inhibitors, these authors have provided data to suggest that CCEFs act through a calmodulin-dependent pathway. Stevens's group in the UK has shown upregulation of mRNA for bone morphogenetic proteins (BMP)-2 and -4 with PEMF in osteoblast cultures. The major limitation of the Stevens study on BMPs is the short duration of PEMF exposure in this study, because the clinical benefit of PEMF is believed to require 3–10 h/d of exposure. Therefore, the role of BMPs in the action of PEMF or any biophysical stimulation technique is not understood. However enticing, more work needs to be performed to fully understand the role of growth factors in transduction of biophysical stimuli and the clinical relevance.

The signal transduction mechanism underlying the effects of these various electrical and electromagnetic signals has been studied extensively by Brighton's group using the mouse MC3T3-E1 osteoblastic cell line *(37)*. In these studies, MC3T3-E1 cells were exposed to CCEF, CMF, or PEMF; DNA content significantly increased in all stimulation groups at various time points of exposure. The important result from this study was the observation that CCEF signaled through voltage-gated calcium channels, whereas the CMF and PEMF (both inductive coupling techniques) signaled through release of intracellular stores of Ca^{2+}. These results demonstrate that the common signal transduction pathway for these techniques is via calcium signaling, with the final pathway based on elevation of intracellular Ca^{2+} leading to an increase in activated cytoskeletal calmodulin.

A body of excellent work from several groups in Italy has demonstrated significant effects in both *in vivo* and clinical studies. The clinical studies are presented in the subsequent section on clinical studies; at this time a brief overview of the relevant *in vivo* studies will be provided. Cane, Cadossi, and colleagues have used a transcortical defect model for the past 10 yr to address basic histomorphometric and molecular aspects of EMF stimulation. These results have provided important insights, as this *in vivo* model is neither metabolic nor pathological in contrast to the osteopenic model systems

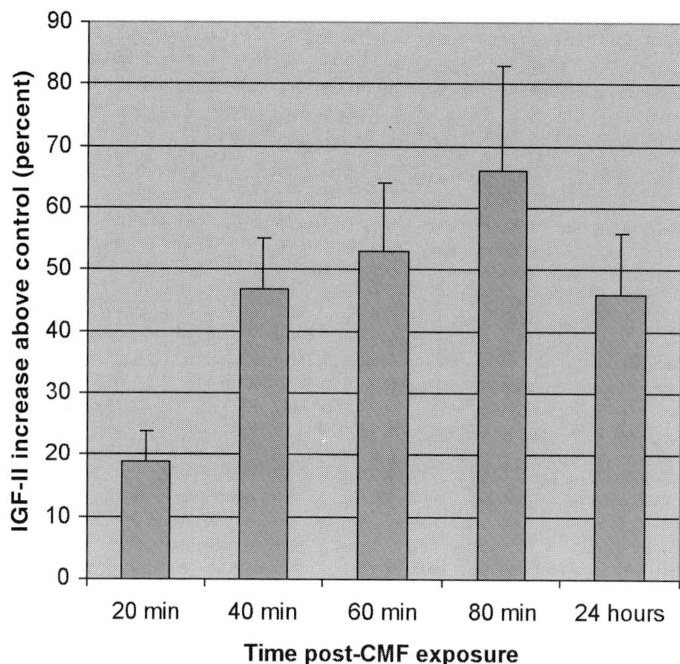

Fig. 1. Combined magnetic field stimulation of IGF-2 release in TE-85 osteoblast-like cells. IGF-2 concentration measured in culture media after CMF exposure *(21)*.

studied by other laboratories. The model used by Cane and coworkers is the bilateral cortical hole defect model in the metacarpal bones in horses, with quantitative histomorphometric methods employed to quantify differences between treated and control limbs. These authors used PEMF stimulation for 24 h/d, with sham-exposed contralateral limbs serving as the nonstimulated controls. In the first study, reported in 1991 *(38)*, PEMF-treated holes showed a statistically significant increase in the amount of new bone formation, ranging from 40% to 120% at 60 d of treatment in diaphyseal defects (**Fig. 2**), with more variable response observed in metaphyseal defects. The follow-up study *(39)* employed dynamic histomorphometric analyses and focused on the effects on osteoblast activity at 30 d of PEMF stimulation using tetracycline double-label technique. These results showed a significant increase in bone formation and mineral apposition rate with PEMF treatment. The authors concluded this effect was due to an increase in osteoblast activity. Caution should be applied to this interpretation, as no effort was reported to rule out any effect on osteoclastic coupling/activity. Recently, this same group reported on the ability of PEMF to stimulate osseointegration into hydroxyapatite implants in a rabbit model *(40)*. Significant increases in affinity index and microhardness values were observed with PEMF treatment, and these authors propose that PEMF may be useful for aiding osseointegration of implants in clinical applications.

Useful information has also been derived from in vivo studies on osteopenic animal models. Brighton et al. *(41,42)* have shown that a low-voltage, high-frequency, capacitively coupled electrical signal can prevent osteopenia due to both sciatic denervation and castration in rat osteopenia models. Skerry et al. *(43)* demonstrated inhibition of bone loss with pulsed electromagnetic fields in an ovariectomized canine model. Their conclusion was this effect was due to inhibition of resorption at the bone surface, not stimulation of new bone formation. Our own work has used the ovariectomized rat model to assess the effects of CMF exposure in reversing osteopenia. In **Fig. 3**, using synchrotron-based X-ray tomography we show that CMF can reverse bone loss due to the hypoestrogenemic state *(44)*, although not to the degree seen with intermittent PTH treatment, an anabolic stimulus.

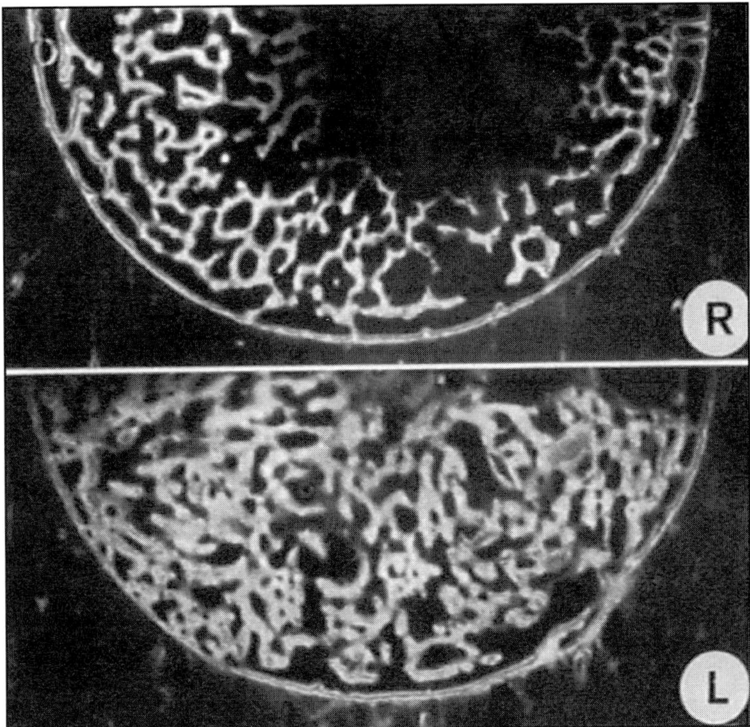

Fig. 2. PEMF stimulation of transcortical bone defect healing. R, control trancortical defect; L, PEMF treated transcortical defect. (From Cane et al. *[38]*, reproduced with permission from *J. Orthopaedic Research.*)

Rubin and McLeod *(45)*, using an avian ulna disuse osteopenia model, showed a significant increase in bone formation in the ulna when treated with pulsed electromagnetic fields. These authors have also used this model to address the frequency dependence of EMF effects, and found the maximal response was observed with low-frequency (15-Hz) sinusoidal EMF generating approx 10 μV/cm^2 in the tissue. These results led these authors to propose that the PEMF signals used clinically are extremely inefficient, and only a small component of the energy output is actually sensed by the healing tissue. Pilla et al. *(46)* had reached this same conclusion several years before, however, not based on in vivo studies using bone specific models.

Recent studies have also looked at clinically relevant models such as osteotomy gap healing and distraction osteogenesis models. Chao's lab has used a well-characterized canine osteotomy gap model to assess the effects of a new low-frequency PEMF signal *(47)*. The features of this signal are a daily treatment time of only 1 h/d, and stimulation was carried out for 4 wk beginning at 4 wk postsurgery. The results showed a statistically significant increase in periosteal callus area, as well as significant increases in torque to failure and torsional stiffness.

In summary, these studies emphasize that basic cellular biochemical control processes are affected by applied EF/EMF, and these cellular- and tissue-level effects provide support for further clinical applications.

Clinical Studies

Nonunion fracture repair has the longest history in the clinical application of electrical and electromagnetic fields. As stated above, various methods exist for EF/EMF stimulation, with specific signal parameters, device configurations, and daily prescribed treatment times. The specific types of technol-

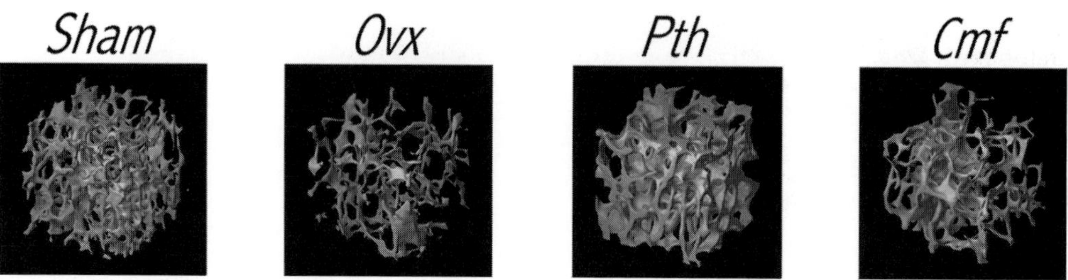

Fig. 3. Three-dimensional reconstructions of trabecular bone cubes taken from the rat proximal tibial metaphysis at 9 μm resolution by synchrotron-based X-ray microtomography *(26)*. Sham, control; Ovx, ovariectomized; Pth, parathyroid hormone; Cmf, combined magnetic field. (Color illustration in insert following p. 212.)

ogies, commonly referred to as bone growth stimulators, have been recently reviewed *(12)* and are summarized here. Three different methods of EF/EMF bone growth stimulation are presently FDA-approved for use in the United States. These are capacitive coupling stimulation using electrodes placed on the skin (noninvasive, manufactured by Biolectron/EBI), direct current stimulation using implanted electrodes (invasive, EBI), and electromagnetic stimulation by inductive coupling using time-varying magnetic fields (noninvasive). The latter category has two different technologies currently FDA-approved for clinical applications: pulsed electromagnetic fields (PEMF, EBI and Orthofix) and combined magnetic fields (CMF, OrthoLogic, IDJ, Regentek).

Evaluation of the clinical benefit of these devices contributed greatly to subsequent clinical trial design in fracture healing. Precedents set by these studies on nonunion treatment include large, multicenter, prospective clinical trials; the use of blinded radiographic panel assessment; monitoring of nonunions for 3 mo prior to study enrollment, ensuring that fracture healing had ceased; and no surgical intervention for 3 mo prior to treatment *(12)*.

The first system developed for clinical treatment of nonunions was direct current stimulation with implanted electrodes. This technique produces a localized electrical current (E field) between electrodes inserted at the fracture site, and is predominantly used clinically today for augmentation of spine fusion. This technique was developed concurrently by Friedenberg and Brighton in the United States *(48)* and Patterson *(49)* in Australia. The relative success rates for nonunion treatment in these prospective clinical studies ranged from 78% to 86%, respectively. Reasons for the limited clinical acceptance of the implanted technology for nonunion treatment were the subsequent availability of noninvasive methods of EF/EMF treatment, limited electric field exposure metrics, and complication rate. The physical mechanism of direct current stimulation is the topic of an excellent monograph by Black *(50)*.

The first noninvasive system approved by the FDA used pulsed electromagnetic fields. This technique was developed by Bassett, Pilla, and Ryaby (the author's father) *(51,52)*, and uses an external coil to produce a complex asymmetric signal of pulses repeating at 15 Hz. The clinical studies demonstrating efficacy of PEMF were conducted by Bassett's group, with follow-up studies by Heckman et al. *(53)*. In one prospective series by Bassett *(54)*, 127 tibial diaphyseal delayed unions or nonunions were exposed to PEMF for 10 h/d. At the study conclusion, 87% were noted to be healed, with a median healing time of 5.2 mo. No long-term follow-up data were provided in this study, therefore the absolute success rates are not known. A follow-up, multicenter, prospective study by Heckman et al. *(53)* showed a success rate of 64% in a series of 149 patients with PEMF treatment. Stratification of the data showed that in the responsive population, 85% healed within 3–6 mo after treatment initiation. Clinical studies using PEMF are the subject of comprehensive reviews by Bassett *(55)*, Gossling et al. *(56)*, and Hinsenkamp *(57)*.

Noninvasive capacitive coupling (CCEF), as developed by Brighton and Pollack *(58)*, uses disk electrodes coupled to the skin via a conductive gel to produce a broad uniform E field in the fracture

Table 2
**Randomized, Double-Blind, Placebo-Controlled Trials Performed
to Assess Safety and Efficacy of Biophysical Stimulation Techniques (Selected)**

Authors	Year (ref.)	Method	Indication
Borsalino et al.	1988 *(62)*	PEMF	Femoral Osteotomies
Sharrard	1990 *(63)*	PEMF	Delayed Union
Mooney	1990 *(72)*	PEMF	Spine Fusion
Mammi et al.	1993 *(64)*	PEMF	Tibial Osteotomies
Scott and King	1994 *(65)*	CCEF	Non-union
Heckman et al.	1994 *(104)*	Ultrasound	Tibial Fractures
Kristansen et al.	1997 *(105)*	Ultrasound	Distal Radius Fractures
Goodwin et al.	1999 *(73)*	CCEF	Spine Fusion
Linovitz et al.	2002 *(75)*	CMF	Spine Fusion
Simonis et al.	2003 *(66)*	PEMF	Tibial Non-union

site. The device produces a 60-kHz symmetrical sine wave, which produces a 5-V peak-to-peak current with approximately 7 µA root mean square at the skin level. The daily treatment time recommended is 24 h/d. The first nonunion study reported an overall efficacy of 77%, with a mean time to healing of 23 wk in a series of 22 nonunions. This study included 17 recalcitrant nonunions, which had failed to heal with bone graft or prior electrical stimulation (the technology was not specified).

The newest technique, combined magnetic fields, was first evaluated as a treatment for fracture non-unions in the mid-1990s *(59)*. The CMF device (an inductive-based method) employs an external pair of coils oriented parallel to one another, which produce two parallel low-energy magnetic fields. The alternating magnetic field is a sinusoidal wave of 76.6 Hz and amplitude of 40 µT peak to peak, with the static field set at 20 µT *(12)*. The study group consisted of 84 nonunions. Nonunions were defined by strict entrance criteria, which was a minimum of 9 mo postinjury and no surgical treatment for 3 mo prior to enrollment. A blinded radiographic review panel verified the presence of nonunion by analyzing radiographs taken a minimum of 3 mo apart prior to study enrollment. The protocol provided for one 30-min treatment dosage per day with the CMF device until healed or for a maximum of 9 mo. The results noted by the blinded radiographic review panel showed 51 nonunions healed (61%) and 33 nonunions did not heal (39%), with a mean healing time of 5.8 mo. Stratification of tibial nonunions demonstrated healing in 31 of 41 nonunions, representing an efficacy of 75.6%. Nonunions that were less than or equal to 24 mo postinjury also healed at 75%, with long-standing nonunions ranging up to 10 yr more resistant to CMF stimulation. The most important feature of the study was the 2-yr posttreatment follow-up. All nonunions determined healed at the end of the study remained healed at the 2-yr posttreatment follow-up. The conclusion drawn from this study was that CMF demonstrates clinical efficacy with a 30-min treatment dosage for the treatment of fracture nonunions.

Randomized, Double-Blind Studies

The use of EF/EMF for nonunion treatment does not have universal acceptance in the orthopedic community. The following double-blind, placebo-controlled trials were designed to use the most rigorous criteria possible for assessment of the beneficial effects of EF/EMF on bone healing (**Table 1**). Laupacis et al. *(60)* and Freedman et al. *(61)* have recently provided excellent reviews on the use of randomized trials in orthopedics. The reader is referred to these articles for information on clinical trial design. These double-blind studies of EF/EMF in orthopedic fracture and bone repair have been recently reviewed *(12)* and are summarized here and in **Table 2**.

The first successful prospective, randomized, double-blind, placebo-controlled trial was performed by Borsalino et al. *(62)* in 1988. These investigators studied the effects of PEMF on 31 femoral intertrochanteric osteotomies with degenerative joint disease of the hip. Patients were randomized to receive either an active or placebo electromagnetic device for 8 h/d over a period of 3 mo. Radiographic assessment was performed by a blinded radiographic review panel 40 and 90 d postosteotomy. In active-device patients, increases in bone density and trabecular bridging of 45% and 64% were observed, respectively, as compared to the control (placebo) group.

The first placebo-controlled, double-blind, randomized trial on fracture healing was reported by Sharrard in 1990 *(63)*. This study assessed the effect of PEMF on healing of tibial delayed unions. Fifty-one patients were randomized to receive either active or placebo devices with a treatment time of 12 h/d for 12 wk. The orthopedic surgeon and a musculoskeletal radiologist performed blinded radiographic assessment. Results of this study showed a significant effect of the active device on healing, with the surgeon's assessment more favorable than the radiologist's. According to the orthopedic surgeon, 45% of the active-device patients healed, compared to only 14% of the placebo patients, a statistically significant difference ($p < 0.02$). The conclusion drawn from this study, that progress to union is significantly affected by electromagnetic stimulation, is valid and supported by the data.

Other double-blind, placebo-controlled studies have included the study by Mammi et al. *(64)*, who reported on the treatment of tibial osteotomies with PEMF for degenerative arthrosis. A blinded panel devised a progressive scale of stage one to four healing based on consecutive serial radiographs. The results showed that use of the active device for 8 h/d increased the percentage of patients at late-stage healing (3/4) to 72%, compared to 26% in the placebo controls ($p < 0.006$).

Finally, the first prospective, randomized, double-blind, placebo-controlled study on nonunions was reported by Scott and King in the United Kingdom *(65)*. Capacitively coupled electrical fields were used on long-bone nonunions for 24 h/d for a maximum of 6 mo. Sixty percent of the active-device patients healed, with a mean time of 21 wk, compared to none in the placebo-device group, a statistically significant effect by Fisher's exact test ($p < 0.004$). Very recently, another double-blind, randomized clinical trial on electrical treatment of tibial nonunions has been published by Simonis et al. *(66)*. Each nonunion was surgically treated with an oblique fibular osteotomy and unilateral external fixator, and randomized to either an active or placebo EMF device. The EMF signal used in this study is not FDA approved in the United States. Thirty-four nonunions were studies in this trial, with a high proportion of smokers (21/34) and atrophic nonunions (16/34). Eighty-nine percent of the active group showed bony union, compared to 50% in the placebo group, which was statistically significant, and smoking was shown to impair healing of nonunions in both placebo- and active-treated patients.

Double-blind clinical trials have also demonstrated therapeutic efficacy of electrical and electromagnetic fields for treatment of spinal fusions. This is the subject of a recent meta-analysis *(67)*, and these spine fusion clinical trials will be comprehensively summarized in the next section. In summary, these trials on surgical bone repair and nonunions have all demonstrated effectiveness of various forms of electrical and electromagnetic stimulation devices. Additional indications for these devices remains to be further defined.

Spine Fusion

Three randomized, double-blind, placebo-controlled clinical trials have been performed addressing the use of bone growth stimulation technologies as an adjunct to spine fusion. **Table 3** provides a comparison of the different technologies that are FDA-approved for use in the United States. These trials form the basis for FDA approval of these technologies as adjunctive stimulation devices for the enhancement of spine fusion.

Initially, the use of bone growth stimulation in spine fusion was limited to surgically implantable direct-current stimulation devices, as reported by Dwyer in 1974 *(68)*. Following this report was the randomized study by Kane in 1989, who utilized an implantable DC stimulator *(69)*. This device uses electrodes that are surgically placed lateral to the fusion site and powered by a battery pack to deliver

Table 3
Comparison of FDA-Approved Technologies for Stimulation of Spine Fusion

	EBI SP F II™ stimulator	ORTHOFIX SPINAL STIM LITE™	BIOLECTRON/ EBI SPINALPAK™	ORTHOLOGIC SPINALOGIC™
Technology	Direct Current (DC)	Pulsed Electromagnetic Fields (PEMF)	Capacitively Coupled Electrical Fields (CCEF)	Combined Magnetic Fields (CMF)
Reference	69	72	73	75
Device Type	Implanted Electrodes/ Battery, Invasive	External, Non-invasive	External, Non-invasive	External, Non-invasive
Percent Change	27	18	20	21
Active	81	83	85	64
Placebo	54*	65	65	43
Clinical Trial	Randomized, *not placebo controlled	Double-blind, placebo controlled	Double-blind, placebo controlled	Double-blind, placebo controlled
Study Population	Instrumented/ non-instrumented fusions	Instrumented/ non-instrumented fusions	Instrumented/ non-instrumented fusions	Non-instrumented fusions
Study Population Age (years)	N/A	38	45	58
FDA Approval Date	1987	1990	1999	1999
Daily Treatment Time	Continuous	Minimum 2 hours	Recommended 24 hours	30 minutes

a current of 20 μA/cm^2. There were three components to this report. The most important component was the randomized trial, which was a small trial compared to the standards of today but was precedent-setting in the field of spine fusion. This study comprised 59 total patients, both male and female, 28 control and 31 active. The stimulated group healed with a percentage of 81%, compared to 54% in the control group, a statistically significant increase. The major caveat in this study was the lack of a placebo control, therefore this was not a blinded study. Second, no discussion of the fusion procedure or use of instrumentation was provided. Third, the study had a high dropout rate; only 59 patients out of 99 enrolled patients were included in the data analysis. This was ascribed to not filling the block randomization of four patients at each site. The other two components of this trial were a retrospective and prospective component. The retrospective component studied the effect of DC stimulation on 82 patients in comparison to a historical cohort of 159 patients of comparable diagnosis. The effect of DC showed an overall success rate of 92%, compared to 81% in the nonstimulated group. The author concluded that the implanted DC stimulation technique is a useful adjunct to the spine fusion process. Meril *(70)* reported on the use of DC in anterior and posterior interbody fusion procedures. The overall results were higher in stimulated patients, 93%, compared to 75% for the control. However, the study design was not prospective and did not use a placebo control. Finally, a recent study by Tejano et al. *(71)* showed a fusion percentage of 92% in their series of 118 patients in a prospective trial. The strengths of this study were (1) the use of no instrumentation, (2) all patients received autograft, and (3) the long-term follow-up of these patients. The limitation of this study was the lack of placebo control, making it difficult to assess the true effect of the adjunctive stimulation.

The first use of a noninvasive electromagnetic technology was the study by Mooney, who reported on the use of pulsed electromagnetic fields for stimulation of interbody fusions in 1990 (Orthofix

SpinalStim) *(72)*. This was a multicenter, prospective, placebo-controlled, randomized double-blind trial, and analysis was ultimately performed on 195 patients with a mean age of 38 yr. Patients were fitted with electromagnetic coils in a brace and instructed to use the electromagnetic device for a minimum of 8 h/d for 12 mo. Patients and surgeons were unaware of whether the brace was an active or placebo device. Two additional strengths of this study were the use of a confirmatory reading of fusion success by a blinded radiologist, and for a two-level fusion both levels had to be fused to be a success. The data were stratified into consistent (≥8 h/d) and inconsistent users (<4 h/d). In the 117 patients who were consistent users, the active-device patients achieved a fusion success rate of 92%, compared to the placebo success rate of 68%, a statistically significant difference. Patients who used the device inconsistently had the same success rate in the active and placebo groups, 65% and 61%, respectively. These results were the first to show a dose response for noninvasive electromagnetic treatment. Some limitations of the study include no control for use of instrumentation, daily treatment time, or type of graft (autograft or allograft).

The second noninvasive technology for stimulation of spine fusion is capacitively coupled electrical fields (CCEF, Biolectron, SpinalPak), as reported by Goodwin in 1999 *(73)*. The study design was a multicenter, prospective, randomized, double-blind, placebo-controlled trial that reported on 179 patients, with a mean age of 43 yr. Patients were randomized to receive either an active or a placebo device within 3 wk postsurgery. The daily device treatment time was 24 h/d using two electrodes placed laterally 10 cm apart at the fusion site, connected to the signal generator. This study used a blinded radiographic and clinical review, and the study end point was 9 mo. The results showed that 85% of the active-device patients fused, compared to 65% of the placebo patients, a statistically significant difference. Posterolateral fusion patients comprised this group, as the anterio- and posterio-lateral interbody fusion groups did not reach statistical significance due to low patient numbers. Limitations of the study included no control for use of instrumentation, daily treatment time (average patient use was 16 h/d), or type of graft (autograft, allograft, or a combination). The one puzzling outcome was that the noninstrumented patient population fused at a higher success rate than the instrumented patients, which was unexpected in reference to the literature *(74)*.

The third noninvasive technology for adjunctive stimulation of spine fusion is combined magnetic fields, as reported by Linovitz et al. *(75)*. The clinical study conducted was a prospective, randomized, double-blind, placebo-controlled trial on primary, uninstrumented lumbar spine fusion. Patients had one- or two-level fusions (between L3 and S1) without instrumentation, with either autograft alone or in combination with allograft. The combined magnetic field device uses a single posterior coil, centered over the fusion site, with one 30-min treatment per day for 9 mo. The primary end point was assessment of fusion at 9 mo, based on radiographic evaluation by a blinded panel consisting of the treating physician, a musculoskeletal radiologist, and a spine surgeon. The difference in this panel evaluation was that the treating surgeon's assessment of fusion could be overruled by the blinded panel. This is the largest study to date, with 201 patients evaluated. Among all active-device patients, 64% had healed at 9 mo, compared to 43% of placebo-device patients, a statistically significant difference. This was the first study to stratify by gender. The results showed 67% of active-device females fused, compared to 35% of placebo-device females ($p = 0.001$ by Fisher's exact test). For the overall patient population, repeated-measures analysis of fusion outcomes (by generalized estimating equations [GEE]) showed a main effect of treatment, favoring the active treatment ($p = 0.030$). For the first time, GEE analysis also showed a significant time by treatment interaction ($p = 0.024$), indicating acceleration of healing. The lower overall fusion rates in this study may be due to (1) the high-risk patient group, with an average age of 57 yr, (2) the use of noninstrumented technique with posterolateral fusion only, and (3) reliance on extremely critical blinded radiographic and clinical criteria for fusion assessment, without surgical confirmation.

The major differences between the implantable and noninvasive electrical and electromagnetic technologies are the need for surgical implantation and explantation and patient compliance with use of the device. A comparison is provided in **Table 3**. The implantable DC stimulator does not pose a compliance issue if the electrodes are not in contact with internal fixation. The noninvasive devices require the patient to comply with the treatment protocol. The combined magnetic field technology requires only 30 min of treatment per day. Pulsed electromagnetic fields (Mooney study, *[72]*) required a minimum of 8 h/d of treatment to show a statistically significant effect. In the study of capacitively coupled field stimulation *(73)* by Goodwin et al., patients used the device approximately 16 h/d. Clearly it is difficult to predict patient compliance in daily clinical practice based on these extended daily treatment times.

To date there are no studies on any of the EF/EMF technologies with fusion cages, nor are there studies comparing the benefits of the devices using an outcome instrument such as the Oswestry score. However, the above studies do provide strong support for the adjunctive use of EF/EMF in spine fusion patients.

Registry Data

Registry data have been proposed by the author in a recent review article *(12)* to provide an important assessment of noninvasive bone growth stimulation technologies, as these data demonstrate efficacy at the practitioner/community level. However, registry data are not available for all the biophysical stimulation technologies. As an example, clinical outcome success rate for CMF (OrthoLogic 1000) was calculated by performing a prospective analysis of data provided by treating physicians. The information collected included age, diagnosis, type of fracture, period of time from fracture to initiation of treatment, duration of treatment, nonunion healing success rates, and time to healing. All patients were treated with the CMF device for 30 min/d. Outcome was determined by the treating physician based on radiographic and clinical evaluation, with success criteria limited to healed (success) and not healed (failure). Data were tabulated for each fracture site, and respective outcome rates were calculated.

The results are based on a total of 4100 patients, on whom complete follow-up was obtained on 2370 patients (58%). One limitation of a registry process is securing the compliance of both the patient and physician, as no clinical monitoring is performed in a registry study, thereby leaving conclusive determination of efficacy open to question. Of the 2370 patients, the overall results showed that 75% of the nonunion fractures healed in an average of 4.9 mo. The median time from injury to the initiation of treatment for the total patient population was 6.3 mo. In **Table 4**, the outcome data is stratified by anatomical site.

As can be seen in **Table 4**, outcome varied from a low of 57.2% for the humerus to a high of 89.7% for the carpal metacarpal. The mean time to healing also showed a range of 3.4 mo for the phalanx to 6.4 mo for the femur. In the most prevalent nonunion sites (femur, metatarsal, and tibia), the percent success was 75.4%, matching the overall study mean success rates.

The registry data are plagued by several limitations besides the compliance rate. These limitations include no independent diagnosis of nonunion, no blinded radiographic review for determination of outcome, and no control of patients who received CMF treatment. With regard to the latter, only randomized trials in which patients are randomized to either receive CMF or surgery would uncover any underlying bias. These registry results do show, however, that the treatment of nonunions with CMF can yield a positive outcome of benefit to the patient. Additionally, there is a clear benefit to intervention with CMF in terms of both healing and time to heal.

ULTRASOUND STIMULATION

Scientific Basis

Ultrasound has found a wide range of applications in medicine, as both a diagnostic tool as well as a therapeutic modality. The clinical use of ultrasound for diagnostic imaging is well documented and has been extensively accepted worldwide. Therapeutic use of low-energy, nonthermal ultrasound for stimulation of bone repair is the subject of an excellent recent review *(76)*. Ultrasound is also widely

Table 4
Outcomes of CMF Treatment of Nonunions by Anatomical (Fracture) Site

By site (efficacy)	Healed no./total no. *n/N*	Outcome rate	Average healing time (mo)
ANKLE	110/145	75.9%	4.7
CARPAL NAVICULAR	154/218	70.6%	3.9
CARPAL/ METACARPAL	35/39	89.7%	5.3
CLAVICLE	79/114	69.3%	5.1
FEMUR	160/250	64.0%	6.4
FIBULA	58/68	85.3%	4.3
HUMERUS	103/180	57.2%	5.5
METATARSAL	408/477	85.5%	3.8
PHALANX (FINGER)	21/24	87.5%	3.4
PHALANX (TOE)	22/29	75.9%	3.7
RADIUS	81/96	84.4%	5.0
RADIUS/ULNA	14/17	82.4%	5.3
TARSAL	51/77	66.2%	4.3
TIBIA	285/372	76.6%	6.2
TIBIA/FIBULA	122/154	79.2%	5.8
ULNA	77/110	70.0%	5.0

used therapeutically for localized, site-specific treatment of musculoskeletal soft tissue injuries. In this mode, the mechanism of therapeutic action is assumed to be a localized, transient increase in temperature that leads to enhanced blood flow *(77)*. Recently, ultrasound has gained considerable attention for a new clinical therapeutic application, acceleration of the rate of bone fracture healing. In 1994, ultrasound was approved for treatment of fresh fracture healing by the FDA, with nonunion approval granted in 2000. This section reviews and summarizes current knowledge on the use of nonthermal ultrasound for stimulation of bone growth and repair.

Ultrasound is acoustic (mechanical) energy at frequencies above 18 kHz. Although the basic physics of acoustic propagation are well understood, the biophysical interaction of ultrasound with biological tissue is extremely complex. The nonthermal mechanisms of ultrasound stimulation are believed to be due to the forces associated with the applied ultrasound energy. These forces are mechanical forces associated with the acoustic wave and radiation pressure *(78)*, and associated acoustic streaming. It is accurate to state that on this level the effect of ultrasound is mechanical and differentiated from that of EF/EMF. One drawback to ultrasound that is not shared by EMF devices is that the ultrasonic wave is attenuated by soft tissue. For example, in a fracture where there is 1 cm of overlying soft tissue, the relative proportion of ultrasound intensity that *actually enters the bone* is approx 60% of its initial value *(79)*.

Ultrasound has also been demonstrated to have effects in various animal models. In two studies, pulsed and continuous ultrasound at a frequency of 3.5 MHz and intensity of 100 mW/cm^2 (SATA) was shown to promote soft tissue wound healing *(80)*. In 1953, one of the earliest reported animal studies on the effect of ultrasound on bone repair *(81)* demonstrated that treatment for 5 min/d for a period of 15 d at an intensity of 1.5 W/cm^2 accelerated healing compared to contralateral controls. Similar results were reported in the early 1970s by other investigators *(82,83)*. In one study, an intensity of 200 mW/cm^2 was shown to enhance the healing of fibular fractures in rabbits *(84)*. In a similar study, pulsed ultrasound at an intensity of 100 mW/cm^2 (SATA) and a daily exposure of 5 min was shown to enhance fracture healing in the rat fibula *(85)*. Duarte performed a study on rabbits using intensities of 50 and 57 mW/cm^2 (SATA) applied for 15 min daily *(86)*. Comparisons of the planar projected callus area as measured from a photograph of the dissected bone suggested that ultrasound stimulated fracture healing.

The results of a large study on fracture healing using rabbits were reported by Pilla et al. in 1990 *(87)*. In this study, a 30-mW/cm^2 (SATA), 1.5-MHz pulsed ultrasound signal was applied for 20 min daily to a highly reproducible fibular osteotomy model. This study was the first to utilize biomechanical torsion testing to assess the effects of ultrasound. Results demonstrated a statistically significant stimulation by the ultrasound exposure throughout the healing period, which was shown to result in an acceleration of healing by a factor of 1.7. A power-intensity dosimetry study was subsequently conducted using the same animal model *(88)*. In this study the ultrasound intensity varied from 1 to 45 mW/cm^2 for 20 min daily, and the biomechanical data analyzed on d 17 postfracture showed a statistically significant stimulation at all doses except the 1-mW/cm^2 intensity. An independently performed study using the same power intensity (30 mW/cm^2 SATA) as Pilla et al. *(87)* was carried out using the standard rat femoral fracture model *(89)* by Bolander et al. *(90)*. The results demonstrated a 22% increase in maximum torque in comparison to the contralateral limb at 21 d postfracture. Signals with both 0.5- and 1.5-MHz carrier frequencies were demonstrated to be effective. In a subsequent investigation, a higher power intensity was used (50 mW/cm^2 SATA) and demonstrated increases in maximum torque and torsional stiffness of 29% and 37%, respectively, at 21 d postfracture *(90)*.

What biological mechanisms underlie the observed effects on fracture healing? Early studies on ultrasound effects on fibroblasts in culture suspensions demonstrated structural changes and an increased rate of protein synthesis *(91)*. Using the same ultrasound signal as now approved by the FDA for fracture and nonunion repair, increased proteoglycan synthesis in chondrocytes *(92)* and adenylate cyclase activity in osteoblasts *(93)* have been reported. These authors proposed that the dynamic mechanical forces associated with the ultrasound input, perhaps through acoustic microstreaming, served as the physical stimulus for the observed response.

The most extensive work on cell- and tissue-level mechanisms has been performed by Bolander and colleagues *(94)*. These investigators performed quantitative analysis of mRNA after ultrasound exposure in the rat closed femoral fracture model. Statistically significant increases in aggrecan gene expression at d 7 postfracture were observed in the ultrasound-treated group. No effects of ultrasound were observed on α1(I) or (II) procollagen, bone gla protein, alkaline phosphatase, or TGF-β mRNA levels. The authors concluded that the differentiation of cartilage and cartilage hypertrophy is stimulated earlier with ultrasound exposure, accelerating the process of endochondral bone formation.

These authors completed a follow-up study on chondrocytes in culture *(95)*. Ultrasound was shown to increase aggrecan mRNA levels and proteoglycan synthesis after 3 d of exposure for 10 min/d. How does chondrogenesis get affected? This group has recently shown *(96)* that chondrocyte calcium signaling is directly affected by ultrasound, perhaps through acoustic microstreaming, which causes upregulation of aggrecan gene expression. An alternative view is that the production of growth factors in the mesenchymal microenvironment may be stimulated with ultrasonic exposure. Using different ultrasound exposure conditions, studies have shown both fibroblast growth factor and vascular endothelial growth factor production to be stimulated in cell culture *(97)*; also, in co-cultures of osteoblasts and endothelial cells, platelet-derived growth factor is stimulated *(98)*.

Ultrasound has also been investigated as an aide to distraction osteogenesis. Shimazaki et al. showed that ultrasound accelerated bone maturation in a rabbit distraction osteogenesis model *(99)*. Mayr, using a sheep model of distraction osteogenesis, also showed effects of ultrasound on maturation of the bone regenerate *(100)*. In contrast, Chao's group applied ultrasound during the consolidation phase of distraction osteogenesis, finding larger callus area, but no effect on mechanical strength or bone mineral density *(101)*.

In summary, the above cellular and animal studies suggest that ultrasound has the potential to be an important noninvasive stimulus for bone formation and repair. The low intensities shown to have a positive effect on fracture healing (e.g., as low as 5 mW/cm^2) suggest that a nonthermal mechanism of action is involved. However, it is not possible to rule out small, localized increases in temperature, which could have some effect on the healing process.

Clinical Applications

The use of ultrasound in the enhancement of fracture healing in patients was reported as early as 1959, when it was shown to have positive effects in the treatment of delayed unions in 181 patients *(102)*. Xavier and Duarte in Brazil also suggested that ultrasound could have a beneficial effect on fracture healing in a clinical setting *(103)*, and this study set the precedent for the development of the Exogen SAFHS technology. Recently, results of the two prospective double-blind, randomized, placebo-controlled clinical trials conducted in the United States and Israel have been published *(104, 105)*. These studies provided the basis for ultrasound FDA approval in the United States. It should be noted the ultrasound signal and treatment regimen are identical to that reported by Pilla et al. in their rabbit studies *(87,88)*. The first study, by Heckman et al. *(104)*, assessed the efficacy of pulsed ultrasound on fresh fracture healing in 67 closed or grade I open fractures. Both clinical and radiographic assessment demonstrated statistically significant acceleration of fresh fracture healing. Based on clinical assessment (pain and tenderness at the fracture site), the fractures healed 25% faster (86 d for the ultrasound-treated group, compared to 114 d for the placebo group; $p < 0.01$). Clinical and radiographic evaluation demonstrated a 38% reduction in healing time (96 d for the ultrasound group, compared to 154 d for the placebo group; $p < 0.001$). The major limitation of this study is that only closed and grade 1 tibial fractures were studied.

The second study, by Kristiansen et al. *(105)*, investigated ultrasound effects on healing of 61 fresh distal radius fractures using radiographic criteria. Ultrasound stimulation decreased the fracture healing time from 98 d in the placebo group, compared to 61 d in the ultrasound-treated group, a 37% decrease in healing time. Loss of reduction was also decreased in the ultrasound-treated group, from 43% to 20%. However, no data were provided on clinical assessment of healing, due to difficulties in comparing results between investigative sites. The major limitation of this study is the clinical literature describing Colles fracture healing times to be of the order of 45–60 d, essentially the same as the ultrasound-treated group. Therefore, in the clinical setting, the usefulness of ultrasound on distal radius fracture healing is unclear.

One recent prospective double-blind, randomized, placebo-controlled study by Larsson's group in Sweden investigated the use of ultrasound to enhance the healing of intramedullary fixed fractures *(106)*. Patients with tibial fractures treated by static locked intramedullary nailing were randomized to receive active or placebo devices. Active-device patients showed a slight increase in healing time as assessed by the radiologist. The orthopedist's assessment noted no difference between active and placebo patients. The authors concluded that there was no effect of ultrasound on fracture healing in this study. However, the key difference in this study was the limitation of ultrasound for the first 75 d of healing, in comparison to its continuous use throughout healing in the Heckman *(104)* and Kristiansen *(105)* studies.

In summary, ultrasound has been demonstrated to have a stimulatory effect on the rate of bone healing. Several *in vitro* animal and clinical studies have shown significant biological and biomechanical effects of ultrasound stimulation. Other potential clinical indications for noninvasive ultrasound technology exist. These include enhanced osseointegration of orthopedic prosthetic components *(107)* and enhancement of spinal fusion *(108)*. Further studies are necessary to elucidate the mechanism(s) by which ultrasound modulates the process of bone healing. It is reasonable to assume that such further studies will lead to more effective signals and a broader range of applications in clinical orthopedic practice.

ADDITIONAL CLINICAL APPLICATIONS

Two recent papers describe effects of biophysical stimulation techniques on Charcot neuroarthropathy. The first was a clinical trial that assessed the effect of CMF on treatment of Charcot neuroarthropathy by Hanft and colleagues *(109)*. The trial was a prospective, randomized pilot study on acute, phase 1 Charcot patients. The study design initially randomized 21 patients, 10 to the control group and 11 to the CMF group. Patients were followed weekly and treated until consolidation, with CMF treatment time of 30 min/d. Statistical analysis of this initial group revealed a statistical benefit for the CMF treatment group. Subsequently, an additional 10 patients were enrolled in the CMF treatment group. The final results showed that the mean time to consolidation in the control group was 23.2 ± 7.7 wk. In contrast, treatment with the CMF device decreased the time to consolidation to 11.1 ± 3.2 wk, a statistically significant difference ($p < 0.001$). There was no statistically significant difference in entry criteria between the control and CMF groups; and the authors concluded that the CMF treatment significantly accelerated the process of consolidation in this study. The second paper, a case report by Strauss and Gonya *(110)*, described the effect of ultrasound on ankle arthodesis in two patients with severe Charcot neuroarthropathy. Both patients healed after treatment with ultrasound, demonstrating that even these difficult conditions may be amenable to treatment with these biophysical techniques.

Well-conducted trials have also been performed on venous ulcer healing in humans *(111,112)*. The current FDA approval status for this indication is not known at the time of this writing.

DISCUSSION AND CONCLUSION

There remain many open, unanswered questions in the clinical applications of electrical, electromagnetic, and ultrasonic fields. When are these biophysical technologies indicated? Under what conditions do these technologies offer patient benefit? Are there subgroups of patients (based on, i.e., age, gender, fracture type) that benefit more than others? When using these biophysical techniques for primary treatment (i.e., nonunions, pseudoarthosis of spine), how do the outcomes compare to those of standard surgical procedures?

To answer these remaining questions, several approaches could be widely used to design future clinical studies. The first would be to conduct additional double-blind clinical trials to determine if biophysical stimulation can affect the healing rate and outcome for the intended orthopedic indication. Second, outcome studies could be performed, randomizing patients in two- or three-arm clinical trials comparing different treatment regimes to biophysical stimulation. Third, well-designed registry studies may be useful in expansion of clinical indications for which there already exist FDA-approved indications. For example, this might include an expansion of indication of EF/EMF techniques to all spine fusions, not the current limitation of lumbar fusion only.

In conclusion, electrical, electromagnetic, and ultrasonic devices have been demonstrated to positively affect the healing process in fresh fractures, delayed and nonunions, osteotomies, and spine fusion. These outcomes have been validated by well-designed and statistically powered double-blind clinical trials. The FDA-approved indications for these biophysical stimulation devices are limited at present to these indications. Based on these findings, biophysical stimulation technologies provide an additional arm to current treatment management strategies for these conditions. Future delineation of additional clinical indication(s) for musculoskeletal conditions awaits further basic scientific, pre-clinical, and clinical research.

REFERENCES

1. Fukada, E. and Yasuda, I. (1957) On the piezoelectric effect of bone. *J. Phys. Soc. Japan* **12,** 121–128.
2. Bassett, C. A. L. and Becker, R. O. (1962) Generation of electric potentials in bone in response to mechanical stress. *Science* **137,** 1063–1064.

 3. Friedenberg, Z. B. and Brighton, C. T. (1966) Bioelectric potentials in bone. *J. Bone Joint Surg.* **48A,** 915–923.
 4. Shamos, M. H. and Lavine, L. S. (1967) Piezoelectricity as a fundamental property of biological tissues. *Nature* **212,** 267–268.
 5. Williams, W. S. and Perletz, L. (1971) P-n junctions and the piezoelectric response of bone. *Nat. New. Biol.* **233,** 58–59.
 6. Anderson, J. C. and Eriksson, C. (1968) Electrical properties of wet collagen. *Nature* **227,** 166–168.
 7. Bassett, C. A. L. and Pawluk, R. J. (1974) Electrical behavior of cartilage during loading. *Science* **814,** 575–577.
 8. Grodzinsky, A. J., Lipshitz, H., and Glimcher, M. J. (1978) Electromechanical properties of articular cartilage during compression and stress relaxation. *Nature* **275,** 448–450.
 9. Duncan, R. L. and Turner, C. H. (1995) Mechanotransduction and the functional response of bone to mechanical strain. *Calcif. Tissue Int.* **57,** 344–358.
10. Luben, R. A. (1991) Effects of low-energy electromagnetic fields (pulsed and DC) on membrane signal transduction processes in biological systems. *Health Phys.* **61,** 15–28.
11. Pilla, A. A. and Markov, M. S. (1994) Bioeffects of weak electromagnetic fields. *Rev. Environ. Health* **10,** 155–169.
12. Ryaby, J. T. (1998) Clinical effects of electromagnetic and electrical fields on fracture healing. *Clin. Orthop. Rel. Res.* **355S,** S205–S215.
13. Aaron, R. K., Lennox, D., Bunce, G. E., and Ebert, T. (1989) The conservative treatment of osteonecrosis of the femoral head. A comparison of core decompression and pulsing electromagnetic fields. *Clin. Orthop. Rel. Res.* **249,** 209–218.
14. Steinberg, M. E., Brighton, C. T., Corces, A., et al. (1989) Osteonecrosis of the femoral head. Results of core decompression and grafting with and without electrical stimulation. *Clin. Orthop. Rel. Res.* **249,** 199–208.
15. Binder, A., Parr, G., Hazelman, B., and Fitton-Jackson, S. (1984) Pulsed electromagnetic field therapy of persistent rotator cuff tendinitis: a double blind controlled assessment. *Lancet* **1(8379),** 695–697.
16. Zizik, T. M., Hoffman, K. C., Holt, P. A., et al. (1995) The treatment of osteoarthritis of the knee with pulsed electrical stimulation. *J. Rheumat.* **22,** 1757–1761.
17. Otter, M. W., McLeod, K. J., and Rubin, C. T. (1988) Effects of electromagnetic fields in experimental fracture repair. *Clin. Orthop. Rel. Res.* **355S,** 90–104.
18. McLeod, B. R. and Liboff, A. R. (1987) Cyclotron resonance in cell membranes: the theory of the mechanism, in *Mechanistic Approaches to Interactions of Electromagnetic Fields with Living Systems* (Blank, M. J. and Findl, E., eds.), Plenum, New York, pp. 97–108.
19. Fitzsimmons, R. J., Strong, D., Mohan, S., and Baylink, D. J. (1992) Low amplitude, low frequency electric field-stimulated bone cell proliferation may in part be mediated by increased IGF-II release. *J. Cell. Physiol.* **150,** 84–89.
20. Fitzsimmons, R. J., Ryaby, J. T., Magee, F. P., and Baylink, D. J. (1994) Combined magnetic fields increase net calcium flux in bone cells. *Calcif. Tissue Int.* **55,** 376–380.
21. Fitzsimmons, R. J., Baylink, D. J., Ryaby, J. T., and Magee, F. P. (1993) EMF-stimulated bone cell proliferation, in *Electricity and Magnetism in Biology and Medicine* (Blank, M. J., ed.), San Francisco Press, San Francisco, pp. 899–902.
22. Fitzsimmons, R. J., Ryaby, J. T., Mohan, S., Magee, F. P., and Baylink, D. J. (1995) Combined magnetic fields increase IGF-II in TE-85 human bone cell cultures. *Endocrinology* **136,** 3100–3106.
23. Fitzsimmons, R. J., Ryaby, J. T., Magee, F. P., and Baylink, D. J. (1995) IGF-II receptor number is increased in TE-85 cells by low-amplitude, low-frequency combined magnetic field (CMF) exposure. *J. Bone Miner. Res.* **10,** 812–819.
24. Ryaby, J. T., Fitzsimmons, R. J., Khin, N. A., et al. (1994) The role of insulin-like growth factor in magnetic field regulation of bone formation. *Bioelectrochem. Bioenerg.* **35,** 87–91.
25. Diebert, M. C., McLeod, B. R., Smith, S. D., and Liboff, A. R. (1994) Ion resonance electromagnetic field stimulation of fracture healing in rabbits with a fibular ostectomy. *J. Orthop. Res.* **12,** 878–885.
26. Ryaby, J. T., Magee, F. P., Haupt, D. L., and Kinney, J. H. (1996) Reversal of osteopenia in ovariectomized rats with combined magnetic fields as assessed by x-ray tomographic microscopy. *J. Bone Miner. Res.* **11,** S564.
27. Ryaby, J. T., Cai, F. F., and DiDonato, J. A. (1997) Combined magnetic fields inhibit IL-1 and TNF-dependent NF-κB activation in osteoblast-like cells. *Trans. Orthop. Res. Soc.* **22,** 180.
28. Aaron, R. K., Wang, S., and Ciombor, D. M. (2002) Upregulation of basal TGFβ1 levels by EMF coincident with chondrogenesis—implications for skeletal repair and tissue engineering. *J. Orthop. Res.* **20,** 233–240.
29. Aaron, R. K. and Ciombor, D. M. (1996) Acceleration of experimental endochondral ossification by biophysical stimulation of the progenitor cell pool. *J. Orthop. Res.* **14,** 582–589.
30. Ciombor, D. M., Lester, G., Aaron, R. K., Neame, P., and Caterson, B. (2002) Low frequency EMF regulates chondrocyte differentiation and expression of matrix proteins. *J. Orthop. Res.* **20,** 40–50.
31. Aaron, R. K., Ciombor, D. M., and Jolly, G. (1989) Stimulation of experimental endochondral ossification by low-energy pulsing electromagnetic fields. *J. Bone Miner. Res.* **4,** 227–233.
32. Lohmann, C. H., Schwartz, Z., Liu, Y., et al. (2000) Pulsed electromagnetic field stimulation of MG63 osteoblast-like cells affects differentiation and local factor production. *J. Orthop. Res.* **18,** 637–646.
33. Lohmann, C. H., Schwartz, Z., Liu, Y., et al. (2003) Pulsed electromagnetic fields affect phenotype and connexin 43 protein expression in MLO-Y4 osteocyte-like cells and ROS 17/2.8 osteoblast-like cells. *J. Orthop. Res.* **21,** 326–334.

34. Guerkov, H. H., Lohmann, C. H., Liu, Y., et al. (2001) Pulsed electromagnetic fields increase growth factor release by nonunion cells. *Clin. Orthop. Rel. Res.* **180,** 265–279.

35. Zhuang, H., Wang, W., Seldes, R. M., Tahernia, A. D., Fan, H., and Brighton, C. T. (1997) Electrical stimulation induces the level of TGF-B1 mRNA in osteoblastic cells by a mechanism involving calcium/calmodulin pathway. *Biochem. Biophys. Res. Commun.* **237,** 225–229.

36. Bodamyali, T., Bhatt, B., Hughes, F. J., et al. (1998) Pulsed electromagnetic fields simultaneously induce osteogenesis and upregulate transcription of bone morphogenetic proteins 2 and 4 in rat osteoblasts in vitro. *Biochem. Biophys. Res. Commun.* **250,** 458–461.

37. Brighton, C. T., Wang, W., Seldes, R., Zhang, G., and Pollack, S. R. (2001) Signal transduction in electrically stimulated bone cells. *J. Bone Joint Surg.* **83A,** 1514–1523.

38. Cane, V., Botti, P., Farnetti, P., and Soana, S. (1991) Electromagnetic stimulation of bone repair: a histomorphometric study. *J. Orthop. Res.* **9,** 908–917.

39. Cane, V., Botti, P., and Soana, S. (1993) Pulsed magnetic fields improve osteoblast activity during the repair of an experimental osseous defect. *J. Orthop. Res.* **11,** 664–670.

40. Fini, M., Cadossi, R., Cane, V., et al. (2002) The effect of pulsed electromagnetic fields on the osseointegration of hydroxyapatite implants in cancellous bone: a morphologic and microstructural in vivo study. *J. Orthop. Res.* **20,** 756–763.

41. Brighton, C. T., Katz, M. J., Goll, S. R., Nichols, C. E., 3rd, and Pollack, S. R. (1985) Prevention and treatment of sciatic denervation disuse osteoporosis in the rat tibia with capacitively coupled electrical stimulation. *Bone* **6,** 87–97.

42. Brighton, C. T., Luessenhop, C. P., Pollack, S. R., Steinberg, D. R., Petrik, M. E., and Kaplan, F. S. (1989) Treatment of castration induced osteoporosis by a capacitively coupled electrical signal in rat vertebrae. *J. Bone Joint Surg.* **71A,** 228–236.

43. Skerry, T. M., Pead, M. J., and Lanyon, L. E. (1991) Modulation of bone loss during disuse by pulsed electromagnetic fields. *J. Orthop. Res.* **9,** 600–608.

44. Ryaby, J. T., Haupt, D. L., and Kinney, J. H. (1996) Reversal of osteopenia in ovariectomized rats with combined magnetic fields as assessed by x-ray tomographic microscopy. *J. Bone Miner. Res.* **11,** S231.

45. McLeod, K. J. and Rubin, C. T. (1992) The effect of low-frequency electrical fields on osteogenesis. *J. Bone Joint Surg.* **74A,** 920–929.

46. Pilla, A. A., Kaufman, J. J., and Ryaby, J. T. (1987) Electrochemical kinetics at the cell membrane: the physiochemical link for electromagnetic bioeffects, in *Mechanistic Approaches to Interactions of Electric and Electromagnetic Fields with Living Systems* (Blank, M. and Findl, E., eds.), Plenum, New York, pp. 39–53.

47. Inoue, N., Ohnishi, I., Chen, D., Deitz, L. W., Schwardt, J. D., and Chao, E. Y. S. (2002) Effect of a pulsed electromagnetic fields (PEMF) on late-phase osteotomy gap healing in a canine tibial model. *J. Orthop. Res.* **20,** 1106–1114.

48. Brighton, C. T., Friedenberg, Z. B., Mitchell, E. I., and Booth, R. E. (1977) Treatment of nonunion with constant direct current. *Clin. Orthop. Rel. Res.* **124,** 106–123.

49. Patterson, D. (1984) Treatment of nonunion with a constant direct current: a totally implantable system. *Orthop. Clin. N. Am.* **15,** 47–59.

50. Black, J. (1987) *Electrical Stimulation: Its Role in Growth, Repair, and Remodeling of the Musculoskeletal System.* Praeger, New York.

51. Bassett, C. A. L., Pawluk, R. J., and Pilla, A. A. (1974) Augmentation of bone repair by inductively coupled electromagnetic fields. *Science* **184,** 575–577.

52. Hinsenkamp, M., Ryaby, J., and Burny, F. (1985) Treatment of non-union by pulsing electromagnetic fields: European multicenter study of 308 cases. *Reconstr. Surg. Traumatol.* **19,** 147–156.

53. Heckman, J. D., Ingram, A. J., Lloyd, R. D., Luck, J. V., and Mayer, P. W. (1981) Nonunion treatment with pulsed electromagnetic fields. *Clin. Orthop. Rel. Res.* **161,** 58–66.

54. Bassett, C. A. L., Mitchell, S. N., and Gaston, S. R. (1981) Treatment of ununited tibial diaphyseal fractures with pulsing electromagnetic fields. *J. Bone Joint Surg.* **63A,** 511–523.

55. Bassett, C. A. L. (1989) Fundamental and practical aspects of therapeutic uses of pulsed electromagnetic fields (PEMFS). *CRC Crit. Rev. Biomed. Eng.* **17,** 451–529.

56. Gossling, H. R., Bernstein, R. A., and Abbott, J. (1992) Treatment of ununited tibial fractures: a comparison of surgery and pulsed electromagnetic fields. *Orthopedics* **16,** 711–717.

57. Hinsenkamp, M. (1982) Treatment of non-unions by electromagnetic stimulation. *Acta Orthop. Scand. Suppl.* **196,** 63–79.

58. Brighton, C. T. and Pollack, S. R. (1985) Treatment of recalcitrant nonunion with a capacitively coupled electric field. *J. Bone Joint Surg.* **67A,** 577–585.

59. Longo, J. A. (1998) The management of recalcitrant nonunions with combined magnetic field stimulation. *Orthop. Trans.* **22,** 408–409.

60. Laupacis, A., Rorabeck, C. H., Bourne, R. B., Feeny, D., Tugwell, P., and Sim, D. A. (1989) Randomized trials in orthopaedics: why, how, and when? *J. Bone Joint Surg.* **71A,** 535–543.

61. Freedman, K. B., Back, S., and Bernstein, J. (2001) Sample size and statistical power of randomised, controlled trials in orthopaedics. *J. Bone Joint Surg.* **83B,** 397–402.

62. Borsalino, G., Bagnacani, M., Bettati, E., et al. (1988) Electrical stimulation of human femoral intertrochanteric osteotomies: double blind study. *Clin. Orthop. Rel. Res.* **237**, 256–263.

63. Sharrard, W. J. W. (1990) A double-blind trial of pulsed electromagnetic fields for delayed union of tibial fractures. *J. Bone Joint Surg.* **72B**, 347–355.

64. Mammi, G. I., Rocchi, R., Cadossi, R., and Traina, G. C. (1993) Effect of PEMF on the healing of human tibial osteotomies: a double blind study. *Clin. Orthop. Rel. Res.* **288**, 246–253.

65. Scott, G. and King, J. B. (1994) A prospective double blind trial of electrical capacitive coupling in the treatment of non-union of long bones. *J. Bone Joint Surg.* **76A**, 820–826.

66. Simonis, R. B., Parnell, E. J., Ray, P. S., and Peacock, J. L. (2003) Electrical treatment of tibial non-union: a prospective, randomized, double-blind trial. *Injury* **34**, 357–362.

67. Akai, M., Kawashima, N., Kimura, T., and Hayashi, K. (2002) Electrical stimulation as an adjunct to spinal fusion: a meta-analysis of controlled clinical trials. *Bioelectromagnetics* **23**, 496–504.

68. Dwyer, A. F., Yau, A. C., and Jeffcoat, K. W. (1974) The use of direct current in spine fusion. *J. Bone Joint Surg.* **56A**, 442–446.

69. Kane, W. J. (1988) Direct current electrical bone growth stimulation for spinal fusion. *Spine* **13**, 363–365.

70. Meril, A. J. (1994) Direct current stimulation of allograft in anterior and posterior interbody fusions. *Spine* **19**, 2393–2398.

71. Tejano, N. A., Puno, R., and Ignacio, J. M. P. (1996) The use of implantable direct current stimulation in multilevel fusion without instrumentation. *Spine* **21**, 1904–1908.

72. Mooney, V. (1990) A randomized double blind prospective study of the efficacy of pulsed electromagnetic fields for interbody lumbar fusions. *Spine* **15**, 708–715.

73. Goodwin, C. B., Brighton, C. T., Guyer, R. D., Johnson, J. R., Light, K. I., and Yuan, H. A. (1999) A double blind study of capacitively coupled electrical stimulation as an adjunct to lumbar spinal fusions. *Spine* **24**, 1349–1357.

74. Zdeblick, T. D. (1993) A prospective, randomized study of lumbar fusion: preliminary results. *Spine* **18**, 983–991.

75. Linovitz, R., Pathria, M., Bernhardt, M., et al. (2002) Combined magnetic fields accelerate and increase spine fusion: a double-blind, randomized, placebo controlled study. *Spine* **27**, 1383–1389.

76. Rubin, C. T., Bolander, M., Ryaby, J. P., and Hadjiargyrou, M. (2001) The use of low-intensity ultrasound to accelerate the healing of fractures. *J. Bone Joint Surg.* **83A**, 259–270.

77. Lehmann, J. F., DeLateur, B. J., Warren, C. G., and Stonebridge, J. S. (1968) Heating produced by ultrasound in bone and soft tissue. *Arch. Phys. Med. Rehab.* **48A**, 397–401.

78. Coakley, W. T. (1978) Biophysical effects of ultrasound at therapeutic intensities. *Physiotherapy* **64**, 166–169.

79. Hill, C. R., ed. (1986) *Physical Principles of Medical Ultrasonics*. Halstead, New York.

80. Dyson, M. and Pond, J. B. (1970) The effect of pulsed ultrasound on tissue regeneration. *Physiotherapy* **56**, 136–142.

81. Corradi, C. and Cozzolino, A. (1953) Gli ultrasuoni e l'evoluzione delle fratture sperimentali dei conigli. *Arch. Ortop.* **66**, 77–98.

82. Knoch, H. G., Dominok, G. W., and Schramm, H. (1971) Distant action of ultrasound upon callous tissue. *Z. Exp. Chir.* **4**, 93–99.

83. Goldblat, V. I. (1971) Effect of various methods of ultrasonic treatment on bone tissue regeneration (an experimental study). *Ortop. Travmatol. Protez.* **32**, 59–63.

84. Klug, W., Franke, W. G., and Knoch, H. G. (1986) Scintigraphic control of bone-fracture healing under ultrasonic stimulation: an animal experimental study. *Eur. J. Nuclear Med.* **11**, 494–497.

85. Dyson, M. and Brookes, M. (1983) Stimulation of bone repair by ultrasound. *Ultrasound Med. Biol.* **8(S2)**, 61–66.

86. Duarte, L. R. (1983) The stimulation of bone growth by ultrasound. *Arch. Orthop. Trauma Surg.* **101**, 153–159.

87. Pilla, A. A., Mont, M. A., Nasser, P. R., et al. (1990). Non-invasive low-intensity pulsed ultrasound accelerates bone healing in the rabbit. *J. Orthop. Trauma* **4**, 246–253.

88. Pilla, A. A., Figueiredo, M., Nasser, P., et al. (1991) Acceleration of bone repair by pulsed sine wave ultrasound: animal, clinical, and mechanistic studies, in *Electromagnetics in Biology and Medicine* (Brighton, C. T. and Pollack, S. R., eds.), San Francisco Press, San Francisco, pp. 331–341.

89. Bonnarens, F. and Einhorn, T. A. (1984) Production of a standard closed fracture in laboratory animal bone. *J. Orthop. Res.* **2**, 97–101.

90. Wang, S. J., Lewallen, D. G., Bolander, M. E., Chao, E. Y. S., Ilstrup, D. M., and Greenleaf, J. F. (1994) Low-intensity ultrasound treatment increases strength in a rat femoral fracture model. *J. Orthop. Res.* **12**, 40–47.

91. Webster, D. F., Harvey, W., Dyson, M., and Pond, J. B. (1980) The role of ultrasound-induced cavitation in the "in vitro" stimulation of collagen synthesis in human fibroblasts. *Ultrasonics* **18**, 33–37.

92. Ryaby, J. T., Cai, F. F., Kaufman, J. J., and Lippiello, L. (1998) Mechanical stimulation of cartilage by ultrasound, in *Electricity and Magnetism in Biology and Medicine* (Bersani, F., ed.), Plenum, New York, pp. 947–950.

93. Ryaby, J. T., Duarte-Alves, P., Mathew, J., and Pilla, A. A. (1991) Low intensity pulsed ultrasound modulates adenylate cyclase activity and transforming growth factor beta synthesis, in *Electromagnetics in Biology and Medicine* (Brighton, C. T. and Pollack, S. R., eds.), San Francisco Press, San Francisco, pp. 95–100.

94. Yang, K. H., Parvizi, J., Wang, S. J., et al. (1996) Exposure to low-intensity ultrasound increases aggrecan gene expression in a rat femur fracture model. *J. Orthop. Res.* **14,** 802–809.

95. Parvizi, J., Wu, C. C., Lewallen, D. G., Greenleaf, J. F., and Bolander, M. E. (1999) Low-intensity ultrasound stimulates proteoglycan synthesis in rat chondrocytes by increasing aggrecan gene expression. *J. Orthop. Res.* **17,** 488–494.

96. Parvizi, J., Parpura, V., Kinnick, R. R., Greenleaf, J. F., and Bolander, M. E. (1997) Low-intensity ultrasound increases intracellular concentration of calcium in chondrocytes. *Trans. Orthop. Res. Soc.* **22,** 465.

97. Reher, P., Doan, N., Bradnock, B., Meghji, S., and Harris, M. (1999) Effect of ultrasound on the production of IL-8, basic FGF and VEGF. *Cytokine* **11,** 416–423.

98. Ito, M., Azuma, Y., Ohta, T., and Komoriya, K. (2000) Effects of ultrasound and 1,25-dihydroxyvitamin D3 on growth factor secretion in co-cultures of osteoblasts and endothelial cells. *Ultrasound Med. Biol.* **26,** 161–166.

99. Shimazaki, A., Inui, K., Azuma, Y., Nishimura, N., and Yamano, Y. (2000) Low intensity pulsed ultrasound accelerates bone maturation in distraction osteogenesis in rabbits. *J. Bone Joint Surg.* **82B,** 1077–1082.

100. Mayr, E., Laule, A., Suger, G., Ruter, A., and Claes, L. (2002) Radiographic results of callus distraction aided by pulsed low-intensity ultrasound. *J. Orthop. Trauma* **15,** 407–414.

101. Tis, J. E., Meffert, R. H., Inoue, N., et al. (2002) The effect of low intensity pulsed ultrasound applied to rabbit tibiae during the consolidation phase of distraction osteogenesis. *J. Orthop. Res.* **20,** 793–800.

102. Hippe, P. and Uhlmann, J. (1959) Die Anwendung des Ultraschalls bei schlecht heilenden Fracturen. *Zent. Chirug.* **28,** 1105–1110.

103. Xavier, C. A. M. and Duarte, L. (1983) Estimulaca ultra-sonica de calo osseo: applicaca clinica. *Rev. Brasil. Ortop.* **18,** 73–80.

104. Heckman, J. D., Ryaby, J. P., McCabe, J., Frey, J. J., and Kilcoyne, R. F. (1994) Acceleration of tibial fracture-healing by non-invasive, low-intensity pulsed ultrasound. *J. Bone Joint Surg.* **76A,** 26–34.

105. Kristiansen, T. K., Ryaby, J. P., McCabe, J., Frey, J. J., and Roe, L. R. (1997) Accelerated healing of distal radial fractures with the use of specific, low-intensity ultrasound. A multicenter, prospective, randomized, double-blind, placebo-controlled study. *J. Bone Joint Surg.* **79A,** 961–973.

106. Emami, A., Petren-Mallmin, M., and Larsson, S. (1999) No effect of low-intensity ultrasound on healing time of intramedullary fixed tibial fractures. *J. Orthop. Trauma* **13,** 252–257.

107. Tanzer, M., Harvey, E., Kay, A., Morton, P., and Bobyn, J. D. (1996) Effect of noninvasive low intensity ultrasound on bone growth into porous-coated implants. *J. Orthop. Res.* **14,** 901–906.

108. Glazer, P. A., Heilmann, M. R., Lotz, J. C., and Bradford, D. S. (1998) Use of ultrasound in spinal arthrodesis. A rabbit model. *Spine* **23,** 1142–1148.

109. Hanft, J. R., Goggin, J. P., Landsman, A., and Surprenant, M. (1998) The role of combined magnetic field bone growth stimulation as an adjunct in the treatment of neuroarthropathy/Charcot joint: an expanded pilot study. *J. Foot Ankle Surg.* **37,** 510–515.

110. Strauss, E. and Gonya, G. (1998) Adjunct low intensity ultrasound in Charcot neuroarthropathy. *Clin. Orthop. Rel. Res.* **349,** 132–138.

111. Ieran, M., Zaffuto, S., Bagnacani, M., Annovi, M., Moratti, A., and Cadossi, R. (1990) Effect of low frequency pulsing electromagnetic fields on skin ulcers of venous origin in humans: a double-blind study. *J. Orthop. Res.* **8,** 276–282.

112. Stiller, M. J., Pak, G. H., Shupack, J. L., Thaler, S., Kenny, C., and Jondreau, L. (1992) A portable pulsed electromagnetic field (PEMF) device to enhance healing of recalcitrant venous ulcers: a double-blind, placebo-controlled clinical trial. *Br. J. Dermatol.* **127,** 147–154.

16

Vascularized Fibula Grafts

Clinical Applications

Richard S. Gilbert, MD and Scott W. Wolfe, MD

INTRODUCTION

The reconstruction of large skeletal defects has posed a challenging problem to the orthopedic surgeon. Such defects may be a result of trauma, infection, tumor resection, or reconstruction of congenital differences. Moore, Weiland, and Daniel have shown that for skeletal defects less than 6 cm, conventional cortical or cancellous bone grafts may prove satisfactory *(1)*. However, for larger defects, or in a poorly vascularized tissue bed, conventional bone grafting results in an unacceptably high rate of complications. These include fatigue fracture, failure of incorporation, and nonunion *(1)*. Such complications often lead to multiple surgical procedures and the need for prolonged immobilization. To prevent such complications, Moore et al. *(1)* and others *(2–5)* have recommend employing a microvascular bone transfer when reconstructing skeletal defects greater than approximately 6 cm in length, or in poorly vascularized tissue beds.

The technique of bone grafting for the reconstruction of skeletal defects was first introduced by Barth in the late nineteenth century *(6)*. However, it was not until the advances in microsurgical instruments and technique beginning in the 1960s *(7)*, that it would become a possibility to transfer autogenous bone on a vascular pedicle. These microsurgical advances were pioneered by the work of Jacobson and Suarez in 1960, who reported a 100% patency rate anastomosing arteries 1.4 mm in diameter *(7)*. These and subsequent refinements led McKee to perform the first clinically successful vacularized bone graft of an osteocutaneous rib flap to a mandible in 1970 *(8)*. The first description of a free vascularized fibula transfer was by Taylor et al. in 1975 *(9)*. They successfully transferred a 22-cm segment of vascularized fibula to reconstruct a contralateral tibial defect. Union occurred at 10 mo, and by 12 mo the graft had hypertrophied to a size approaching that of the tibia. Since these early reports in the 1970s, the field of microvascular bone transfer in general, and free vascularized fibula grafting in particular, has rapidly expanded.

BIOLOGICAL ADVANTAGES OF VASCULARIZED BONE GRAFTS

The biological advantages of vascularized bone grafting over conventional grafts include more rapid and predictable union, less graft resorption, lower rate of infections, fewer fatigue fractures, graft hypertrophy, and the ability to respond to biomechanical loads similar to living bone *(1,2,4,10)*. These advantages result from the differing manner in which conventional and vascularized bone grafts incorporate into the recipient bed. Nonvascularized bone grafts heal via "creeping substitution" *(6)*. This process was first described by Barth in 1895 *(6)*, and the term was coined by Phemister in 1914 *(11)*. In this process, the bone graft serves as a necrotic trabecular scaffolding onto which new bone formation occurs. Host capillaries invade the avascular graft and bring in osteogenic cells. Graft resorption occurs before new bone forms. This involves a protracted course of incorporation and strength is

From: *Bone Regeneration and Repair: Biology and Clinical Applications*
Edited by: J. R. Lieberman and G. E. Friedlaender © Humana Press Inc., Totowa, NJ

significantly diminished during the revascularization phase *(12–14)*. In large defects, this often leads to nonunion, graft resorption, or fatigue fracture *(1,15)*.

In a vascularized bone graft, microvascular anastomosis to the recipient vessels preserves circulation to the graft and allows osteoblasts and osteocytes to survive. Nutrient blood supply is preserved, and thus the graft does not undergo necrosis. Healing occurs in a manner similar to that of a segmental fracture, without the need for "creeping substitution." Healing is more rapid and the incidence of nonunion is minimized *(16–20)*. With its vascular supply intact, the graft can remodel in response to biomechanical stresses, and will hypertrophy when axially loaded *(10,19,21–24)*. This increases the strength and stiffness of the transferred graft, shortens the postoperative immobilization period, and lowers the incidence of fatigue fractures *(24,25)*.

THE VASCULARIZED FREE FIBULA GRAFT

Since the first description of a free vascularized fibula transfer by Taylor et al. in 1975 *(9)*, the indications have expanded and the technique has been refined. Today, vascularized fibula grafts are employed for the reconstruction of extensive long bone and composite bone and soft tissue defects following trauma *(4,26–37)*, tumor resection *(1,32,38–45)*, and infection *(4,46–54)*. In addition, the free fibula graft has been used in the revascularization of osteonecrosis of the femoral head *(55–62)*, for joint and spine arthrodesis *(40,42,44,63–70)*, congenital tibial *(71–78)* and forearm *(79–82)* pseudarthrosis reconstruction, and for free epiphyseal transfer for congenital differences and pediatric trauma *(14,20,83–85)*. (Applications in head and neck surgery are discussed Chapters 17 and 18.)

The fibula is the ideal bone for microvascular reconstruction of extensive segmental long bone defects. It has a high density of cortical bone, is straight and tubular, and has a triangular cross section. This results in a high resistance to angular and torsional stresses *(15,86)*. In the adult, a straight length of up to 22–26 cm of fibula can be harvested for vascularized graft *(36)*.

The pedicle of the vascularized fibula flap is relatively consistent and is based on the peroneal artery, the largest branch of the posterior tibial artery, together with its accompanying two or three venae comitantes *(14,15,20)*. The peroneal vessels maintain both the periosteal and endosteal circulation to the fibula *(15,87)*. The pedicle has a variable length of 4–8 cm and enters the fibula at the junction of the proximal and middle thirds of the bone *(14,20,88)*. The pedicle is predictably identified through a relatively uncomplicated surgical approach *(36,89–91)* (see surgical technique section). The vessels are of sufficient diameter (arterial diameter of 1.5–2.5 mm and venous diameter of 2–3 mm) to make the microvascular anastomosis relatively straightforward *(15,22)*.

The size and shape of the fibula is a close match for the radius or ulna, and when doweled, it fits into the intramedullary canals of the humerus, femur, and tibia *(37,92,93)*. It can also be employed as a "double-barrel" graft to reconstruct defects involving a large cross-sectional area *(22,94–97)*. During the healing phase, the fibula hypertrophies and has the potential to take on the contour of the bone to which it is transferred *(15)*. It can be transferred as a free bone, or with an accompanying fasciocutaneous and/or muscular flap based on the same pedicle *(26–28,30,31,34,35)*. This permits concomitant reconstruction of an associated soft tissue defect in a single procedure. Finally, in the pediatric population, the proximal epiphysis and physeal plate can be incorporated into the transfer to provide for potential longitudinal growth *(14,20,83–85)*.

INDICATIONS FOR VASCULARIZED FIBULA GRAFTS

Traumatic Bone Defects and Nonunions

The successful treatment of traumatic long bone defects is predicated on achieving bony stabilization and union, while preventing the development of infection. This is accomplished by thorough debridement of all nonviable bone and soft tissue, stabilization of the bone ends, and some form of bone graft to bridge the defect. Adequate debridement often results in a large bony defect. As discussed previously,

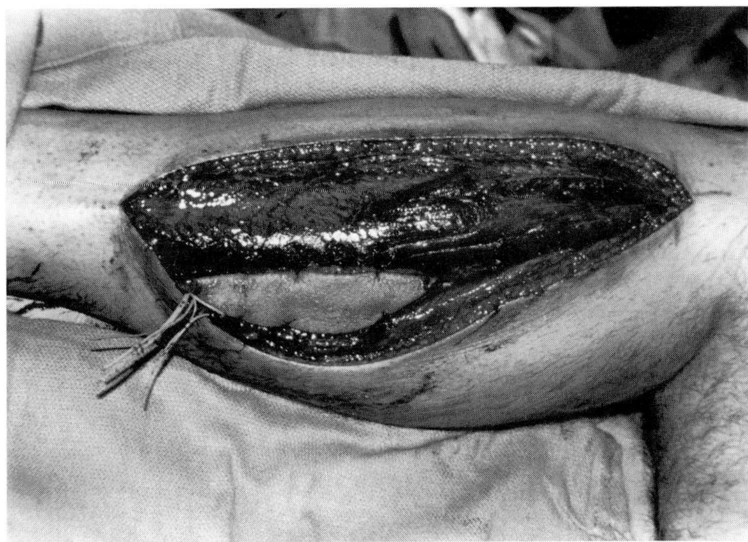

Fig. 1. Intraoperative photogragh demonstrating the harvest of an osteofasciocutaneous vascularized fibula graft to treat a combined traumatic bone and soft tissue defect.

skeletal defects greater than 6 cm in length treated with conventional bone grafts result in unacceptably high rates of nonunion, failure of incorporation, and fatigue fractures *(1)*. Moreover, when traumatic skeletal defects of this caliber are encountered, they are, with few exceptions, associated with a significant injury to the local soft tissues. This leads to a relative avascular zone surrounding the skeletal defect, making nonvascularized graft incorporation via "creeping substitution" even more improbable *(32,33)*.

The vascularized fibula graft has a role in both the acute and late treatment of traumatic long bone defects. Acutely, it is employed to bridge large defects and stabilize the fracture fragments *(37)*. In severe trauma associated with extensive bony injury or loss, there is often concomitant damage to the skin and overlying subcutaneous tissue and muscle. When there is an associated soft tissue defect, the fibula can be transferred together with skin, subcutaneous tissue, and/or muscle *(26–28,30,31,34,35)* (see **Fig. 1**). This will provide for restoration of the bone and soft tissue defect in a single operative procedure. The accompanying skin can also serve to monitor the adequacy of the circulation to the transferred fibula postoperatively *(28,30,33,34)*.

In late traumatic nonunions there is usually a significant skeletal defect associated with an extensively impaired vascular bed. Often the patient has undergone multiple procedures to the area, resulting in significant scarring and the loss of local vascularity, further precluding successful incorporation of a nonvascularized graft *(29,32,33,37)*. Vascularized fibula grafting allows for extensive debridement of all nonviable bone fragments, fracture stabilization, and predictable healing in an avascular tissue envelope. Its high density of cortical bone enhances the intrinsic stability of the nonunion *(29)*. Overall, vascularized fibula transfer for trauma can be expected to heal in greater than 90% of patients *(4,29,30,35–37)*. The highest rates of union employing vascularized fibula transfer can be expected in those patients with an etiology of traumatic nonunion *(4)*.

Tumor Reconstruction

Vascularized fibula transfer allows for immediate or delayed reconstruction following the resection of locally aggressive benign or low-grade malignant bone tumors *(1,32,38–45)*. For defects less

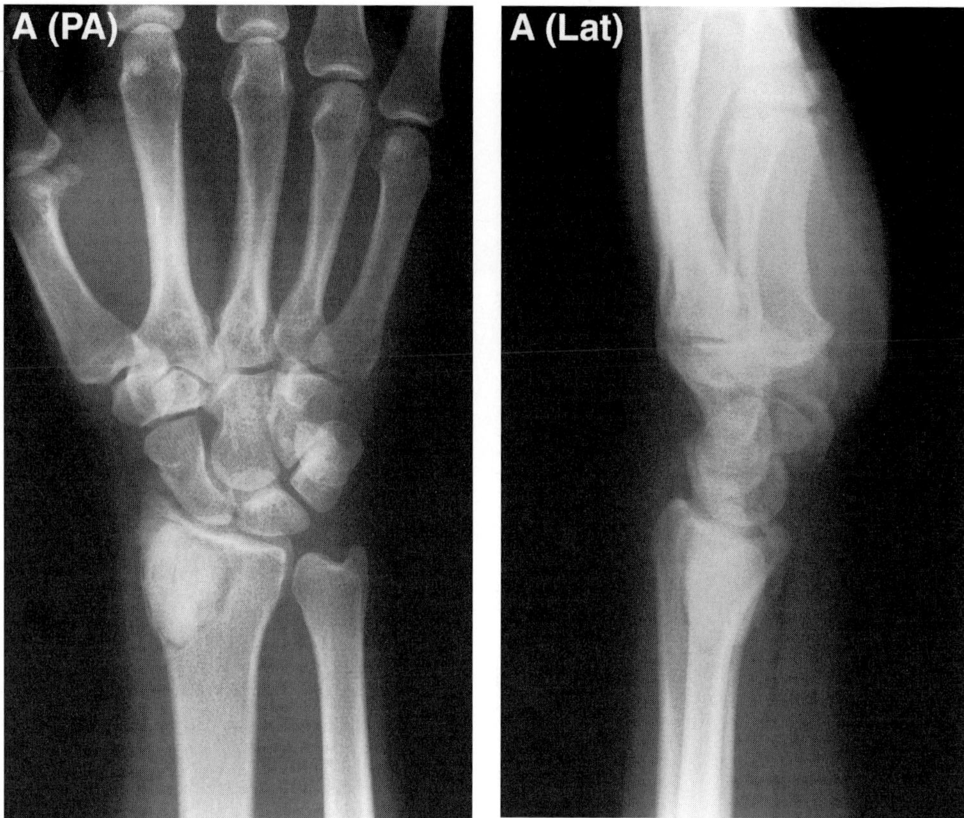

Fig. 2. Radiographs of the wrist of a 29-yr-old male with a recurrent giant cell tumor of the distal radius. **(A)** Patient was initially treated with currettage and cement fixation. Lucency surrounding the cement represents tumor recurrence 2 yr after the primary surgery.

than 6 cm, conventional bone grafts are usually successful *(1)*. In larger tumors, vascularized fibula transfer allows for a radical resection of the neoplasm, without compromising pathologic margins for the purpose of limb salvage. Reconstruction with nonvascularized bone grafting often requires multiple procedures and a prolonged period of immobilization, with a significant chance of nonunion and infection *(1)* (see **Fig. 2**). Often the tumor bed has been irradiated preoperatively and subjected to previous surgical procedures. This results in significant scarring and impaired local perfusion, further jeopardizing conventional bone grafting procedures *(36)*. The free fibula graft, with its own inherent vascular supply, can survive and heal in such compromised tissue beds. In addition, adjuvant chemotherapy does not appear to impair the incorporation of a vascularized fibula graft *(32,40,45,98)*, as has been documented in conventional grafts *(99)*. Most series reporting on vascularized fibula transfer for tumor reconstruction report graft healing and incorporation in over 80% of patients, with low rates of tumor recurrence *(32,40,45)*.

In addition to its use as an intercalary graft for tumor reconstruction, the vascularized fibula has also been used extensively to reconstruct the distal radius following resection of advanced giant cell tumors *(39,42–44)*. Pho first described this technique in 1979 in a 23-yr-old female *(43)*. At 6 mo postoperative, the fibula had healed without resorption, and the patient had a 65° flexion–extension arc and a 60° pronation-supination arc at the wrist. Anatomically, the articular surface of the proximal fibula is similar to that of the distal end of the radius *(39,43)*. Nonvascularized proximal fibula grafts

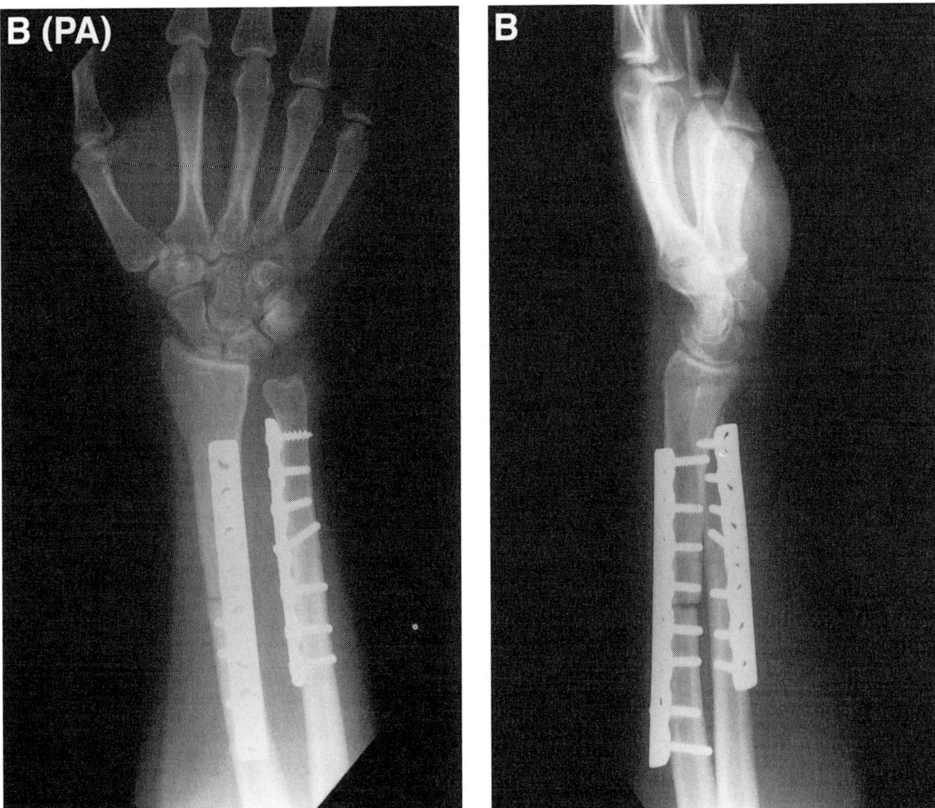

Fig. 2. (B) Patient was treated with resection, distal radius allograft reconstruction, and ulna-shortening osteotomy.

have been similarly employed in the past. A high rate of complications including nonunion, stress fracture, deformity, and degenerative change at the fibulo–carpal joint have been reported *(43,44)*. These complications often result in the need for wrist arthrodesis *(43)*. Vascularized fibula grafting allows for an aggressive *en bloc* resection of the tumor and immediate reconstruction of the resultant defect, usually without the need for wrist fusion *(44)*. Most series report painless and adequate wrist motion, with low recurrence rates *(39,42,44)*.

Osteomyelitis and Infected Nonunions

Treatment of extensive osteomyelitis and septic nonunions pose a particularly challenging problem to the orthopedic surgeon. Treatment is based on adequate debridement of all infected and nonviable bone and soft tissue, and subsequent reconstruction and stabilization of the remaining skeletal defect. This may be done in one or multiple stages. Conventional grafts do not heal well in infected, necrotic, and hypovascular tissue beds *(32,50,52,54)*. In addition, after extensive debridement, the skeletal defect is often too large to be bridged by a nonvascularized graft *(1)*. Instability and persistent infection are common end results *(46)*.

The vascularized fibula transfer is an effective procedure for reconstructing such defects *(46–48, 50–54)* (see **Fig. 3**). It permits for a more aggressive debridement of all infected and nonviable tissue, essentially without concern for the length of the resultant defect. The vascularized graft does not depend on the compromised local soft tissue bed in order to incorporate and heal *(46,50,53)*. The vascularity

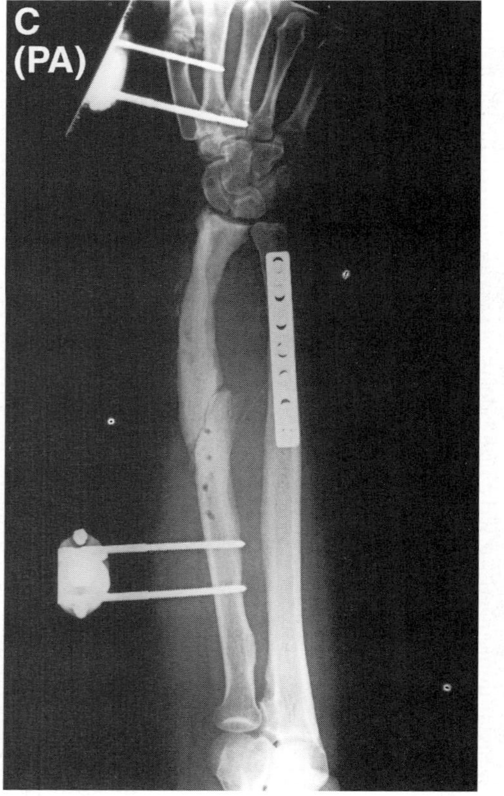
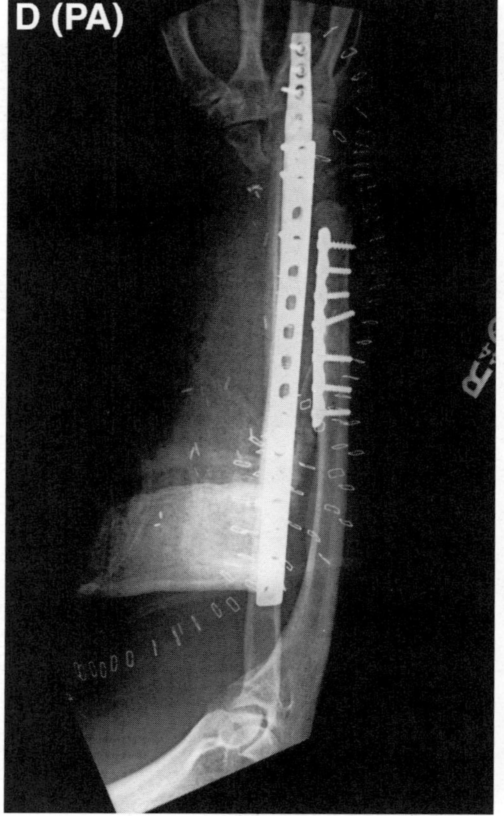
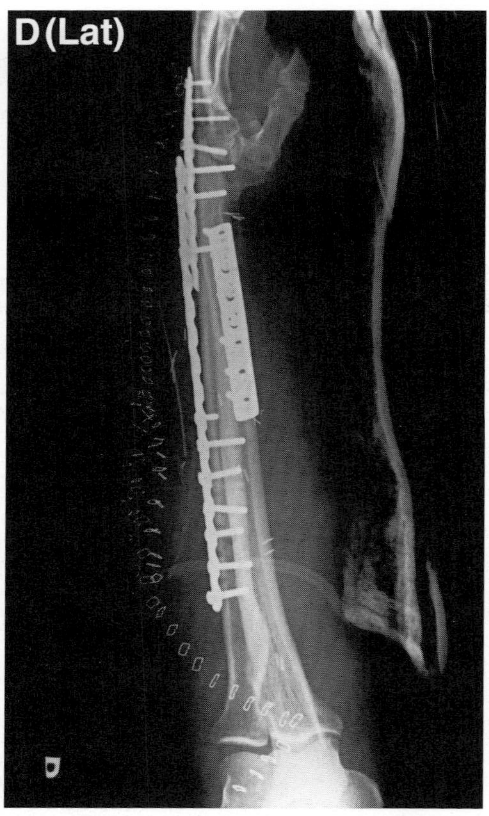

Fig. 2. (**C**) Eighteen months later, patient developed an infection of the allograft, which was treated with removal of hardware, debridement, and external fixation. (**D**) After repeated debridements and intravenous antibiotics, patient was treated with wrist arthrodesis employing vascularized fibula transfer and plate fixation.

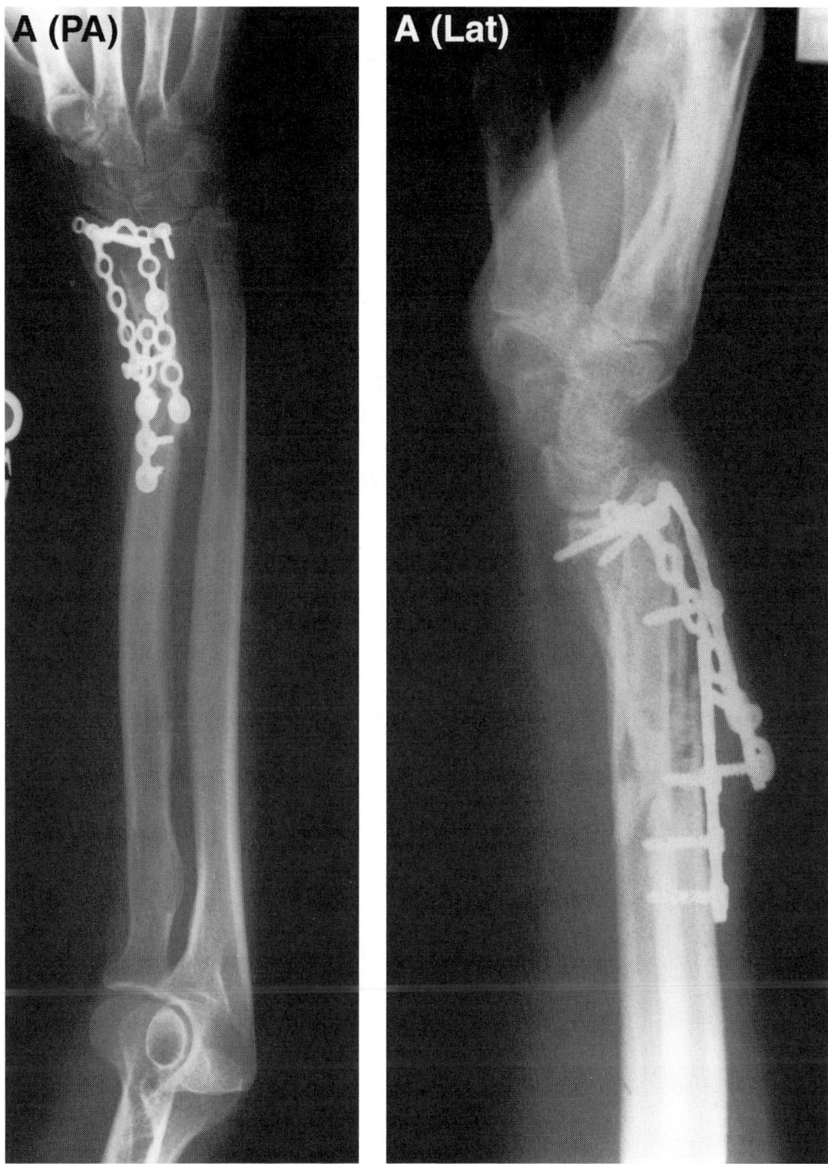

Fig. 3. Radiographs of the forearm of 46-yr-old female with an infected nonunion of the distal radius. **(A)** Patient was referred after she developed an infected nonunion of the distal radius 2 mo after open reduction and internal fixation of an extraarticular fracture.

of the graft also provides an inherent resistance against infection and infectious rejection of the grafted bone *(46)*. Moreover, with successful reanastomosis, the transferred fibula provides for enhanced delivery of antibiotics into the infected tissues *(46,47,49,54)*. This aids in eradicating any residual infection that remains after debridement.

A number of series have reported successful eradication of the infection and ultimate healing of the nonunion in 80–90% of patients treated *(47,50,54)*. This often requires additional surgical procedures, less commonly in the upper than the lower extremities. Overall, results of the transfer for infection are inferior to those reported for other indications, such as trauma, tumor, and congenital reconstruction

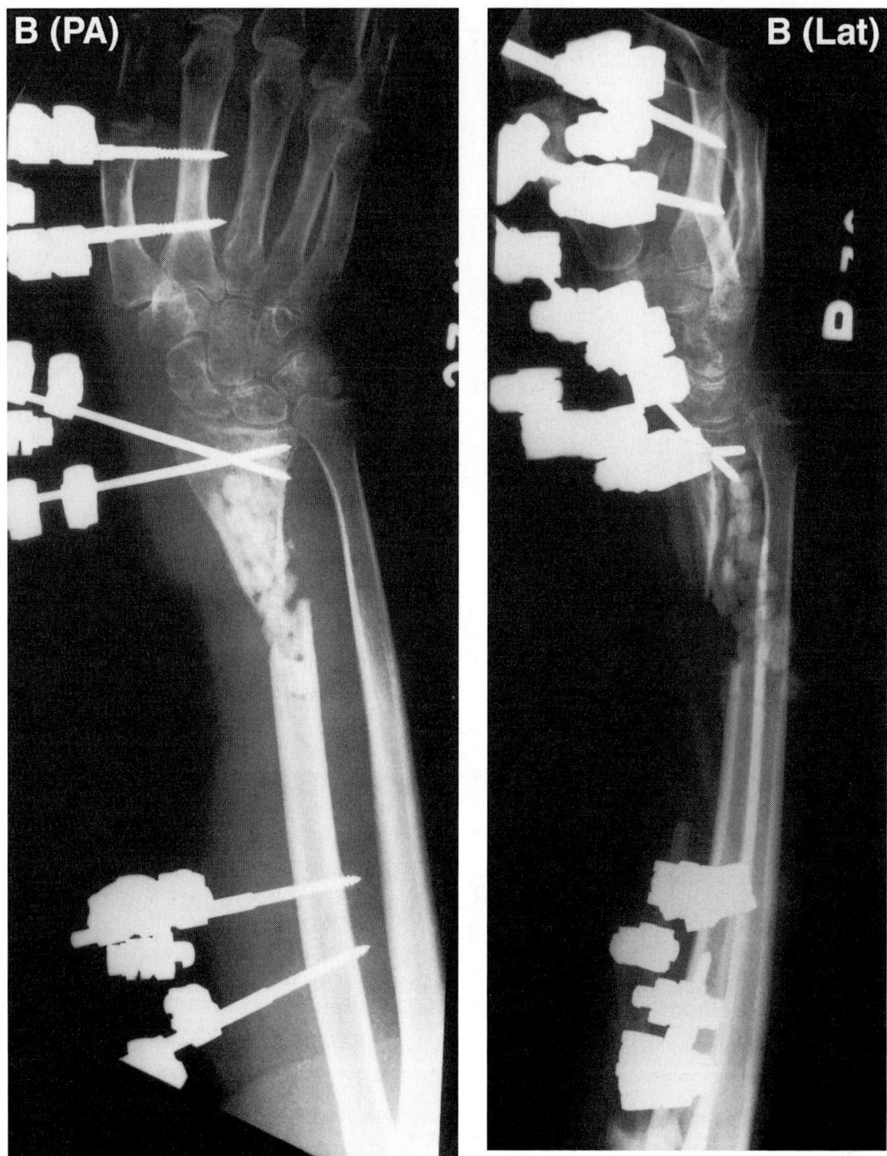

Fig. 3. (B) Patient was initially treated with extensive debridement, external fixation, and placement of antibiotic impregnated cement beads.

(4,100,101). De Boer et al. reported a higher nonunion rate for patients treated with vascularized fibula graft for a diagnosis of osteomyelitis, as compared to other diagnoses *(101)*. This is not surprising, considering the amount of fibrosis and necrosis that occurs in the infected tissue bed. However, in many of these patients, amputation would have been the alternative treatment option *(4)*.

Osteonecrosis of the Femoral Head

Osteonecrosis of the femoral head is a debilitating disease that primarily affects patients in the third through fifth decades of life *(55)*. It is the result of multiple etiologies, most commonly alcoholism, exposure to prolonged systemic steroid administration, or trauma *(59,60)*. Left untreated, it progres-

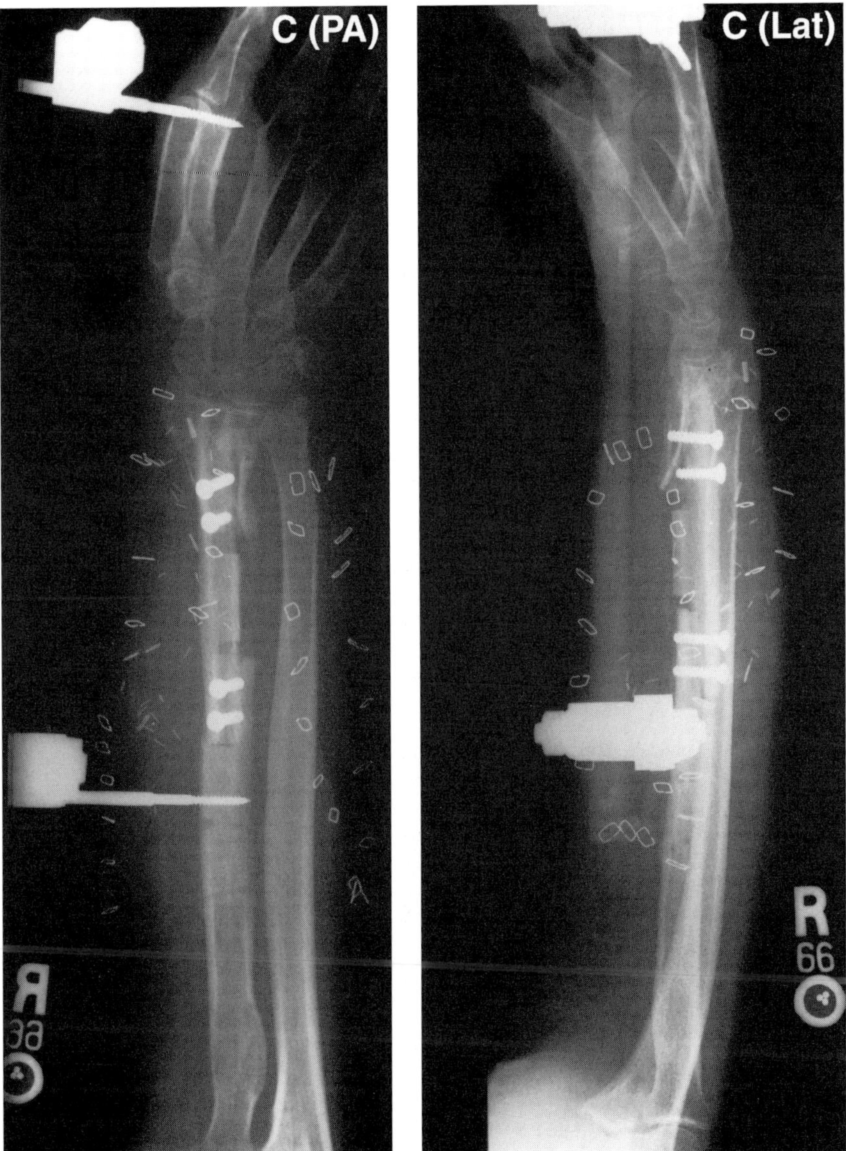

Fig. 3. (C) After repeated debridements and intravenous antibiotics, patient was treated with vascularized fibula transfer.

sively leads to articular incongruity and subsequent osteoarthrosis of the hip joint *(55,58,60)*. Osteonecrosis accounts for approximately 18% of total hip replacements in Western countries *(61)*. Because it affects relatively younger patients, numerous interventions have been employed in an attempt to avoid total joint arthroplasty. These have included restricted weight bearing, core decompression, osteotomy, nonvascularized structural grafts, and electrical stimulation *(58,59,62)*. Overall, the results of these interventions have been unsatisfactory, particularly in the more advanced stages *(58,60)*. Progression of the disease and articular collapse are common sequelae.

Vascularized fibula grafting provides for a source of vascularity and osteocytes to enhance osteogenesis in the femoral head. It also serves as a cortical structural graft that supports the subchondral

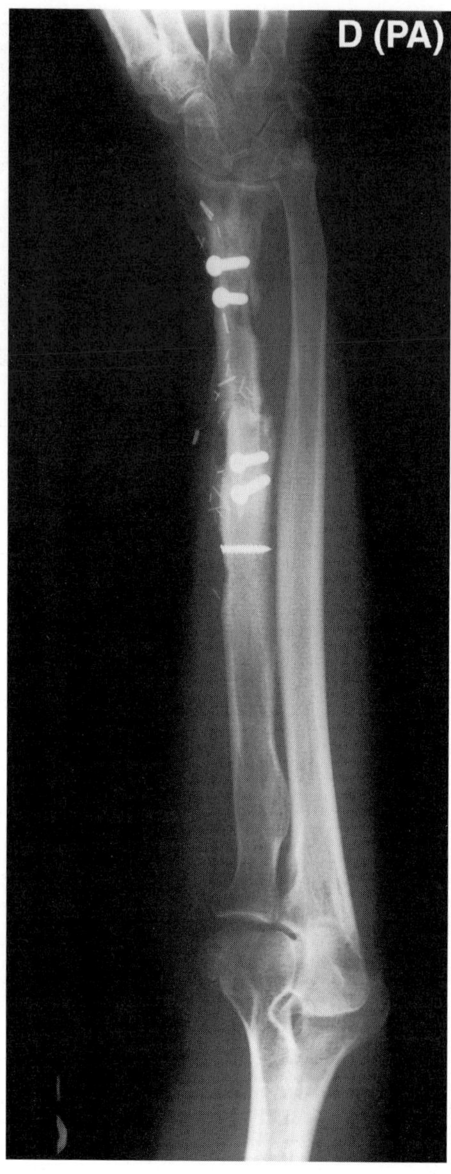

Fig. 3. (D) At 4 mo postoperative there is full incorporation of the fibula proximally and distally, with no evidence of recurrence of the infection.

articular surface *(55–60,62)*. The femoral head is preserved, and the presence of the fibular graft does not preclude later conversion to a total hip arthroplasty, if required *(60)*. Treatment consists of removing all necrotic bone beneath the articular surface of the femoral head. This region is augmented with cancellous bone graft, and then buttressed with the vascularized fibula graft *(60,61)*. The goal of this procedure is to either delay or prevent the progression of osteonecrosis, thereby avoiding the need for total joint arthroplasty *(58)* (see **Fig. 4**). Urbaniak and colleagues have had the widest experience with treating osteonecrosis of the femoral head with vascularized fibula transfer *(58,60,61)*. In a series of 103 consecutive patients, at a minimum follow-up of 5 yr, the procedure was successful in avoiding conversion to total hip arthroplasy in more than 80% of precollapse hips and 70% of hips that pre-

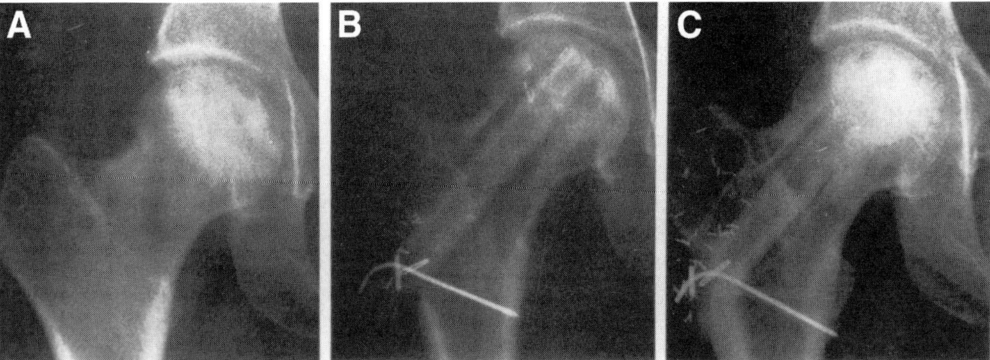

Fig. 4. Anteroposterior radiographs of the hip of a 35-yr-old woman who had stage III avascular necrosis of the femoral head. **(A)** Preoperative radiograph demonstrating evidence of subchondral collapse (crescent sign). **(B)** Six weeks after treatment with vascularized fibula grafting. **(C)** Eight years postoperative demonstrating maintenance of articular congruity. (From Urbaniak, J. R., Coogan, P. G., Gunneson, E. B., and Nunley, J. A. [1995] Treatment of osteonecrosis of the femoral head with free vascularized fibular grafting. A long-term follow-up study of one hundred and three hips. *J. Bone Joing Surg.* **77A,** 681–694. Reprinted with permission.)

operatively demonstrated articular collapse *(60)*. They advocate the procedure for patients less than 50 yr old with stage 1–4 disease *(61)*.

Arthrodesis

Vascularized fibula grafting has been employed to facilitate arthrodesis in the upper and lower extremities, as well as the spine *(40,42,44,63–70)* (see **Fig. 5**). The largest number of series have been reports involving fusion of the knee joint and spine *(63–70)*. In the knee, vascularized fibula transfer is indicated for arthrodesis in patients with a large bony defect, a failed arthrodesis, or a substantial avascular segment *(65,69,70)*. These are most commonly encountered at the site of a previously infected or failed total knee arthroplasty *(69,70)*. The fibula can be used as either an ipsilateral pedicled graft based on antegrade perfusion, or as a single- or double-strut free transfer *(65,69)*. A pedicled transfer is often limited in range by the relatively short peroneal vascular pedicle *(65)*. An intramedullary rod or external fixator is usually employed in conjunction with the fibula transfer *(69,70)*. The Mayo Clinic group reported a solid fusion and a successful result in 12 of 13 patients who underwent knee arthrodesis with vascularized free or pedicled fibula transfer for a variety of diagnoses *(69)*. The average time to union was 7 mo, and none of the patients required secondary grafting procedures.

In the spinal column, the vascularized fibula graft has been employed to fuse high-grade kyphotic deformities, segmental spinal defects, and multiple (greater than three) cervical vertebral levels *(63, 64,66–68)*. It has been most widely used to facilitate anterior arthrodesis in patients with severe kyphotic deformities *(66–68)*. Classically, anterior spinal fusion for kyphosis is accomplished with the use of a nonvascularized rib or fibula strut graft *(66)*. Incorporation may take up to 2 yr *(68)*. In high-grade curves, there is a significant risk of fracture and resultant loss of anterior stabilization during the graft resorption phase *(66,68,102)*. Bradford reported this complication in 4 of 23 patients using a nonvascularized fibula for anterior fusion of kyphotic curves *(103)*. Pedicled rib grafts have also been employed; however, they are mechanically weak, curved, and limited by the short intercostal vascular pedicle *(66)*. A vascularized fibula graft is mechanically stronger than a rib, and can be used to manage a kyphosis of any length or angle throughout the spinal column *(68)*. Studies have demonstrated reliably rapid and solid bony incorporation of the vascularized fibula graft, without evidence of pseudarthrosis *(66–68)*.

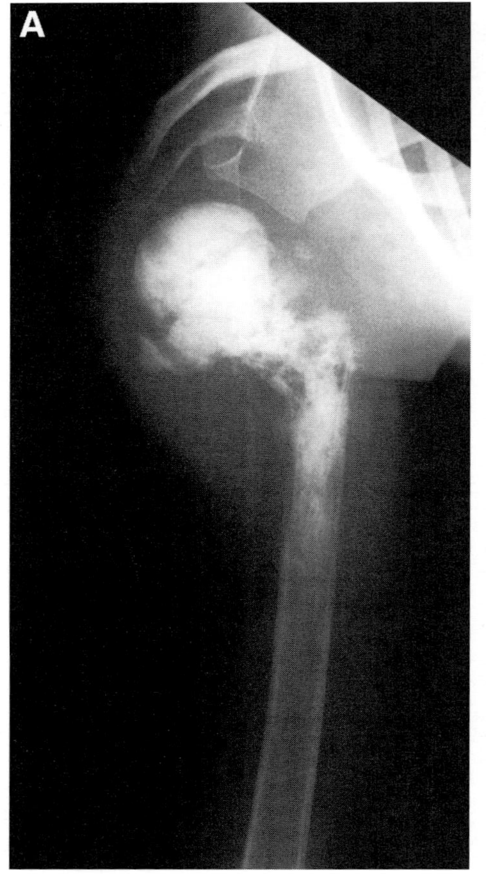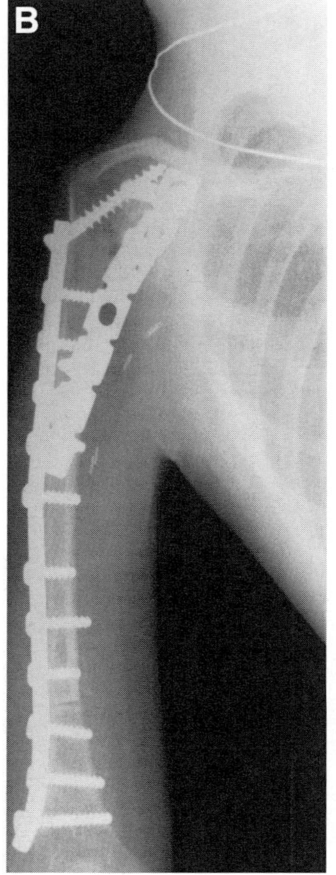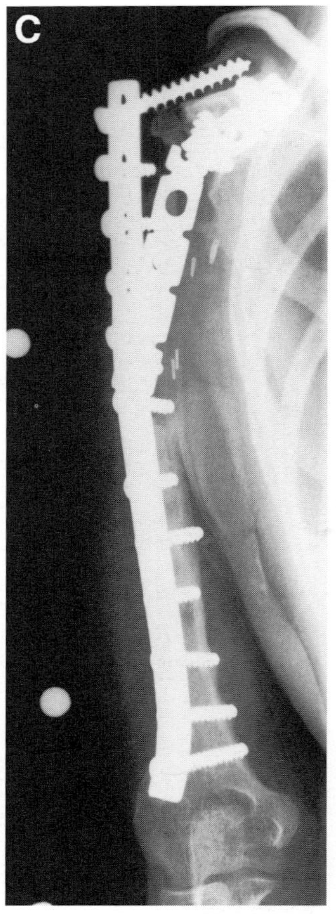

Fig. 5. Anteroposterior radiographs of the proximal humerus of an 18-yr-old female who developed a nonunion of her glenohumeral joint after the resection of an osteosarcoma. (**A**) Preoperative radiograph demonstrating the extent of the tumor and pathological fracture. (**B**) Patient was initially treated with resection of the tumor and shoulder arthrodesis with allograft. (**C**) Radiograph 7 yr later demonstrates complete resorption of the allograft and breakage of the hardware.

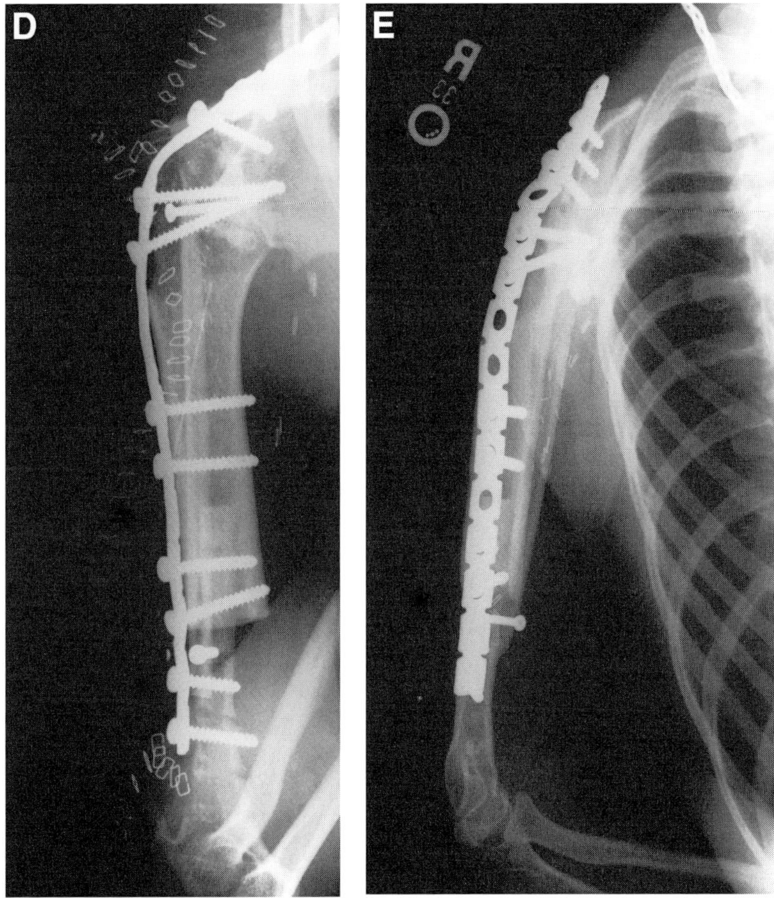

Fig. 5. (D) Patient was treated with removal of hardware and revision of the arthrodesis with vascularized fibula graft, allograft, iliac crest bone graft, and plate fixation. **(E)** Radiograph 2 yr postoperative demonstrating incorporation of the fibula graft and successful fusion of the shoulder joint.

Congenital and Pediatric Reconstruction

Congenital Tibial Pseudarthrosis

Congenital pseudarthrosis of the tibia is a rare disorder that historically represents one of the most challenging reconstructive problems for the orthopedic surgeon *(72,75)*. The etiology is unknown, although it is frequently associated with neurofibromatosis *(77)*. It has remained resistant to most forms of treatment aimed at promoting healing *(76,78)*. Results of conventional onlay grafts, pedicle grafts, bypass grafts, reverse osteotomy, and intramedullary rods have been disappointing, particularly when the tibial defect is greater than 3 cm *(76–78)*. Morrissy et al. reported a nonunion rate of 45% employing conventional bone grafting in a variety of different procedures *(104)*. The graft is frequently resorbed and often results in fracture, nonunion, and multiple surgical procedures. Moreover, severe shortening, ankle deformities, and ultimately, below-knee amputations are not infrequent end results *(77,78,105)*. Some series report amputation rates as high as 40–50% using these treatment modalities *(1,106)*. More recently, electrical stimulation has been employed in an effort to enhance healing.

Overall results, however, have been less than satisfactory in the more severe forms, or when the defect is greater than 3 cm *(73,74,76,78,107)*.

The use of a free vascularized fibula graft in the treatment of congenital tibial pseudarthrosis was first described by Judet et al. in 1978 *(74)*. Its use is indicated when the tibial defect is greater than 3 cm, when the leg length discrepancy is 5 cm or greater, or when the condition has remained refractory to other treatment modalities *(76,78)*. It allows the orthopedist to completely excise all pathological avascular tissue, essentially preventing recurrence, without concern for the length of the residual skeletal defect *(71,75)*. The transferred fibula permits for correction of the angular deformity and the leg length discrepancy in a single procedure *(71,75)*. Moreover, the vascularized fibula graft, unlike conventional grafting techniques, will not resorb *(72,77)*.

Results of treating congenital tibial pseudarthrosis with vascularized fibula transfer have surpassed those of other treatment options. Weiland et al. reported an ultimate union rate of 95% in 19 patients at average follow-up of 6.3 yr *(78)*. Similarly, Gilbert and Brockman reported a healing rate of 94% in 29 patients at skeletal maturity *(73)*. It should be noted that 41% of the patients in Gilbert and Brockman's series and 26% of the patients in Weiland's series required secondary surgical procedures to achieve ultimate union. In addition, residual tibial malalignment and leg length discrepancy were not uncommon sequelae. Still, their ultimate functional results were superior to those of other treatment options currently available.

Congenital Forearm Pseudarthrosis

Congenital pseudarthrosis of one or both forearm forearm bones is a much rarer entity than congenital tibial pseudarthrosis, with approximately 60 cases being reported in the English-language literature *(79,82)*. Neurofibromatosis has been cited as an etiological factor in approximately 80% of cases *(80)*. Similar to its tibial counterpart, it is resistant to standard forms of treatment *(82)*. Numerous procedures have been described, including conventional bone grafting, Ilizarov distraction lengthening, creation of a one-bone forearm, and electrical stimulation *(82)*. These procedures have been met with varying degrees of success *(81,82)*. Their limitations are similar to those already discussed with regard to congenital tibial pseudarthrosis. Treatment with vascularized fibula transfer was first reported by Allieu et al. in 1981 *(79)*. It permits wide resection of the pathologic fibrous tissue and reconstruction of the resultant defect. Its size and shape closely matches those of the shafts of the radius and ulna *(79–82)*. A recent review of the literature found vascularized fibular grafting to achieve the highest union rate among all reported procedures, with overall excellent results *(82)*.

Epiphyseal Transfer

Free vascularized proximal fibula epiphyseal transfer has been employed in the reconstruction of the distal radius for radial clubhand, pediatric tumors, and physeal arrest secondary to trauma or infection *(14,20,83–85)*. This transfer potentially allows for continued growth of the limb to which it is transferred, through the open physeal plate. Moreover, in a young child, the fibula may remodel and conform to the configuration of the proximal carpal row *(14)*. The proximal end of the fibula is transferred with its vascular pedicle consisting of the lateral inferior geniculate artery and vein, usually branching from the popliteal vessels *(20)*. This preserves the vascularity to both the articular surface and epiphyseal plate of the fibula *(14)*. The peroneal artery is also sometimes included in the transfer *(85)*.

To date, reported results have been variable. Early reports from the first several cases performed by Weiland et al. were encouraging *(14)*. However, in a larger series, Wei Tsai et al. reported less favorable results *(85)*. In eight cases of vascularized fibular epiphyseal transfer to the upper extremity for a variety of pathologies, four demonstrated premature physeal closure and only one of the eight showed continued longitudinal growth. At present, the utility of vascularized epiphyseal transfer remains uncertain. Further research is required to determine how a transplanted growth plate will react when transferred to a new anatomical site and exposed to different stress loads *(85)*.

PREOPERATIVE EVALUATION

Numerous factors must be taken into consideration before proceeding with a vascularized fibula graft. Age, comorbidities, and history of previous trauma or surgery to the donor and recipient sites will factor into the decision-making process. A preoperative physical examination of the donor and recipient extremities, with particular regard for distal pulses and soft tissue status, is imperative *(108)*. The bony, soft tissue, and vascular status of the recipient site must be assessed. At a minimum, the recipient site must be evaluated with plain X-rays to assess the dimensions and characteristics of the skeletal defect. The method of fixation of the fibula to the recipient bone can usually be determined with plain radiographs. Further workup may include magnetic resonance imaging (MRI), computerized tomography (CT), or bone scan, depending on the particular circumstances.

Most authors advocate preoperative imaging of the recipient site with angiography to map out the vascular anatomy in the recipient bed *(36,109)*. Considerable debate exists, however, with regard to preoperative imaging of the donor site. Many authors do not recommend routine donor-site angiography, unless there are absent pedal pulses on physical exam, a history of vascular disease, or a history of previous leg trauma or surgery *(108–111)*. They claim that, unless indicated by history or examination, angiography will not add any relevant new information. Much of the literature, however, supports preoperative angiography of the donor fibula to identify possible vascular abnormalities secondary to anatomic variants, congenital malformations, or prior trauma to the leg *(36,87,112)*. The length of the fibular pedicle is highly variable *(113)*. Preoperative angiography will demonstrate those patients who have an inadequate peroneal vascular pedicle, which would preclude successful vascularized transfer and reanastomosis *(110)*. Moreover, in 5–7% of the population, the peroneal artery has a dominant role in the circulation of the foot *(112,114)*. Harvesting a fibula graft with its peroneal pedicle in such patients may jeopardize the perfusion to the foot *(112,113)*. Young et al. found that preoperative angiography altered the surgical plan in 7 of 28 patients *(25%) (115)*. More recently, a number of reports in the literature have recommend less invasive preoperative vascular imaging, such as MRI *(113,114)* or noninvasive color duplex imaging *(116)*. These modalities are gaining support and do not have any associated morbidity, as does angiography *(108,113)*.

SURGICAL TECHNIQUE

This surgical technique is based on that described by Weiland *(36)*. During harvesting of the fibula graft, the patient is in the supine position with the knee flexed 135° and the hip flexed 60°. The surgery is performed under pneumatic tourniquet. The fibula is harvested through a lateral approach (see **Fig. 6**). The length of the incision depends on the length of fibula required at the recipient site. The skin on the lateral border of the fibula is incised through a straight incision between the fibular head and the lateral malleolus. The interval between the peroneus longus and soleus muscles is identified. The fascia between these two muscles is split longitudinally along the course of the incision. The peroneus longus muscle is dissected off the anterior fibula and the soleus muscle is dissected off the fibula posteriorly. All muscular dissections are performed extraperiosteally. There are three perforating vessels to the skin that must be identified posteriorly in the fascia that overlies the soleus. These vessels must be ligated, unless an osteofasciocutaneous flap is to be harvested *(89–91)*.

In a proximal-to-distal direction, the peroneus longus and brevis muscles are extraperiosteally dissected off the anterior fibula. The peroneal nerve is protected proximally. The anterior crural septum is identified and divided longitudinally along the length of fibula to be harvested. The extensor muscle group is dissected off the anterior aspect of the interosseous membrane. The anterior tibial neurovascular bundle should be identified and preserved during this dissection. The posterior crural membrane is then identified and incised longitudinally along the length of fibula graft. The soleus and flexor hallucis longus muscles are dissected off the posterior aspect of the fibula. The peroneal vessels are identified and protected on the posterior surface of the intermuscular membrane. Two or

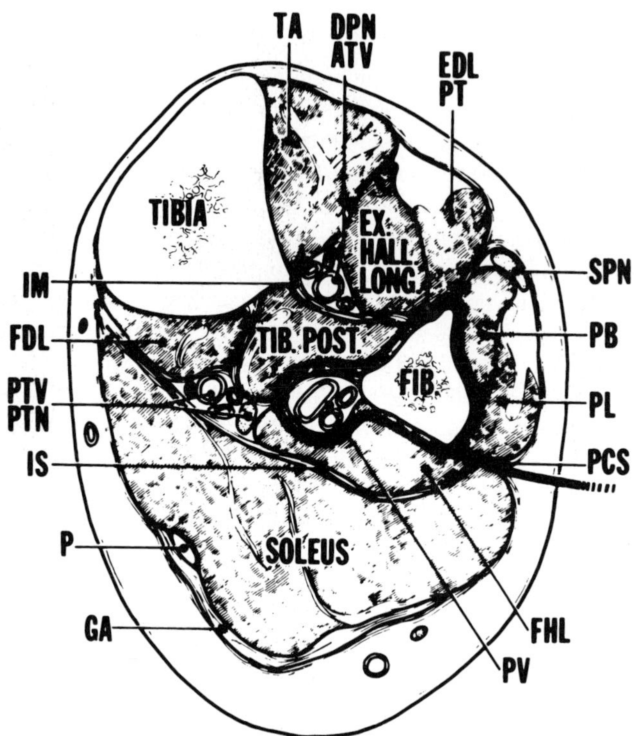

Fig. 6. Cross-sectional diagram of the leg depicting the plane of dissection for harvesting a vascularized fibula graft through the lateral approach (see darkened line). TA, tibialis anterior; DPN, deep peroneal nerve; ATV, anterior tibial vessels; Ex. Hall. Long., extensor hallucis longus; EDL, extensor digitorum longus; PT, peroneus tertius; SPN, superficial peroneal nerve; PB, peroneus brevis; PL, peroneus longus; PCS, posterior crural septum; FHL, flexor hallucis longus; PV, peroneal vessels; GA, gastrocnemius aponeurosis; P, plantaris; IS, intermuscular septum; PTV, posterior tibial vessels; PTN, posterior tibial nerve; FDL, flexor digitorum longus; IM, interosseous membrane; Tib. Post., tibialis posterior. (From Bishop, A. T. [1999] Vascularized bone grafting, in *Green's Operative Hand Surgery,* 4th ed. (Green, D. P., Hotchkiss, R. N., and Pederson, W. C., eds.), Churchill Livingstone, Philadelphia, pp. 1221–1250. Reprinted with permission.)

three peroneal artery branches to the soleus muscle will be encountered. These need to be ligated, uness an osteomuscular flap including the soleus muscle is to be harvested *(91).*

The length of fibula graft to be harvested is then measured and marked with methylene blue. The proximal and distal 6 cm should not be included in the graft, to maintain knee and ankle stability (see **Fig. 7**). As discussed previously, the proximal fibula may be employed to reconstruct defects of the distal end of the radius *(14,20,39,42–44,83–85).* In these cases, the lateral collateral ligament that inserts into the fibular head should be reconstructed to prevent instability of the knee joint *(43,85).* Distally, in children with open physes, a distal tibio-fibular synostosis proximal to the physis should be performed to prevent the subsequent development of ankle valgus instability *(45,78,117,118).*

The distal osteotomy is performed first using a Gigli saw. The peroneal vessels, which lie posteriorly, are protected. The proximal osteotomy is similarly performed, again protecting the peroneal vessels. The distal peroneal vessels at the distal end of the graft are then ligated and divided. The distal aspect of the graft is retracted posterolaterally, and the interosseous membrane is incised longitudinally in a distal to proximal direction. The fibula is then retracted anteriorly and the remaining muscle, the tibialis posterior muscle, is dissected off of the posterior middle third of the fibula (see **Fig. 8**).

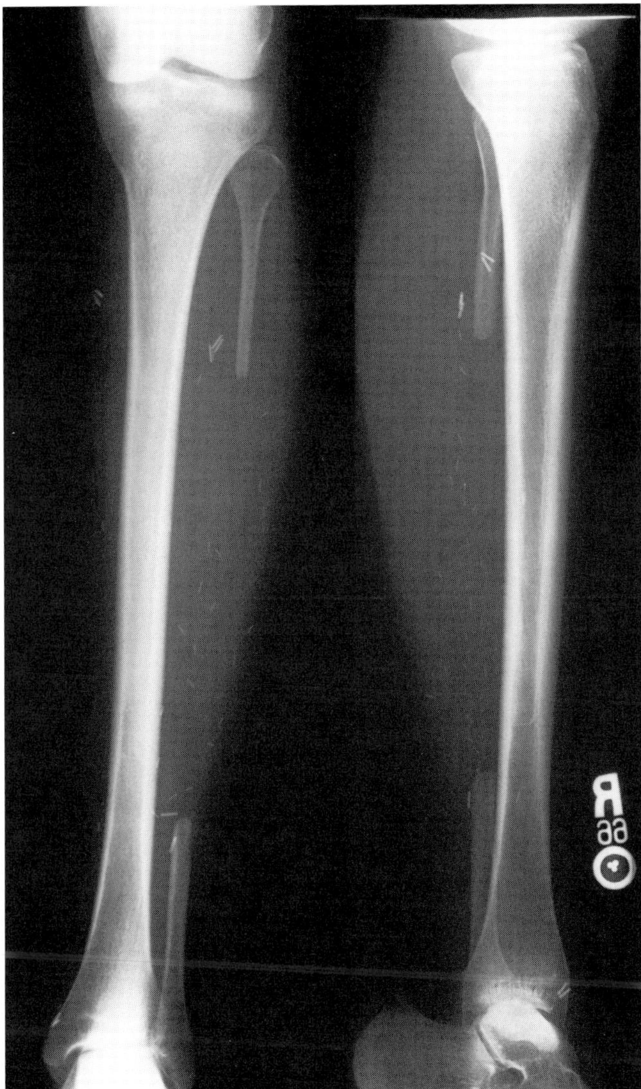

Fig. 7. Anteroposterior and lateral radiographs demostrating the osseous defect after vascularized fibula harvest. Note that the proximal and distal portions of the fibula have been retained in order to maintain knee and ankle stability, respectively.

The peroneal artery and its venae comitantes are then dissected proximally to the point at which the artery divides off of the posterior tibial artery. The fibula is then placed back into its tissue bed. At this point, the tourniquet is deflated to perfuse the graft. Careful hemostasis is obtained. The recipient bed is then prepared, if not previously prepared by a second surgical team. Once the recipient bed is fully prepared, the peroneal vessels are ligated and divided as far proximal as possible. The graft is placed into its recipient bed. Skeletal fixation is then completed, using plates and screws, an external fixation device, an intramedullary rod, or some combination thereof. Microvascular anastomoses of the peroneal artery and vein to their recipient vessels are then performed. The subcutaneous layer and skin are closed over suction drains.

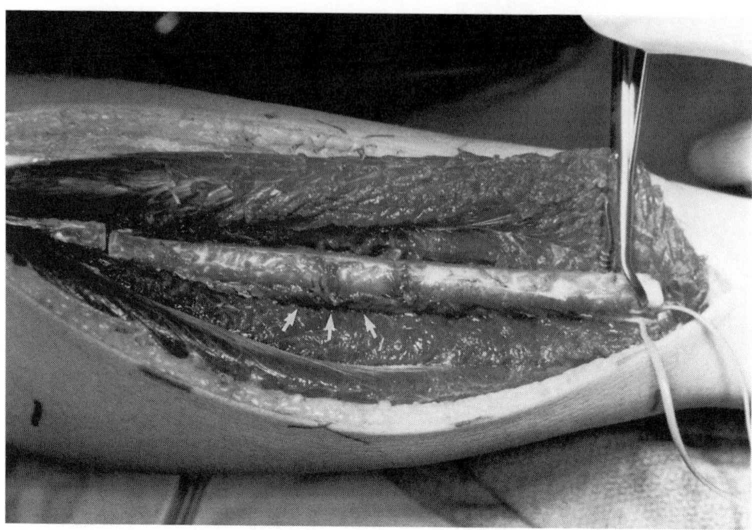

Fig. 8. Intraoperative photograph demonstrating the vasularized fibula graft in its tissue bed after the proximal and distal osteotomies have been completed. The clamp is on the distal aspect of the graft and the arrows are pointing to the peroneal vascular pedicle.

POSTOPERATIVE MONITORING

Monitoring of the circulation to the vascularized fibula flap in the immediate postoperative period is a controversial subject. The graft is subcutaneous and is therefore not visible for direct monitoring *(37)*. Some authors believe that postoperative vascular monitoring is not indicated *(32,114)*. They reason that even if a test revealed failure of the vascular anastomosis, surgical revision of the anastomosis may not be feasible *(114)*. Moreover, by the time a failure of the pedicle anastomosis is detected, it may be too late to restore blood flow to the graft *(15,37)*. The fibula would then simply serve as a nonvascularized graft *(14)*.

In contrast, numerous reports in the literature advocate some form of postoperative vascular monitoring *(4,5,27,37,119–124)*. Bone scintigraphy using technetium-99m methylene diphosphonate is the most widely advocated method in the immediate postoperative period *(4,5,37,119)*. A positive bone scan within the first postoperative week has been correlated clinically and experimentally with patency of the microvascular anastomosis and viability of the graft *(15,125)*. A positive bone scan later than 1 wk postoperative, however, does not necessarily indicate that the anastomosis is patent, or that the fibula is viable. After 1 wk, experimental studies have demonstrated that a positive bone scan may also represent activity secondary to "creeping substitution" on the surface of a nonviable graft *(5,15,18,125)*.

Some authors advocate incorporating a small "buoy flap" of skin with the vascularized fibula graft to be used for monitoring of the circulation to the graft *(27,124)* (see **Fig. 9**). The vascular supply to the "buoy flap" is via perforating cutaneous branches of the peroneal artery, and is therefore in continuity with that of the fibula *(27,55,124)*. By constantly observing the color of the skin island, it is possible to determine immediately whether the anastomosis has become thrombosed. Because this can be observed immediately, some form of surgical intervention could theoretically salvage the vascularity of the fibula graft *(124)*. This is an advantage over bone scanning, which gives information at only one point in time. Others report that the use of such a monitoring flap is unreliable because the quality of the perforating branches may be insufficient *(22,126)*. Moreover, the circulation to the monitoring flap may not fully correspond with that of the transferred fibula *(34)*.

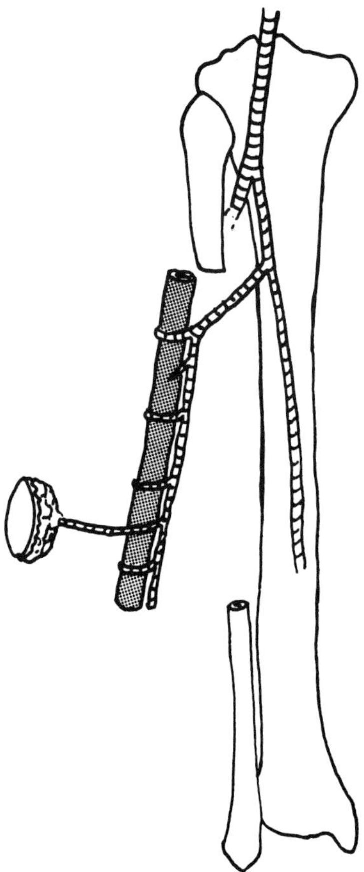

Fig. 9. Diagram depicting a vascularized fibula graft isolated on its peroneal vascular pedicle with a "buoy flap." (From Yoshimura, M., Shimamura, K., Iwai, Y., Yamauchi, S., and Ueno, T. [1983] Free vascularized fibular transplant. A new method for monitoring circulation of the grafted fibula. *J. Bone Joint Surg.* **65A(9),** 1295–1301. Reprinted with permission.)

Various other methods for postoperative monitoring of the circulation to the transferred fibula have been advocated, including laser Doppler flowmetry *(121)*, Doppler color-flow imaging *(123)*, implanted thermocouple probes *(120)*, and measurement of hydrogen washout *(122)*. These methods allow for continuous monitoring of the flap, without the limitations associated with the "buoy flap." In addition, they do not require an additional surgical step, as does the incorporation of a "buoy flap" into the transferred fibula. In our recent practice, we have not routinely employed the previously discussed methods for postoperative monitoring of the graft, and rarely harvest a "buoy flap" for postoperative vascular monitoring. Evidence of early callus formation, healing at the graft junctions, and graft hypertrophy are used as indirect evidence of vessel patency.

COMPLICATIONS

Stress Fracture

Complications secondary to vascularized fibula transfer include stress fracture *(10,28,94,98,101, 119,127–130)*, delayed and nonunion *(4,89,101,127,131)*, thrombosis *(15,121,124)*, infection *(49,127, 132)*, and those related to the fibula donor site *(4,75,78,93,111,117,118,133–136)*. Stress fracture of the graft after union, particularly in the lower extremity, is the most commonly reported complication in the literature *(10,127)*. De Boer and Wood studied 62 cases of vascularized transfer and reported a 25% stress fracture rate, occurring at an average of 8 mo postoperative *(10)*. Overall, reported stress fracture rates vary from 20% to 40% *(10,128–130)*, the majority occurring within the first postoperative year *(10,98,101,117)*.

Stress fracture is significantly less common in transfers to the upper extremity, perhaps due to lower applied loads *(10,40,98,137)*. Vascularized fibula transfers in the upper extremity usually hypertrophy and incorporate rapidly *(10,114)*. In de Boer and Wood's study, fractures occurred only in the grafts transferred to the lower extremity *(10)*. Stress fractures are a result of excessive loading during the hypertrophy phase, before adequate incorporation has occurred *(10)*. Most occur within the middle of the transferred fibula, rather than at the junction sites *(28)*. Once fracture has occurred, provided the graft is adequately vascularized, with proper immobilization and protection, exuberant callus and hypertrophy usually results *(10,28)*. Secondary bone grafting procedures are sometimes required *(98)*.

To limit the incidence of stress fracture in the transferred fibula, the graft should be protected from excessive mechanical loading until hypertrophy is well established *(10,40,98)*. This usually occurs by 1 yr, and can be followed by serial radiographs *(4,10,32)*. Limited mechanical loading, however, will enhance hypertrophy and remodeling *(10)*. Stress fractures are particularly prevalent in vascularized fibula transfer to reconstruct the femur, because of the disparity between the cross-sectional area of the femoral and fibular shafts *(94)*. These can potentially be avoided by dividing the fibula into two struts as a "double-barrel" graft, preserving the vascular supply to both *(22,94–97)*.

Delayed and Nonunion

Delayed or nonunion at one or both junctions of a vascularized fibula transfer is not uncommon. Rates in the literature vary, but nonunion generally is reported to occur in 10–20% of cases, when patients who had secondary grafting procedures are included *(4,127,131)*. A review of 478 vascularized fibula grafts performed for all indications documented a primary union rate of 68% and an overall rate of 82% after supplemental bone grafting procedures *(89)*. The Mayo Clinic reported a primary union rate of 62% of 132 vascularized fibula transfers *(4)*. After secondary grafting procedures, they reported an overall union rate of 80%, at an average follow-up of 42 mo. Weiland reviewed 123 vascularized fibula grafts and reported an ultimate union rate of 87%, with 10% of the patients requiring supplemental bone grafts *(131)*.

The incidence of nonunion differs depending on the underlying pathology of the patient. The results for osteomyelitis are much less favorable than those for tumor, trauma, or nonunion reconstruction *(101)*. De Boer et al. reported an overall union rate of 93% in patients who underwent vascularized fibula transfer for a diagnosis of tumor or trauma, compared to a 59% union rate for those whose underlying diagnosis was osteomyelitis *(101)*. Nonunions are also more common in fibula grafts transferred to the lower extremity, as compared to the upper extremity *(4)*. Stable initial fixation, most commonly with plates and/or screws, has been shown by some to correlate with higher rates of union, as compared to other fixation methods, such as external fixation *(4,101)*. The addition of nonvascularized bone graft at the fibula–recipient junctions at the time of transfer has also been demonstrated to increase primary union rates *(101,138)*. Nonunions are treated with secondary bone grafting procedures, which lead to eventual healing in most instances *(101,132)*.

Thrombosis

Thrombosis occurs in approximately 10% of vascularized fibula grafts in the early postoperative period, as diagnosed by continual laser Doppler flowmetry and confirmed by surgical exploration *(121)*. Whether or not surgical exploration of a thrombosed vessel of the pedicle is indicated remains controversial *(15,124)*. Experimentally, Siegert and Wood demonstrated that the viability of a thrombosed vascularized bone graft is less than that of a conventional nonvascularized graft *(139)*.

Infection

The incidence of infection in vascularized fibula transfer ranges from approximately a 14% deep infection rate to a 33% superficial infection rate *(127)*. Deep infection appears to be more common following reconstruction for the diagnosis of osteomyelitis or tumor *(132)*. Experimentally, vascularized bone grafts have been shown to become infected less often than do conventional nonvascularized bone grafts *(140)*. When fibula grafts do become infected, the infection is easier to eradicate in a successfully vascularized graft, as compared to a nonviable transfer *(49)*. Infection is usually more widespread in nonviable grafts. When deep infection does occur, the response of a viable vascularized fibula graft is similar to that of normal cortical bone. Treatment should consist of intravenous antibiotics with debridement as necessary *(49)*.

Fibula Donor-Site Morbidity

Weiland, Jupiter, and others have documented minimal or no morbidity at the fibula donor site *(1, 22,141)*. Others, however, have found a number of associated complications *(4,93,111,134–136)*. Most commonly, these include residual paresthesias *(134,136)*, occasional pain and cramps *(135,136)*, altered gait *(93,135,136)*, weakness *(93,136)*, reduced walking distance *(134)*, and cold intolerance *(134)*. Gore et al. reported on fibula donor-site morbidity in 41 patients at an average of 27 mo postoperative *(135)*. They found that 42% had pain, 7% complained of muscle pain on exertion, 10% complained of a tired, weak feeling associated with vigorous activity, and 2% had trouble with balance wearing high-healed shoes. A review of 132 vascularized fibula grafts performed at the Mayo Clinic demonstrated donor-site complications in 8% of the patients *(4)*. These included flexor hallucis longus contracture, transient peroneal nerve palsy, compartment syndrome of the leg, and stress fracture of the ipsilateral tibia. Youdas et al. evaluated the gait mechanics of 11 patients who had vascularized fibula transfer to the upper extremity *(93)*. They found muscle strength, especially foot inversion and eversion, to be significantly impaired. There existed an inverse relationship between the length of the resected fibula and the strength of the evertor muscles of the ankle.

The development of an ankle valgus deformity after vascularized fibula graft harvest in patients with open physes is a complication which is well documented in the literature *(75,78,117,118,133)*. This has not been demonstrated to occur in the adult, provided that more than 6 cm of the distal fibula is retained *(22,136)*. In children, this deformity can be prevented by performing a distal tibio-fibular synostosis proximal to the physis at the time of fibula harvest *(45,117,118)*. Deformity has not been demonstrated to occur proximally when the proximal fibular epiphysis is transferred in children *(85, 136)*. The lateral collateral ligament which inserts into the fibular head should be reconstructed, however, to prevent instability of the knee joint *(85)*.

CONCLUSION

Since the first report of a vascularized fibula transfer by Taylor et al. in 1975 *(9)*, the indications for this procedure have expanded widely. Today, it has become one of the established modalities for the orthopedic surgeon in the reconstruction of extensive long bone defects following trauma, tumor resection, and infection. Moreover, it is now widely employed in the treatment of osteonecrosis of the femoral head, congenital tibial and forearm pseudarthrosis, congenital differences and pediatric trauma, and to facilitate spine and joint arthrodesis. Although vascularized fibula transfer is a procedure associated with a number of well-documented complications, these are far outweighed by its ultimate clinical benefits. Future refinements in the use of the fibula as a free epiphyseal transfer and in the area of postoperative monitoring are still needed.

REFERENCES

1. Moore, J. R., Weiland, A. J., and Daniel, R. K. (1983) Use of free vascularized bone grafts in the treatment of bone tumors. *Clin. Orthop.* **175,** 37–44.

2. Brunelli, G., Vigasio, B., Battiston, B., Di Rosa, F., and Brunelli, G. J. (1991) Free microvascular fibular versus conventional bone grafts. *Int. Surg.* **76,** 33–42.

3. Dell, P. C., Burckhardt, H., and Glowczewski, F. P. (1985) A roentgenographic, biomechanical, and histological evaluation of vascularized and nonvascularized segmental fibular canine autografts. *J. Bone Joint Surg.* **67A(1),** 105–112.

4. Han, C. S., Wood, M. B., Bishop, A. T., and Cooney, W. P. III. (1992) Vascularized bone transfer. *J. Bone Joint Surg.* **74A,** 1441–1449.

5. Osterman, A. L. and Bora, F. W. (1984) Free vascularized bone grafting for large-gap nonunion of long bones. *Orthop. Clin. N. Am.* **15,** 131–142.

6. Barth, H. (1895) Histologische Untersuchungen uber Knochen Transplantation. *Beitr. Parthol. Anat Allg. Pathol.* **17,** 65–142.

7. Jacobson, J. H. II and Suarez, E. L. (1960) Microsurgery in anastomosis of small vessels. *Surg. Forum* **11,** 243–245.

8. McKee, D. M. (1978) Microvascular bone transplantation. *Clin. Plast. Surg.* **5,** 283–292.

9. Taylor, G. I., Miller, G. D., and Ham, F. J. (1975) The free vascularized bone graft. A clinical extension of microvascular techniques. *Plast. Reconstr. Surg.* **55,** 533–544.

10. de Boer, H. H. and Wood, M. B. (1989) Bone changes in the vascularized fibular graft. *J. Bone Joint Surg.* **71B,** 374–378.

11. Phemister, D. B. (1914) The fate of transplanted bone and regenerative power of its various constituents. *Surg. Gynecol. Obstet.* **19,** 303–333.

12. Abbott, L. C., Schottslaedt, E. R., Saunders, J. B. D M., and Bost, F. C. (1947) The evaluation of cortical and cancellous bone as grafting material: a clinical and experimental study. *J. Bone Joint Surg.* **29,** 381–414.

13. Enneking, W. F., Burchardt, H., Puhl, J. J., and Piotrowski, G. (1975) Physical and biological aspects of repair in dog cortical-bone transplants. *J. Bone Joint Surg.* **57A,** 237–252.

14. Weiland, A. J., Kleinert, H. E., Kutz, J. E., and Daniel, R. K. (1979) Free vascularized bone grafts in surgery of the upper extremity. *J. Hand Surg.* **4(2),** 129–144.

15. Sowa, D. T. and Weiland, A. J. (1987) Clinical applications of vascularized bone autografts. *Orthop. Clin. N. Am.* **18,** 257–273.

16. Arata, M. A., Wood, M. B., and Cooney, W. P. III. (1984) Revascularized segmental diaphyseal bone transfers in the canine. An analysis of viability. *J. Reconstr. Microsurg.* **1,** 11–19.

17. Berggren, A., Weiland, A. J., and Dorfman, H. (1982) The effect of prolonged ischemia time on osteocyte and osteoblast survival in composite bone grafts revascularized by microvascular anastomoses. *Plast. Reconstr. Surg.* **69,** 290–298.

18. Bos, K. E. (1979) Bone Scintigraphy of experimental composite bone grafts revascularized by microvascular anastomoses. *Plast. Reconstr. Surg.* **64,** 353–360.

19. Doi, K., Tominaga, S., and Shibata, T. (1977) Bone grafts with microvascular anastomoses of vascular pedicles: an experimental study in dogs. *J. Bone Joint Surg.* **59A,** 809–815.

20. Weiland, A. J. (1981) Current concepts review: vascularized free bone transplants. *J. Bone Joint Surg.* **63A(1),** 166–169.

21. Fujimaki, A. and Suda, H. (1994) Experimental stud and clinical observations on hypertrophy of vascularized bone grafts. *Microsurgery* **15,** 726–732.

22. Jupiter, J. B., Bour, C. J., and May, J. W. J. (1987) The reconstruction of defects in the femoral shaft with vascularized transfers of fibular bone. *J. Bone Joint Surg.* **69A,** 365–374.

23. Ostrup, L. T. and Fredrickson, J. M. (1974) Distant transfer of a free, living bone graft by microvascular anastomoses. An experimental study. *Plast. Reconstr. Surg.* **54,** 274–285.

24. Shaffer, J. W., Field, G. A., Goldberg, V. M., and Davy, D. T. (1985) Fate of vascularized and non-vascularized autografts. *Clin. Orthop.* **197,** 32–43.

25. Davis, P. K., Mazur, J. M., and Coleman, G. N. (1982) A torsional strength comparison of vascularized and nonvascularized bone grafts. *J. Biomech.* **15,** 875–880.

26. Baudet, J., Panconi, B., Cai, P., Schoofs, M., Amarante, J., and Kaddoura, R. (1982) The composite fibula and soleus transfer. *Int. J. Microsurg.* **4,** 10–26.

27. Chen, Z. W. and Yan, W. (1983) The study and clinical applications of the osteocutaneous flap of fibula. *Microsurgery* **4,** 11–16.

28. Harrison, D. H. (1986) The osteocutaneous free fibular graft. *J. Bone Joint Surg.* **68B,** 804–807.

29. Jupiter, J. B. (1990) Complex non-union of the humeral diaphysis: treatment with a medial approach, an anterior plate, and a vascularized fibular graft. *J. Bone Joint Surg.* **72A(5),** 701–707.

30. Jupiter, J. B., Gerhard, H. J., Guerrero, J., Nunley, J. A., and Levin, L. S. (1997) Treatment of segmental defects of the radius with use of the vascularized osteoseptocutaneous fibular autogenous graft. *J. Bone Joint Surg.* **79A(4),** 542–550.

31. Koshima, I., Higaki, H., and Soeda, S. (1991) Combined vascularized fibula and peroneal composite-flap transfer for severe heat-press injury of the forearm. *Plast. Reconstr. Surg.* **88,** 338–341.

32. Malizos, K. N., Nunley, J. A., Goldner, R. D., Urbaniak, J. R., and Harrelson, J. M. (1993) Free vascularized fibula in traumatic long bone defects and in limb salvaging following tumor resection: comparative study. *Microsurgery* **14,** 368–374.

33. Newington, D. P. and Sykes, P. J. (1991) The versatility of the free fibula flap in the management of traumatic long bone defects. *Injury* **22,** 275–281.
34. Wei, F. C., Chen, H. C., Chuang, C. C., and Noordhoff, M. S. (1986) Fibular osteoseptocutaneous flap: anatomic study and clinical application. *Plast. Reconstr. Surg.* **78(2),** 191–199.
35. Wei, F. C., El-Gammal, T. A., Lin, C. H., and Ueng, W. N. (1997) Free fibular osteoseptocutaneous graft for reconstruction of segmental femoral shaft defects. *J. Trauma* **43(5),** 784–792.
36. Weiland, A. J. (1984) Vascularized bone transfers. *Instruct. Course Lect.* **33,** 446–460.
37. Weiland, A. J., Moore, J. R., and Daniel, R. K. (1983) Vascularized bone autografts. Experience with 41 cases. *Clin. Orthop.* **174,** 87–95.
38. Aberg, M., Rydholm, A., Holmberg, J., and Wieslander, J. B. (1988) Reconstruction with a free vascularized fibular graft for malignant bone tumor. *Acta Orthop. Scand.* **59,** 430–437.
39. Bajec, J. and Gang, R. K. (1993) Bone reconstruction with a free vascularized fibular graft after giant cell tumour resection. *J. Hand Surg.* **18B(5),** 565–567.
40. Hsu, R. W. W., Wood, M. B., Sim, F. H., and Chao, E. Y. S. (1997) Free vascularized fibular grafting for reconstruction after tumour resection. *J. Bone Joint Surg.* **19B(1),** 36–42.
41. Leung, P. C. and Chan, K. T. (1986) Giant cell tumor of the distal end of the radius treated by the resection and free vascularized iliac crest graft. *Clin. Orthop.* **202,** 232–236.
42. Ono, H., Yajima, H., Mizumoto, S., Miyauchi, Y., Mii, Y., and Tamai, S. (1997) Vascularized fibular graft for reconstruction of the wrist after excision of giant cell tumor. *Plast. Reconstr. Surg.* **99(4),** 1086–1093.
43. Pho, R. W. (1979) Free vascularized fibular transplant for replacement of the lower radius. *J. Bone Joint Surg.* **61B(3),** 362–365.
44. Pho, R. W. (1981) Malignant giant-cell tumor of the distal end of the radius treated by a free vascularized fibular transplant. *J. Bone Joint Surg.* **63A(6),** 877–884.
45. Shea, K. G., Coleman, D. A., Scott, S. M., Coleman, S. S., and Christianson, M. (1997) Microvascularized free fibular grafts for reconstruction of skeletal defects after tumor resection. *J. Pediatr. Orthop.* **17(4),** 424–432.
46. Dell, P. C. and Sheppard, J. E. (1984) Vascularized bone grafts in the treatment of infected forearm nonunions. *J. Hand Surg.* **9A,** 653–658.
47. Doi, K., Kawakami, F., Hiura, Y., Oda, T., Sakai, K., and Kawai, S. (1995) One-stage treatment of infected bone defects of the tibia with skin loss by free vascularized osteocutaneous grafts. *Microsurgery* **16,** 704–712.
48. Lee, K. S., Chung, H. K., and Kim, K. H. (1991) Vascularized osteocutaneous fibular transfer to the tibia. *Int. Orthop.* **15,** 199–203.
49. Low, C. K., Pho, R. W. H., Kour, A. K., Satku, K., and Kumar, V. P. (1996) Infection of vascularized fibular grafts. *Clin. Orthop.* **323,** 163–172.
50. Mattar, R., Azze, R. J., Ferreira, M. C., Starck, R., and Canedo, A. C. (1994) Vascularized fibular graft for management of severe osteomyelitis of the upper extremity. *Microsurgery* **15,** 22–27.
51. Nonnenmacher, J., Bahm, J., and Moui, Y. (1995) The free vascularised fibular transfer as a definitive treatment in femoral septic non-unions. *Microsurgery* **16,** 383–387.
52. Vitkus, K. and Vitkus, M. (1992) Reconstruction of large infected tibia defects. *Ann. Plast. Surg.* **29(2),** 97–108.
53. Wood, M. B. and Cooney, W. P. III. (1984) Vascularized bone segment transfers for management of chronic osteomyelitis. *Orthop. Clin. N. Am.* **15,** 461–472.
54. Yajima, H., Tamai, S., Mizumoto, S., and Inada, Y. (1993) Vascularized fibular grafts in the treatment of osteomyelitis and infected nonunion. *Clin. Orthop.* **293,** 256–264.
55. Cho, B. C., Kim, S. Y., Lee, J. H., Ramasastry, S. S., Weinweig, N., and Baik, B. S. (1998) Treatment of osteonecrosis of the femoral head with free vascularized fibular transfer. *Ann. Plastic Surg.* **40(6),** 586–593.
56. Malizos, K. N., Soucacos, P. N., and Beris, A. E. (1995) Osteonecrosis of the femoral head: hip salvaging with implantation of a vascularized fibular graft. *Clin. Orthop.* **314,** 67–75.
57. Malizos, K. N., Soucacos, P. N., Beris, A. E., Korobilias, A. B., and Xenakis, T. A. (1994) Osteonecrosis of the femoral head in immunosuppressed patients: hip salvaging with implantation of a vascularized fibular graft. *Microsurgery* **15,** 485–491.
58. Scully, S. P., Aaron, R. K., and Urbaniak, J. R. (1998) Survival analysis of hips treated with core decompression or vascularized fibular grafting because of avascular necrosis. *J. Bone Joint Surg.* **80A(9),** 1270–1275.
59. Sotereanos, D. G., Plakseychuk, A. Y., and Rubash, H. E. (1997) Free vascularized fibula grafting for the treatment of osteonecrosis of the femoral head. *Clin. Orthop.* **344,** 243–256.
60. Urbaniak, J. R., Coogan, P. G., Gunneson, E. B., and Nunley, J. A. (1995) Treatment of osteonecrosis of the femoral head with free vascularized fibular grafting. A long-term follow-up study of one hundred and three hips. *J. Bone Joint Surg.* **77A,** 681–694.
61. Urbaniak, J. R. and Harvey, E. J. (1998) Revascularization of the femoral head in osteonecrosis. *J. Am. Acad. Orthop. Surg.* **6(1),** 44–54.
62. Yoo, M. C., Chung, D. W., and Hahn, C. S. (1992) Free vascularized fibula grafting for the treatment of osteonecrosis of the femoral head. *Clin. Orthop.* **277,** 128–138.

63. Doi, K., Kawai, S., Sumiura, S., and Sakai, K. (1988) Anterior cervical fusion using the free vascularized fibular graft. *Spine* **13(11)**, 1239–1244.
64. Freidberg, S. R., Gumley, G. J., Pfeifer, B. A., and Hybels, R. L. (1989) Vascularized fibular graft to replace resected cervical vertebral bodies: case report. *J. Neurosurg.* **71**, 283–286.
65. Hallock, G. G. (1997) The role of pedicled or free fibular grafts in knee arthrodesis. *Ann. Plast. Surg.* **39(1)**, 60–63.
66. Hubbard, L. F., Herndon, J. H., and Buonanno, R. (1985) Free vascularized fibula transfer for stabilization of the thoracolumbar spine: a case report. *Spine* **10(10)**, 891–893.
67. Kaneda, K., Kurakami, C., and Minami, A. (1988) Free vascularized fibular strut graft in the treatment of kyphosis. *Spine* **13(11)**, 1273–1277.
68. Minami, A., Kaneda, K., Satoh, S., Abumi, K., and Kutsumi, K. (1997) Free vascularized fibular strut graft for anterior spinal fusion. *J. Bone Joint Surg.* **79B(1)**, 43–47.
69. Rasmussen, M. R., Bishop, A. T., and Wood, M. B. (1995) Arthrodesis of the knee with a vascularized fibular rotatory graft. *J. Bone Joint Surg.* **77A(5)**, 751–759.
70. Usui, M., Ischii, S., Naito, T., et al. (1996) Arthrodesis of knee joint by vascularized fibular graft. *Microsurgery* **17**, 2–8.
71. Bos, K. E., Besselaar, P. P., Van Der Eyken, J. W., Taminiau, A. H. M., and Verbout, A. J. (1993) Reconstruction of congenital tibial pseudarthrosis by revascularized fibular transplants. *Microsurgery* **14**, 558–562.
72. Dormans, J. P., Krajbich, J. I., Zuker, R., and Demuynk, M. (1990) Congenital pseudarthrosis of the tibia: treatment with free vascularized fibular grafts. *J. Pediatr. Orthop.* **10(5)**, 623–628.
73. Gilbert, A. and Brockman, R. (1995) Congenital pseudarthrosis of the tibia: long-term followup of 29 cases treated by microvascular bone transfer. *Clin. Orthop.* **314**, 37–44.
74. Judet, J., Gilbert, Judet H., and Servan, M. (1978) Apport de la micro-chirurgie a la chirurgie osseuse. *Chirurgie* **104 (10)**, 921–924.
75. Paterson, D. (1989) Congenital pseudarthrosis of the tibia: an overview. *Clin. Orthop.* **247**, 44–54.
76. Simonis, R. B., Shirali, H. R., and Mayou, B. (1991) Free vascularized fibular grafts for congenital pseudarthrosis of the tibia. *J. Bone Joint Surg.* **73B(2)**, 211–215.
77. Uchida, Y., Kojima, T., and Sugioka, Y. (1991) Vascularized fibular graft for congenital pseudarthrosis of the tibia. *J. Bone Joint Surg.* **73B(5)**, 846–850.
78. Weiland, A. J., Weiss, A. P. C., Moore, J. R., and Tolo, V. T. (1990) Vascularized fibular grafts in the treatment of congenital pseudarthrosis of the tibia. *J. Bone Joint Surg.* **72A(5)**, 654–662.
79. Allieu, Y., Gomis, R., Yoshimura, M., Dimeglio, A., and Bonnel, F. (1981) Congenital pseudarthrosis of the forearm —two cases treated by free vascularized fibular graft. *J. Hand Surg.* **6(5)**, 475–481.
80. Allieu, Y., zu Reckendorf, G. M., Chammas, M., and Gomis, R. (1999) Congenital pseudarthrosis of both forearm bones: long-term results of two cases managed by free vascularized fibular graft. *J. Hand Surg.* **42A(3)**, 604–608.
81. Mathoulin, C., Gilbert, A., and Azze, R. G. (1993) Congenital pseudarthrosis of the forearm: treatment of six cases with vascularized fibular graft and a review of the literature. *Microsurgery* **14**, 252–259.
82. Witoonchart, K., Uerpairojkit, C., Leechavengvongs, S., and Thuvasethakul, P. (1999) Congenital pseudarthrosis of the forearm treated by free vascularized fibular graft: a report of three cases and a review of the literature. *J. Hand Surg.* **24A(5)**, 1045–1055.
83. Pho, R. W. H., Patterson, M. H., Kour, A. K., and Kumar, V. P. (1988) Free vascularized epiphyseal transplantation in upper extremity reconstruction. *J. Hand Surg.* **13B(4)**, 440–447.
84. Shea, K. G., Coleman, S. S., and Coleman, D. A. (1997) Growth of the proximal fibular physis and remodeling of the epiphysis after microvascular transfer. *J. Bone Joint Surg.* **79A(4)**, 583–596.
85. Tsai, T. M., Ludwig, L., and Tonkin, M. (1986) Vascularized fibular epiphyseal transfer. A clinical study. *Clin. Orthop.* **210**, 228–234.
86. Taylor, G. I. (1977) Microvascular free bone transfer. *Orthop. Clin. N. Am.* **8**, 425–446.
87. Mckee, N. H., Haw, P., and Vettesse, T. (1984) Anatomic study of the nutrient foramen in the shaft of the fibula. *Clin. Orthop.* **184**, 141–144.
88. van Twisk, R., Pavlov, P. W., and Sonneveld, J. (1988) Reconstruction of bone and soft tissue defects with free fibula transfer. *Ann. Plastic Surg.* **21(6)**, 555–558.
89. Bishop, A. T. (1999) Vascularized bone grafting, in *Green's Operative Hand Surgery,* 4th ed. (Green, D. P., Hotchkiss, R. N., and Pederson, W. C., eds.), Churchill Livingstone, Philadelphia, pp. 1221–1250.
90. Bowen, C. V. and Tomaino, M. (1996) Vascularized bone transfers, in *Surgery of the Hand and Upper Extremity* (Peimer, C. A., ed.), McGraw-Hill, New York, pp. 1941–1974.
91. Serafin, D. (1996) *Atlas of Microsurgical Composite Tissue Transplantation.* Saunders, Philadelphia, pp. 547–573.
92. Taylor, G. I. (1983) The current status of free vascularized bone grafts. *Clin. Plast. Surg.* **10**, 185–209.
93. Youdas, J. W., Wood, M. B., Cahalan, T. D., and Chao, E. Y. S. (1988) A quantitative analysis of donor site morbidity after vascularized fibula transfer. *J. Orthop. Res.* **6(5)**, 621–629.
94. Jones, N. F., Swartz, W. M., Mears, D. C., Jupiter, J. B., and Grossman, A. (1988) The "double barrel" free vascularized fibular bone graft. *Plast. Reconstr. Surg.* **81(3)**, 378–385.

95. O'Brien, B. M., Gumley, G. J., Dooley, B. J., and Pribaz, J. J. (1988) Folded free vascularized fibula transfer. *Plast. Reconstr. Surg.* **82(2)**, 311–318.

96. Tomita, Y., Murota, K., Takahashi, F., Moriyama, M., and Beppu, M. (1994) Postoperative results of vascularized double fibula grafts for femoral pseudoarthrosis with large bony defect. *Microsurgery* **15**, 316–321.

97. Yajima, H. and Tamai, S. (1994) Twin-barrelled vascularized fibular grafting to the pelvis and lower extremity. *Clin. Orthop.* **303**, 178–184.

98. Kasashima, T., Minami, A., and Kutsumi, K. (1998) Late fracture of vascularized fibular grafts. *Microsurgery* **18**, 337–343.

99. Friedlaender, G. E., Tross, R. B., Doganis, A. C., Kirkwood, J. M., and Baron, R. (1984) Effects of chemotherapeutic agents on bone. I. Short-term methotrexate and doxorubicin (adriamycin) treatment in a rat model. *J. Bone Joint Surg.* **66A(4)**, 602–607.

100. Chew, W. Y. C., Low, C. K., and Tan, S. K. (1995) Long-term results of free vascularized fibular graft. A clinical and radiographic evaluation. *Clin. Orthop.* **311**, 258–261.

101. de Boer, H. H., Wood, M. B., and Hermans, J. (1990) Reconstruction of large skeletal defects by vascularized fibula transfer. Factors that influenced the outcome of union in 62 cases. *Int. Orthop.* **14**, 121–128.

102. Streitz, W., Brown, J. C., and Bonnett, C. A. (1977) Anterior fibular strut grafting in the treatment of kyphosis. *Clin. Orthop.* **128**, 140–148.

103. Bradford, D. S. (1980) Anterior vascular pedicle bone grafting for the treatment of kyphosis. *Spine* **5**, 318–323.

104. Morrissy, R. T., Riseborough, E. J., and Hall, J. E. (1981) Congenital pseudarthrosis of the tibia. *J. Bone Joint Surg.* **63B(3)**, 367–375.

105. Zumiotti, A. and Ferreira, M. C. (1994) Treatment of congenital pseudarthrosis of the tibia by microsurgical fibula transfer. *Microsurgery* **15**, 37–43.

106. Hagan, K. F. and Buncke, H. J. (1982) Treatment of congenital pseudarthrosis of the tibia with free vascularized bone graft. *Clin. Orthop.* **166**, 34–44.

107. Kort, J. S., Schink, M. M., Mitchel, S. N., and Bassett, C. A. L. (1962) Congenital pseudarthrosis of the tibia: treatment with pulsing electromagnetic fields. *Clin. Orthop.* **165**, 124–137.

108. Disa, J. J. and Cordeiro, P. G. (1998) The current role of preoperative arteriography in free fibula flaps. *Plast. Reconstr. Surg.* **102(4)**, 1083–1088.

109. Lutz, B. S., Ng, S. H., Cabailo, R., Lin, C. H., and Wei, F. C. (1998) Value of routine angiography before traumatic lower-limb reconstruction with microvascular free tissue transplantation. *J. Trauma* **44(4)**, 682–686.

110. Lutz, B. S., Wei, F. C., Ng, S. H., Chen, I. H., and Chen, S. H. T. (1999) Routine donor leg angiography before vascularized free fibula transplantation is not necessary: a prospective study in 120 clinical cases. *Plast. Reconstr. Surg.* **103(1)**, 121–127.

111. Vail, T. P. and Urbaniak, J. R. (1996) Donor-site morbidity with use of vascularized autogenous fibular grafts. *J. Bone Joint Surg.* **78A(2)**, 204–211.

112. Carroll, W. R. and Esclamado, R. (1996) Preoperative vascular imaging for the fibular osteocutaneous flap. *Arch. Otolaryngol. Head Neck Surg.* **122**, 708–712.

113. Manaster, B. J., Coleman, D. A., and Bell, D. A. (1990) Magnetic resonance imaging of vascular anatomy before vascularized fibular grafting. *J. Bone Joint Surg.* **72A(3)**, 409–414.

114. Manaster, B. J., Coleman, D. A., and Bell, D. A. (1990) Pre- and postoperative imaging of vascularized fibular grafts. *Radiology* **176(1)**, 161–166.

115. Young, D. M., Trabulsky, P. P., and Anthony, J. P. (1994) The need for preoperative leg angiography in fibula free flaps. *J. Reconstr. Microsurg.* **10**, 283–287.

116. Hallock, G. G. (1994) Evaluation of fasciocutaneous perforators using color duplex imaging. *Plast. Reconstr. Surg.* **94**, 644–651.

117. Minami, A., Kaneda, K., Itoga, H., and Usui, M. (1989) Free vascularized fibular grafts. *J. Reconstr. Microsurg.* **5**, 37–43.

118. Omokawa, S., Tamai, S., Takakura, Y., Yajima, H., and Kawanishi, K. (1996) A long-term study of the donor-site ankle after vascularized fibula grafts in children. *Microsurgery* **17**, 162–166.

119. Itoh, K., Minami, A., Sakuma, T., and Furudate, M. (1989) The use of three-phase bone imaging in vascularized fibular and iliac bone grafts. *Clin. Nuclear Med.* **14(7)**, 494–500.

120. May, J. W., Lukash, F. N., Gallico, G. G. 3rd, and Stirrat, C. R. (1983) Removable thermocouple probe microvascular patency monitor: an experimental and clinical study. *Plast. Reconstr. Surg.* **72**, 366–379.

121. Schuurman, A. H., Bos, K. E., and Van Nus, Y. H. (1987) Laser doppler bone probe in vascularized fibula transfers: a preliminary report. *Microsurgery* **8**, 186–189.

122. Shima, I., Yamauchi, S., Matsumoto, T., et al. (1985) A new method for monitoring circulation of grafted bone by use of electrochemically generated hydrogen. *Clin. Orthop.* **198**, 244–249.

123. Stevenson, T. R., Rubin, J. M., and Herzenberg, J. E. (1988) Vascular patency of fibular free graft: assessment by doppler color-flow imager. A case report. *J. Reconstr. Microsurg.* **4**, 409–413.

124. Yoshimura, M., Shimamura, K., Iwai, Y., Yamauchi, S., and Ueno, T. (1983) Free vascularized fibular transplant. A new method for monitoring circulation of the grafted fibula. *J. Bone Joint Surg.* **65A(9)**, 1295–1301.
125. Berggren, A., Weiland, A. J., and Ostrup, L. T. (1982) Bone scintigraphy in evaluating the viability of composite bone grafts revascularized by microvascular anastomoses, conventional autogenous bone grafts, and free nonrevascularized periosteal grafts. *J. Bone Joint Surg.* **64A**, 799–809.
126. Yajima, H., Tamai, S., Mizumoto, S., and Ono, H. (1993) Vascularized fibular grafts for reconstruction of the femur. *J. Bone Joint Surg.* **75B(1)**, 123–128.
127. Coghlan, B. A. and Townsend, P. L. G. (1993) The morbidity of the free vascularised fibula flap. *Br. J. Plast. Surg.* **46**, 466–469.
128. Ihara, K., Doi, K., Sakai, K., Kuwata, N., and Kawai, S. (1992) Fracture of free vascularized fibular grafts. *J. Jpn. Soc. Reconstr. Microsurg.* **5**, 86–95.
129. Minami, A., Kimura, T., Matsumoto, O., and Kutsumi, K. (1993) Fracture through united vascularized bone grafts. *J. Reconstr. Microsurg.* **9**, 227–232.
130. Tamai, S. (1991) Treatment of traumatic extensive bony defect in the lower leg: with vascularized fibula graft. *J. Jpn. Soc. Reconstr. Microsurg.* **4**, 152–155.
131. Weiland, A. J. (1991) Clinical applications of vascularized bone grafts, in *Bone and Cartilage Allografts* (Friedlaender, G. E. and Goldberg, V. M., eds.), American Academy of Orthopaedic Surgeons, Park Ridge, IL, pp. 239–245.
132. Moran, C. G. and Wood, M. B. (1993) Vascularized bone autografts. *Orthop. Rev.* **22**, 187–197.
133. Coleman, S. S. and Coleman, D. A. (1994) Congenital pseudoarthrosis of the tibia: treatment by transfer of the ipsilateral fibula with vascular pedicle. *J. Pediatr. Orthop.* **14**, 156–160.
134. Goodacre, T. E. E., Walker, C. J., Jawad, A. S., Jackson, A. M., and Brough, M. D. (1990) Donor site morbidity following osteocutaneous free fibula transfer. *Br. J. Plast. Surg.* **43**, 410–412.
135. Gore, D. R., Gardner, G. M., Sepic, S. B., Mollinger, L. A., and Murray, M. P. (1987) Function following partial fibulectomy. *Clin. Orthop.* **220**, 206–210.
136. Lee, E. H., Goh, J. C. H., Helm, R., and Pho, R. W. H. (1990) Donor site morbidity following resection of the fibula. *J. Bone Joint Surg.* **72B(1)**, 129–131.
137. Olekas, J. and Guobys, A. (1991) Vascularised bone transfer for defects and pseudarthroses of forearm bones. *J. Hand Surg.* **16B(4)**, 406–408.
138. Pirela-Cruz, M. A. and DeCoster, T. A. (1994) Vascularized bone grafts. *Orthopedics* **17(5)**, 407–412.
139. Siegert, J. J. and Wood, M. B. (1987) Thrombosed vascularized bone graft: viability compared with a conventional bone graft. *J. Reconstr. Microsurg.* **3**, 99–103.
140. Haw, C. S., O'Brien, B. C., and Kurata, T. (1978) The microsurgical revascularization of resected segments of tibia in the dog. *J. Bone Joint Surg.* **60B**, 266–269.
141. Tang, C. L., Mahoney, J. L., McKee, M. D., et al. (1998) Donor site morbidity following vascularized fibular grafting. *Microsurgery* **18**, 383–386.

17

Craniofacial Repair

Bruce A. Doll, DDS, PhD, Charles Sfeir, DDS, PhD, Kodi Azari, MD,
Sarah Holland, MD, and Jeffrey O. Hollinger, DDS, PhD

INTRODUCTION

Annually, skeletal injury and specifically craniofacial injury total approx 12.2 million people in the United States *(1)*. Advances in craniofacial therapy, founded on developing knowledge of the molecular signals and intercellular communication, has greatly improved the restoration of form and function. Fracture healing is a complex physiological process. Cellular and biochemical processes that occur during fracture healing parallel those that take place in the growth plate during development, except in fracture healing these processes occur on a temporal scale *(2–4)*. Similarities in the processes occurring at the growth plate and at the fracture site permit some knowledge from growth-plate analysis to comprehend events in fracture healing. Fracture healing involves a series of distinct cellular responses. Specific paracrine and autocrine intercellular signaling pathways control cellular and osseous tissue mineralization (**Fig. 1**). However, extrapolation of knowledge of growth-plate molecular dynamics is insufficient to achieve consistently optimal bone regeneration during primary and secondary fracture healing.

Fracture healing has been divided into primary fracture healing and secondary fracture healing. Attempts by the cortex to reestablish itself once it has become interrupted characterize primary fracture healing *(5,6)*. Responses in the periosteum and external soft tissues lead to callus formation during secondary healing. Bone on one side of the cortex unites with bone on the other side of the cortex to reestablish mechanical continuity. Anatomical restoration of the fracture is favored when the fragments are coapted and stable *(7)*. Under these conditions, bone-resorbing cells on one side of the fracture undergo a tunneling resorptive response whereby they reestablish new Haversian systems by providing pathways for the penetration by blood vessels. These new blood vessels are accompanied by endothelial cells and perivascular mesenchymal cells, which become the osteoprogenitor cells for osteoblasts.

The regeneration of the bone form and function appears to have limits. Some fractures heal slowly or not at all. Destruction of a critical mass of osseous topography, i.e., a critical-sized defect (CSD), does not regenerate completely. A complex series of molecular cues temporally and spatially influence healing. A *critical-sized defect* has been defined as an intraosseous deficiency that will not heal with more than 10% new bone formation within the life expectancy of the patient (where a patient may be human or nonhuman) *(8)*. A critical-sized defect heals with scar formation—a fibrous unification of osseous cortical plates.

Overcoming the predisposition for skeletal nonunion requires supplemental treatments. Surgeons may circumvent scar formation by numerous approaches, some based on empirical evidence *(9–11)*. Accounts richly detail treatments that surgeons have used to augment fracture healing and continuity defect regeneration (reviewed in ref. *12*). Present treatments include bone grafts, alloplasts, electrical

From: *Bone Regeneration and Repair: Biology and Clinical Applications*
Edited by: J. R. Lieberman and G. E. Friedlaender © Humana Press Inc., Totowa, NJ

FRACTURE HEALING

Injury Phase

PLATELETS: Clotting, growth factors
Complement Activation
INFLAMATION: neutrophils, lymphocytes, monocytes, mast cells, macrophages
RESORPTION: osteoclasts, macrophages
FACTORS: TGF-β, PDGF, VEGF, FGF, EGF, cytokines, interleukins
ADDITIONAL CELLS: pericytes, undifferentiated cells

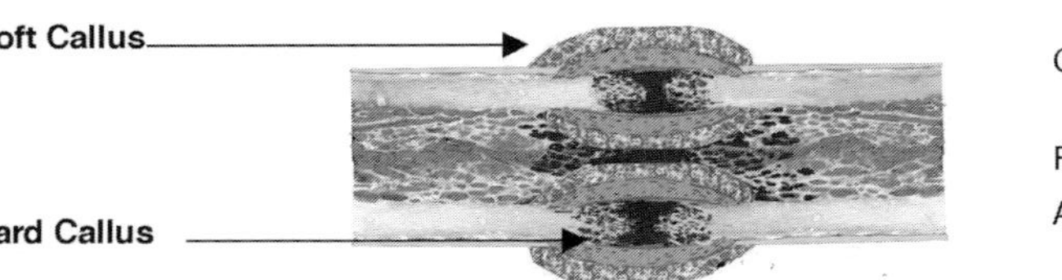

Proliferation Phase

GRANULATION TISSUE: endothelial cells, Neovascularization, collagens (I, II, III, IV, IX, X), fibroblasts
FACTORS: TGF-β, PDGF, IGF, VEGF, BMPs
ADDITIONAL CELLS: macrophages, endothelial cells, chondrocytes, osteoclasts

Early Remodeling

Woven bone develops into lmellar bone
FACTORS: TGF-β, CT, IGF, PTH, GH, BMPs
CELLS: osteoclasts, osteoblasts
Type I collagen predominates

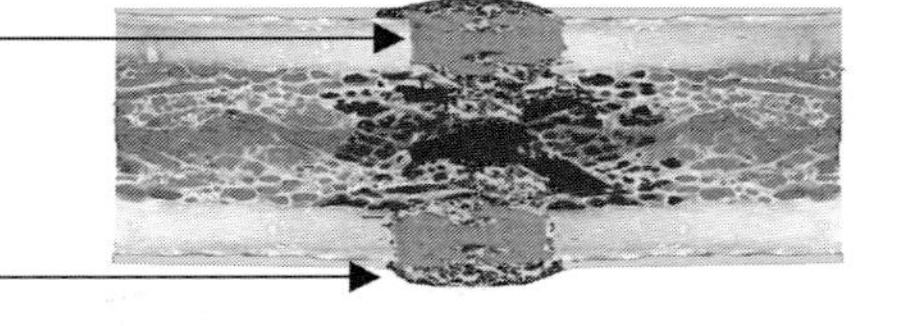

Late Remodeling

Lamellar bone, haversian bone
Bone contour restored

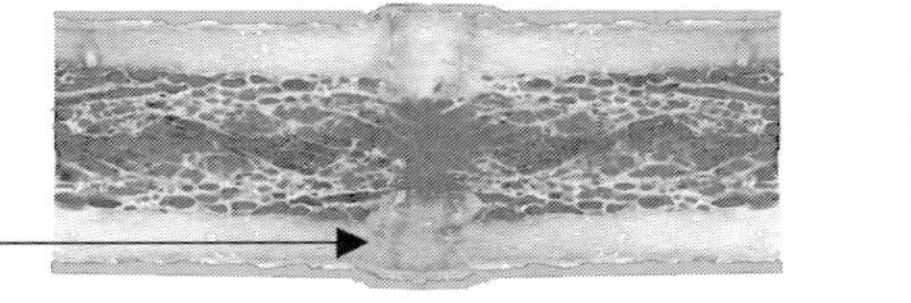

338

Fig. 1. Four phases of fracture healing are noted. A controlled temporal and spatial cascade controls the behavior of cellulars elements. Each phase is characterized by bone formation and remodeling, exhibiting the coupling of osteoblastic and osteoclastic activities. (Color illustration in insert following p. 212.)

stimulation, distraction osteogenesis, guided bone regeneration, and growth factors, whereas targeted-cells and gene delivery represent a relatively new approach for enhanced bone growth (reviewed in refs. *13–17*).

Comprehensive reviews summarizing historical origins of enhanced bone healing are available and are cited for completeness. However, the present review emphasizes selected bone regeneration options that may be new to many surgeons, offering potential for establishing new standards of care in the near future. The review emphasizes contemporary therapies in the context of an ongoing elaboration of bone biology and healing. Present therapies must focus on the controlled delivery of a single agent or rely on the presence of numerous factors purported for bone auto- and allografts. Effectiveness of grafting is attributed to the host of factors released from the graft and the host's support of the graft. The graft lacks temporal, spatial, and stoichiometric precision for dispersion of the factors encouraging bone growth. Therefore, fine predictability for graft success is not currently possible. Successful therapies anticipate the necessity for delivery vehicles *(18)*. The review underscores the important affiliation at several levels—scaffold and factor, chemist, engineer, biologist, and clinician—to achieve a predictable, regenerative therapy.

CELL AND MOLECULAR BIOLOGY OF FRACTURE HEALING

Treatment of most fractures accepts a degree of motion *(19,20)*. The majority of fractures heal by secondary facture healing involving a combination of intramembranous and endochondral ossification. Both processes contribute to repair in an orchestrated sequence of four or five phases of healing. Hematoma and inflammation precede angiogenesis *(18)* and chondrogenesis. Cartilage is removed and osteoprogenitor cells induce bone formation and remodeling in concert with osteoclastogenesis *(21)*.

Multiple events occur during bone injury. Fracture is an injury that incites an inflammatory response, activation of complement ensues, and vascular damage leads to fluid extravasation (**Fig. 1**). At the fracture, there is a decrease in pH to 4–5 (the acidotic state), monocyte and macrophage recruitment, platelet degranulation, and disruption of bone marrow architecture *(22)*. Proteolytic degradation of extracellular matrix (ECM) produces chemotactic remnants attracting monocytes and macrophages to the wound. Chemokines at the fracture site establish selective migration gradients for polymorphonuclear leukocytes (PMNs), and growth factors released from the alpha granules of degranulating platelets attract additional PMNs, as well as lymphocytes, monocytes, and macrophages. Activation of macrophages elicits secretion of fibroblast growth factor (FGF) and vascular endothelial growth factor (VEGF), stimulating endothelial cells to express plasminogen activator and procollagenase *(22)*.

Hematoma forms from extravasated blood and establishes a hemostatic plug. The hematoma may be a source of signaling molecules that have the capacity to initiate the cascades of cellular events critical to fracture healing. Blood-volume depletion is minimized. Aggregated platelets provide hemostasis control and mediator-signaling through isoforms of platelet-derived growth factor (PDGF), transforming growth factor-β (TGF-β), insulin-like growth factors (IGFs), and fibroblast growth factors (FGFs) *(23)*. Inflammatory cells that secrete cytokines such as interleukin-1 (IL-1) and IL-6 may be important in regulating the early events in the fracture-healing process.

Macrophages remove cellular and tissue remnants. Macrophages can develop into polykaryon, multinucleated giant cells to manage a protracted bacterial presence. Macrophages synthesize cytokines, interleukins (e.g., IL-1, IL-5, IL-6), tumor necrosis factor (TNF), and macrophage colony-stimulating factor, in addition to PDGF and TGF-β isoforms that stimulate cell activity, recruit cells, and provoke mitogenesis and chemotaxis. Il-5 can induce ectopic ossification. During the first 24–36 h, the environment is characterized by acidotic, hypoxic conditions favorable to PMNs and macrophage activities. PMNs remove microbes and cellular debris *(24)*.

Approximately 3–5 d after fracture, a blastema develops *(25)*. The blastema is similar to an embryological environment: new blood vessels, collagen isotypes, pluripotential cells, supportive ECM, and associated signaling molecules such as growth factors, chemokines, cytokines, and interleukins. Within the blastema, preferential, selective binding of growth factors to collagens (e.g., bone morphogenetic protein and type IV collagen) may localize, protect, and position growth factors to optimize cell interactions. Inflamatory cells secrete cytokines (Il-1, IL-6) and may contribute to regulation of early fracture healing. The blastema-rich collagen facilitates molecular interactions with receptive cells and offers a provisional solid-state matrix for differential cell attachment and promotion of cell transductive mechanisms. Undifferentiated cells traversing neovasculature and osteoprogenitor cells localized to periosteum and endosteum, guided to the fracture site by chemotactic signals (e.g., TGF-β, bone morphogenic proteins [BMPs]), anchor to the collagen-ECM and differentiate into chondrocytes and osteoblasts. The orchestration of cell anchorage, mechanotransduction, and cell-factor interaction promotes cell differentiation to specific phenotypes to favor fracture-wound healing *(26)*. The periosteum undergoes an intramembranous bone formation response, and histological evidence shows formation of woven bone opposed to the cortex within a few millimeters from the site of the fracture during the first 7–10 d. Concurrently, chondrogenesis commences within the callus overlying the fracture site.

Cell differentiation in the adjacent periosteum and external soft tissues, accumulation of their expression products, and maturation of ECM leads to callus formation. The process extends over several weeks. Callus formation is survival-linked. Fracture chondrogenesis and callus provide rapid stabilization of unstable skeletal parts. Callus components include vascular elements, a community of cells (such as, chondrocytes, chondroclasts, fibroblasts, endothelial cells, smooth muscle cells, preosteoblasts, and pluripotential cells), cartilage, and bone. Cartilage is a normal event during fracture healing in the endochondrally derived appendicular skeleton. However, in the intramembranous flat bones of the craniofacial complex, cartilage during fracture healing is indicative of an unstable fracture. A probable etiology for cartilage in the wound is localized motion of unstable bone that provokes cell-shape alterations *(26–31)*. Possibly, the pluripotent cellular population susceptible to mechanotransduction progresses to a cartilage phenotype with minimal vascularity.

Hypertrophic chondrocytes become embedded in a calcified matrix. Tissue and cells are removed as woven bone develops. The sequence of tissue and cell removal is consistent. The removal of chondrocytes during endochondral fracture healing probably involves an ordered process of programmed cell death (apoptosis) *(32)*. Elongated proliferative chondrocytes undergo mitosis and divide approximately 9 d after fracture. Shortly, cell proliferation decreases and hypertrophic chondrocytes become the dominant cell type in the callus. Membrane structures bud from the hypertrophic chondrocytes to form vesicularized bodies, known as matrix vesicles. The matrix vesicles migrate to the extracellular matrix and participate in the regulation of calcification *(21)*. The mitochondria in these cells store and release calcium for transport by matrix vesicles *(33)*. Einhorn and coworkers demonstrated that the matrix vesicles contain enzymes important in proteolytic degradation of the matrix, a necessary step in the preparation of the callus for calcification *(34)*. In addition, matrix vesicles possess phosphatases needed to degrade matrix phosphodiesters to release phosphate ions for precipitation with calcium. A peak in all types of neutral proteases occurs at approx 14 d after fracture, with the peak in alkaline phosphatase occurring at approx 17 d *(34)*. Importantly, a temporal and spatial distribution of enzymes is supportive of the concept that large proteins and proteoglycans in the extracellular matrix of the callus may inhibit calcification until they are degraded sufficiently.

Cellular migration into the wound site is dependent on a scaffold of molecules that mediate adhesion and migration in fracture repair. Fibroblasts, chondrocytes, and osteoblasts produce fibronectin in the callus. Fibronectin was detected in the fracture hematoma during the first 3 d after fracture. Fibronectin was distributed in the fibrous portions of the matrices and, to a lesser extent, in cartilage matrix. Subperiosteal woven bone did not contain fibronectin. *In situ* hybridization showed a moderate signal in poorly differentiated mesenchymal cells and immature chondrocytes a week after fracture. There was no evidence of this signal in the periosteum or in osteoblasts and osteocytes of periosteal woven bone. Northern hybridization showed low levels of fibronectin mRNA in intact bone but marked elevations in expression in the soft callus within 3 d after fracture. These levels increased with time, reaching a maximum 2 wk postfracture. Fibronectin appears to be present throughout the fracture-repair process, but the production of fibronectin by cells associated with the callus appears to be greatest in the earliest stages of healing. The provisional fibers in cartilaginous matrices are synthesized during the early phase of wound healing *(21)*.

Although several extracellular matrix proteins are involved in bone repair and regeneration, most studies have focused on osteopontin, osteonectin, osteocalcin, and several minor fibrillar collagens *(35)*. Studies of temporal expression of mRNAs for matrix proteins in callus demonstrated a peak osteonectin expression on d 9 in the soft callus. A prolonged peak in osteonectin expression in the hard callus was apparent from d 9 to d 15 *(36,37)*. Osteocalcin was detected only in hard callus. Osteocalcin expression commenced between d 9 and d 11. Osteocalcin is a possible marker for osteoblast phenotype. Peak expression was observed at approximately d 15. *In situ* hybridization showed widespread distribution of osteonectin within the fracture callus from d 3 through d 28. Other studies have detected osteonectin mRNA throughout the healing process *(38–40)*. The osteonectin signal was strongest in osteoblastic cells in subperiosteal woven bone where intramembranous ossification was progressing during d 4–7. Type I and type V procollagen signals were intense in the osteonectin-positive cells.

Osteocalcin demonstrated a weak signal and was detected in the deeper zones of the hard callus. Subperiosteal detection of osteonectin signal had diminished by d 10 at the endochondral ossification front. A similar alteration of the spatial expression of type I and type V collagen was detected. Osteonectin was not detected in hypertrophic chondrocytes. Osteonectin was weakly detected in proliferative chondrocytes at all times and sites. Immunohistochemical analysis showed intracellular staining in proliferating and early hypertrophic chondrocytes. Matrices surrounding the chondrocytes contained type II collagen *(35)*. Osteonectin may play a regulatory role in the early stages of ossification. Fibril-forming collagens (types I and V) were concurrently expressed with osteonectin. Osteonectin may regulate tissue morphogenesis in conjunction with other matrix components *(41,42)*.

Fracture healing progresses to bone formation in an immobilized environment. The vascularogenesis supports reconstitution of the craniofacial complex. Proangiogenic modulators (FGF-1, VEGFs) prevent chondrogenesis *(43)*. Alternatively, FGF-2 is important to the capacity for bone regeneration *(44,45)*. Replacement of cartilaginous foci within the fracture by woven bone is observed. Neovascularization is accompanied by endothelial cells and perivascular mesenchymal cells. Osteoprogenitor cells apparently arise from the mesenchymal cells. Eventually, osteoblasts are present in discrete bone-forming loci. Woven bone is cellular, randomly oriented spicules of immature bone that advances to less cellular lamellar bone and consists of bony sheets, about 100 μm thick, directed to buttress fracture fragments. Remants of lamellar bone persist within the remodeled bone. Haversian systems characterize the mature bone. Cells and associated growth factors and hormones (e.g., PTH, TGF-β, FGFs, VEGFs, the BMPs, PDGFs, IGFs *(46,47)* orchestrate this dynamic process. Generally, fracture healing is completed by about 6–8 wk.

Whereas complete restoration of the original anatomy is apparent in children, remodeling of the newly formed bone in adults also leads to a mechanically stable lamellar structure. Regeneration of structure comparable to a preinjured tissue is crafted by red blood cells, leukocytes, fibroblasts, endothelial cells, committed pluripotential cells, and stem cells. Osteoblasts and osteoclasts constitute the community of cells that remodel bone and must be quantitatively and qualitatively stimulated for the healing task. Basal-level activity is inadequate to address the acute challenge of fracture healing. Consequently, cells must be recruited, expanded in number, and acted on by the proper combination of growth factors. Furthermore, cell renewal is crucial for bone regeneration and homeostasis.

Osteoclasts and osteoblasts are expended through several fates: traumatic or pathological loss of supporting stroma and exhaustion of vital cellular components. Osteoclasts have a life expectancy of about 2 wk. A majority of osteoblasts undergo apoptosis (cell death) after approximately 4 wk. Some become lining cells along trabeculae, while others are engulfed in calcified matrix and are referred to as osteocytes. Therefore, osteoblast–osteoclast renewal is essential to sustain the dynamics of bone wound healing over the 8–10 wk period. The source for cells that will differentiate into osteoclasts and osteoblasts, as well as the molecular signals controlling differentiation, are being elucidated.

Osteocalcin, bone sialoprotein, and γ-carboxyglutamic acid may enhance attraction of cells to the skeletal defect. Molecular attractants for pericytes, preosteoblast cells from endosteum, periosteum, and marrow may include fragments of collagen and tissue debris sustained during injury *(48)*.

Bone marrow stromal cells can undergo asymmetric division, with one daughter cell retaining progenitor capability and the other differentiating to end-state cells. Evidence suggests that FGF and PDGF can stimulate mitogenesis *in vitro,* and they may be operational at fracture healing.

The process of cell recruitment to the fracture site is followed by proliferation and, finally, differentiation. Pluripotent cells in the site retain capacity to differentiate along several pathways. Adipogenesis, fibrogenesis, myogenesis, neurogenesis, chondrogenesis, and osteogenesis represent possible cell fates dictated by an appropriate molecular milieu *(49–56)*.

Determinants for cell fate (i.e., differentiation) are being elucidated. Determinants for differentiation may include extracellular signals, the BMPs, and TGF-β *(57)*. Extracellular signaling molecules represent a refined means to address bone regeneration. Success depends not only on their identification but also on a suitable delivery vehicle for correct temporal and spatial placement of osteoinductive agent.

The osteoblast, osteoclast, and osteocyte retain both commonly shared and lineage-specific signal pathways. Temporal and spatial restrictions to the availability and utility of intracellular signaling pathways are inherent during the differentiative cell phases. The pathways of intracellular signaling may be temporally restricted to a cell's developmental phase and exhibit receptivity to threshold stimulus levels with duration and concentration limits. Comprehending the initiation, processing, and consequence of intracellular signaling in bone cells is complicated and intriguing. Furthermore, it offers a potential precise method of inducing cell fate along the osteogenic path.

Intracellular signaling molecules have been identified that either sustain the action of extracellular cues or by themselves can induce differentiation. The differentiation program within the cell cytoplasm is sustained by Smads (a contraction of *Caenorhabditus elegans* Sma and *Drosophila melanogaster* Mad) and STAT (Signal Transducers and Activators of Transduction) *(58–63)*. The Smads shuttle signals received from receptor interaction with TGF-β and BMPs to the nucleus, where another set of signals either restricts or enables transcription of specific genes (**Fig. 2**). A strategy to direct cell fate to the osteoblast phenotype could involve bypassing the BMP receptor with a Smad therapy.

Within the cell nucleus, an activated Smad complex can stimulate nuclear transcriptional activity encoded by DNA. The process includes certain nuclear transcriptional factors that can stimulate expression of specific genes. However, Smads exhibit a range of activities, including transmission of growth-inhibitory signals originating from TGF-β and activation of the Smad/hFAST-1-mediated transcription *(64)*. Recent evidence has identified a unique nuclear transcription factor for osteoblast differentiation previously known as core binding factor A, cbfa (reviewed in refs. 62 and 65). The transcription factor RUNX2 plays a specific, crucial role in the formation of the mineralized skeleton during embryogenesis and regulates maturation of the osteoblast phenotype. RUNX2 (also known as Cbfa1, AML-3, and PEBP2alphaA) supports commitment and differentiation of progenitor cells to osteoblasts. It is hypothesized that activation of RUNX2 results from a chain of events beginning with BMP-initiated receptor interaction, followed by intracellular Smad signaling *(66)*. Smad-activated complexes transit the cell cytoplasm, cross the nuclear membrane, and bind to DNA, where they induce a transcriptional response for RUNX2 *(64,67,68)*. RUNX2 binds to the osteocalcin transcription promoter, culminating in osteoblast differentiation *(69)*. Osteocalcin and RUNX2 appear to be unique to the osteoblast. Could direct therapeutic interdiction with RUNX2 be an alternative to BMPs? In contrast to BMP therapies, RUNX2 may yield a precise targeted induction to primarily osteoblast differentiation (**Fig. 2**).

Osteoclasts, in addition to osteoblasts, are required for fractures to progress through a dynamic series of processes. Osteoclasts are relatively short-lived cells (about 2 wk), and therefore must be

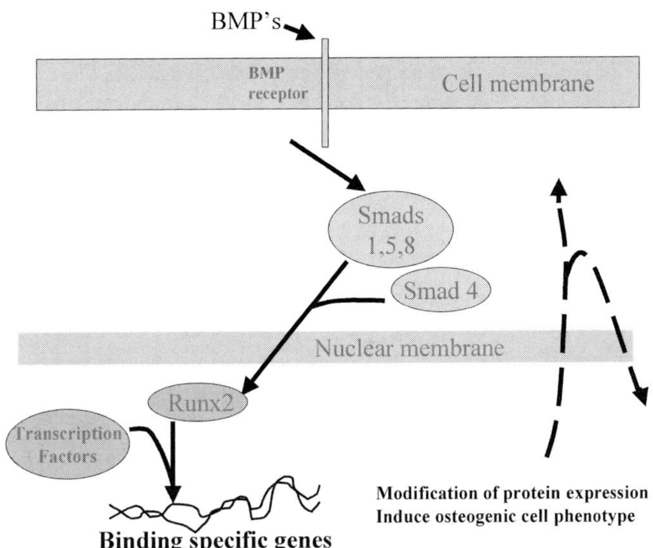

Fig. 2. Interaction of BMP with osteogenic precursor cell and BMP binding the receptor. Subsequent stimulus of Smad 1, -5, -8, interactions with Smad-4, and transport across the nuclear membrane. Complex interaction with Runx2 and other transcription factors may enhance or inhibit the gene transcription. Cell may be induced toward osteogenic phenotype by these and other undefined events.

renewed. Fragments of fibronectin (a ubiquitous attachment factor) and ECM degradation products attract monocytes to the healing wound. Moreover, local preosteoblasts secrete TRANCE (a member of the TNF family), which activates preosteoclast-like cells through the receptor RANK *(48,56)*. Differentiated osteoblasts express interleukins (IL-1, IL-6, and IL-11) that push osteoclast differentiation *(19)*. Furthermore, macrophages at the wound vicinity express FGF and VEGFs, enabling blood permeability and neoangiogenesis, thereby providing transit conduits for additional monocytes to replenish those lost to injury.

Osteoblasts and osteoclasts participate in a coupling mechanism to sustain the wound-healing cascade. Osteoblasts express an osteoclast-inhibiting molecule, osteoprotegerin (OPG) *(70,71)*. OPG is a decoy receptor for TRANCE; it interferes with RANK binding, abrogating additional osteoclast formation *(48)*. Osteoblasts also express IL-1, IL-6, and IL-11 and these factors promote osteoclast differentiation, although the interactions require further study *(72)*.

When osteoblast lineage cells and osteoblasts are quantitatively and qualitatively deficient, as well as operationally dysfunctional (indicative of the aged, osteoporotic condition), corruption in bone-regeneration dynamics may be predicted due to the decrement in directing cues and impairment in the operational status of the craftsmen for regeneration (reviewed in ref. *13*).

IMPORTANCE OF ELUCIDATING THE FRACTURE HEALING MODEL

Fracture-healing components provide the physiological context for rational therapy. Cells, ECM, blood vessels, and signaling molecules (extracellular and intracellular) must be accurately organized. Components of the fracture-healing continuum can be exploited for a rational clinical therapy. Potential therapeutic candidates include pluripotential stem cells, ECM, and extracellular and intracellular signaling molecules. Cell and gene-based therapies have potential for the practice of craniofacial bone regeneration.

CURRENT THERAPIES

Evidence-based clinical data underscore the value of established therapies for bone regeneration. By analytically, dispassionately tabulating virtues and liabilities of autografts and bank bone formulations, the astute clinician can develop a logical argument supporting decisions for alternatives.

Autographs and Allogeneic Banked Bone Preparations

Autografts and allogeneic bank bone preparations offer patients with bone deficiencies the possiblity for regeneration. Successful clinical outcomes from autografts can exceed 80%, whereas allogeneic preparations are slightly less successful *(73,74)*. Reasons for this clinical outcome have been reviewed extensively *(74–77)*.

Autogenous grafts are nonimmunogenic. Surgical intervention and postoperative healing are usually unremarkable. ECM, signaling molecules, and compatible cells bolster healing.

In the absence of infection, regeneration in appendicular and craniofacial locales with autografts is the safest, most predictable, and most reliable form of osseous therapy. Allogeneic bank preparations are less successful than autografts. Immunogenicity (less than 4%) and absence of viable transplanted cells that can become osteoblasts are possible explanations.

Several factors influence clinical outcome. Putative variables influencing autogenous and allogeneic bone bank preparations can include patient health, age, surgical expertise, facilities, personnel factors, transplant embryology, format, and preparation of the transplanted tissue.

Does embryogenesis of either the transplanted bone or recipient bone influence treatment outcome? Evidence of superiority for either endochondral or intramembranous bone grafts is arguable. Biochemical and cellular parity exists between the two embryological derivatives. Morphologically, the intramembranous bone has more robust cortices than endochondral bone. Consequently, intramembranous bone grafts appear to prompt a more enduring result than endochondral ones.

Anatomically, there are biofunctional and vascular differences between the two embryological derivatives. In general, the craniofacial complex is more highly vascularized than appendicular bones, and the intramembranous bones of the calvaria sustain significantly less biofunctional challenge than long bones. Vascularity and functionality could affect the recipient bed interaction with an implanted material. Intuitively, the importance of vascularity and cell viability underscores a more rapid positive response to grafting in the craniofacial complex vs distal extremities. There is no evidence that intramembranous bone inherently possesses more angiogenic signaling molecules (FGFs, VEGFs) than endochondral bone for accelerating and sustaining a vascular response.

Do physical properties of the transplanted bone inspire different clinical outcomes? When we look at bone clinically, cortical and cancellous compartments are different. Cancellous autografts are more cellular, have more signaling factors, and promote a more intense osteotropic healing response than cortical bone. The trade-off is that structurally, cancellous grafts are weaker than cortical. For the craniofacial skeleton, strength, in general, is not the deciding selection criterion.

Can bone banks determine clinical outcome? Yes. The US Food and Drug Administration (FDA) has issued documents for determining donor bone selection. The documents establish guidelines subject to voluntary adherence by each bank. However, more formal FDA requirements are pending. The American Association of Tissue Banking (AATB) does have stringent preparation protocols; however, all banks are not members of the AATB and some do not follow important guidelines. Consequently, procurement and preparation standards vary.

Donor variability (e.g., age, gender) can influence outcome. Several studies revealed that osteoblasts from aged animals were dysfunctional compared to those from adults; osteoblast precursors were fewer in aged than in adult donors; and bone derivatives from aged donors were less potent than derivatives from adult donors (reviewed in ref. *13*).

Some banks gamma-irradiate, ethylene oxide-sterilize, or do neither. While contradictory data exist on how each of these factors affects bank bone outcome, evidence indicates that sterilization may

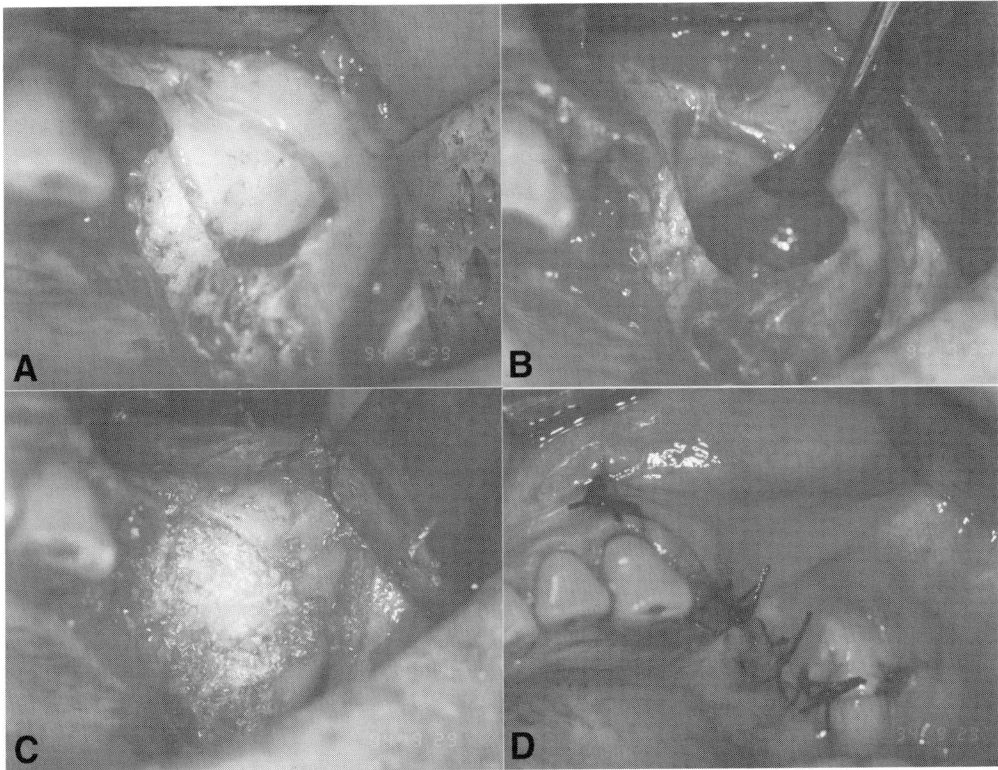

Fig. 3. Maxillary sinus graft: (**A**) exposed maxilla and outline of sinus door; (**B**) intrusion of the osseous door and visualization of the sinus; (**C**) placement of graft—demineralized bone matrix; (**D**) closure of soft tissue after membrane placement. (Photographs courtesy of Dr. F. Gotia.)

decrease bone regeneration potency. The question is whether the decrement is clinically significant to offset the peril from nonsterilized donor material. Significantly, donor screening at reputable banks is sufficiently stringent to satisfy most concerned clinicians and their patients. Nevertheless, for the percentage of individuals who are concerned about the transmissibility of disease and infection from allogeneic bank bone products, laboratory-engineered alternatives need to be considered.

Maxillary Sinus Augmentation

Sinus grafting provides an example of the use of autogenous and allogeneic graft material (**Fig. 3**). Once the sinus is exposed, a combination of both graft materials is used to enhance bone growth (*78–80*). A derived natural product, platelet-rich plasma (PRP), has been employed for sinus grafting. Results were equivocal as to the role of PRP in a combinatorial therapy with freeze-dried bone allograft (*81,82*). Recombinant human BMP-7 compared unfavorably with autogenous grafting of the sinus (*83*). Tarnow recommends the use of barrier membranes to enhance graft viability (*159*). Krauser demonstrated enhanced bone healing for PepGen P-15 placement compared with anorganic bovine bone in the sinus (*84*). Recombinant BMP-2 and marrow mesenchymal stem cells support bone growth in sinus augmentation (*85*). Numerous combination therapies have been attempted, though none is an optimal grafting material for the maxillary sinus. Many are case studies and do not support general conclusions about ideal therapy.

Mandibular Distraction Osteogenesis

Unlike sinus graft, another surgical procedure illustrates the use of bone's capacity to bridge a skeletal gap *(86)*. Sometimes denoted as hard tissue engineering, distraction osteogenesis can be a controlled separation of mandible bone while maintaining an environment supporting bone formation between the two pieces *(87,88)* (**Fig. 4**). Several animal studies support the predictable process of bone formation *(87,89,90)*. However, relapse is possible. For both sinus augmentation and osteogenesis, treatment therapies must be validated by long-term human studies in the future.

Orbital Fractures

The primary goals of orbital fracture repair are to preserve ocular muscle mobility, avoid exopthalmos, and maintain structural support for the globe. Both axial and coronal computerized tomography (CT) scans have become instrumental in defining the bony and soft tissue anatomy of orbital injuries. Thorough physical examination is required to assess visual acuity, muscle mobility, numbness, and any changes in orbital volume. Orbital fractures require operative intervention for ocular muscle entrapment, enopthalmos, and large bony defects (>2 cm^2) that may lead to enopthalmos over time.

Treatment modalities for orbital trauma depend on the specific injury. Fractures can be plated with metallic miniplates and screws and biodegradable plates. Resorbable plates induce bone remodeling *(91)* and theoretically have less long-term risk of becoming infected. In cases of bone defects, both alloplastic materials and autogenous tissues are available for the reconstruction effort. Autogenous tissues are preferred given their lower rate of infection and extrusion, but availability and donor-site morbidity can be issues. Ribs, iliac crest, calvarial bone *(92)*, maxillary antral bone *(93)*, nasoseptal cartilage *(32)*, and conchal cartilage have all been employed in orbital repairs. Dural membranes have been used successfully to repair small orbital fractures and comminuted fractures without bone loss *(94)*.

Cleft Palate

Initial repair of cleft palate should be done at 10–12 mo of age, prior to speech development. Often these patients require bone grafts to fill the cleft of the alveolar ridge, behind which orol-naslal fistulae tend to develop. This must be done prior to the eruption of permanent teeth (8–12 yr), to provide solid support for the teeth on either side of the cleft as well as a strong scaffold for the alar base of the nose. Iliac crest provides the best cancelous bone and, with a strong foundation for the dental ridge, orthodontics can complete the occlusion.

Velopharyngeal incompetence is ubiquitous in the cleft palate population. There are many surgical procedures to elongate the posterior portion of a cleft palate to prevent velopharyngeal incompetence. The Furlow, Von Langenbeck, and Veau-Wardill-Kilner palatoplasty techniques all involve manipulation of the soft tissues exclusively. Some groups have looked at distraction ostoegenesis as a way to lengthen the palate, and in animal studies this has been successful *(95)*. Distraction osteogenesis may be the future of cleft palate repair.

SOME AVAILABLE (PUTATIVE) BONE GRAFTING AGENTS

The Food and Drug Administration has approved a number of bone substitutes for selected clinical applications. The clinician of the 21st century is compelled to seek information and to understand bone substitutes and the various bone bank preparations.

Fig. 4. (*Opposite page*) Distraction osteogenesis of the mandibular naterior sextant: (**A**) preoperative gingival contour of mandible; (**B**) gingival flap retracted to expose osseous contour; (**C**) vertical and horizontal mobilization of anterior osseous segment; (**D**) securing guide apparatus to bone; (**E**) radiograph to confirm orientation following placement; (**F**) 2 wk postsurgery; (**G**) clinically hard tissue interposed at the distraction site. (Photographs courtesy of Dr. F. Gotia.)

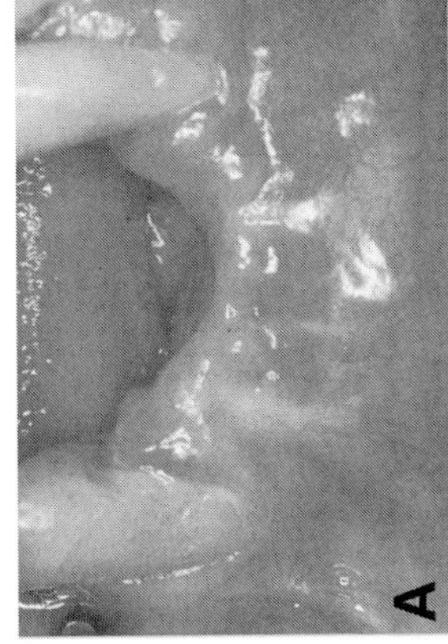

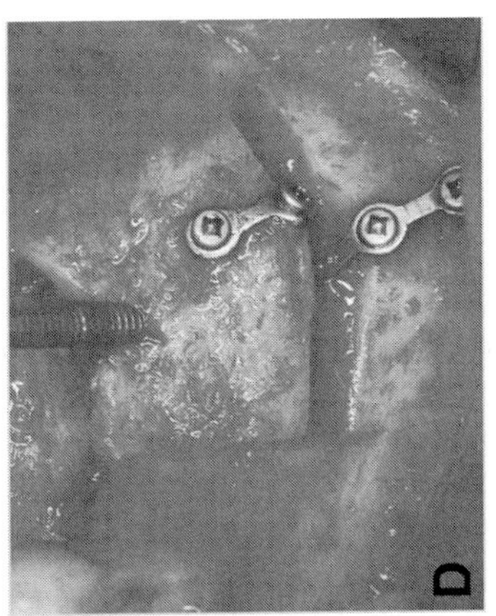

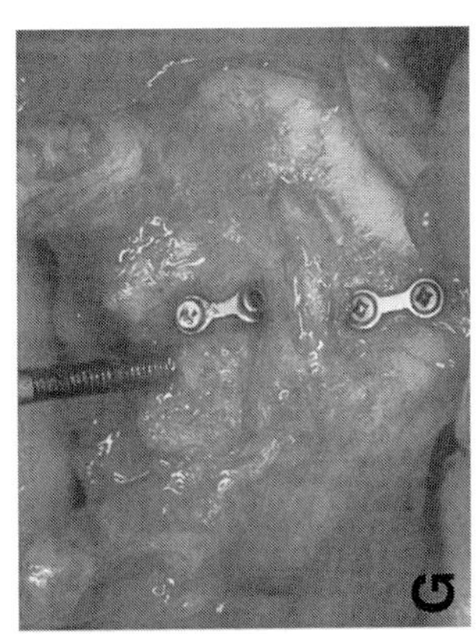

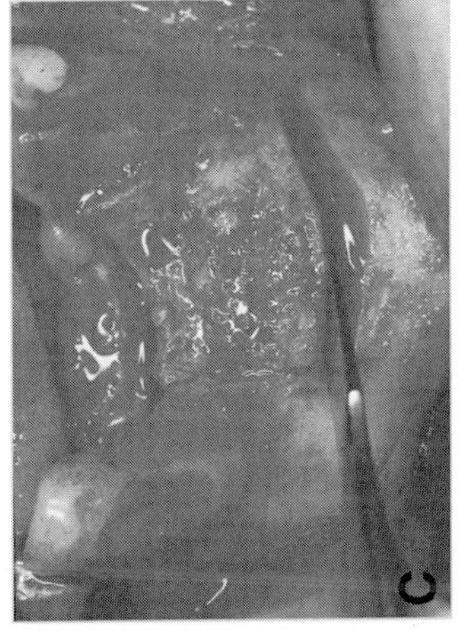

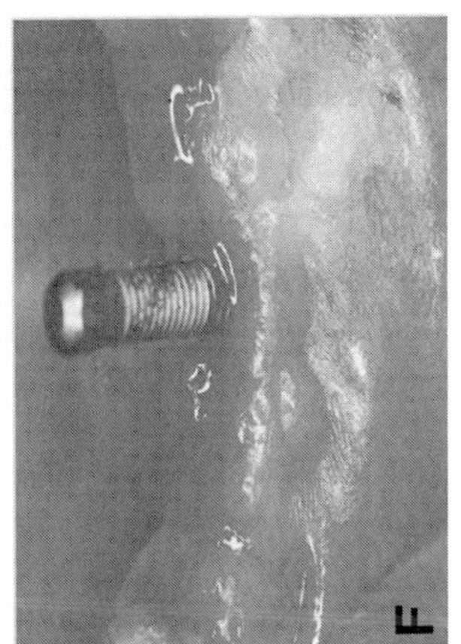

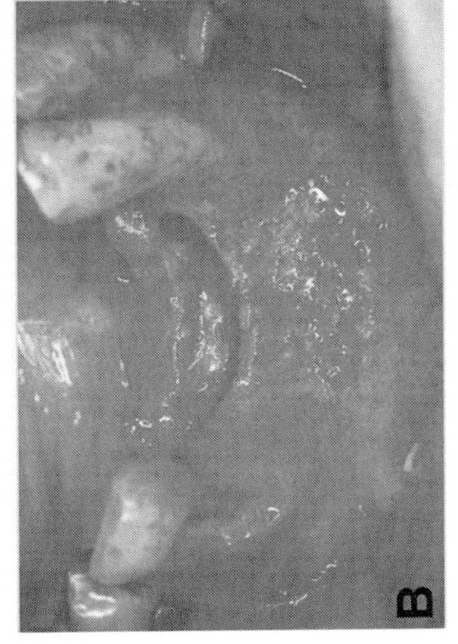

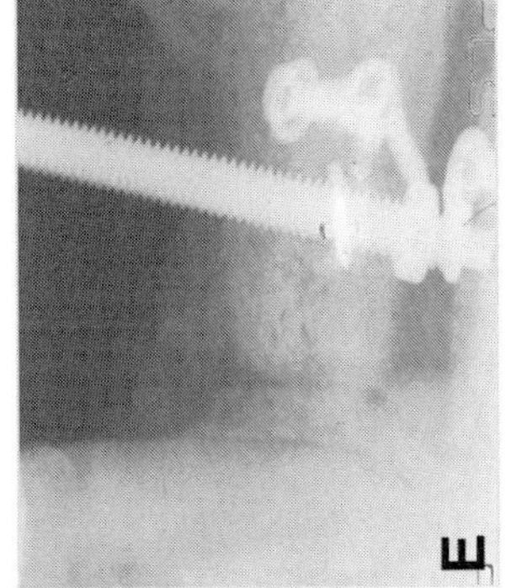

Polymers

Polymethylmethacrylate (PMMA) *(96,97)*, polytetrafluoroethylene (PTFE), and PMMA/polyhydroxy-ethylmethacrylate (PHEMA) *(96,98)* may be described as alloplastic, synthetic, nonbiodegradable polymers. PMMA has considerable versatility. It is used for dentures, arthroplasties, cranioplasties, and as a cement for many orthopedic prostheses.

Guided bone regeneration employs PTFE for augmentation (the Gore material SAM, subcutaneous augmentation material) of periodontal defects *(99)*. The principle of guided bone regeneration is that new bone formation will occur by providing a passageway for osteoblast lineage cells and osteoblasts. Preventing soft tissue prolapse into an osseous deficit creates the environment to facilitate preferential cell growth and differentiation into osteoblasts. Physiologically, fibroblasts are more likely to populate an intraosseous deficit than osteoblasts; therefore, by sustaining a zone for migration of osteoblast lineage cells, the clinical outcome should be bone and not connective tissue scar.

Hard tissue replacement (HTR-MFI) is PMMA/PHEMA prepared as blocks and particulates. The block format is for augmentation, whereas particulates have periodontal applications to restore deficient alveolar bone.

Collagen and Demineralized Bone Materials

The organic phase of bone is principally type I collagen. When bone is demineralized with hydrochloric acid, the method used by most commercial venders, the bone derivative is largely type I collagen and a minimal percent mixture of cell debris, a soup of soluble signaling molecules that is resistant to acidic demineralization, and residue ECM components. The format for the demineralized bone (DBM) can be either a range of particulate matter, blocks, or strips

Some clinicians voice laudatory praise for DBM products, while others have abandoned the notion that DBM is worthwhile. The variables influencing clinical outcome from DBM therapy include nonstandardized procurement and preparation techniques, donor age and gender, reagents and quality control, and sterilization (e.g., gamma-irradiate, ethylene oxide-sterilize). Clinical reports on combinations of DBM and autograft favor this composition over DBM alone and underscore DBM as an autograft expander.

Fundamental to DBM preparations is the principle of *bone induction (osteoinduction)*, a term promulgated by Urist and Strates in 1971 *(100)*. The working definition of osteoinduction based on Urist and Reddi embodies the recruitment of cells, differentiation to chondrocytes, osteoblasts, osteoclasts, and marrow cells, and the product of this activity is an *ossicle* (i.e., a small bone) *(100,101)*. Implicit to the Urist-Reddi definition of osteoinduction is bone formation induced at a nonbony locale (i.e., the promotion of heterotopic bone formation).

Hydroxyapatites

Hydroxyapatite (HA) is a major component of the inorganic compartment of bone. HA prepared commercially is biocompatible, with biodegradability either absent or extended over years. There has been an increasing interest in clinical applications of HA, inspiring clinical applications and spawning penetrating reviews *(102,103)*. A convenient categorization of HA includes laboratory-modified coralline forms, laboratory-modified bone-derived forms and laboratory-produced synthetic forms.

Laboratory-Modified, Bovine Bone-Derived HA

Bovine bone-derived products are available from several commercial sources. The deorganified bovine bone product has found a niche in the dental market, where it is used to restore deficient alveolar bone caused by periodontal disease.

Laboratory-Produced, Synthetic HA

Particulate HA has been synthesized in two versions: OsteoGraf/LD and OsteoGraf/D. The denominator identifies low density and dense, respectively. OsteoGraf/LD purportedly biodegrades more rapidly than OsteoGraf/D. We could not locate data in the literature quantitating products and degradation time.

A recent report with laboratory-produced HA from a research facility indicated that, combined with rhBMP-2 and implanted subcutaneously in rats, optimum pore size for osteoconduction ranged from 300 to 400 μm *(104)*. This pore size is consistent with the Haversian systems *(25)* in mature bone and could provide a capacity for self-renewal that is often absent with HA. Applications to the craniofacial skeleton of rhBMP-2/HA combinations may provide therapeutic options for onlays, which are inherently susceptible to resorption.

Injectable HAs have been proposed for craniofacial applications *(105)*. A critical review of these materials labeled them calcium phosphate cements (CPCs) *(102)*. CPCs are available from several commercial sources. CPCs are biocompatible and they have the capacity to be molded *in situ* and subsequently harden at the osseous site in 15–20 min *(106)*. Alpha-BSM can slowly biodegrade and become replaced with bone *(107)*.

Physiological mechanisms for CPCs relevant to bone regeneration are obscure and will need to be determined. The bone-promoting effect from HA materials such as the CPCs may be caused by the adsorption of signaling cues to the implanted material. The notion of local adsorption of growth factors is derived from the purification protocols of the 1980s, when HA affinity chromatography was used to purify BMPs from bone *(108,109)*.

CPCs may become part of the craniofacial repertoire. However, it is crucial for any commercially marketed product to face stringent assessments and be reported in peer-reviewed journals so that clinicians can have an unbiased, propaganda-free opportunity to judge for themselves.

OPTIONS FOR THE 21ST CENTURY

Tissue Engineering

For most regenerative applications, the autogenous graft is a benchmark against which alternatives should be judged. There is a bubbling passion among a cadre of dedicated researchers to develop options for surgeons that will rival the autograft. Individuals answering the challenge represent a diverse spectrum of disciplines. This diversity is the sinew for the field known as tissue engineering, which is an amalgam of clinical specialties and basic scientists. The power of this unity is that clinical conditions can be rationally addressed by clinical experts combining expertise with developmental biologists, cell and molecular biologists, polymer chemists, and so on.

In the area of the craniofacial complex, signaling molecules, cells, and delivery systems have been studied with the goal of assembling key ingredients into a clinical therapy. The opportunities and challenges have prompted an array of provocative options.

Final commentary emphasizes novel promising tissue engineering options that could affect treatment of osseous deficits in the craniofacial skeleton.

A device needs to be engineered to carry either the cells or signaling molecule, together or individually, to the surgical site. The term *delivery system* is not accurate. A suitable term must be developed to describe the functional roles, as well as to provide clinical convenience for the surgeon, to localize and protect the payload, to release that payload in a predictable and time-controlled fashion, to permit cell ingrowth, and to act transiently as an ECM until sufficient cell performance constructs a new substratum *(110–112)*. The concept we like is a biomimetic ECM, in light of the physiological roles provided by the ECM.

Signaling Molecules

Critical to any therapy are signaling molecules, defined as hormones, growth factors, and cytokines. This confusing array of molecules are the principal extracellular regulators of all physiological function, including embryogenesis, tissue postnatal growth, differentiation, maintenance, and repair and remodeling. The current understanding of the interaction of the temporal and spatial relationships of these molecules remains in its infancy. These molecules can be provided *ex vivo*, manipulated *in vivo*, and/or the host's own physiology can be relied on to provide these molecules. Our accumulated knowledge remains so inadequate that current selection criteria for possible therapies remains within the realm of experience and prejudices of the individual researcher. Specific rationales for any given signaling molecule are further limited by the limitations of translating the various sources of information into a coherent story. This is further exasperated by the failure of continuity in results between *in vitro*, *in vivo* animal and clinical applications.

Selected molecules that according to the literature and our estimation appear to offer probable clinical impact on bone regeneration for the craniofacial skeleton are discussed. Prejudices that sway decisions about therapy composition can be formidable deterrents to selection. A recent, insightful review on bone formation and growth factors presented a number of examples where data could support selection of platelet-derived growth factor to regenerate bone rather than insulin-like growth factor-1, but in certain circumstances, fibroblast growth factor-2 was better. It is relatively easy and convenient to argue for one factor vs another. Hundreds of literature citations and our experiences with growth factors underscores this statement. We feel strongly enough about this predicament that we are singling out bone morphogenetic protein to explain why we would not select it.

One therapeutic option that is highly visible to clinicians is recombinant human (rh) BMPs. In combination with either collagen or calcium-phosphate, rhBMP-2 appears to be an effective spine fusion option *(111,113,114)*. However, rhBMP-2 has had limited success for maxillary sinus lifts *(115)* and alveolar bone preservation *(116)*. Furthermore, clinical studies report the need for milligram doses of rhBMP-2 *(117)*. OP-1 is presently available in Europe, Canada, Australia, and under certain circumstances, the United States. BMPs are powerful morphogens capable of stimulating differentiation of the osteoblast lineage that may lead to bone regeneration. However multiple effects of BMPs, some nonosteogenic, are expected.

Potential options for therapy include other growth factors. Fibroblast growth factor (FGF) comprises a group of nine distinct members; FGF-1 (acidic FGF-2 may be a candidate agent for bone regeneration) is angiogenic and osteogenic *(118,119)*.

The biological activities of FGF-2 include mitogenesis for osteoblast-like phenotypes and angiogenesis *(120)*, suggesting a strong involvement in bone formation, fracture repair, and bone regeneration. Moreover, FGF-2 has a potent influence on bone formation *(121–124)*, especially in cranial osteogenesis *(125)*. Furthermore, the angiogenic capacity of FGF-2 is a noteworthy advantage for bone healing. There is a strong correlation between the temporal sequence of vascular reestablishment and viability and incorporation of a bone graft: revascularization supports graft incorporation and survival *(126–128)*.

The transcription factor Runx2 (previously known as Cbfa1, core binding factor alpha 1, and as Osf2, osteoblast specific factor 2) is a critical regulator of osteoblast differentiation *(56,69,129)*. Runx2 is the only osteoblast-specific transcription factor identified to date. Molecular and genetic evidence have demonstrated that it acts as an activator of osteoblast differentiation during embryonic development in mouse and human.

Osteoinductive factors such the BMPs regulate Runx2 expression in osteoblasts *(69)*. However, Runx2 is different from BMPs. BMPs are powerful morphogens that can induce bone formation through embryogenic recapitulation of several cell phenotypes. In contrast, Runx2 functions downstream of BMP to promote osteoblast differentiation *(130)* and control matrix deposition *(129)*. Evidence for this function was determined in mice that were homozygous for a targeted deletion of Runx2 that were absent of osteoblasts and lacked bone *(131,132)*.

The term *signaling molecules* is not specific. However, the term has crept into the literature and broadly includes molecules that provoke an effect. Signaling molecules can be either extracellular or intracellular. It would exceed the scope of this chapter to address the expanding list of signaling molecules and bone. Therefore, we have specified bone morphogenic proteins.

Bone Morphogenetic Proteins

Bone morphogenic proteins are signaling molecules that are attracting considerable attention for bone regeneration *(133–138)*. Contemporary literature has reported 16 BMPs (reviewed in refs. *65* and *139*) and while the B in BMP refers to bone, BMPs do more than prompt bone regeneration. Moreover, not all BMPs are osteoinductive. The extraordinary spectrum of activities includes developmental and functional organization of limbs, digits, liver, brain, kidney, muscle, and spleen *(140–148)*.

Except for BMP-1, the BMPs are members of the transforming growth factor-β (TGF-β) clan and are highly conserved proteins structurally and functionally related to regulatory gene products expressed by the invertebrate fruit fly (*D. melanogaster*) and primitive vertebrate newt (*C. elegans*). Specifically, BMP-2 and -4 are mammalian homologs to fruit fly gene *decapentaplegic* (*dpp*), and BMP-5, -7, and -8 have striking homology to fruit fly *Vgr/60A*. By determining functional relationships of BMP gene products in *D. melanogaster* and *C. elegans* to developing embryogenesis in these organisms, scientists can derive homologous functional significance to human conditions.

An additional subgroup of BMPs are the growth and differentiating factors (GDFs). The GDFs include GDF-1 (also known as cartilage-derived morphogenetic protein-1, CDMP-1), and at least seven other GDFs (reviewed in refs. *65, 149,* and *150*). Data indicate that the GDFs are implicated with cartilage development.

At this time, BMPs and GDFs have not been singled out as etiopathological factors for craniofacial developmental deficiencies. In contrast, mice with mutations encoding for GDF5 are associated with brachypodism (reduction in length of bones in the limbs, altered formation of bones and joints in the sternum, and reduction in the number of bones in the digits *[150]*). Furthermore, the condition in mice known as the *short ear* mouse is related to a defect in the BMP-5 gene. The *short ear* mouse has deficiently sized ears as well as shapes of ribs, sternum, and vertebral processes *(141)*. Limbs are not affected. These observations underscore the positional differences in regulators for the appendicular and axial skeleton. Moreover, the human condition known as cleidocranial dysplasia (abnormal or absent clavicles and supernumerary teeth) is associated with humans who are heterozygotes for the *cbfa1* gene mutation *(151)*.

Creative tissue engineering technologies exploiting GDFs and BMPs could be invoked to induce cartilage and bone formation preferentially in selective domains of the temporomandibular joint. Accomplishing this goal may involve several of the factors and signaling molecules discussed in this chapter, including pro-angiogenic molecules (e.g., VEGF, FGF), anti-BMPs, Smads, and cbfa.

Despite membership in the common clan of TGF-β, TGF-β molecules and BMPs have different influences on osteoblast lineage cells. TGF-β increases expression of collagen type I, osteopontin, and osteonectin, but reduces alkaline phosphatase and osteocalcin expression, and BMP increases alkaline phosphatase and osteocalcin expression *(152)*. Variability in mechanisms of action among these molecules, as well as developmental and maturational differences among cells of the same phenotype, may be responsible for gene expression consequences noted. Therefore, unanticipated clinical outcomes from therapeutic application of powerful signaling molecules should not be a surprise. We described in the fracture model how the dynamics of a healing wound embody a heterogenous cell population at several maturational levels. In contrast, embryogenesis may be viewed as a less dynamic, more refined, silky process.

The clinical appeal for BMPs is their capacity to promote bone regeneration. Prior to clinical testing of BMPs, a frenzy of preclinical studies determined that BMP-2 and -4 through -7 were osteoinductive (reviewed in ref. *139*). However, a dose–response parity among BMPs was not consistent. Furthermore, preclinical testing clearly indicated that a significant ramping up in dosing requirements for BMPs was necessary as testing animal models ascended a phylogenetic ladder. A dosing relationship for BMPs revealed that when progressing from rat to rabbit, a 3× dose was needed; from rabbit to dog, 10×; and from dog to nonhuman primate and clinical studies in patients, 3–5× *(74,115,153–155)*.

Consideration of the clinical location must also be made, in which the craniofacial complex may require a different dose of BMP than the distal extremity or spine.

Limited clinical testing has been reported to date *(115,116,156)* including the use of OP-1 in tibial nonunion and BMPs in the spine. In reported studies, a local pharmacological effect required superphysiological BMP doses ranging from 1.7 to 3.4 mg. As a consequence of pharmacological dosing, are there effects distant from the local administration site? A legitimate concern is whether BMP transit from the local osseous wound to other tissues may either induce unwanted bone formation or inspire an unexpected reaction. The concept to underscore is that BMP is a *morphogen*, and morphogens exert profound cellular effects that are dose- and cell phenotype-dependent.

BMPs promote an effect with responding cells through membrane-bound transmembrane receptors (serine-threonine kinases), designated types IA, IB, and II (reviewed in refs. *62* and *157*). Following receptor binding, a series of intracellular events is initiated, known as signal transduction. Signal transduction for BMPs is modulated by intracellular signaling molecules called Smads (noted earlier in the chapter) (**Fig. 2**). Outside the cell, regulation of BMP-receptor binding may be abrogated through proteins that either bind to the receptor and block BMP-receptor interaction or else bind directly to the BMPs, thereby interfering and short-circuiting ligand–receptor interaction.

A number of anti-BMPs have been identified; examples include chordin, noggin, follistatin, DAN, and cerbrus (reviewed in ref. *62*). Osteoblasts express BMPs, noggin and chordin *(158)*. Therefore, a self-regulatory, autocrine loop could enable an osteoblast either to control itself (i.e., autocrine) or neighboring osteoblasts (i.e., paracrine).

Binding proteins for growth factors are the rule in nature, where TGF-β and IGFs are expressed in an inactive, protein-bound form requiring activation. BMPs are secreted in an active form, and curtailing activity must be accomplished through either proteolytic degradation or anti-BMPs. The anti-BMPs may be upregulated as bone regeneration nears completion, thereby shutting down an overabundance of unnecessary growth. While most craniofacial clinicians envision the power of BMP and bone regeneration, the opportunities to use an anti-BMP strategy could benefit the orthopedic surgeon in preventing heterotopic bone formation that may occur with hip arthroplasties.

Delivery Systems

Available systems will be developed to use with cells, molecules, or DNA. The vehicle must be biocompatible, supportive of cell viability, and protective against undesireable protein and DNA degradation. The device must be capable of temporal and spatial release of the osteoinductive agent *(110–112)*.

CONCLUSIONS

New therapies incorporating newly identified osteoinductive molecules and materials will enhance regeneration of the bone. Autogenous graft, allogeneic bank bone, and various alloplastics, as well as technical improvements in distraction osteogenesis, will continue to provide beneficial clinical treatment to correct craniofacial osseous deficiencies.

REFERENCES

1. Furner, S. E. and Kozak, L. J. (1993) Health data on older Americans: United States, 1992. Acute care. *Vital Health Stat* 3, 115–142.
2. Böstrom, M. P. (1998) Expression of bone mrophogenetic proteins in fracture healing. *Clin. Orthop. Rel. Res.* **355S**, S116–S123.
3. Johari, A. N. and Sinha, M. (1999) Remodeling of forearm fractures in children. *J. Pediatr. Orthop. B* **8(2)**, 84–87.
4. Ekholm, E. C., Ravanti, L., Kahari, V., Paavolainen, P., and Penttinen, R. P. (2000) Expression of extracellular matrix genes: transforming growth factor (TGF)-beta1 and ras in tibial fracture healing of lathyritic rats. *Bone* **27(4)**, 551–557.
5. Kusuzaki, K., Kageyama, N., Shinjo, H., et al. (2000) Development of bone canaliculi during bone repair. *Bone* **27(5)**, 655–659.
6. Bebchuk, T. N., Degner, D. A., Walshaw, R., et al. (2000) Evaluation of a free vascularized medial tibial bone graft in dogs. *Vet. Surg.* **29(2)**, 128–144.

7. McKibbin, B. (1978) The biology of fracure healing in long bones. *J. Bone Joint Surg.* **60B(2)**, 150–162.
8. Kleinschmidt, J. and Hollinger, J. O. (1992) Animal models in bone research, in *Bone Grafts and Bone Substitutes* (Habal, M. and Reddi, A. H., eds.), Saunders, Philadelphia, pp. 133–147.
9. Weinzweig, J., Pantaloni, M., Spangenberger, A., Marler, J., and Zienowicz, R. J. (2000) Osteochondral reconstruction of a non-weight-bearing joint using a high-density porous polyethylene implant. *Plast. Reconstr. Surg.* **106(7)**, 1547–1554.
10. Craft, P. D. and Sargent, L. A. (1989) Membranous bone healing and techniques in calvarial bone grafting. *Clin. Plast. Surg.* **16(1)**, 11–19.
11. Verdi, G. D. and Alpert, B. (1986) Bone grafts in orthognathic and craniofacial surgery. *Ear Nose Throat J.* **65(11)**, 506–511.
12. Reddi, A. H. (1998) Initiation of fracture repair by bone morphogenetic proteins. *Clin. Orthop. Rel. Res.* **355S**, S66–S72.
13. Burgess, E. A., Mayer, M. H., and Hollinger, J. O. (2000) Bone grafting and substitutes, in *Plastic Surgery: Indications Operations Outcomes* (Coleman, J. J., Vander Kolk, C. A., and Achauer, B. M., eds.), Mosby-Year Book, Philadelphia, in press.
14. Adami, R. C., Collard, W. T., Gupta, S. A., Kwok, K. Y., Bonadio, J., and Rice, K. G. (1998) Stability of peptide-condensed plasmid DNA formulations. *J. Pharm. Sci.* **87**, 678–683.
15. Bonadio, J., Smiley, E., Patil, P., and Goldstein, S. (1999) Localized, direct plasmid gene therapy *in vivo*: prolonged therapy results in reproducible tissue regeneration. *Nat. Med.* **5(7)**, 753–759.
16. Perry, C. (1999) Bone repair techniques, bone graft, and bone graft substitutes. *Clin. Orthop. Rel. Res.* **360**, 71–86.
17. Baltzer, A., Lattermann, C., Whalen, J., et al. (2000) Potential role of direct adenoviral gene transfer in enhancing fracture repair. *Clin. Orthop. Rel. Res.* **S379(3)**, S120–S125.
18. Chow, K. M. and Rabie, A. B. (2000) Vascular endothelial growth pattern of endochondral bone graft in the presence of demineralized intramembranous bone matrix—quantitative analysis. *Cleft Palate Craniofac. J.* **37(4)**, 385–394.
19. Ahn, D. K., Sims, C. D., Randolph, M. A., et al. (1997) Craniofacial skeletal fixation using biodegradable plates and cyanoacrylate glue. *Plast. Reconstr. Surg.* **99(6)**, 1508–1515; discussion 16–17.
20. Bensinger, W. I., Maloney, D., and Storb, R. (2001) Allogeneic hematopoietic cell transplantation for multiple myeloma. *Semin. Hematol.* **38(3)**, 243–249.
21. Einhorn, T. A. (1998) The cell and molecular biology of fracture healing. *Clin. Orthop.* **355(Suppl)**, S7–S21.
22. Clark, R. A. F. (1996) *The Molecular and Cellular Biology of Wound Repair.* 22nd ed. Plenum, New York.
23. Bolander, M. E. (1992) Regulation of fracture repair by growth factors. *Proc. Soc. Exp. Biol. Med.* **200**, 165–170.
24. Macias, M. P., Fitzpatrick, L. A., Brenneise, I., McGarry, M. P., Lee, J. J., and Lee, N. A. (2001) Expression of IL-5 alters bone metabolism and induces ossification of the spleen in transgenic mice. *J. Clin. Invest.* **107(8)**, 949–959.
25. Hollinger, J. O., Buck, D. C., and Bruder, S. (1998) Biology of bone healing: its impact on clinical therapy, in *Tissue Engineering: Applications in Maxillofacial Surgery and Periodontics* (Lynch, S., Marx, R., and Genco, R., eds.), Quintessence, San Diego, pp. 17–53.
26. Le, A. X., Miclau, T., Hu, D., and Helms, J. A. (2001) Molecular aspects of healing in stabilized and non-stabilized fractures. *J. Orthop. Res.* **19(1)**, 78–84.
27. Alhopuro, S. (1978) Premature fusion of facial sutures with free periosteal grafts. An experimental study with special reference to bone formation with free periosteal grafts from the tibia, the scapula and the calvarium. *Scand. J. Plast. Reconstr. Surg. Suppl.* **17**, 1–68.
28. Jones, D., Leivseth, G., and Tenbosch, J. (1995) Mechano-reception in osteoblast-like cells. *Biochem. Cell Biol.* **73 (7–8)**, 525–534.
29. Harter, L. V., Hruska, K. A., and Duncan, R. L. (1995) Human osteoblast-like cells respond to mechanical strain with increased bone matrix protein production independent of hormonal regulation. *Endocrinology* **136(2)**, 528–535.
30. Schierle, H. P. and Hausamen, J. E. (1997) [Modern principles in treatment of complex injuries of the facial bones]. *Unfallchirurg* **100(5)**, 330–337.
31. Mathog, R. H., Toma, V., Clayman, L., and Wolf, S. (2000) Nonunion of the mandible: an analysis of contributing factors. *J. Oral Maxillofac. Surg.* **58(7)**, 746–752; discussion 52–53.
32. Caplanis, N., Lee, M. B., Zimmerman, G. J., Selvig, K. A., and Wikesjo, U. M. (1998) Effect of allogeneic freeze-dried demineralized bone matrix on regeneration of alveolar bone and periodontal attachment in dogs. *J. Clin. Periodontol.* **25(10)**, 801–806.
33. Brighton, C. T. and Hunt, R. M. (1986) Histochemical localization of calcium in the fracture callus with potassium pyroantimonate. Possible role of chondrocyte mitochondrial calcium in callus calcification. *J. Bone Joint Surg.* **68(5)**, 703–715.
34. Einhorn, T. (1995) Current concepts review. Enhancement of fracture healing. *J. Bone Joint Surg.* **77A(6)**, 940–956.
35. Yamazaki, M., Majeska, R. J., Yoshioka, H., Moriya, H., and Einhorn, T. A. (1997) Spatial and temporal expression of fibril-forming minor collagen genes (types V and XI) during fracture healing. *J. Orthop. Res.* **15(5)**, 757–764.
36. Jingushi, S., Heydemann, A., Kana, S. K., Macey, L. R., and Bolander, M. E. (1990) Acidic fibroblast growth factor (aFGF) injection stimulates cartilage enlargement and inhibits cartilage gene expression in rat fracture healing. *J. Orthop. Res.* **8(3)**, 364–371.

37. Jingushi, S., Joyce, M. E., and Bolander, M. E. (1992) Genetic expression of extracellular matrix proteins correlates with histologic changes during fracture repair. *J. Bone Miner. Res.* **7(9)**, 1045–1055.
38. Hiltunen, A., Metsaranta, M., Perala, M., Saamanen, A. M., Aro, H. T., and Vuorio, E. (1995) Expression of type VI, IX and XI collagen genes and alternative splicing of type II collagen transcripts in fracture callus tissue in mice. *FEBS Lett.* **364(2)**, 171–174.
39. Hiltunen, A., Metsaranta, M., Virolainen, P., Aro, H. T., and Vuorio, E. (1994) Retarded chondrogenesis in transgenic mice with a type II collagen defect results in fracture healing abnormalities. *Dev. Dyn.* **200(4)**, 340–349.
40. Hiltunen, A., Aro, H. T., and Vuorio, E. (1993) Regulation of extracellular matrix genes during fracture healing in mice. *Clin. Orthop.* **297**, 23–27.
41. Sugimoto, M., Hirota, S., Sato, M., et al. (1998) Impaired expression of noncollagenous bone matrix protein mRNAs during fracture healing in ascorbic acid-deficient rats. *J. Bone Miner. Res.* **13(2)**, 271–278.
42. Termine, J. D. (1990) Cellular activity, matrix proteins, and aging bone. *Exp. Gerontol.* **25**, 217–221.
43. Chen, J. H., Wang, X. C., Kan, M., and Sato, J. D. (2001) Effect of FGF-1 and FGF-2 on VEGF binding to human umbilical vein endothelial cells. *Cell Biol. Int.* **25(3)**, 257–260.
44. Montero, A., Okada, Y., Tomita, M., et al. (2000) Disruption of the fibroblast growth factor-2 gene results in decreased bone mass and bone formation. *J. Clin. Invest.* **105(8)**, 1085–1093.
45. Hosokawa, R., Kikuzaki, K., Kimoto, T., et al. (2000) Controlled local application of basic fibroblast growth factor (FGF-2) accelerates the healing of GBR. An experimental study in beagle dogs. *Clin. Oral Implants Res.* **11(4)**, 345–353.
46. Fitzpatrick, L. A. and Bilezkian, J. P. (1996) Actions of parathyroid hormone, in *Principles of Bone Biology* (Bilezkian, J. P., Raisz, L. G., and Rodan, G. A., eds.), Academic, New York, pp. 339–346.
47. Hill, P. A., Tumber, A., and Meikle, M. C. (1997) Multiple extracellular signals promote osteoblast survival and apoptosis. *Endocrinology* **138(9)**, 3849–3858.
48. Rodan, G. (1998) Control of bone formation and resorption: biological and clinical perspective. *J. Cell Biochem.* **Suppl. 30**, 55–61.
49. Ahrens, M., Ankenbauer, T., Schröder, D., Hollnagel, A., Mayer, H., and Gross, G. (1993) Expression of human morphogenetic proteins-2 or -4 in murine mesenchymal progenitor C3H10T1/2 cells induces differentiation into distinct mesenchymal cell lineages. *DNA Cell Biol.* **12(10)**, 871–880.
50. Chan, Y. S., Ueng, S. W., Wang, C. J., Lee, S. S., Chao, E. K., and Shin, C. H. (1998) Management of small infected tibial defects with antibiotic-impregnated autogenic cancellous bone grafting. *J. Trauma* **45(4)**, 758–764.
51. Celeste, A. J., Song, J. J., Cox, K., Rosen, V., and Wozney, J. M. (1994) Bone morphogenetic protein-9, a new member of the TGF-beta superfamily. *J. Bone Miner. Res.* **9(1)**, S137.
52. Amedee, J., Bareille, R., Rouais, F., Cunningham, N., Reddi, N., and Hamand, M. F. (1994) Osteogenin (bone morphogenetic protein-3) inhibits proliferation and stimulates differentiation of osteoprogenitors in human bone marrow. *Differentiation* **58**, 157–164.
53. Ikeda, T., Nagai, A., Yamaguchi, A., Yokose, S., and Yoshiki, S. (1995) Age-related reduction in bone matrix protein mRNA expression in rat bone tissues: application of histomorphometry to in situ hybridization. *Bone* **16(1)**, 17–23.
54. Asahina, I., Sampath, T. K., and Hauschka, P. V. (1996) Human osteogenic protein-1 induces chondroblastic, osteoblastic, and/or adipocytic differentiation. *Exp. Cell Res.* **222**, 38–47.
55. Reissmann, E., Ernsberger, U., Francis-West, P. H., Rueger, D., Brickell, P. M., and Rohrer, H. (1996) Involvement of bone morphogenetic protein-4 and bone morphogenetic protein-7 in the differentiation of the adrenergic phenotype in developing sympathetic neurons. *Development* **122**, 2079–2088.
56. Ducy, P. and Karsenty, G. (1998) Genetic control of cell differentiation in the skeleton. *Curr. Opin. Cell Biol.* **10**, 614–619.
57. Uusitalo, H., Hiltunen, A., Ahonen, M., Kahari, V., Aro, H., and Vuorio, E. (2001) Induction of periosteal callus formation by bone morphogenetic protein-2 employing adenovirus-mediated gene delivery. *Matrix Biol.* **20(2)**, 123–127.
58. Derynck, R., Zhang, Y., and Feng, X. (1998) Smads: transcription activatros of TGF-beta responses. *Cell* **95**, 737–740.
59. Barnes, G., Kostenuik, P., Geerstenfeld, L., and Einhorn, T. (1999) Growth factor regulation of fracture repair. *J. Bone Miner. Res.* **14(11)**, 1805–1815.
60. Hirano, T. (1999) Molecular basis underlying functional pleiotropy of cytokines and growth factors. *Biochem. Biophys. Res. Commun.* **260**, 303–308.
61. Sakou, T., Onishi, T., Yamamoto, T., Nagamine, T., Sampath, K., and Ten Dijke, P. (1999) Localization of Smads, the TGF-beta family intracellular signaling components during endochondral ossification. *J. Bone Miner. Res.* **14(7)**, 1145–1152.
62. Schmitt, J. M., Hwang, K., Winn, S. R., and Hollinger, J. O. (1999) Bone morphogenetic proteins: an update on basic biology and clinical relevance. *J. Orthop. Res.* **17(2)**, 269–278.
63. Weinstein, M., Yang, X., and Deng, C. (2000) Functions of mammalian Smad genes as revealed by targeted gene disruption in mice. *Cytokine Growth Factor Rev.* **11(1–2)**, 49–58.
64. Yanagisawa, K., Uchida, K., Nagatake, M., et al. (2000) Heterogeneities in the biological and biochemical functions of Smad2 and Smad4 mutants naturally occurring in human lung cancers. *Oncogene* **19(19)**, 2305–2311.

65. Mont, M. A., Jones, L. C., Einhorn, T. A., Hungerford, D. S., and Reddi, A. H. (1998) Osteonecrosis of the femoral head. Potential treatment with growth and differentiation factors. *Clin. Orthop.* **355(Suppl)**, S314–S335.
66. Hay, E., Lemonnier, J., Fromigue, O., and Marie, P. J. (2001) Bone morphogenetic protein-2 promotes osteoblast apoptosis through a Smad-independent, protein kinase C-dependent signaling pathway. *J. Biol. Chem.* **276(31)**, 29028–29036.
67. Lee, K. S., Kim, H. J., Li, Q. L., et al. (2000) Runx2 is a common target of transforming growth factor beta1 and bone morphogenetic protein 2, and cooperation between Runx2 and Smad5 induces osteoblast-specific gene expression in the pluripotent mesenchymal precursor cell line C2C12. *Mol. Cell Biol.* **20(23)**, 8783–8792.
68. Zhang, Y. W., Yasui, N., Ito, K., et al. (2000) A RUNX2/PEBP2alpha A/CBFA1 mutation displaying impaired transactivation and Smad interaction in cleidocranial dysplasia. *Proc. Natl. Acad. Sci. USA* **97(19)**, 10549–10554.
69. Ducy, P., Zhang, R., Geoffroy, V., Ridall, A., and Karsenty, G. (1997) Osf2/Cbfa1: a transcriptional activator of osteoblast differentiation. *Cell* **89**, 747–754.
70. Simonet, W., Lacey, D., Dunstan, C., et al. (1997) Osteoprotegerin: a novel secreted protein involved in the regulation of bone density. *Cell* **89**, 309–319.
71. Itonaga, I., Sabokbar, A., Murray, D. W., and Athanasou, N. A. (2000) Effect of osteoprotegerin and osteoprotegerin ligand on osteoclast formation by arthroplasty membrane derived macrophages. *Ann. Rheum. Dis.* **59(1)**, 26–31.
72. Mentaverri, R., Lorget, F., Wattel, A., Maamer, M., Kamel, S., and Brazier, M. (2000) [Osteoblastic regulation of osteoclast survival: effect of calcitriol]. *C. R. Acad. Sci. III* **323(11)**, 951–957.
73. Mayer, M. H. (1994) Clinical perspectives on bone grafting: avoiding the complications of developmental malformations, in *Bone Repair and Regeneration* (Seyfer, A. and Hollinger, J., eds.), Saunders, Philadelphia, pp. 365–376.
74. Hollinger, J. O., Mayer, M., Buck, D., et al. (1996) Poly(alpha-hydroxy acid) carrier for delivering recombinant human bone morphogenetic protein-2 for bone regeneration. *J. Controlled Rel.* **39**, 287–304.
75. Hanamura, H., Higuchi, Y., Nakagawa, M., Iwata, H., Nogami, H., and Urist, M. R. (1980) Solubilized bone morphogenetic protein (BMP) from mouse osteosarcoma and rat demineralized bone matrix. *Clin. Orthop. Rel. Res.* **148**, 281–290.
76. Goldberg, V. M. and Stevenson, S. (1987) Natural history of autografts and allografts. *Clin. Orthop. Rel. Res.* **225**, 7–16.
77. Stevenson, S. (1998) Enhancement of fracture healing with autogenous and allogeneic bone grafts. *Clin. Orthop. Rel. Res.* **355S**, S239–S246.
78. Nishibori, M., Betts, N. J., Salama, H., and Listgarten, M. A. (1994) Short-term healing of autogenous and allogeneic bone grafts after sinus augmentation: a report of 2 cases. *J. Periodontol.* **65(10)**, 958–966.
79. Chanavaz, M. (2000) Sinus graft procedures and implant dentistry: a review of 21 years of surgical experience (1979–2000). *Implant Dent.* **9(3)**, 197–206.
80. Hanisch, O., Lozada, J. L., Holmes, R. E., Calhoun, C. J., Kan, J. Y., and Spiekermann, H. (1999) Maxillary sinus augmentation prior to placement of endosseous implants: a histomorphometric analysis. *Int. J. Oral Maxillofac. Implants* **14(3)**, 329–336.
81. Kassolis, J. D., Rosen, P. S., and Reynolds, M. A. (2000) Alveolar ridge and sinus augmentation utilizing platelet-rich plasma in combination with freeze-dried bone allograft: case series. *J. Periodontol.* **71(10)**, 1654–1661.
82. Lozada, J. L., Caplanis, N., Proussaefs, P., Willardsen, J., and Kammeyer, G. (2001) Platelet-rich plasma application in sinus graft surgery: Part I—Background and processing techniques. *J. Oral Implantol.* **27(1)**, 38–42.
83. van den Bergh, J. P., ten Bruggenkate, C. M., Groeneveld, H. H., Burger, E. H., and Tuinzing, D. B. (2000) Recombinant human bone morphogenetic protein-7 in maxillary sinus floor elevation surgery in 3 patients compared to autogenous bone grafts. A clinical pilot study. *J. Clin. Periodontol.* **27(9)**, 627–636.
84. Krauser, J. T., Rohrer, M. D., and Wallace, S. S. (2000) Human histologic and histomorphometric analysis comparing OsteoGraf/N with PepGen P-15 in the maxillary sinus elevation procedure: a case report. *Implant Dent.* **9(4)**, 298–302.
85. Ueda, M., Tohnai, I., and Nakai, H. (2001) Tissue engineering research in oral implant surgery. *Artif. Organs* **25(3)**, 164–171.
86. Costantino, P. D., Friedman, C. D., Shindo, M. L., Houston, G., and Sisson, G. A., Sr. (1993) Experimental mandibular regrowth by distraction osteogenesis. Long-term results. *Arch. Otolaryngol. Head Neck Surg.* **119(5)**, 511–516.
87. Wiltfang, J., Kessler, P., Merten, H. A., and Neukam, F. W. (2001) Continuous and intermittent bone distraction using a microhydraulic cylinder: an experimental study in minipigs. *Br. J. Oral Maxillofac. Surg.* **39(1)**, 2–7.
88. Vu, T., Shipley, J., Bergers, G., et al. (1998) MMP-9/gelatinase B is a key regulator of growth plates angiogenesis and apoptosis of hypertrophic chondrocytes. *Cell* **93**, 411–422.
89. Al Ruhaimi, K. A. (2001) Bone graft substitutes: a comparative qualitative histologic review of current osteoconductive grafting materials. *Int. J. Oral Maxillofac. Implants* **16(1)**, 105–114.
90. Jorgensen, C., Noel, D., Apparailly, F., and Sany, J. (2001) Stem cells for repair of cartilage and bone: the next challenge in osteoarthritis and rheumatoid arthritis. *Ann. Rheum. Dis.* **60(4)**, 305–309.
91. Rozema, F. R., Bos, R. R., Pennings, A. J., and Jansen, H. W. (1990) Poly(L-lactide) implants in repair of defects of the orbital floor: an animal study. *J. Oral Maxillofac. Surg.* **48(12)**, 1305–1309; 1310.
92. Maves, M. D. and Matt, B. H. (1986) Calvarial bone grafting of facial defects. *Otolaryngol. Head Neck Surg.* **95(4)**, 464–470.

93. Copeland, M. and Meisner, J. (1991) Maxillary antral bone grafts for repair of orbital fractures. *J. Craniofac. Surg.* **2(1)**, 18–21.

94. Iatrou, I., Theologie-Lygidakis, N., and Angelopoulos, A. (2001) Use of membrane and bone grafts in the reconstruction of orbital fractures. *Oral Surg. Oral Med. Oral. Pathol. Oral Radiol. Endodont.* **91(3)**, 281–286.

95. Ascherman, J. A., Marin, V. P., Rogers, L., and Prisant, N. (2000) Palatal distraction in a canine cleft palate model. *Plast. Reconstr. Surg.* **105(5)**, 1687–1694.

96. Yukna, R. A. and Yukna, C. N. (1997) Six-year clinical evaluation of HTR synthetic bone grafts in human grade II molar furcations. *J. Periodontal. Res.* **32(8)**, 627–633.

97. Yukna, R. A. (1994) Clinical evaluation of HTR polymer bone replacement grafts in human mandibular Class II molar furcations. *J. Periodontol.* **65(4)**, 342–349.

98. Trombelli, L., Lee, M. B., Promsudthi, A., Guglielmoni, P. G., and Wikesjo, U. M. (1999) Periodontal repair in dogs: histologic observations of guided tissue regeneration with a prostaglandin E1 analog/methacrylate composite. *J. Clin. Periodontol.* **26(6)**, 381–387.

99. Eickholz, P., Kim, T. S., Holle, R., and Hausmann, E. (2001) Long-term results of guided tissue regeneration therapy with non-resorbable and bioabsorbable barriers. I. Class II furcations. *J. Periodontol.* **72(1)**, 35–42.

100. Urist, M. R. and Strates, B. S. (1971) Bone morphogenetic protein. *J. Dent. Res.* **50**, 1392–1406.

101. Carrington, J. L., Roberts, A. B., Flanders, K. C., Roche, N. S., and Reddi, A. H. (1988) Accumulation, localization, and compartmentation of transforming growth factor β during endochondral bone development. *J. Cell Biol.* **107**, 1969–1975.

102. Schmitz, J. P., Hollinger, J. O., and Milam, S. B. (1999) Reconstruction of bone using calcium phosphate bone cements: a critical review. *J. Oral Maxillofac. Surg.* **57**, 1122–1126.

103. Burgess, E. A. and Hollinger, J. O. (1999) Options for engineering bone, in *Frontiers in Tissue Engineering* (Patrick, C. W., ed.), Elsevier, pp. 383–399.

104. Tsuruga, E., Takita, H., Itoh, H., Wakisaka, Y., and Kuboki, Y. (1997) Pore size of porous hydroxyapatite as the cell-substratum controls BMP-induced osteogenesis. *J. Biochem.* **121(2)**, 317–324.

105. Chow, L., Takagi, S., Constantino, P. D., and Friedman, C. D. (1991) Self-setting calcium phosphate cements. *Mater. Res. Soc. Symp. Proc.* **179**, 3–24.

106. Knaack, D., Goad, M. E., Aiolova, M., et al. (1998) Resorbable calcium phosphate bone substitute. *J. Biomed. Mater. Res.* **43(4)**, 399–409.

107. Baltzer, A. W., Lattermann, C., Whalen, J. D., Braunstein, S., Robbins, P. D., and Evans, C. H. (1999) A gene therapy approach to accelerating bone healing. Evaluation of gene expression in a New Zealand white rabbit model. *Knee Surg. Sports Traumatol. Arthrosc.* **7(3)**, 197–202.

108. Syftestad, G. T., Triffit, J. T., Urist, M. R., and Caplan, A. I. (1984) An osteo-inductive bone matrix extract stimulates the in vitro conversion of mesenchyme into chondrocytes. *Calcif. Tissue Int.* **36**, 625–627.

109. Sampath, T. K., Muthukumaran, N., and Reddi, A. H. (1987) Isolation of osteogenin, an extracellular matrix-associated bone-inductive protein, by heparin affinity chromatography. *Proc. Natl. Acad. Sci. USA* **84**, 7109–113.

110. Hollinger, J. O. (1997) What's new in bone biology? *J. Histotech.* **20(3)**, 235–240.

111. Boden, S. D.. Hair, G. A., Viggeswarapu, M., Liu, Y., and Titus, L. (2000) Gene therapy for spine fusion. *Clin. Orthop.* **379(Suppl)**, S225–S233.

112. Lee, Y. M., Seol, Y. J., Lim, Y. T., et al. (2001) Tissue-engineered growth of bone by marrow cell transplantation using porous calcium metaphosphate matrices. *J. Biomed. Mater. Res.* **54(2)**, 216–223.

113. Sandhu, H. and Boden, S. (1998) Biologic enhancement of spine fusion. *Orthop. Clin. N. Am.* **29(4)**, 621–631.

114. Boden, S. (2000) Biology of lumbar spine fusion and use of bone graft substitutes: present, future, and next generation. *Tissue Eng.* **6(4)**, 383–400.

115. Boyne, P. J., Marx, R. E., Nevins, M., et al. (1997) A feasibility study on evaluation rhBMP-2/absorbable collagen sponge for maxillary sinus augmentation. *Int. J. Periodont. Restor. Dent.* **17(1)**, 11–25.

116. Howell, T. H., Fiorellini, J., Jones, A., et al. (1997) A feasibility study evaluating rhBMP-2/absorbable collagen sponge device for local alveolar ridge preservation of augmentation. *Int. J. Periodont. Restor. Dent.* **17(2)**, 125–139.

117. Riedel, G. E. and Valentin-Opran, A. (1999) Clinical evaluation of rhBMP-2/ACS in orthopedic trauma: a progress report. *Orthopedics* **22(7)**, 663–665.

118. Rodan, S. B. and Rodan, G. A. (1992) Fibroblast growth factor and platelet derived growth factor, in *Cytokines and Bone and Metabolism* (Gowan, M., ed.), CRC Press, Boca Raton, FL, pp. 116–140.

119. Dunstan, C. R., Boyce, B. F., Boyce, R., et al. (1999) Systemic administration of acidic fibroblast growth factor (FGF-1) prevents bone loss and increases new bone formation in ovariectomized rats. *J. Bone Miner. Res.* **14**, 953–959.

120. Martin, P. (1997) Wound healing-aiming for the perfect skin. *Science* **276**, 75–81.

121. Nakamura, T., Hanada, K., Tamura, M., et al. (1995) Stimulation of endosteal bone formation by systemic injection of recombinant basic fibroblastic growth factor in rats. *Endocrinology* **136(3)**, 1276–1284.

122. Nakamura, K., Kawaguchi, H., Aoyama, I., et al. (1997) Stimulation of bone formation by intraosseous application of recombinant basic fibroblast growth factor in normal and ovariectomized rabbits. *J. Bone Joint Surg.* **15**, 307–313.

123. Kato, T., Kawaguchi, H., Hanada, K., et al. (1998) Single local injection of recombinant fibroblast growth factor-2 stimulates healing of segmental bone defects in rabbits. *J. Orthop. Res.* **16,** 654–659.

124. Nakamura, T., Hara, Y., Tagawa, M., et al. (1998) Recombinant human fibroblast growth factor accelerates fracture healing by enhancing callus remodeling in experimental dog tibial fracture. *J. Bone Miner. Res.* **13(6),** 942–949.

125. Debiais, F., Hott, M., Graulet, A. M., and Marie, P. J. (1998) The effects of fibroblast growth factor-2 on human neonatal calvaria osteoblastic cells are differentiation stage specific. *J. Bone Miner. Res.* **13(4),** 645–654.

126. Eppley, B. L., Delfino, J. J., Connolly, D. T., and Feder, J. (1988) Angiogenic enhancement in bone graft healing by basic fibroblast growth factor. *Clin. Res.* **36,** A851.

127. Lu, S., Zhang, Z., and Wang, J. (1996) Guided bone regeneration in long bone. An experimental study. *Chin. Med. J. (Engl.)* **109(7),** 551–554.

128. Wang, J. and Aspenberg, P. (1996) Basic fibroblast growth factor enhances bone-graft incorporation: dose and time dependence in rats. *J. Orthop. Res.* **34,** 316–323.

129. Ducy, P., Schinke, T., and Karsenty, G. (2000) The osteoblast: a sophisticated fibroblast under central surveillance. *Science* **289,** 1501–1504.

130. Harada, H., Tagashira, S., Fujiwara, M, et al. (1999) Cbfa1 isoforms exert functional differences in osteoblast differentiation. *J. Biol. Chem.* **274(11),** 6972–6978.

131. Komori, T. and Ksihimoto, T. (1998) Cbfa1 in bone development. *Curr. Opin. Genes Dev.* **8,** 494–499.

132. Yamaguchi, A., Komori, T., and Suda, T. (2000) Regulation of osteoblast differentiation mediated by bone morphogenetic proteins, hedgehogs, and Cbfa1. *Endocr. Rev.* **21(4),** 393–411.

133. Fujimura, K., Bessho, K., Kusumoto, K., Konishi, Y., Ogawa, Y., and Iizuka, T. (2001) Experimental osteoinduction by recombinant human bone morphogenetic protein 2 in tissue with low blood flow: a study in rats. *Br. J. Oral Maxillofac. Surg.* **39(4),** 294–300.

134. Linkhart, T. A., Mohan, S., and Baylink, D. J. (1996) Growth factors for bone growth and repair: IGF, TGF beta, and BMP. *Bone* **19(1),** 1–12.

135. Becker, W., Urist, M., Becker, B. E., et al. (1996) Clinical and histologic observations of sites implanted with intraoral autologous bone grafts or allografts. 15 human case reports. *J. Periodontol.* **67(10),** 1025–1033.

136. Gao, T. J., Lindholm, T. S., Kommonen, B., et al. (1997) The use of a coral composite implant containing bone morphogenetic protein to repair a segmental tibial defect in sheep. *Int. Orthop.* **21(3),** 194–200.

137. Viljanen, V. V. and Lindholm, T. S. (1997) The search for new members of the BMP/TGF-beta family, in *Skeletal Reconstruction and Bioimplantation: Demineralized Bone Matrix, Non-collagenous, Native, and Recombinant Bone Morphogenetic Proteins* (Lindholm, T. S., ed.), Academic Press, New York, pp. 241–248.

138. Riedel, G. E. and Valentin-Opran, A. (1999) Preliminary report: new technology. Clinical evaluation of rhBMP-2/ACS in orthopedic trauma: a progress report. *Orthopedics* **22(7),** 663–665.

139. Welch, R. D., Jones, A. L., Bucholz, R. W., et al. (1998) Effect of recombinant human bone morphogenetic protein-2 on fracture healing in a goat tibial fracture model. *J. Bone Miner. Res.* **13(9),** 1483–1490.

140. Kingsley, D. (1994) What do BMPs do in mammals? Clues from the mouse short-ear mutation. *Trends Genet.* **10(1),** 16–21.

141. Storm, E. E., Huynh, T. V., Copeland, N. G., Jenkins, N. A., Kingsley, D. M., and Lee, S. (1994) Limb alterations in brachypodism mice due to mutations in a new member of the TGF-beta superfamily. *Nature* **368,** 639–643.

142. Francis, P. H., Richardson, M. K., Brickell, P. M., and Tickle, C. (1994) Bone morphogenetic proteins and a signaling pathway that controls patterning in the developing chick limb. *Development* **120,** 209–218.

143. Kingsley, D. M. (1995) Genes that define the number and shape of bones in the mouse skeleton, in *Portland Bone Symposium* (Hollinger, J. O. and Seyfer, A. E., eds.), Portland, OR, pp. 505–516.

144. Song, J. J., Celeste, A., Kong, F. M., Jirtle, R. L., Rosen, V., and Thies, R. S. (1995) Bone morphogenetic protein-9 binds to liver cells and stimulates proliferation. *Endocrinology* **136(10),** 4293–4297.

145. Tomizawa, K., Matsui, H., Kondo, E., et al. (1995) Developmental alteration and neuron-specific expression of bone morphogenetic protein-6 (BMP-6) mRNA in rodent brain. *Molec. Brain Res.* **28,** 122–128.

146. Chang, S. C., Hoang, B., Thomas, J. T., et al. (1994) Cartilage-derived morphogenetic proteins. *J. Biol. Chem.* **269,** 28227–28234.

147. Lindholm, T. S. (1996) *Bone Morphogenetic Proteins: Biology, Biochemistry and Reconstructive Surgery.* Academic Press, San Diego, CA.

148. Basic, N., Basic, V., Bulic, K., et al. (1996) TGF-beta and basement membrane matrigel stimulate the chondrogenic phenotype in osteoblast cells derived from fetal rat calvaria. *J. Bone Miner. Res.* **11(3),** 384–391.

149. Brunet, L., McMahon, J., McMahon, A., and Harland, R. (1998) Noggin, cartilage morphogenesis, and joint formation in the mammalian skeleton. *Science* **280,** 1455–1457.

150. Fonseca, M., Storm, G., Hennik, W., Gerritsen, W., and Haisma, H. (1999) Cationic polymeric gene delivery of beta-glucouronidase for doxorubicin pro-drug therapy. *J. Genet. Med.* **1(6),** 407–414.

151. Mundlos, S., Engel, H., Michel-Behnke, I., and Zabel, B. (1990) Distribution of type I and type II collagen gene expression during the development of human long bones. *Bone* **11,** 275–279.

152. Rodan, G. (1991) Perspectives: mechanical loading, estrogen deficiency, and the coupling of bone formation to bone resorption. *J. Bone Miner. Res.* **6,** 527–530.

153. Kenley, R., Marden, L., Turek, T., Jin, L., Ron, E., and Hollinger, J. (1994) Osseous regeneration in the rat calvarium using novel delivery systems for recombinant human bone morphogenetic protein-2 (rhBMP-2). *J. Biomed. Mater. Res.* **28(10),** 1139–1147.

154. Bostrom, M., Lane, J., Tomin, E., et al. (1996) Use of bone morphogenetic protein-2 in the rabbit ulnar nonunion model. *Clin. Orthop. Rel. Res.* **327,** 272–282.

155. Zegzula, H. D., Buck, D., Brekke, J., Wozney, J., and Hollinger, J. O. (1997) Bone formation with use of rhBMP-2 (recombinant human bone morphogenetic protein-2). *J. Bone Joint Surg.* **79A(12),** 1778–1790.

156. Geesink, R. G., Hoefnagels, N. H., and Bulstra, S. K. (1999) Osteogenic activity of OP-1 bone morphogenetic protein (BMP-7) in a human fibular defect. *J. Bone Joint Surg.* **81B(4),** 710–718.

157. Hollinger, J. O., Joh, S. P., Suh, K. W., Buck, D. C., Schmitt, J., and Zegzula, H. D. (1998) Regenerating the radius in a rabbit CSD model with recombinant human bone morphogenetic protein-2 and a collagen carrier. *J. Appl. Biomater.* **43,** 356–364.

158. Gazzerro, E., Gangji, V., and Canalis, E. (1998) Bone morphogenetic proteins induce the expression of noggin, which limits their activity in cultured rat osteoblasts. *J. Clin. Invest.* **102(12),** 2106–2114.

159. Tarnow, D. P., Wallace, S. S., Froum, S. J., Rohrer, M. D., and Cho, S. C. (2000) Histologic and clinical comparison of bilateral sinus floor elevations with and without barrier membrane placement in 12 patients: Part 3 of an ongoing prospective study. *Int. J. Periodontics Restorative Dent.* **20,** 117–125.

18

Bone Regeneration Techniques in the Orofacial Region

Samuel E. Lynch, DMD, DMSc

INTRODUCTION

There are numerous indications for bone regeneration materials and techniques in the orofacial region. Examples of some typical indications, as illustrated in **Fig. 1**, include:

1. Periodontal bone defects, i.e., osseous defects around teeth
2. Periimplant defects, i.e., osseous defects around endosseous dental implants
3. Large, extraction defects following tooth extraction, especially where the buccal plate of bone is missing or damaged
4. Large defects at the tooth root apex resulting from pulpal infection or a failed root canal procedure
5. Resorbed or atrophic alveolar ridges
6. Tumor resection or trauma resulting in osseous defects
7. Developmental abnormalities

Of interest is the frequency of bone reconstructive procedures in the orofacial area. For example, approximately 2.1 million periodontal surgeries are performed annually for the treatment of moderate to advanced periodontal disease (**Fig. 1**). To put this number in perspective, it is likely the most frequently performed surgery on the human body. In addition, there are over 10 million "surgical" tooth extractions involving the resection of bone to facilitate tooth removal (*1*). Clearly there is a need for large numbers of bone augmentation procedures to be performed by clinicians in this field.

To satisfy the need to restore bone architecture in oral and maxillofacial indications, a number of bone regeneration materials and procedures have been developed. Some of these materials and techniques are similar to those developed for treatment of bone deficiencies in other skeletal sites, while some materials and procedures are unique to the orofacial field (**Table 1**). The techniques and materials utilized for the periodontal and maxillofacial fields are reviewed below. For a more detailed description, the reader is referred to ref. *2*.

Bone regeneration materials and techniques commonly utilized in periodontal and oral and maxillofacial bone grafting include:

Autogenous bone grafts from either intraoral sites (commonly the symphysis of the chin or ascending ramus), or extraoral sites (iliac crest or head of the tibia)
Bone allografts, including mineralized and demineralized freeze-dried bone
Alloplasts, such as coral-derived materials, "bioactive" glasses, calcium sulfates, and calcium phosphates
Bone xenografts, primarily purified deproteinized bovine bone mineral
Membraneous sheets of materials to facilitate selective cell repopulation of the wound (also commonly referred to as guided tissue regeneration)
Distraction osteogenesis
Growth factors and morphogens, both natural and recombinant, such as platelet-derived growth factor (PDGF) and bone morphogenetic proteins (BMP-2 and -7); the recombinant growth factors are currently under clinical development but not yet approved by the US Food and Drug Administration (FDA).

The use of each of these materials in the orofacial region is summarized below. However, first it is important to appreciate some of the unique aspects of the biology of the periodontium.

From: *Bone Regeneration and Repair: Biology and Clinical Applications*
Edited by: J. R. Lieberman and G. E. Friedlaender © Humana Press Inc., Totowa, NJ

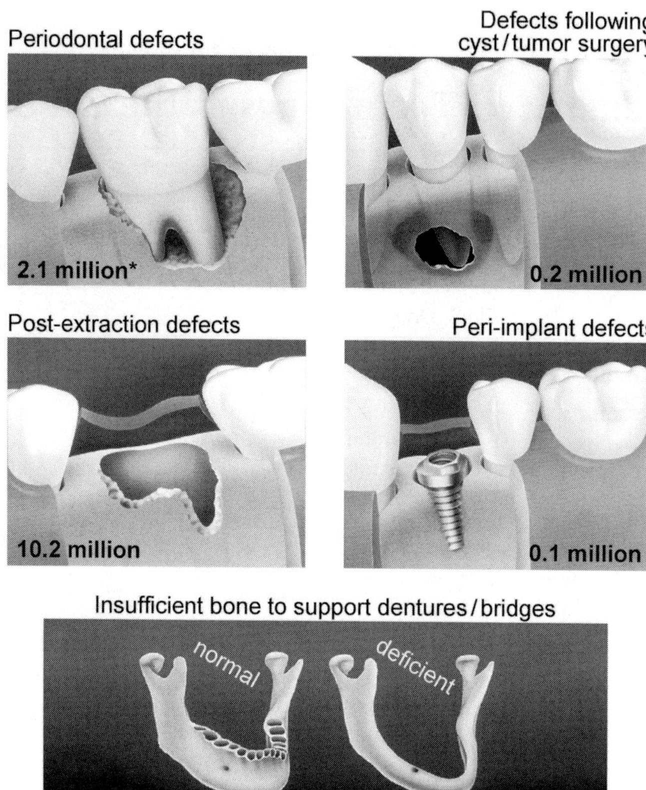

Fig. 1. Illustration of some of the more common indications for bone grafting in the orofacial region. The estimated number of each procedure performed annually is shown at the bottom of each panel. Clearly, a very large number of bone grafting procedures are performed annually in this field.

BIOLOGY OF PERIODONTAL WOUND HEALING

Anatomy of the Periodontium

The biological and anatomic considerations of the periodontal attachment structures present some unique challenges to the periodontal surgeon whose ultimate goal is to regenerate these structures. The periodontium is composed of the gingival epithelium and connective tissue attachment to the tooth, the cementum or outer layer of the tooth root, the periodontal ligament (PDL), which is a narrow band of connective tissue that connects the cementum to the alveolar bone, and the alveolar bone (**Fig. 2**). Reconstruction of these tissues, following destruction by periodontal disease, to their original physiological orientation and spatial relationship has proven difficult. In fact, it was not until the past decade that regeneration of the periodontium was considered possible.

The reasons for difficulty in achieving regeneration of the periodontal tissues lie in the anatomy of the site and cellular responses following conventional surgery. The anatomic challenges are (1) difficult access to, and visualization of, the damaged tissue (this is especially true of sites in the posterior region of the mouth, in deep, narrow intraosseous lesions, and in deep furcations, i.e., the region between two roots of the same tooth); (2) poor vascular supply (again, especially true in the furcation region where three sides bordering the bone defect are essentially avascular); (3) potential for bacterial contam-

Table 1
Currently Available Bone Grafts/Substitutes for Orofacial Indications[a]

Product type	Companies	Collective market share
Allograft	Osteotech, DePuy/Gensci	30%
(human cadaver bone)	Numerous hospitals and local tissue banks	
Xenografts	Osteohealth	25%
(bovine bone and porcine	Biora	
enamel matrix deriv.)	CereMed	
Bone substitutes	Implant Innovations/Orthovita	10%
(synthetic)	Block Drug	
	LifeCore	
Membranes for GTR	Osteohealth	35%
(synthetic and natural)	Implant Innovations/W. L. Gore	
	Block Drug/Atrix	
	Sulzer	
PDGF and bone	BMPI	0%[b]
morphogenetic proteins	Stryker	
(osteoinductive proteins)	Genetics Institute	
	Sulzer	

[a]The approximate market share of each product is shown to the right. Other than autograft, the most widely used material in orofacial bone reconstruction is allograft, because of the proposed presence of growth factors. Cell-occlusive barrier membranes for guided tissue regeneration (GTR) represent a group of materials that is also often utilized.

[b]Not yet FDA-approved.

ination of the surgical site from the oral cavity (it is often difficult to obtain true primary closure around and between teeth in maxillary posterior sites); and (4) micromovement of the wound tissues and clot due to masticatory forces.

Cellular responses that normally follow periodontal surgery also contribute to the common result of repair of the wound with little regeneration of the original, healthy architecture of the site. Repair following surgical treatment results in wound closure with scarring instead of the desired tissue relationship that existed prior to the diseased state. Specifically, there is often little or no bone fill of the osseous defect following periodontal surgery without grafting. Instead, the osseous lesion is generally occupied with gingival soft connective tissue and the root surface is lined by a long junctional epithelium rather than a connective tissue attachment or periodontal ligament.

This repair (not regeneration) process results from several responses at the cellular level. The epithelium migrates at a rate of 0.5 mm/day, considerably faster than the rate of migration of the periodontal ligament and bone-forming cells. Therefore, the epithelium covers the exposed root surface prior to the fibroblasts and osteoblasts repopulating the wound site. Once the long junctional epithelium forms it does not remodel, even if bone forms in the adjacent connective tissue compartment. This type of healing can result in newly formed bone being separated from the root surface by the epithelium rather than being connected to the root through a physiological periodontal ligament.

Importance of the Periodontal Ligament

During normal healing following periodontal surgery, the cells most critical in the regeneration process are the periodontal ligament fibroblasts, osteoblasts, and cementoblasts. These cells must proliferate, migrate into the periodontal defect, and sythesize the appropriate matrix in the proper position (**Figs. 2** and **3**). Unfortunately, these processes do not occur efficiently following treatment. It has been reported that active proliferation of PDL and bone cells occurs in only a narrow band of about 200 μm adjacent to the periodontal bone defect following surgery *(3,4)*.

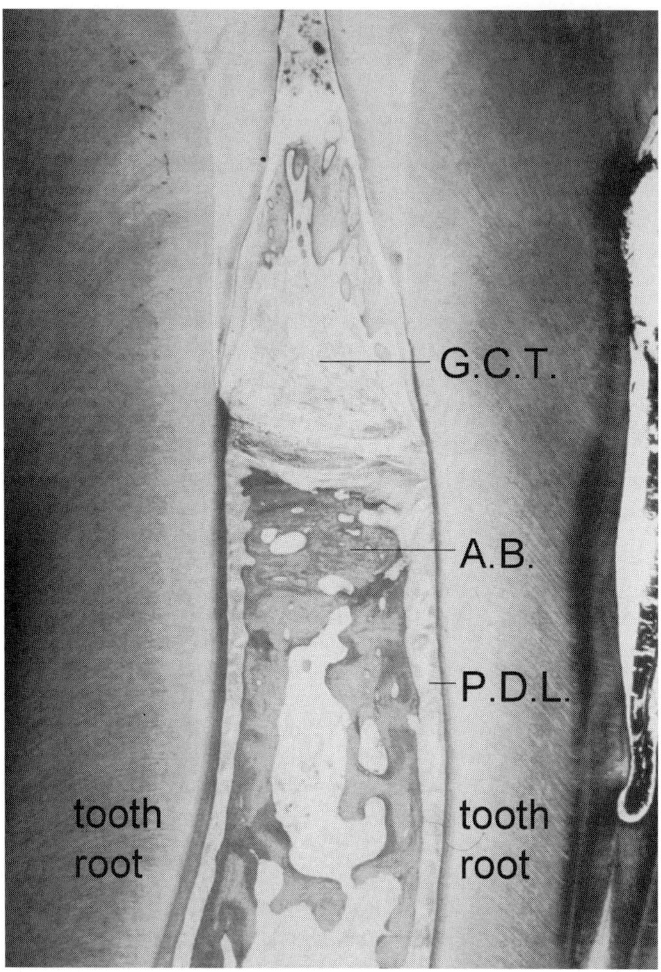

Fig. 2. Photomicrograph of the periodontal attachment structures. The unique anatomy of the periodontium presents substantial challenges to those hoping to achieve regeneration of the periodontium by grafting. Materials used in this indication must not only facilitate bone and cementum formation, at the same time they must also allow for formation of the periodontal ligament. Use of iliac crest autograft, the graft material of choice for most bone reconstructive procedures, is contraindicated around teeth because of increased root resorption. Graft materials for periodontal use must also be able to perform well in the potential presence of bacterial contamination from the oral environment and micromovement from masticatory forces. (GCT, gingival connective tissue; AB, alveolar bone; PDL, periodontal ligament. (Photomicrograph courtesy of Dr. Robert Schenk.)

To appreciate the significance of such a limited zone of cellular activity, one must recognize that periodontal osseous defects can range from 1 or 2 mm up to 10 mm or more in depth and width. The defects can be horizontal or flat in nature or intrabony, meaning that the periodontal infection has created a hole in the bone immediately adjacent to the tooth root, but a few millimeters away the bone may be present at its normal height. These intrabony defects may have one, two, or three bone walls. Horizontal or flat bone defects in which the bone resorption has occurred relatively uniformly across large segments of the jaw (alveolar bone) are considered to have no bony walls. The greater the number of bone walls, the greater the potential for some amount of regeneration to occur. The reason for this phenomenon appears to be related to the diminutive amount of cellular activity within the bone and

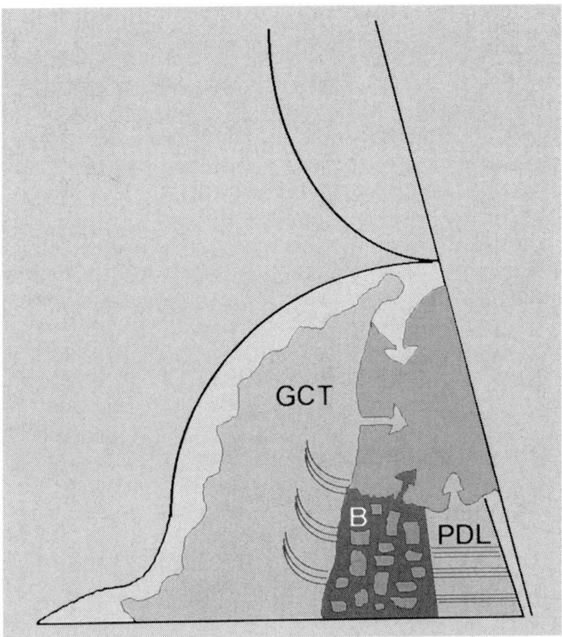

Fig. 3. Schematic of cellular events necessary for periodontal regeneration to occur. The periodontal ligament (PDL) and bone (B) cells must proliferate and migrate coronally prior to the defect being occupied by the junctional epithelium and gingival connective tissue (GCT). Normally, PDL cells are capable of migrating only a short distance, with or without placement of a physical barrier to exclude the epithelium and connective tissue. Growth factors such as PDGF-BB can increase recruitment of bone and PDL cells into the defect, leading to increased regeneration.

PDL adjacent to the osseous defect, as discussed above. Given that the zone of cell proliferation within the PDL and bone appears to be limited to within 200 μm of the borders of the defect, it is unlikely that the cellular response is robust enough to regenerate tissues up to a centimeter away. In such large, broad lesions, regeneration is also limited by the more rapidly proliferating gingival connective tissue fibroblasts and epithelium. On the other hand, if the osseous defect is a deep intrabony three-wall lesion such that a source of PDL and bone cells surrounds most of the defect, the distance that these cells must migrate to reach the center of the defect is limited. Consequently, at least partial regeneration often occurs in deep, narrow defects after surgery.

CURRENT MATERIALS FOR BONE REGENERATION

In an attempt to achieve more predictable bone regeneration in osseous defects and regeneration of the complete periodontal attachment apparatus around teeth, many bone grafting materials and techniques are available.

Autogenous Bone Grafts

Overview of Use in Periodontal and Periimplant Surgery

Similar to other indications in the skeleton, the material of choice for most oral and maxillofacial surgeons and periodontists from a purely biological and wound-healing perspective is currently autogenous bone. The preference for autogenous bone is due to the belief that it provides the most predictable outcome. The more favorable clinical outcome appears to be due to the presence of osteoblasts and osteoprogenitor cells within the graft, the natural presence of growth factors and morphogens, and the osteoconductive effects of the graft. While the biological advantages of autogenous bone are well recognized, so are the clinical disadvantages. The clinical disadvantages of autografts relate to the increased morbidity of the patient due to the harvest site, increased potential for postoperative complications, limited availability especially from intraoral sites, and the added cost to the patient and time for the surgeon.

Harvesting Techniques and Locations

Autogenous bone of two types is used, depending on the indication and preference of the surgeon: (1) cortical block; or (2) particulate marrow and cancellous bone (PMCB). If possible, autogenous bone is harvested from intraoral sites. The most common sites of harvesting intraoral bone are the symphysis of the chin, the retromolar region, the ascending ramus of the mandible, and the edentulous areas of the alveolar ridge, if present *(5)*. Block grafts are utilized most commonly in non-space-maintaining defects, where space maintenance is critical to the success of the procedure. An example of a clinical indication requiring space maintenance is alveolar ridge augmentation, both lateral and vertical augmentation. Although PMCB can be used for ridge augmentation, it requires that the graft be placed into a metal framework or mesh to stabilize it at the graft site and maintain the space. Without the protective effect of the titanium mesh, the force of the soft tissues will often compress the graft, resulting in a decreased volume of bone formation. Additionally, the masticatory forces can cause micromovement of an unprotected PMCB graft, also resulting in poor bone formation *(5)*. For these reasons, ridge augmentations are often performed with cortico-cancellous block bone.

The use of PMCB grafts is generally preferable in naturally space-maintaining defects and in reconstructing large ablative osseous defects when used in conjunction with a stabilizing framework. The exception to this guideline is the use of PMCB from the iliac crest around teeth. It has been reported that use of this type of graft can result in significant ankylosis and root resorption *(6,7)*. This observation, coupled with the significant morbidity associated with harvesting iliac crest autograft, has resulted in general avoidance of its use around natural teeth. Consequently, cortico-cancellous chips harvested from intraoral sites are preferred, either alone or in conjunction with block grafts. Examples of indications for PMCB are treatment of alveolar clefts, augmentation of the maxillary sinus floor, periimplant defects, and reconstruction of large defects following removal of cysts or tumors.

Bone Allografts

Overview of Use in Periodontal and Periimplant Surgery

When autograft is not readily available or the side effects of the autograft harvest are not justified, alternative materials must be utilized to treat significant orofacial osseous defects. At this time the most frequently utilized alternative to autograft is allograft, or bone harvested from human cadavers. Decalcified freeze-dried bone allograft (DFDBA) is the most widely used form of allograft, although mineralized freeze-dried allograft (FDBA) has been recommended for use in sinus floor augmentations.

Critical Review of Literature

In a series of studies, Bowers and coworkers obtained human biopsy specimens of teeth roots and adjacent tissues following periodontal flap surgery with and without DFDBA. Histological evaluation of these specimens showed that regeneration of the periodontal attachment structures, including the PDL and bone, is possible following the use of DFDBA but does not occur following open-flap debridement surgery without grafting *(8,9)*. Another human clinical trial in periodontal defects found that sites treated with DFDBA have a mean bone fill of 65%, compared to 38% in the control nongrafted group *(10)*. In this study, 78% of sites treated by grafting exhibited more than 50% bone fill, compared to 40% of control sites with at least 50% fill. DFDBA has also been reported to improve the long-

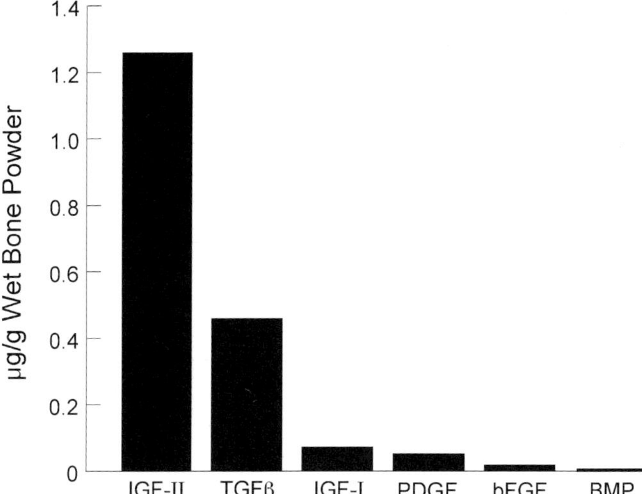

Fig. 4. Many growth factors are found in bone matrix. The most abundant growth factor, based on analysis of matrix proteins extracted from bone, is insulin-like growth factor-2 (IGF-2), followed by TGF-β, IGF-1, and PDGF. BMPs are found only in very low levels in bone matrix (19).

term benefits of guided tissue regeneration (GTR) in periodontal defects *(11)*. Shorter-term studies have reported mixed results regarding the benefit of combining DFDBA with barrier membranes for guided tissue regeneration (GTR) *(12–15)*. Some investigators have reported little or no bone formation with DFDBA *(16,17)*.

Numerous studies have also evaluated the use of allografts for repairing bone defects around titanium endosseous dental implants. When allografts are to be used to regenerate bone into which endosseous implants will be placed, use of FDBA (i.e., mineralized allograft) is generally recommended *(18)*. Preference for the use of mineralized allografts around endosseous implants stems from the observation that bone formed using this material tends to be more dense than bone formed using DFDBA.

Mineralized vs Demineralized Allografts

Despite its clinical prevalence, the use of allograft, and its clinical and biological benefits, remains controversial to some degree. The conflicting views surrounding the use of DFDBA may be the result of variability of the osteoinductive potential of this material. Osteoinduction is the process of inducing bone formation by stimulating the differentiation of pluripotential stem cells into cartilage and bone-forming cells. It is well established that bone, particularly cortical bone, contains a number of growth factors and morphogens *(19)*. The level of some growth factors, such as insulin-like growth factors, is substantial. However, the level of morphogens, such as bone morphogenetic proteins, is extremely low (**Fig. 4**) and dependent on the age and general health status of the donor *(20)*. Tests in immunologically neutral mice have shown wide variability among commercial bone bank preparations of DFDBA and its ability to induce new bone formation *(21)*. In fact, although demineralization has been shown to be necessary for the inductive potential of bone matrix to be realized, a study comparing DFDBA to FDBA in periodontal lesions showed no difference between the two with regard to the amount of bone formation *(22)*. This finding suggests that insufficient bone-inductive proteins are present in the small quantity of DFDBA placed into the defect to produce a clinically visible difference. It is therefore likely that DFDBA as well as FDBA may be stimulating bone formation primarily by osteoconduction (i.e., by providing a scaffolding for cell ingrowth and stabilization of the clot) rather than osteoinduction.

One approach to improving the predictability of allografts would be to add recombinant growth factors or morphogens to increase the concentration of these important cell-modulating proteins. Such an approach might result in an improved osteogenic response due to the combined effect of recombinant growth factors actively stimulating cell growth that is facilitated by the presence of an osteoconductive matrix, which may supply some osteoinductive properties as well.

Alloplasts (Bone Substitutes)

Examples of alloplastic materials used in orofacial reconstructive procedures include coralline-derived materials, so-called bioactive glasses, and medical-grade calcium sulfate. Currently, there is little evidence to justify the use of these materials by themselves for regeneration of bone in the orofacial region. Calcium sulfate has been used with some clinical success as a binder for allograft, although there are few well-controlled, rigorous scientific studies to support its benefits. On the other hand, if filling of the defect space by fibrous encapsulation of the graft is acceptable, then an alloplastic material may suffice *(23,24)*.

Although there is little evidence that use of alloplastic materials by themselves results in any substantial amount of bone formation, there is considerable interest in using these materials as carriers for growth factors and morphogens. Alloplastic materials have the advantage of being synthetic, which may reduce the perceived risk of disease transmission compared to allografts or xenografts (albeit the true risk for these materials appears to be exceedingly small; in fact, there has never been a reported case of HIV or hepatitis transmission from DFDBA use in the orofacial region) and increase patient acceptance. To be used as carriers for growth factors, it is important that the alloplast meet the following minimal criteria: (1) bind and release the proteins being delivered in a biologically active form; (2) be resorbed or remodeled over time in such a way that does not interfere with the bone formation and natural healing processes; (3) truly act as an osteoconductive scaffolding by facilitating cell growth and migration; (4) allow for, and ideally encourage angiogenesis and neovacularization of the wound, by possessing biologically acceptable porosity and surface area; and (5) have clinically acceptable handling properties, including a cohesiveness that limits migration from the osseous defect and a rigidity sufficient to maintain the space and prevent prolapse of soft tissue (scar tissue) into the defect. Several research studies in animals have demonstrated favorable results by combining alloplastic materials with growth factors or morphogens *(25–29)*. However, more work is needed to translate this preclinical work into advances in patient care.

Xenografts

Until recently, xenografts were not considered acceptable therapy in orofacial indications. This opinion has changed within the last 5 yr as purified anorganic (deproteinized) bovine bone has become increasingly popular for certain grafting procedures. Although at least two types of deproteinized bovine bone mineral are commercially available, the predominance of the data are for the highly porous, nonscintered form. This material has been reported to stimulate regeneration of the periodontal attachment structures, including new bone, PDL, and cementum, when used in combination with a collagen membrane *(30,31)* to treat intrabony periodontal defects. These human studies have clearly shown that the material is osteoconductive and becomes incorporated into the host bone. These human histological studies also demonstrated that the new generation of deproteinized bovine bone products were highly biocompatible and elicited no histologically detectable immunological response.

The porous bovine bone mineral material has also become popular as a graft for augmenting the floor of the maxillary sinus. Numerous studies have demonstrated that porous, nonscintered anorganic bovine bone provides an osteoconductive scaffold that facilitates bone formation in this indication *(32–34)*.

While the osteoconductive nature of this material is well documented in animals as well as humans, limiting factors to its use are the lack of inductive proteins or growth factors, resulting in the material acting purely by a conductive mechanism, and the apparent slow resorption time. Some studies suggest that the porous nonscintered material is included in the natural remodeling process, while other reports demonstrate the presence of the material even after 5 yr in humans *(32,35,36)*. Because the material incorporates into bone, and appears to remain incorporated over time, the question then becomes whether the slow resorption has any adverse biological or clinical effects. At this time no detrimental effects of the long-term presence of the material have been reported.

The preferable resorption rate of a bone substitute material for orofacial use has not been well defined. In general, the substitute should remain intact at the site long enough to maintain the space during the critical early phases of bone formation and provide a conductive scaffold until the defect is filled with relatively mature bone. While there are many different views on the precise time necessary for these events to occur, it is generally suggested that the substitute material should remain mostly intact for 3–6 mo and be completely remodeled within 18–24 mo. When the bone substitute is used as a carrier for growth factors, it may be acceptable for the resorption rate to be somewhat faster because the growth factors are presumably accelerating the rate and amount of bone formation. However, it is important that the resorptive process of the matrix material not interfere with neoosteogenesis.

GUIDED TISSUE REGENERATION (SELECTIVE CELL REPOPULATION)

Guided tissue regeneration (GTR) and its corollary, guided bone regeneration (GBR), is the technique of attempting to select the cells repopulating a wound site by placing a thin barrier membrane between the osseous defect and the overlying soft tissues (**Fig. 5**). By placing a barrier between the bone defect and the soft tissues, fibroblasts and epithelial cells from the soft tissues are prevented from migrating into the bone defect, thereby allowing more time for cells from the bone tissue compartment to recolonize the defect. Presumably, by allowing only cells from the bone to repopulate the defect (and the bone graft placed into the defect), bone formation is facilitated.

This process has been used with some success around teeth to facilitate a new connective tissue attachment. However, around teeth it is impossible to obtain a true exclusion of all tissue types other than bone, because of the presence of the periodontal ligament between the bone and tooth root. The amount of bone formation following GTR around teeth is therefore minimal. GBR has been used successfully to augment bone formation when used around titanium endosseous implants or for augmentation of the alveolar ridge *(37,38)*. Generally, GBR is used in conjunction with a bone graft, such as allograft or bone substitute. The presence of the bone substitute prevents collapse of the barrier, aids in clot stabilization, and provides scaffolding for the new bone growth.

Membrane Materials (Collagens and Synthetic Polymers)

Initially, nonresorbable barriers made from expanded poytetrafluoroethelyene (ePTFE) were utilized in the GTR/GBR procedure. The materials functioned reasonably well, especially for GBR indications. However, use of these materials is limited by the technique sensitivity, the increased potential for postoperative complications such as infection and tissue recession, and the need for a second surgical procedure to remove the barrier. Newer generations of materials have been developed that address many of the concerns with the ePTFE barriers. The newer materials are resorbable and less prone to infection and tissue sloughing. Many of the new materials are made from either collagens or resorbable synthetic polymers. The collagen membranes in particular appear to have improved tissue compatibility, although they may be limited by lack of rigidity (allowing micromovement and therefore limiting bone formation), and some types may have resorption that is too fast to allow complete bone formation. The malleable collagen barriers in particular must be used in conjunction with a bone graft or substitute.

An additional benefit to the use of collagen GBR barriers is improved graft containment. Even when using autograft, it is important to contain the graft in the defect site and prevent its migration into neighboring connective or neuronal tissues. The collagen barrier membranes appear to be very useful for this purpose.

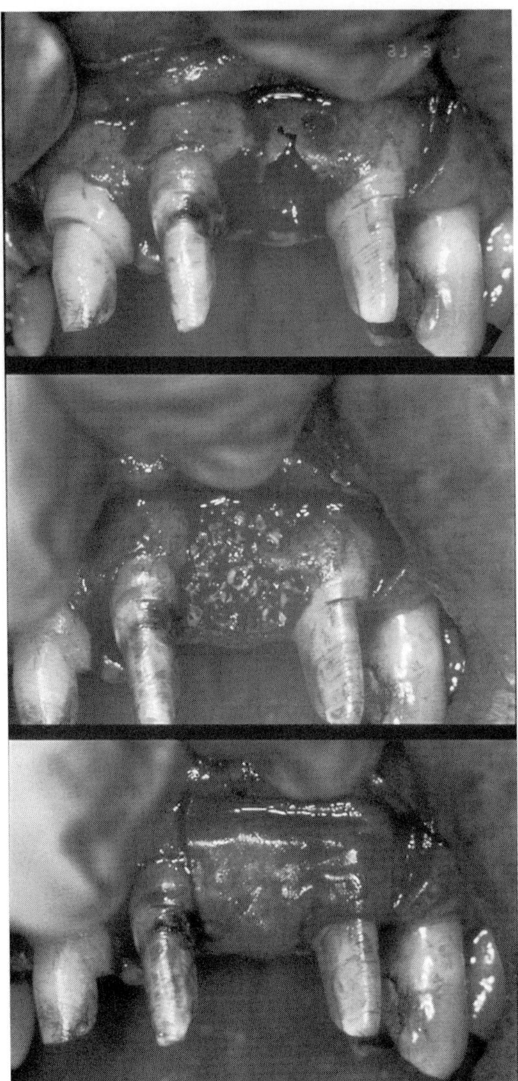

Fig. 5. Guided tissue (bone) regeneration involves placement of a physical barrier between the bone and surrounding soft tissue in an attempt to enrich the percentage of bone cells recolonizing an osseous defect. In the above case, extraction of the maxillary incisor resulted in fracture of the buccal bone plate (*top*). Such a fracture, especially combined with the thin adjacent bone, would normally lead to a significant lose of bone upon healing. Lose of alveolar bone subsequently leads to a depression of the alveolar ridge, resulting in compromised esthetics and function. To prevent such bone lose and ridge collapse from occurring, the site is grafted (*center*) and a thin barrier membrane is placed over the graft and surrounding bone (*bottom*). The soft tissues are then sutured to achieve primary closure. The barrier membrane prevents the soft fibrous connective tissue from the gingiva from growing into the bone defect. By preventing ingrowth of the soft tissue there is more time for the bone, a slower-healing tissue, to fill the defect. (From surgical case by the author.)

DISTRACTION OSTEOGENESIS

Distraction osteogenesis is a form of in vivo tissue engineering in which the gradual separation of cut bone edges results in the generation of new bone. The use of distraction osteogenesis for correcting orofacial skeletal abnormalities has received considerable interest over the past several years. This technique involves: (1) creation of a surgical fracture; (2) gradual, controlled distraction (pulling apart) of the two separated bone segments; (3) new bone formation between the two distracted segments, similar to that which occurs during fracture healing (although histologically some differences are present); and (4) stabilization of the distracted bone segments for sufficient time to allow the new bone to mature to a point that it will be stable following removal of the distraction device.

These clinical steps, when correctly applied, result in the three sequential biological phases of distraction, which are latency, distraction, and consolidation *(39)*. During the latency phase, a callus forms by a process similar to that which occurs after a fracture. In osteodistraction, however, traction forces are applied during the process of callus formation. These tensional forces stimulate changes at both the cellular and subcelluar level that longitudinally orient the fibrous components of the soft callus along the axis of distraction. By the end of the second week of distraction, distinct zones of biological processes are apparent, resulting in a well-characterized continuous bone formation process *(39)*. After distraction ceases, the center fibrous zone mineralizes to form a continuous bridge of calcified tissue. The area of distraction subsequently gradually remodels in a process similar to the remodeling of a fracture callus. The biological phases of distraction of membranous bones of the craniofacial complex are similar to those of long bones *(40–42)*.

Clinical Applications in the Jaw

Distraction has been evaluated in the craniofacial region since at least the 1970s, when the first experimental studies in animals were reported demonstrating the technique for lengthening the mandible using distraction *(43,44)*. The principals of distraction were first applied clinically for rapid palatal expansion and lengthening of the mandibular symphysis *(45–47)*. Osteodistraction was subsequently utlilized in patients with congenital cranio-facial abnormalities *(48,49)*. It is currently utilized in main indications: (1) lengthening of the mandible; (2) advancement of the midface at the Le Fort III level; and (3) reconstruction of the alveolar ridge for dental implant placement. The clinical techniques for distraction of orofacial structures have been reviewed by Chin *(49)*.

Distraction offers the significant advantage of expanding soft tissues at the same time that the bone abnormality is being corrected. The technique also has the advantage, when used in conjunction with dental implants, of transporting the mature lamellar bone into the site to receive the greatest load. This is in contrast to grafting or GBR procedures, in which new bone formation is required in the site to be loaded.

GROWTH FACTORS AND MORPHOGENS IN OROFACIAL BONE REGENERATION

While guided bone regeneration is an accepted surgical method to increase the quantity and quality of host bone, the lack of predictability in osseous regenerative procedures using various grafting materials suggests that enhancement of the osteogenic properties of these materials is a highly desirable goal. Similar to the orthopedic field, growth factors and morphogens have received a great deal of attention in the periodontal and cranio-maxillofacial fields as clinicians continue to seek an "off-the-shelf" material that could replace the need for autograft and provide better, more consistent results than current bone substitutes. The two categories of molecules that have received the greatest attention are the growth factors, which are primarily mitogens and chemotactic agents, and morphogens that act primarily by osteoinduction, i.e., stimulating the differentiation of stem cells into bone-forming cells. Within each of these broad categories, the molecules that have received the most attention and are being developed clinically are platelet-derived growth factor (PDGF) and the bone morphogenetic proteins (BMPs).

PDGF and Other Growth Factors

Isoforms of both PDGF and BMPs are present in bone matrix and both are produced locally at fracture sites *(19,50–52)*. Further suggesting the role of these proteins in normal fracture healing, cell surface receptors for PDGF, for example, are present and, in fact, upregulated during fracture healing *(51,53)*.

PDGF is often regarded as nature's "wound-healing hormone." It is contained within the α-granules of blood platelets, where it is delivered precisely to sites of injury. Once released from the platelets (or the recombinant protein is applied), it has been shown to stimulate rapid cell proliferation (mitogenesis) of osteoblasts and periodontal ligament fibroblasts as well as directing their migration (chemotaxis) and subsequent protein synthesis by binding to well-characterized cell-surface receptors. PDGF has been shown to improve the effectiveness of autografts as well as stimulate new bone formation when applied directly into the bone defect or onto tooth root surfaces exposed as a result of periodontitis.

Current knowledge of the scientific principles of the host response to autogenous cortical and cancellous bone grafts suggests that the process begins with the formation of a hematoma that includes the implanted bone. The role of the hematoma in this process is not fully understood but clearly involves the release of bioactive molecules such as growth factors and cytokines from degranulated platelets in the graft *(54)*. The growth factors include the BB isoform of PDGF (PDGF-BB), which has been shown to stimulate mitogenesis of the marrow stem cells as well as osteoblasts transferred in the autograft, thereby increasing their numbers by several orders of magnitude *(55)*. PDGF also initiates angiogenesis leading to capillary budding into the graft by inducing endothelial cell mitosis.

Studies with transgenic mice have underscored the physiological role of PDGF as essential for bone development and repair *(56,57)*. For example, spontaneous deletion of the α-receptors for PDGF results in maldevelopment of craniofacial bones and vertebrae, demonstrating the importance of PDGF in skeletal development *(58)*.

A number of in vitro and in vivo studies, including those described earlier, have been conducted with respect to the effects of PDGF on bone. PDGF in both its AB and BB isoforms has been shown to be a potent stimulator of migration and DNA synthesis in the fetal rat calvarial system as well as in cultures of cells from adult human bone explants *(59,60)*. Several in vitro studies on the effect of PDGF on parameters of osteoblastic cell differentiation suggest that the effects of the growth factor are complex and dependent on its concentration and incubation conditions. Although earlier studies reported that PDGF reduces alkaline phosphatase, a marker of osteoblastic cell differentiation *(61)*, and inhibits bone matrix formation *(62)*, the most recent extensive studies of Hsieh and Graves *(63)* reveal that the in vitro osteoblastic cell cultures effects are biphasic. Multiple brief exposures to PDGF enhance formation of a mineralized matrix, while continuous longer-term exposure is inhibitory. The study explains how exogenous PDGF could enhance osseous healing in vivo despite the reported inhibitory effects of the mitogen on osteoblastic cell differentiation in vitro.

Preclinical animal studies by several research groups have shown that PDGF-BB (alone and in combination with insulin-like growth factor-1 (IGF-1), and basic fibroblast growth factor (bFGF), have the capacity to stimulate new bone as well as periodontal ligament and cementum formation in periodontal lesions in dogs and nonhuman primates *(64–69)* (**Figs. 6** and **7**). Basic FGF-applied sites exhibited significant regeneration as represented by new bone formation rate, rate of new trabecular bone, and cementum formation.

In the nonorofacial region, PDGF has been shown to have significant effects on bone repair in tibial osteotomies in rabbits and to increase bone density in the rodent skeleton *(70,71)*. In addition, systemic administration of PDGF-BB in ovarectomized rats has been shown to promote significant increases in osteoblast cell number and trebecular bone volume in both long bones and vertebrae, whole skeletal bone mineral density as judged by DEXA and strength of long bones and vertebrae (three-point bending and compression, respectively) *(71)*.

Certainly, PDGF is not the only growth factor that influences bone regeneration. It has been shown in animal models that other growth factors, such as bFGF, IGF-1, and transforming growth factor-β (TGF-β) are expressed and may play different roles in the remodeling phase of mandibular distraction osteogenesis *(72,74,75)*.

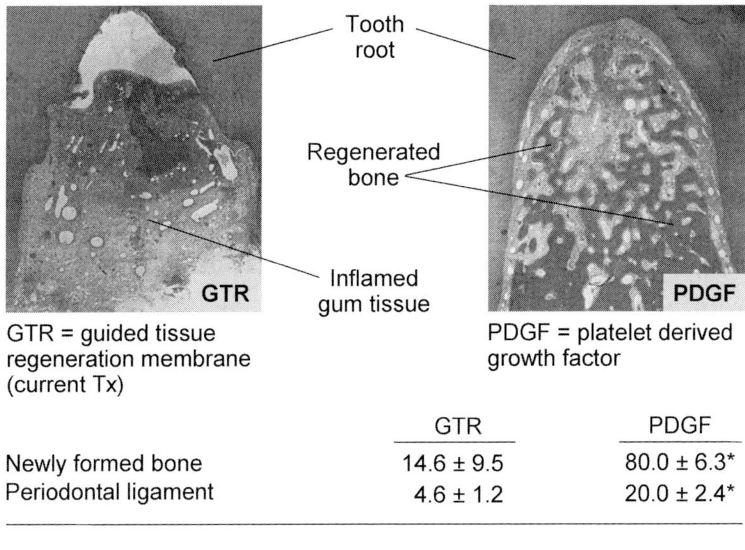

	GTR	PDGF
Newly formed bone	14.6 ± 9.5	80.0 ± 6.3*
Periodontal ligament	4.6 ± 1.2	20.0 ± 2.4*

*p < 0.05

Fig. 6. Comparison of guided tissue regeneration (GTR) with and without PDGF-BB. rhPDGF-BB greatly increases the effectiveness of GTR as assessed by new bone and PDL formation in Class III furcation defects in dogs. By 8 wk postoperative there was complete fill of the defects treated with rhPDGF-BB (80% bone and 20% PDL), compared to only 19.2% fill of defects treated with GTR alone. This is a particularly striking finding because GTR is a popular current therapy. (From Cho, M. I., Lin, W. L., and Genco, R. J. (1995) Platelet-derived growth factor-modulated guided tissue regenerative therapy. *J. Periodontol.* **66,** 522–530.)

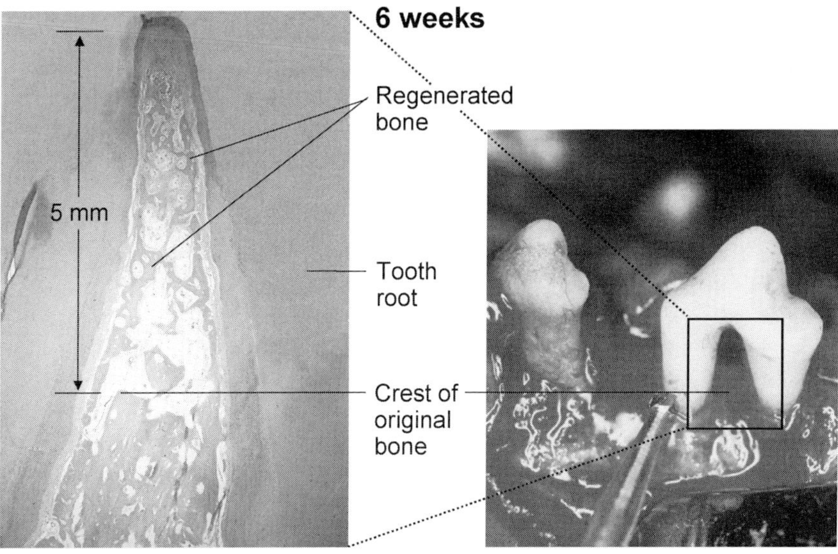

Fig. 7. Clinical appearance and 6-wk postoperative photomicrograph showing the significant amount of supracrestal new bone and PDL that formed in this canine specimen as a result of treatment of the defect with rhPDGF-BB in combination with IGF-1. Nearly 5 mm of supracrestal bone had formed during the 6-wk healing time.

BMPs

The bone morphogenetic proteins have pivotal roles in the regulation of bone induction, maintenance, and repair *(73)*. They act through an autocrine or paracrine mechanism by binding to cell-surface receptors and initiating a sequence of downstream events that have effects on various cell types. Differentiation of osteoprogenitor mesenchymal cells and upregulation of osteoblastic features occur under the influence of cytokines and growth factors that are expressed with the direct or indirect guidance of BMPs acting at the transcriptional level or higher. Moreover, the inflammatory response observed during wound repair and fracture healing results in by-products that interact with BMPs and affect their biological potential. Similar to expression of growth factors, the temporal and spatial pattern of expression of BMP-2, -4, and to lesser extent BMP-7, suggest that these proteins are important mediators of bone remodeling during mandibular distraction osteogenesis *(74–76)*.

REGENERATION OF THE PERIODONTIUM AROUND TEETH

Platelet-derived growth factor has been the most thoroughly studied growth factor in periodontics. Since PDGF was first discovered by Lynch and coworkers to promote regeneration of bone, cementum, and periodontal ligament in the late 1980s *(64)*, more than 100 studies have been published on its effects on periodontal ligament and alveolar bone cells and regeneration of the periodontium in animals and humans. Without exception, all studies demonstrate the potent stimulatory effects of PDGF on the periodontal attachment structures.

Indeed, the pleotrophic effects of PDGF on both the bone and PDL make it distinctively suited for stimulating regeneration of all the components of the periodontium. PDGF stimulates gingival fibroblast hyaluronate synthesis, a prerequisite for the formation of large aggregates of proteoglycans that provide the lattice for the extracellular matrix *(77)*. Several investigators have demonstrated the potent effects all isoforms of PDGF have on the proliferative activity of periodontal ligament fibroblasts in vitro *(78–85)*. Indeed, a recent *in vitro* study demonstrated that PDGF-BB stimulates the proliferation and collagen synthesis of human PDL cells in a time- and dose-dependent manner *(82)*. This suggests that PDGF-BB is an important regulator of the maintenance of the PDL extracellular matrix, and may play an important role in the regeneration of PDL cells.

PDGF has also been shown to have a mitogenic effect on human PDL cells cultured with various types of allografts, including demineralized freeze-dried cortical (DFDBA) and cancellous (DFBA) allografts, as well as nondemineralized freeze-dried allograft (FBA) from cancellous bone *(83)*. Papadopoulos and colleagues recently showed that human PDL cells exhibit significantly greater ($p < 0.05$) proliferative responses to DFDBA and DFDBA supplemented with PDGF-BB compared with these allografts alone *(83)*. In addition, a statistically significant difference in DNA synthesis ($p < 0.05$) was noted when PDGF-BB was added to PDL cells cultured with FBA. These findings demonstrate the beneficial role of PDGF-BB as a synergistic agent with DFDBA, DFBA, and FBA to induce periodontal regeneration, and suggest that DFDBA and DFBA combined with PDGF to stimulate PDL cell proliferation might be a useful adjunct in the treatment of periodontal defects.

In direct comparative studies, numerous *in vitro* investigations have found that PDGF is substantially more potent than any of the other growth factors, or BMPs, in stimulating PDL cell proliferation, migration, and protein synthesis (**Table 2**). For example, a recent *in vitro* study, which evaluated mitogenic responses of human PDL cells and gingival fibroblasts (isolated from adult patients with moderate periodontitis), found that PDGF-BB acts as a strong mitogen for human PDL cells, whereas TGF-β effects were modest and recombinant human BMP-2 (rhBMP-2) has an opposite (negative) effect on cell mitosis *(84)*.

Principles of GTR dictate that one of the goals of therapy is to modulate the wound-healing processes to favor repopulation of the wound with cells derived from the PDL rather than from the gingival tissues. It is noteworthy, therefore, that Mumford et al. recently demonstrated a significantly greater pro-

Table 2
Effects of Several Growth Factors on Periodontal Ligament Cells[a]

Growth factor	Chemotaxis	Proliferation	Collaen synthesis	Differentiation
EGF	Slight	Slight	Inhibitory	Inhibitory
TGF-β	None	Inhibitory	Moderate	ND
IGF-1	Strong	Strong	None	Strong
PDGF	Strong	Strong	Moderate	ND
PDGF + IGF1	Strongest	Strongest	ND	ND

[a]PDGF-BB is the most potent mitogen and chemoattractant for these specialized cells. PDGF-BB also stimulates protein synthesis within the cells, thus satisfying all the prerequisites for stimulating periodontal regeneration, as described in **Fig. 3**.

Source: Adapted from Cho, M. I., Lin, W. L., and Genco, R. J. (1995) Platelet-derived growth factor-modulated guided tissue regenerative therapy. *J. Periodontol.* **66,** 522–530.

liferative response to PDGF-BB ($p < 0.0001$) in PDL cells than gingival fibroblasts *(85)*. In this *in vitro* human wound model, PDL cells exhibited increased levels of proliferation at concentrations of PDGF-BB > 10 ng/mL, whereas gingival fibroblasts displayed no increase in proliferation at any concentration compared to negative controls. Thus, there may be cell-specific differences critical to periodontal wound healing that may be exploited by the use of PDGF.

In the clinical practice of periodontics, one of the greatest challenges is the treatment of multirooted teeth with interradicular loss of periodontium ("furcation defects"). In a Phase I/II clinical trial with 38 human subjects possessing bilateral osseous periodontal defects, PDGF/IGF-I treatment (150 µg/mL of each growth factor, or about 10 µg per osseous defect), was found to significantly increase alveolar bone formation compared to control subjects *(86)*. These investigators suggested that the growth factor therapy may be particularly effective in extensive furcation defects, which may represent "critical-size defects" that fail to respond to conventional periodontal debridement and flap surgery *(87)*.

Recent evidence of positive clinical and histological responses has emerged from human clinical trials using rhPDGF-BB delivered in bone allograft for the treatment osseous periodontal lesions *(88, 89)*. In these studies, DFDBA was saturated with rhPDGF-BB for the treatment of advanced Class II furcation and interproximal defects. Clinical probing depths and attachment levels were obtained presurgically and 9 mo postsurgical, after which the teeth and surrounding tissues were removed *en bloc*. There was substantial improvement in horizontal (mean 3.5 mm) and vertical (mean 4.25 mm) probing depths, as well as clinical attachment levels (CAL) (mean 3.75 mm) *(88)*. Histological evaluation revealed periodontal regeneration, including new bone, cementum, and periodontal ligament, in four of the six interproximal defects and all evaluable (four of four) furcation defects treated with PDGF (**Fig. 8**) *(89)*. Regeneration was present coronal to the original osseous crest. In one case, where an enamel projection extended into the fornix of the furcation, new calcified tissue with new inserting connective tissue fibers was observed over the enamel.

Thus, this clinical study documented a favorable tissue response to rhPDGF-BB treatment at both the clinical and microscopic levels, and provided the first human histological evidence that new calcified tissue with inserting collagen fibers can occur over enamel projections within furcations *(88)*. In addition, this study demonstrated for the first time that complete periodontal regeneration can be achieved in advanced Class II furcation defects using a combination of purified recombinant growth factor and bone allograft.

Morphogens (BMPs) in Periodontal Regeneration

The role of BMPs in bone development and repair has been extensively reviewed by numerous authors *(19,50,90,91)*. The role of BMPs on periodontal and cranio-maxillofacial osseous reconstruction specifically has also been extensively reviewed recently *(92,93)*. Accordingly, rather than providing yet

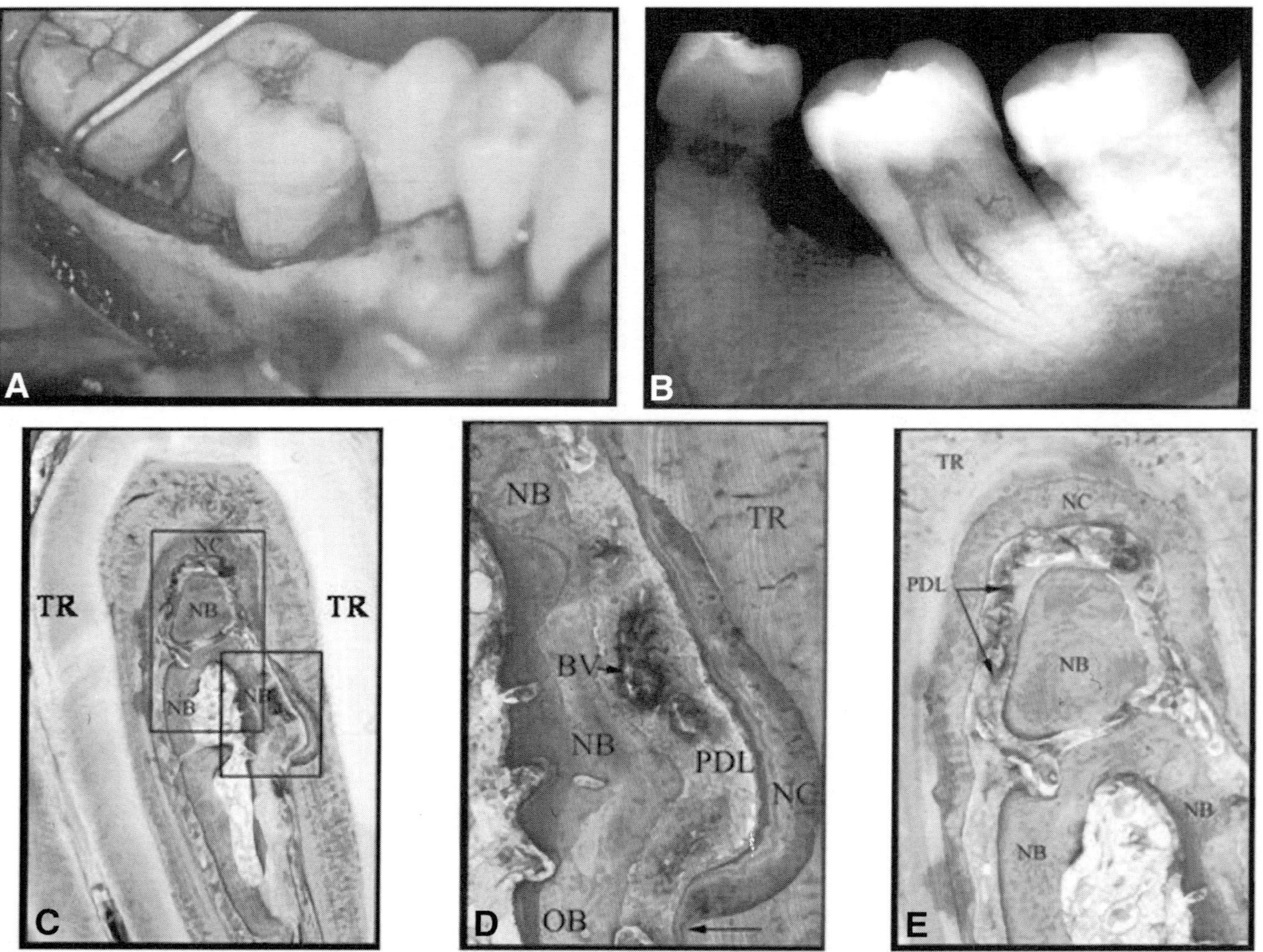

374

another historical review of this subject, only the most recent publications on the effects of BMPs in the orofacial region are summarized below.

Wikesjo and coworkers *(94)* recently evaluated the effects of three different concentrations of recombinant BMP-2 on regeneration of alveolar bone and cementum, and on root resorption and ankylosis in a canine critical-sized periodontal defect model. Three concentrations of BMP-2 (0.05, 0.10, or 0.2 mg/mL) were applied to the periodontal defects in an absorbable collagen sponge (ACS). After 8 wk of healing, block biopsies of the teeth and surrounding bone were taken and examined histologically. All three doses of BMP-2 caused significant increases in alveolar bone formation, comprising 86–96% of the original defect height. Cementum regeneration encompassed 6–8% of the defect height. Root resorption and ankylosis were observed for nearly all teeth for each BMP-2 concentration.

Another, more recent study in a canine surgically induced periodontal defect model similarly concluded that rhBMP-2 elicits a rapid osteoinductive process, although a typical cementum–PDL–alveolar bone relationship is rare *(95)*. Surgical implantation of rhBMP-2 (0.2 mg/mL) in an ACS carrier into large supraalveolar periodontal defects resulted in a variable tissue response. Regenerated bone, exceeding the volume of the normal alveolar process, had formed within 4 wk, and consisted of finely trabeculated woven bone. Marrow spaces exhibited a continuous lining of osteoblasts, osteoclasts, and resting cells. However, a variety of tissue reactions were observed along the root surface, including areas of resorption, areas of hard tissue deposition, and areas without resorptive or appositional activity. In addition, ankylosis was a frequent observation, although areas showing characteristics of a periodontal ligament with a fine layer of acellular fiber cementum and occasional inserting Sharpey's fibers were also observed. Together, these studies highlight one of the concerns of the use of BMPs around teeth. That is, because the BMPs are potent inducers of bone formation, and have little direct effect of the PDL, there is a risk of stimulating ankylosis and root resorption. These events represent an abnormal healing response and, if severe, can jeopardize the survival of the tooth.

Application of BMPs in collagen membranes has recently been reported by two independent groups. In a study in nonhuman primates, a natural preparation of BMP was combined with one or two layers of a fibrous collagen membrane *(96)*. After 12 wk, the single-layer approach showed partial regeneration in the periodontal defects, although it often led to ankylosis (consistent with the observations described above). The double-layer technique gave more favorable results, with new alveolar bone, PDL, and cementum generated along the entire exposed root surface. Ankylosis was rarely observed utilizing this dual-membrane technique. In a 10-d study in rats, BMP-2 treatment resulted in more bone formation when delivered in a collagen gel than in a collagen membrane. However, new cementum formation was greater in the group receiving the BMP-2 in the collagen membrane *(97)*.

Two studies have recently been published comparing the effects of BMPs to PDGF-BB on osteoprogenitor or PDL-derived cells. In a study comparing the effect of a bovine BMP extract to recombinant

---

Fig. 8. (*Opposite page*) **(A)** Intraoperative photograph with probe in place showing a 5-mm, primary horizontal, furcal bone defect on the lingual of tooth #19. **(B)** Presurgical radiographic appearance of horizontal bone class and a Class II furcation lesion on #19. **(C)** Histological section of tooth #19 obtained 9 mo after treatment with rhPDGF-BB mixed with allograft. The notch placed at the apical extent of calculus during the treatment surgery is evident. Complete fill of the original defect area with new bone (NB), periodontal ligament (PDL), and new cementum (NC) is present. The NB is equal in density to the original alveolar bone. There is no epithelial downgrowth into the furcation (long junctional epithelium) even though no membrane was used. (Original magnification ×6.3.) **(D)** Higher-power view of the lower box in **Fig. 8C**, showing the tooth root (TR), NB, new PDL, and NC. The PDL is well organized, with fibers coursing perpendicularly and tangentially between the NC and NB. The new PDL is the same width as the original PDL and contains abundant new blood vessels. (Original magnification ×25.) **(E)** Higher-power view of upper box in **Fig. 8C**, showing the area of the fornix of the furcation. The NB has completely filled the original furcation defect. There is also restoration of a well-organized new PDL throughout the furcation and NC continuously from one root to the other *(89)*. (Original magnification ×25.)

PDGF-BB or IGF-1 on osteoprogenitor cells, PDGF-BB enhanced cell numbers, IGF-1 alone had no significant effect, and the BMPs decreased cell proliferation *(98)*. In this same study, PDGF down-regulated osteopontin and osteocalcin expression, while the BMPs and IGF-1 promoted gene expression of osteopontin and bone sialoprotein. These data indicate that BMPs, PDGF and IGF influence cell activity by different mechanisms. In a second study comparing recombinant BMP-2 to recombinant PDGF-BB, the results showed a significant increase in PDL cell proliferation and protein synthesis in PDGF-BB-treated groups compared to controls and no increase in alkaline phosphatase activity (a marker of osteoblast differentiation). A significant increase in alkaline phosphatase activity was observed in BMP-2-treated groups compared to controls, but protein concentration and cell number were unchanged *(99)*. The conclusion that can be reached from these papers is that PDGF-BB promotes greater cell proliferation while BMP-2 increases cell differentiation toward the osteoblastic lineage.

GROWTH FACTORS AND BMPs IN ALVEOLAR RIDGE AUGMENTATION AND OSSEOINTEGRATION OF TITANIUM IMPLANTS

Alveolar ridge aberrations commonly compromise optimal placement of dental implants. To offset any variance between an aberrant alveolar ridge and prosthetic designs, bone augmentation procedures become necessary. A number of recent studies have shed light on the efficacy of morphogens and growth factors in ridge augmentation and osseointegration of dental implants in canine and non-human primate models, as well as in humans *(95–107)*.

Bone regeneration around implants in dogs, with or without the concurrent application of a combination of PDGF and IGF-1, has been evaluated *(100,101)*. Mandibular premolar extraction sockets were prepared with implant osteotomies (1.25 mm beyond the implant bed in the coronal half). Prior to insertion, implants received a single application of 5 µg/mL of PDGF and IGF-1 delivered in 0.10 mL of 4% methylcellulose gel or the gel alone as a control. Regenerated bone was labeled with a 2% calcein green solution administered by serial intramuscular injections. Substantial benefits were observed with growth factor treatment. The results showed a greater extension of bone–implant contact, a larger percentage of bone area, and greater intensity of bone labeling for test vs control implants ($p < 0.01$) *(101)*. The authors concluded that the combination of PDGF/IGF-1 is an effective alternative for enhancing bone healing around implants.

Some investigators have reported that surgical implantation of BMP-2 is an effective protocol for vertical alveolar ridge augmentation procedures and dental implant osseointegration, though more ambiguous results have also been observed *(103,105–107)*. Wikesjo and colleagues indicated that rhBMP-2 (0.40 and 0.75 mg/mL) in a calcium phosphate cement carrier (alphaBSM) induces substantial augmentation of the alveolar ridge vs controls (alphaBSM alone) in critical size supraalveolar periimplant defects in dogs *(106)*. Superior vertical bone augmentation, new bone area, and new bone–implant contact were observed. Interestingly, other researchers have found that a combination of BMP-2, PDGF, TGF-β, and bFGF in calcium phosphate cement has a significant (synergistic) effect on bone–implant contact and amount of bone per surface area compared with plain cement and (noncement) controls ($p < 0.0009$) in dogs treated with titanium implants placed in prepared sites (1.5 mm beyond the implant bed in the coronal half) *(102)*.

In nonhuman primates, a recent study demonstrated that new bone induced by rhBMP-2 in large surgically created mandibular defects was maintained and functional for at least 1 yr following placement of dental implants into the regenerated bone *(107)*. Excellent remodeling and consolidation of new bone were observed after implant loading. However, another group of researchers reported more ambiguous results using rhBMP-2 in dehiscence defects with nonsubmerged immediate implants *(105)*. In this latter study, Cynomolgus monkeys received dental implants in contralateral extraction socket sites with surgically created buccal dehiscence defects. Contralateral sites received 1.5 mg/mL (0.1 mg per defect) rhBMP-2 in an ACS carrier or served as sham-surgery controls.

After 16 wk, seven of eight defect sites (four of four animals) receiving rhBMP-2 compared to four of eight sites (two of four animals) receiving sham surgery exhibited evidence of osseointegration with newly formed bone in the defect area *(105)*. However, vertical bone gain in rhBMP-2-treated defects did not differ significantly from the sham-surgery control. Nor were there significant differences noted for coronal bone–implant contact, and bone–implant contact within the defect site and within resident bone for the rhBMP-2 and control sites, respectively. In conclusion, the authors questioned the value of rhBMP-2 in an ACS carrier as an osteoinductive biological construct.

Data are less abundant from clinical studies of BMPs in guided bone regeneration in human implant recipients, but rhBMP-2 combined with a xenogenic bone substitute mineral (Bio-Oss) has shown potential in recipients of Branemark implants *(103)*. In a recent study, rhBMP-2 combined with Bio-Oss yielded enhanced maturation of bone regeneration and increased graft–bone contact compared with sites in the same jaw augmented with Bio-Oss without rhBMP-2 *(103)*. Based on these findings, the authors concluded that rhBMP-2 combined with xenogenic bone substitute mineral has the potential to improve and accelerate GBR in human implant recipients.

AUGMENTATION OF THE MAXILLARY SINUS FLOOR

Maxillary sinus floor augmentation with autogenous bone has become a widely accepted procedure in implant dentistry. Given the propensity of BMPs to elicit abnormal healing responses around teeth in the form of increased root resorption and ankylosis, many investigators have evaluated their effects elsewhere in the orofacial region, such as for augmentation of the maxillary sinus floor. Animal and human studies have demonstrated that recombinant BMP-2 and BMP-7 (also referred to as osteogenic protein-1 [OP-1]) are each capable of stimulating bone formation in sinus floor augmentations. However, the appropriate dosage and delivery vehicles for sinus floor augmentation have recently been debated. The reported dosage of rhBMP-2 in a goat study was 3.4 mg per defect *(108)*. In a subsequent human study, the doses ranged from 1.8 to 3.4 mg of rhBMP-2 *(109)*. Although the dosage appeared appropriate in the goat, the human study showed that 3 of 11 patients had inadequate amounts of new bone to allow placement of the endosseous implants, and higher doses were suggested. Nonhuman primate studies using rhOP-1 found that doses below 2.5 mg did not generate sufficient amounts of new bone *(110,111)*.

Some human studies have reported promising results with BMPs for sinus augmentation and alveolar ridge reconstruction, while others have been less enthusiastic. In a recent pilot human study of rhOP-1 in sinus floor augmentation, the investigators found that the results using rhOP-1 were too inconsistent to warrant its clinical use in sinus lifting *(112)*. These authors attributed the inconsistent results at least partially to the collagen carrier, which in one patient appeared to elicit a substantial inflammatory response. A concurrent report suggested that rhOP-1 delivered in deproteinized bovine bone mineral was superior to deproteinized bovine bone mineral alone in sinus floor augmentation with simultaneous implant placement in miniature pigs *(113)*. This report concluded that there was twice the amount of bone to implant contact in the rhOP-1 group compared to the bone substitute alone.

A subsequent clinical study investigated the bone forming potential of rhOP-1 combined with a collagen carrier implanted in the maxillary sinus of three patients compared to a group of three patients treated with sinus floor elevation and autogenous bone grafts *(114)*. Six months after sinus grafting with rhOP-1, well-vascularized bonelike tissue of "good quality" was observed clinically and histologically in one male patient. However, no bone formation was observed in a female patient. Some bonelike formation was seen in another female patient who received bilateral sinus grafts; however, the flexibility of this tissue led to the postponement of implant placement. By contrast, in all five autogenous grafted sinuses, a bone appearance similar to normal maxillary bone was observed clinically as well as histologically, and dental implants could be placed 6 mo after sinus floor elevation surgery. Based on these findings, the authors concluded that the behavior of rhOP-1 delivered via a collagen carrier is insufficiently predictable in this indication area. They recommended further investigation before OP-1 can be successfully used instead of the "gold standard" autogenous bone graft.

In other studies, Boyne *(115)* has reported reconstruction of 2.5-cm hemimandibulectomy defects in nonhuman primates with rhBMP-2. Cochran *(92)* and Howell *(116)* reported favorable safety results in a pilot human study utilizing rhBMP-2 for alveolar ridge augmentation or preservation following tooth extraction. However, it was unclear if the magnitude of the bone response was clinically significant, as augmentation of the alveolar ridge was not observed. It was speculated that the amount of BMP used in the study (0.43 mg/mL) may not have been sufficient for bone augmentation, or that the carrier, a collagen sponge, may not have provided sufficient space maintenance.

PLATELET-RICH PLASMA IN OROFACIAL REPAIR

Naturally concentrated platelet-derived growth factors have been used clinically for the last several years. This procedure involves the production of platelet-rich plasma (PRP) (also referred to as autologous platelet gel after addition of a clotting agent) at the time of the surgical procedure.

Within the last several years, blood cell sorting technology has become clinically accessible. Instruments are now available to isolate and concentrate the various components of whole blood, including platelets, in a completely sterile environment for immediate clinical use. The platelets are concentrated in a small volume of plasma (which is a source of IGF-1 and fibrin) to form platelet-rich plasma. If thrombin and calcium chloride or ITA are added to the PRP, the platelets are activated to release the contents of their α-granule, including PDGF, TGF-β, PD-ECGF, IGF-1, and platelet factor-4 *(117)*. These factors signal the local mesenchymal and epithelial cells to migrate, divide, and increase collagen and matrix synthesis. Meanwhile, the thrombin/calcium or ITA preparations also initiate clotting, including the conversion of fibrinogen to fibrin, resulting in a clinically useful PRP gel that can improve the handling and efficacy of particulate autografts and bone substitutes.

PRP has been suggested for use to increase the rate of bone deposition and quality of bone regeneration when augmenting sites prior to or in conjunction with dental implant placement. Marx and co-workers recently performed a comprehensive clinical and analytical study assessing the mechanism of action and clinical significance of PRP in reconstruction of large mandibular discontinuity defects in humans *(118)*. These investigators confirmed increased platelet counts in PRP, increased PDGF and TGF-β concentrations in the PRP gel, and the presence of PDGF and TGF-β receptors on the surface of cells within iliac crest PMCB autografts. Furthermore, in an 88-patient human clinical trial, they compared the clinical effectiveness of particulate iliac crest autograft alone to autograft plus PRP gel. All 88 patients received major reconstructive surgery resulting in 5-cm or greater discontinuity defects. The 44 defects treated with the PRP/autograft demonstrated increases in both the rate of bone formation and bone density as evaluated clinically, radiographically, and histologically, compared to the 44 patients who received the autograft without PRP.

CARRIERS AND DELIVERY MODALITIES

Much discussion has also focused on the appropriateness of various carrier and delivery modalities for growth factors and morphogens in a variety of treatment settings *(26,27,63,95,106,119–128)* (**Table 3**). Several *in vitro* and *in vivo* studies have investigated delivery of growth factors, including PDGF, via biodegradable carriers that release therapeutic concentrations over a sufficient length of time to achieve optimal effects. For example, Lee and colleagues found that PDGF-BB delivered with a chitosan/tricalcium phosphate (TCP) sponge carrier exhibited good release kinetics in vitro (effective therapeutic PDGF-BB concentration following a high initial burst release), and promoted osseous healing of rat calvarial defects as compared with non-PDGF-BB-treated controls *(121)*. This group of investigators similarly reported that use of PDGF-BB-releasing porous chondroitin-4-sulfate (CS)-chitosan sponge significantly enhanced osteoblast proliferation, and the release rate of PDGF-BB could be controlled by varying the composition of chondroitin-4-sulfate in the sponge or the initial loading content of PDGF-BB *(122)*. Subsequent in vivo research by these investigators in animal calvaria models suggested that PDGF-BB-releasing molded porous poly (L-lactide) (PLLA) membranes improve GBR efficiency in various types of bone defects *(127)*.

Also recently, Bessho et al. reported that the synthetic, biodegradable low-molecular-weight poly (DL-lactide-co-glycolide) (PLGA) copolymer represents a promising slow delivery vehicle for BMPs required for maximal clinical effectiveness *(119)*. In this in vivo study in rats, rhBMP-2 was released in an active form at a soft tissue (calf muscle) implant site during the degradation of the copolymer, resulting in the induction of new bone formation.

Efforts to regenerate tissues (e.g., bone, blood vessels) typically rely on the delivery of single factors, and this may partially explain the limited clinical utility of some current approaches. One constraint on delivering appropriate combinations of factors is a lack of delivery vehicles that allow for a localized and controlled delivery of more than a single factor. However, Richardson et al. recently reported on a new polymeric system that allows for the tissue-specific delivery of two or more growth factors, with controlled dose and rate of delivery *(123)*. The utility of this system was investigated in the context of therapeutic angiogenesis, where it was shown that dual delivery of vascular endothelial growth factor-165 (VEGF-165) and PDGF-BB, each with distinct kinetics, from a single, structural polymer scaffold produces the rapid formation of a mature vascular network. Although the research focus was angiogenesis, this report highlights the benefits of a vehicle capable of delivering multiple factors with distinct kinetics in tissue regeneration and engineering.

A number of studies in the dental and orthopedic settings have investigated the role of mineral materials and bone allograft as matrix for the release of growth factors to enhance bone growth *(88,89)*. As described above, emerging data from human studies indicated that use of purified rhPDGF-BB mixed with bone allograft produce robust periodontal regeneration in both Class II furcations and interproximal intrabony defects *(88,89)*.

Some candidate carriers lack the structural integrity to offset compressive forces that may compromise bone induction, in particular, for challenging onlay indications such as alveolar ridge augmentation. To address this shortcoming, some investigators have studied the potential of rhBMP-2 in a calcium phosphate cement carrier (alphaBSM) to induce augmentation of the alveolar ridge and osseointegration of dental implants *(106)*.

SELECTIVE USES OF GROWTH FACTORS IN OROFACIAL SOFT TISSUE

PDGF has been used to enhance soft tissue as well as bone healing. Moreover, recombinant human PDGF-BB (rhPDGF-BB) is the only recombinant growth factor currently FDA-approved to enhance the healing of soft tissue injuries. Regranex, or rhPDGF-BB has been shown in a series of well-controlled human clinical trials to promote healing of recalcitrant neuropathic dermal ulcers in diabetics. These results suggest that PDGF may provide an important clinical benefit for the healing of soft tissue injuries and deficits in the orofacial region, particularly in healing-compromised patients.

Current therapies for refractory ulcers on the oral soft tissues are often very unsatisfactory. Consequently, studies have been carried out to evaluate the role of growth factors in achieving successful remission of these lesions. For example, a recent study evaluated the effects of systemic administration and topical application of bFGF and epidermal growth factor (EGF) on impaired wound healing of chemically induced gingival ulcers in rabbits injected with cisplatin (CDDP) and peplomycin sulfate *(129)*. According to the investigators, EGF and bFGF promoted proliferation of the fibroblasts, and EGF also promoted proliferation of the keratinocytes isolated from gingival tissue of rabbits in vitro. Systemic injections of EGF and bFGF in rabbits, which had their submandibular glands removed, and topical application of bFGF accelerated healing of the ulcers.

Investigations are also ongoing regarding the use of growth factors in the treatment of other orofacial soft tissue injuries. For instance, some animal studies have shown encouraging effects of EGF, bFGF, and PDGF-AA in accelerating or enhancing healing of acute and chronic tympanic membrane

Table 3
Selected Synopses of Studies Using Various Growth Factor Carriers and Delivery Modalities *(69,120–123,125,128)*

Title, author, and synopsis	Test article	Conclusions
"Pulse application of platelet-derived growth factor enhances formation of a mineralizing matrix while continuous application is inhibitory" *(77)* By S. C. Hsieh and D. T. Graves. 1998 Synopsis: This *in vitro* study evaluates the effect of continuous and pulse treatments with platelet-derived growth factor BB isoform (PDGF-BB) on osteoblast proliferation and nodule formation in mineralizing cultures of fetal rat osteoblastic cells.	Fetal rat osteoblastic cells cultured in mineralizing media with various concentrations of pulsed and continuous PDGF-BB (0 to 50 ng/mL) over a range of time periods.	This *in vitro* study demonstrated that multiple brief exposures to PDGF-BB enhance formation of a mineralized matrix while longer-term exposure is inhibitory. Thus, PDGF-BB may have anabolic as well as inhibitory effects on bone formation, depending on the length of exposure. These findings suggest that multiple, brief exposures to PDGF-BB would enhance bone formation *in vivo* while prolonged exposure to PDGF-BB, which is likely to occur in chronic inflammation, would inhibit differentiated osteoblast function and limit bone regeneration.
"Modification of an osteoconductive anorganic bovine bone mineral matrix with growth factors" *(119)* By D. Jiang, R. Dziak, S. E. Lynch, and E. B. Stephan. August 1999 Synopses: This *in vitro* study investigates the interaction of osteoconductive anorganic bovine bone mineral matrix ("matrix") with platelet-derived growth factor BB isoform (PDGF-BB) and insulin-like growth factor-1 (IGF-1), and determines if the combination of growth factors with the matrix can stimulate proliferation of neonatal rat (Sprague-Dawley) osteoblastic cells.	Anorganic bovine bone matrix incubated with various concentrations of radiolabeled PDGF-BB or IGF-1. To measure proliferation, neonatal rat osteo-blastic cells were cultured with matrix incubated with PDGF-BB (0.4–5.5×10^6 M) or IGF-1 (0.9–14.2×10^6 M). To measure interaction dynamics, radiolabeled PDGF-BB (0.4–5.5×10^6 M) or IGF-1 (0.9–16.06×10^6 M) were incubated with matrix over a range of time periods (1 min to 10 d).	The results of this *in vitro* study indicated that PDGF-BB can be adsorbed to anorganic bovine bone mineral matrix and enhances the osteogenic properties of the matrix. IGF-1 was also adsorbed to the matrix, but was not readily released and did not produce significant effects in osteoblastic cell proliferation. However, the slow release of IGF-1 may be clinically beneficial because IGF-1 is important in long-term bone remolding. It may be clinically feasible to adsorb PDGF-BB to anorganic bovine bone and may have the combination of bone growth factor and matrix has the potential for clinical applications.

"The bone regenerative effect of platelet-derived growth factor-BB delivered with a chitosan/ tricalcium phosphate sponge carrier" *(120)*
By Y. M. Lee, Y. J. Park, S. J. Lee, Y. Ku, S. B. Han, P. R. Klokkevold, and C. P. Chung. 2000
Synopsis: The purpose of this study was to evaluate the bone regenerative effect of PDGF-BB delivered with a chitosan/ tricalcium phosphate (TCP) sponge carrier in a rat calvarial defect model.

Implantation of Chitosan/TCP sponge carriers loaded with/without PDGF-BB into an 8-mm calvarial defect made in Sprague-Dawley rats (15 rats).
Rats were sacrificed at 2 and 4 wk following implantation, and histological and histomorphometrical examinations were performed.
To measure release kinetics of PDGF-BB, ^{125}I-labeled PDGF-BB was loaded onto Chitosan/TCP sponges (plate form) and the concentrations of radioactive labeled PDGF-BB were assayed (over 4 wk)

In vitro examination of loaded sponge carrier revealed that a chitosan/TCP matrix is capable of delivering a consistent dose of PDGF-BB (2–6 ng/d) for up to 21 d.
Histological observations from the rat study indicate that the chitosan/TCP and PDGF-BB loaded chitosan/TCP sponges have osteoconductive and osteoinductive action with spontaneous biodegradation in surgically induced rat calvarial defects.
This provides evidence that a chitosan/TCP matrix is capable of overcoming connective tissue invasion observed with other synthetic graft materials and suggests that a PDGF-BB-loaded chitosan/TCP sponge may have osteogenic applications as a graft material for periodontal and bone regeneration.

"Controlled release of platelet-derived growth factor-BB from chondroitin sulfate-chitosan sponge for guided bone regeneration" *(121)*
By Y. J. Park, Y. M. Lee, J. Y. Lee, Y. J. Seol, C. P. Chung, S. J. Lee. 2000
Synopsis: The aims of this study are to design a way of controlling growth factor delivery by fabrication of a PDGF-BB releasing porous chondroitin-4-sulfate (CS)-chitosan sponge.
1. To achieve steady release of PDGF-BB via ionic interaction between PDGF-BB and CS.
2. To control porosity of the sponge due to ionic aoacervation between chitosan and CS.

Primary cultures of rat calvarial cells were dissected from 21-d-old Sprague-Dawley rat fetus and then striped of periosteum and loosely adherent tissue. Aliquots of 20 µL of cell suspension were seeded on top of rewetted CS-chitosan sponges (density of 10^5 cells/sponge) and incubated for 7 d.

PDGF-BB-loaded CS-chitosan sponge may potentially control PDGF-BB release. Incorporated CS effected controlled release of PDGF-BB from sponge and increased porosity of sponge.

(continued)

Table 3 (Continued)

Title, author, and synopsis	Test article	Conclusions
"Polymeric system for dual growth factor delivery" *(122)* By T. P. Richardson, M. C. Peter, A. B. Ennett, and D. J. Mooney. 2001 Synopsis: This study tests the hypothesis that dual delivery of VEGF and PDGF can direct the formation of a mature vasculature, as compared to the delivery of VEGF or PDGF delivered alone. The polymeric scaffold fabricated from poly(lactide-co-glycolide) (PLG) allowed dual delivery of vascular endothelial growth factor (VEGF)-165 and platelet-derived growth factor (PDGF)-BB, resulting in the rapid formation of a mature vascular network.	*In vitro* studies: Release rates and bioassay using cell proliferation assays with smooth muscle cells and endothelial cells. *In vivo* tests: Lewis rats—subcutaneous implantation. *In vivo* model of therapeutic angiogenesis Nonobese diabetic (NOD) mouse model subjected to femoral and vein ligation.	This study documents that dual delivery of growth factors involved in distinct aspects of vascular development is critical to the rapid formation of mature vasculature. Sustained, localized delivery of two growth factors both initiates formation of a significant number of blood vessels and induces their maturation. The polymeric scaffold allows controlled dual release of two distinct proteins.
"Enhanced bone formation by controlled growth factor delivery from chitosan-based biomaterials" *(123)* By J.-Y. Lee, S.-H. Nam, S.-Y. Im, Y.-J. Park, Y.-M. Lee, Y.-J. Seol, C.-P. Chung, S.-J. Lee. 2002 Synopsis: This study tests the usefulness of chitosan as drug releasing scaffolds and as modification tools for currently used biomaterials to enhance tissue regeneration efficacy. Release of platelet-derived growth factor-BB (PDGF-BB) from these matrices exerted significant osteoinductive effect.	Porous chitosan matrices, chitosan-poly (L-lactide) (PLLA). Release of platelet-derived growth factor-BB (PDGF-BB) from these matrices. Primary cultures of embryonic rat calvarial osteoblasts. *In vivo* studies: Sprague-Dawley rats. A craniotomy defect (8 mm in diameter) was formed by a trephine needle.	Controlled release of PDGF-BB from PLLA-chitosan composite porous matrices significantly promoted bone healing and regeneration. These results suggest the feasibility of a combinative strategy of controlled drug release and tissue engineering in reconstructive therapy in the fields of aerodonetics, orthopedics, and plastic surgery.

382

"Enhancement of osteoblast proliferation *in vitro* by selective enrichment of demineralized freeze-dried bone allograft with specific growth factors" *(124)*

By D. A. Mott, J. Mailhot, M. F. Cuenin, M. Sharawy, and J. Borke. 2002

Synopsis: The purpose of this study is to determine what effect, if any, decalcified freeze-dried bone allografts (DFDBA) enriched with specific growth factors has on the *in vitro* proliferation of murine osteoblasts.

Primary cultured osteoblastic cells harvested from 1-d-old neonatal mice. The calvaria of these mouse pups were digested and frozen for use.

2% residual calcium-DFDBA and supplemented by one or combinations of growth factors; transforming growth factor-β (TGF-β), insulin-like growth factor-1 (IGF-1), platelet-derived growth factor (PDGF), fibroblast growth factors basic (bFGF), or vascular endothelial growth factors (VEGF).

The osteoblasts exposed to 2% DFDBA and enhanced with a combination of growth factors including IGF, TGF-β, and PDGF showed a significant difference at d 7 compared to controls. The DFDBA as a carrier appear to be required for these growth factors to stimulate an increase in cell proliferation.

Individual growth factors IGF-1, PDGF, and TGF-β all showed a noticeable effect and a significant increase in proliferation when adsorbed in DFDBA at d 14.

perforations without significant adverse effects *(129,130)*. However, two reports of clinical trials of topical EGF or bFGF for tympanic membrane perforations have revealed mixed results.

SUMMARY AND FUTURE DIRECTIONS

Today the preferred material for reconstruction of large osseous orofacial defects is autogenous bone from intraoral sources such as the mandibular symphysis. If the volume that can be harvested from intraoral sites is insufficient, autograft is also harvested from the head of the tibia or the iliac crest. For moderate to smaller osseous defects such as periodontal defects, sinus floor augmentation, or most alveolar ridge augmentations, bone allografts or substitutes are being used with increased frequency because of morbidity of autograft harvest. Bone allograft is still the most frequently utilized graft in the United States, although its use in Europe is more limited. Demineralized allograft is used most frequently in periodontal defects, while mineralized allograft is preferred by many clinicians for sinus floor augmentation. Deproteinized bovine bone mineral has gained substantial popularity over the last 5 yr for use in sinus augmentations and periodontal/periimplant defects, because the material has been shown to be biocompatible and highly osteoconductive. The current generation of synthetic alloplastic materials is not widely used, apparently because their osteoconductive capacity is insufficient in most orofacial indications to generate widespread enthusiasm.

The medical field concerned with hard tissue reconstruction eagerly anticipates the commercial availability of bone growth factors and morphogens. Within the orofacial field, the proteins that have received the most enthusiasm and research attention are platelet-derived growth factor and bone morphogenetic proteins. Recombinant human PDGF-BB is currently the only FDA- approved growth factor or morphogen. It is marketed under the brand name Regranex. The formulation of this product, however, is not appropriate for use in treating bone defects. Nevertheless, the widespread commercial use of Regranex (for chronic foot wounds in diabetics) has demonstrated that rhPDGF-BB is a highly safe product. Extensive research focused on the use of PDGF in the periodontal field has elucidated the mechanism of action of PDGF, demonstrating that it is a potent mitogen and chemoattractant for PDL cells and osteoblasts. Animal and initial human trials have further demonstrated that rhPDGF-BB has the capacity to enhance periodontal regeneration. Most recently, emerging data from human studies has shown favorable tissue response to rhPDGF-BB treatment at both the clinical and microscopic levels, and demonstrated for the first time that complete periodontal regeneration can be achieved in advanced Class II furcation defects using a combination of purified recombinant growth factor and bone allograft. Natural platelet-derived growth factors, in the form of platelet-rich plasma, are currently being utilized clinically in conjunction with autograft with good success in reconstruction of large orofacial bone defects.

rhBMP-2 and rhBMP-7 (rhOP-1), as well as a natural bovine-derived bone extract of BMP, are being commercially developed in the orofacial field for augmentation of the sinus floor prior to dental implant placement and, at a somewhat earlier stage, for treatment of periodontal defects. Animal and human studies have demonstrated the capacity for the BMPs to stimulate bone formation in these orofacial defects. The appropriate dosage and, at least in the case of OP-1, the appropriate carrier are being evaluated. A pivotal clinical trial is underway with the use of BMP-2 for sinus floor grafting, for which the results should be available within the next year or two. In the periodontal field, there is clear evidence that both BMP-2 and OP-1 can promote periodontal regeneration in animals. However, the lack of mitogenic and chemotactic effects of the BMPs on PDL cells, coupled with the promotion of osteoblast phenotypic markers in PDL cells treated with BMP, suggest that ankylosis and root resorption may be a sequelae to use of BMPs or OP-1 around teeth. These abnormal healing responses have been observed in most animal studies utilizing the BMPs in periodontal defects, although a study in dogs did not find an increase in either ankylosis or root resorption.

Clearly, bone regeneration research will continue at its current robust pace. Products based on the principles of tissue engineering (**Fig. 9**), combining osteoconductive materials with growth factors or

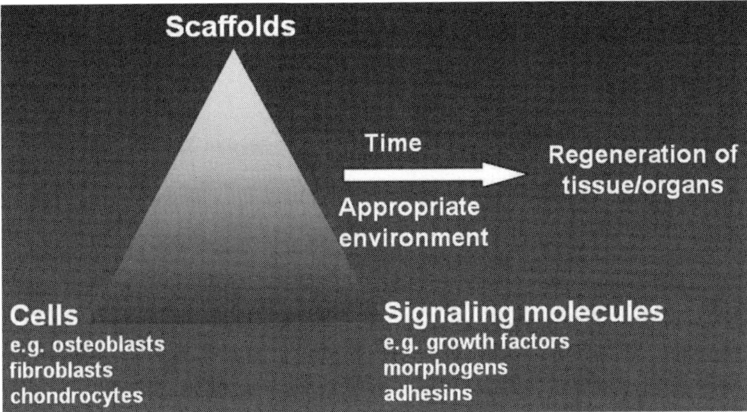

Fig. 9. Tissue engineering triad. The bone regeneration field appears to be moving toward combining recombinant growth factors with osteoconductive scaffolds. In the future, osteoprogenitor stem cells may be added to the graft to further increase effectiveness. No matter what the configuration of the products, they will need to be placed into the appropriate environment, which includes a sufficient vascular supply and quiescent, stable wound to be most effective *(2)*.

morphogens, are being evaluated. Hopefully, sometime in the not-so-distant future, these products will be commercially available and thus obviate or reduce the need to harvest autograft for most reconstructive procedures. Such off-the-shelf therapy would be a great benefit to the patient undergoing reconstructive surgery in the orofacial region.

REFERENCES

1. American Dental Association. (1990) *Survey of Dental Procedures Rendered.*
2. Lynch, S. E., Genco, R. J., and Marx, R. E. (1999) *Tissue Engineering: Applications in Maxillofacial Surgery and Periodontics.* Quintessence, Chicago, pp. 1–297.
3. Aukhil, I. and Iglhaut, J. (1988) Periodontal ligament cell kinetics following experimental regenerative procedures. *J. Clin. Periodontol.* **15,** 374–382.
4. Iglhaut, J., Aukhil, I., Simpson, D. M., Johnston, M. C., and Koch, G. (1988) Progenitor cell kinetics during guided tissue regeneration in experimental periodontal wounds. *J. Periodontol. Res.* **23,** 107–117.
5. Boyne, P. J. *Osseous Reconstruction of the Maxilla and Mandible.* Quintessence, Chicago, pp. 1–105.
6. Schallhorn, R. G. (1972) Postoperative problems associated with iliac transplants. *J. Periodontol.* **43,** 3–9.
7. Dragoo, M. and Sullivan, H. (1973) A clinical and histologic study of autogenous iliac bone grafts in humans. II. External root resorption. *J. Periodontol.* **44,** 614–625.
8. Bowers, G. M., Chadroff, B., Carnevale, R., et al. (1989) Histologic evaluation of new attachment apparatus formation in humans, Part II. *J. Periodontol.* **60,** 674–682.
9. Bowers, G. M., Chadroff, B., Carnevale, R., et al. (1989) Histologic evaluation of new attachment apparatus formation in humans, Part III. *J. Periodontol.* **60,** 683–693.
10. Mellonig, J. T. (1984) Decalcified freeze dried bone allograft as an implant material in human periodontal defects. *Int. J. Periodont. Res. Dent.* **4,** 41–55.
11. McClain, P. and Schallhorn, R. (1993) Long-term assessment of combined osseous composite grafting, root conditioning and guided tissue regeneration. *Int. J. Periodont. Res. Dent.* **13,** 9–27.
12. Anderegg, C. R., Martin, S. J., Gray, J., Mellonig, J. T., and Gher, M. E. (1991) Clinical evaluation of decalcified freeze-dried bone allograft with guided tissue regeneration in the treatment of molar furcation invasions. *J. Periodontol.* **62,** 264–268.
13. Wallace, S., Gellin, R., Miller, M., and Mishkin, M. (1994) Guided tissue regeneration with and without decalcified freeze-dried bone in mandibular class II furcation invasions. *J. Periodontol.* **65,** 244–254.
14. Guillemin, M., Mellonig, J. T., Brunsvold, M., and Steffensen, B. (1993) Healing in periodontal defects treated by decalcified freeze-dried bone allografts in combination with ePTFE membranes. Clinical and SEM analylsis. *J. Clin. Periodontol.* **20,** 528–536.

15. Mellado, J. R., Salkin, L. M., Freedman, A. L., and Stein, M. D. (1995) A comparative study of ePTFE periodontal membranes with and without decalcified freeze-dried bone allografts for the regeneration of interproximal intraosseous defects. *J. Periodontol.* **66,** 751–755.

16. Becker, W., Becker, B. E., and Caffessee, R. (1994) A comparison of demineralized freeze-dried bone and autogenous bone to induce bone formation in human extraction sockets. *J. Periodontol.* **65,** 1128–1133.

17. Becker, W., Urist, M. R., Tucker, L. M., Becker, B. E., and Ochsenbein, C. (1995) Human demineralized freeze-dried bone. Inadequate induced bone formation in athymic mice. A preliminary report. *J. Periodontol.* **66,** 822–828.

18. Nevins, M. and Mellonig, J. T. (1998) *Implant Therapy.* Quintessence, Chicago, p. 74.

19. Mohan, S. and Baylink, D. J. (1996) Therapeutic potential of TGF-β, BMP, and FGF in the treatment of bone loss, in *Principles of Bone Biology.* Academic Press, New York, pp. 1111–1123.

20. Schwartz, Z., Somers, A., Mellonig, J. T., et al. (1998) Ability of commercial demineralized freeze-dried bone allograft to induce new bone formation is dependent on donor age but not gender. *J. Periodontol.* **69,** 470–478.

21. Schwartz, Z., Mellonig, J. T., Carnes, D. L., et al. (1996) Ability of commercial demineralized freeze-dried bone allograft to induce new bone formation. *J. Periodontol.* **67,** 918–926.

22. Rummelhardt, J., Mellonig, J. T., Gray, J., and Towle, H. (1989) Comparison of freeze-dried bone allograft in human periodontal osseous defects. *J. Periodontol.* **60,** 655–663.

23. Mellonig, J. T. (1998) Periodontal regeneration: bone grafts, in *Periodontal Therapy* (Nevins, M. and Mellonig, J. T., eds.), Quintessence, Chicago, pp. 233–248.

24. Garrett, S. (1996) Periodontal regeneration around natural teeth. *Ann. Periodontol.* **1,** 621–666.

25. Lynch, S. E. (1994) The role of growth factors in periodontal repair and regeneration, in *Periodontal Regeneration: Current Status and Directions* (Polson, A. ed.), Quintessence, Chicago, pp. 179–198.

26. Park, Y. J., Lee, Y. M., Park, S. N., Sheen, S. Y., Chung, C. P., and Lee, S. J. (2000) Platelet-derived growth factor releasing chitosan sponge for periodontal bone regeneration. *Biomaterials* **21,** 153–159.

27. Park, Y. J., Ku, Y., Chung, C. P., and Lee, S. J. (1998) Controlled release of platelet-derived growth factor from porous poly (L-lactide) membranes from guided tissue regeneration. *J. Controlled Release* **51,** 201–211.

28. Vikjaer, D., Blom, S., Horting-Hansen, E., and Pinholt, E. M. (1997) Effect of platelet-derived growth factor-BB on bone formatin in calvarial defects: and experimental study in rabbits. *Eur. J. Oral Sci.* **105,** 50–66.

29. Chung, C. P., Kim, D. K., Park, Y. J., Nam, K. H., and Lee, S. J. (1997) Biological effects of drug-loaded biodegradable membranes for guided bone regeneration. *J. Periodont. Res.* **32,** 172–175.

30. Mellonig, J. T. (2000) Human histologic evaluation of a bovine-derived bone xenograft in the treatment of periodontal osseous defects. *Int. J. Periodont. Restor. Dent.* **20,** 19–29.

31. Camelo, M., Nevins, M., Schenk, R., et al. (1998) Clinical, radiographic and histologic evaluation of human periodontal defects treated with Bio-Oss and Bio-Gide. *Int. J. Periodontol. Restor. Dent.* **18,** 321–331.

32. McAllister, B., Margolin, M. D., Cogan, A. G., et al. (1999) Eighteen-month radiographic and histologic evaluation of sinus grafting with anorganic bovine bone in the chimpanzee. *Int. J. Oral Maxillofac. Implants* **14,** 361–368.

33. Valentini, P., Abensur, D., Densari, D., Graziani, J. N., and Hammerle, C. H. F. (1998) Histologic evaluation of Bio-Oss in a 2-stage sinus floor elevation and implant procedure. *Clin. Oral Impl. Res.* **9,** 233–241.

34. Valentini, P. and Abensur, D. (1997) Maxillary sinus floor elevation for implant placement with demineralized freeze-dried bone and bovine bone (Bio-Oss): a clinical study of 20 patients. *Int. J. Periodont. Res. Dent.* **17,** 233–241.

35. Spector, M. (1999) Basic principles of tissue engineering, in *Tissue Engineering: Applications in Maxillofacial Surgery and Periodontics* (Lynch et al., eds.), Quintessence, Chicago, pp. 3–15.

36. Skoglund, A., Hising, P., and Young, C. (1997) A clinical and histologic examination in humans of the osseous response to implanted natural bone mineral. *Int. J. Oral Maxillofac. Implants* **12,** 194–200.

37. Mellonig, J. T. and Nevins, M. (1998) Guided bone regeneration and dental implants, in *Implant Therapy* (Nevins, M. and Mellonig, J. T., eds.). Quintessence, Chicago, pp. 53–81.

38. Shanaman, R. H. (1998) Use of guided bone regeneration to facilitate ideal prosthetic placement of implants, in *Implant Therapy.* Quintessence, Chicago, pp. 83–90.

39. Samchukov, M. L., Cherkashin, A. M., and Cope, J. B. (1999) Distraction osteogenesis: history and biologic basis of new bone formation, in *Tissue Engineering: Applications in Maxillofacial Surgery and Periodontics* (Lynch et al., eds.), Quintessence, Chicago, pp. 131–146.

40. Bell, W. H., Harper, R. H., Fields, R. T., Richardson, S. A., and Samchukov, M. L. (1997) Mandibular widening by osteodistraction after symphyseal osteotomy, *Br. J. Oral Maxillofac. Surg.* **35,** 11–17.

41. Komuro, Y., Takato, T., Harii, K., and Yonemara, Y. (1994) The histologic analysis of distraction osteogenesis of the mandible in rabbits. *Plastic Reconstr. Surg.* **94,** 152–158.

42. Karaharju, E. O., Aalto, K., Kahri, A., et al. (1993) Distraction bone healing. *Clin. Orthop.* **297,** 38–46.

43. Snyder, C. C., Levine, G. A., Swanson, H. M., and Browne, E. Z. (1973) Mandibular lengthening by gradual distraction. *Plastic Reconstr. Surg.* **51,** 506–511.

44. Michieli, S. and Miotti, B. (1977) Lengthening of the mandibular body by gradual surgical-orthodontic distraction. *J. Oral Surg.* **35,** 187.

45. Bell, W. H. and Epker, B. N. (1976) Surgical orthodontic expansion of the maxilla. *Am. J. Orthod.* **70**, 517–521.
46. Guerrero, C. (1990) Expansion rapida mandibular. *Rev. Ven. Orthod.* **1–2**, 48.
47. Guerrero, C. (1993) Espectro de aplicacion de la expansion rapida mandibular. *Rev. Ven. Orthod.* **10**, 170.
48. McCarthy, J. G., Schreiber, J., Karp, N., Thorne, C. H., and Grayson, B. H. (1992) Lengthening the human mandible by gradual distraction. *Plastic Reconstr. Surg.* **89**, 1–9.
49. Chin, M. (1999) Distraction osteogenesis in maxillofacial surgery, in *Tissue Engineering: Applications in Maxillofacial Surgery and Periodontics* (Lynch et al., eds.), Quintessence, Chicago, pp. 147–159.
50. Lane, J. M., Tomin, E., and Bostrom, M. P. G. (1999) Biosynthetic bone grafting. *Clin. Orthop.* **367**, S107–S117.
51. Andrew, J. G., Hoyland, J. A., Freemont, A. J., and Marsh, D. R. (1995) Platelet-derived growth factor expression in normally healing human fractures. *Bone* **16**, 455–460.
52. Bostrom, M. P., Lane, J. M., Berberian, W. S., et al. (1995) Immunolocalization and expression of bone morphogenetic proteins 2 and 4 in fracture healing. *J. Orthop. Res.* **13**, 357–367.
53. Fujii, H., Kitazawa, R., Maeda, S., Mizuno, K., and Kitazawa, S. (1999) Expression of platelet-derived growth factor proteins and their receptor alpha and beta mRNAs during fracture healing in the normal mouse. *Histochem. Cell Biol.* **112**, 131–138.
54. Marx, R. E. (1999) Platelet-rich plasma: a source of autogolous growth factors for bone grafts, in Tissue *Engineering: Applications in Maxillofacial Surgery and Periodontics* (Lynch et al., eds.), Quintessence, Chicago, pp. 71–82.
55. Canalis, E., McCarthy, T. L., Centrella, M. (1989) Effects of platelet-derived growth factor on bone formation in vitro. *J. Cell Physiol.* **140**, 530–537.
56. Bowen-Pope, D. F., van Koppen, A., and Schatteman, G. (1991) Is PDGF really important? Testing the hypotheses. *Trends Genet.* **7**, 411–418.
57. Soriano, P. (1994) Abnormal kidney development and hematological disorders in PDGF beta-receptor mutant mice. *Genes Dev.* **8**, 1888–1896.
58. Schatterman, G. C., Morrison-Graham, K., Van Koppen, A., Weston, J. A., and Bowen-Pope, D. F. (1992) Regulation and role of PDGF receptor alpha subunit expression during embryogenesis. *Development* **115**, 123–131.
59. Centrella, M., McCarthy, T. L., Kusmik, W. F., and Canalis, E. (1992) Relative binding and biochemical effects of heterodimeric and homodimeric isoforms of platelet-derived growth factor in osteoblast-enriched cultures from fetal rat bone. *J. Cell Physiol.* **147**, 420–426.
60. Piche, J. E. and Graves, D. T. (1989) Study of the growth factor requirements of human bone-derived cells: a comparison with human fibroblasts. *Bone* **10**, 131–138.
61. Hock, J. and Canalis, E. (1994) Platelet-derived growth factor enhances bone cell replication, but not differentiated function of osteoblasts. *Endocrinology* **134**, 1423–1428.
62. Canalis, E., Varghese, S., McCarthy, T. L., and Centrella, M. (1992) Role of platelet derived growth factor in bone cell function. *Growth Regul.* **2**, 151–155.
63. Hsieh, S. and Graves, D. (1998) Pulse application of platelet-derived growth factor enhances formation of a mineralizing matrix while continuous application is inhibitory. *J. Cell Biochem.* **69**, 169–180.
64. Lynch, S. E., Williams, R. C., Polson, A. M., et al. (1989) A combination of platelet-derived growth factor and insulin like growth factor enhances periodontal regeneration. *J. Clin. Periodontol.* **16**, 545–548.
65. Lynch, S. E., Castilla, G. R., Williams, R. C., et al. (1991) The effect of short term application of a combination of platelet-derived and insulin-like growth factors on periodontal wound healing. *J. Periodontol.* **62**, 458–467.
66. Giannobile, W. V., Hernandez, R. A., Finkelman, R. D., et al. (1996) Comparative effects of platelet-derived growth factor, insulin-like growth factor, individually and in combination on periodontal regeneration in *Macaca fascicularis. J. Periodont. Res.* **31**, 301–312.
67. Park, J. B., Matsuura, M., Han,, K. Y., et al. (1995) Periodontal regeneration in Class III furcation defects of beagle dogs using guided tissue regenerative therapy with platelet-derived growth factor. *J. Periodontol.* **66**, 462–477.
68. Rutherford, R. B., Niekrash, C. E., Kennedy, J. E., and Charette, M. F. (1992) Platelet-derived and insulin-like growth factors stimulate regeneration of periodontal attachment in monkeys. *J. Periodont. Res.* **27**, 285–290.
69. Murakami, S., Takayama, S., Kitamura, M., et al. (2003) Recombinant human basic fibroblast growth factor (bFGF) stimulates periodontal regeneration in class II furcation defects created in beagle dogs. *J. Periodont. Res.* **38(1)**, 97–103.
70. Nash, T. J., Howlett, C. R., Martin, C., Steele, I., Johnson, K. A., and Hicklin, D. J. (1994) Effect of platelet-derived growth factor on tibial osteotomies in rabbits. *Bone* **15**, 203–208.
71. Mitlak, B. H., Finkelman, R. D., Hill, E. I., et al. (1996) The effect of systemically administered PDGF-BB on the rodent skeleton. *J. Bone Miner. Res.* **11**, 238–247.
72. Si, X., Jin, Y., Yang, L., Tipoe, G. L., and White, F. H. (1997) Expression of BMP-2 and TGF-beta 1 mRNA during healing of the rabbit mandible. *Eur. J. Oral Sci.* **105**, 325–330.
73. Sykaras, N. and Opperman, L. A. (2003) Bone morphogenetic proteins (BMPs): how do they function and what can they offer the clinician? *J. Oral Sci.* **45(2)**, 57–73.

74. Tavakoli, K., Yu, Y., Shahidi, S., Bonar, F., Walsh, W. R., and Poole, M. D. (1999) Expression of growth factors in the mandibular distraction zone: a sheep study. *Br. J. Plastic Surg.* **52(6),** 434–439.

75. Campisi, P., Hamdy, R. C., Lauzier, D., Amako, M., Rauch, F., and Lessard, M. L. (2003) Expression of bone morphogenetic proteins during mandibular distraction osteogenesis. *Plastic Reconstr. Surg.* **111(1),** 201–208; discussion 209–210.

76. Campisi, P., Hamdy, R. C., Lauzier, D., Amako, M., Schloss, M. D., and Lessard, M. L. (2002) Overview of the radiology, histology, and bone morphogenetic protein expression during distraction osteogenesis of the mandible. *J. Otolaryngol.* **31(5),** 281–286.

77. Bartold, P. M. (1993) Platelet-derived growth factor stimulates hyaluronate but not proteoglycan synthesis by human gingival fibroblasts in vitro. *J. Dent. Res.* **72,** 1473–1480.

78. Dennison, D. K., Vallone, D. R., Pinero, G. J., Rittman, B., and Caffesse, R. G. (1994) Differential effect of TGF-beta1 and PDGF on proliferation of periodontal ligament cells and gingival fibroblasts. *J. Periodontol.* **65,** 641–648.

79. McAllister, B. S., Leeb-Lundberg, F., Mellonig, J. T., and Olson, M. S. (1995) The functional interaction of EGF and PDGF with bradykinin in the proliferation of human gingival fibroblasts. *J. Periodontol.* **66,** 429–437.

80. Oates, T. W., Rouse, C. A., and Cochran, D. L. (1993) Mitogenic responses of growth factors on human periodontal ligament cells in vitro. *J. Periodontol.* **64,** 142–148.

81. Piche, J. E. and Graves, D. T. (1989) Study of growth factor requirements of human bone-derived cells. A comparison with human fibroblasts. *Bone* **10,** 131–138.

82. Ojima, Y., Mizuno, M., Kuboki, Y., and Komori, T. (2003) In vitro effect of platelet-derived growth factor-BB on collagen synthesis and proliferation of human periodontal ligament cells. *Oral. Dis.* **9(3),** 144–151.

83. Papadopoulos, C. E., Dereka, X. E., Vavouraki, E. N., and Vrotsos, I. A. (2003) In vitro evaluation of the mitogenic effect of platelet-derived growth factor-BB on human periodontal ligament cells cultured with various bone allografts. *J. Periodontol.* **74(4),** 451–47.

84. Marcopoulou, C. E., Vavouraki, H. N., Dereka, X. E., and Vrotsos, I. A. (2003) Proliferative effect of growth factors TGF-beta1, PDGF-BB and rhBMP-2 on human gingival fibroblasts and periodontal ligament cells. *J. Int. Acad. Periodontol.* **5(3),** 63–70.

85. Mumford, J. H., Carnes, D. L., Cochran, D. L., and Oates, T. W. (2001) The effects of platelet-derived growth factor-BB on periodontal cells in an in vitro wound model. *J. Periodontol.* **72(3),** 331–340.

86. Howell, T. H., Fiorellini, J. P., Paquette, D. W., Offenbacher, S., Giannobile, W. V., and Lynch S. E. (1997) A Phase I/II clinical trial to evaluate a combination of recombinant human platelet-derived growth factor-BB and recombinant human insulin-I like growth factor-I in patients with periodontal disease. *J. Periodontol.* **68,** 1186–1193.

87. Schroer, M., Wahl, T., Hutchens, I., Moriarty, J., and Bergenholtz, B. (1991) Closed versus open debridement of facial grade II molar furcation. *J. Clin. Periodontol.* **18,** 323–329.

88. Camelo, M., Nevins, M. L., Schenk, R. K., Lynch, S. E., and Nevins, M. (2003) Periodontal regeneration in human Class II furcations using purified recombinant human platelet-derived growth factor-BB (rhPDGF-BB) with bone allograft. *Int. J. Periodont. Restor. Dent.* **23(3),** 213–225.

89. Nevins, M., Camelo, M., Nevins, M. L., Schenk, R. K., and Lynch, S. E. (2003) Periodontal regeneration in human Class II furcations using purified recombinant human platelet-derived growth factor-BB (rhPDGF-BB) with bone allograft. *J. Periodontol.* **74(9),** 1282–1292.

90. Hollinger, J. O., Buck, D. C., and Bruder, S. P.. (1999) Biology of bone healing: its impact on clinical therapy, in *Tissue Engineering: Applications in Maxillofacial Surgery and Periodontics* (Lynch et al., eds.), Quintessence, Chicago, pp. 17–53.

91. Wozney, J. M. (1999) Biology and clinical application of rhBMP-2, in *Tissue Engineering: Applications in Maxillofacial Surgery and Periodontics* (Lynch et al., eds.), Quintessence, Chicago, pp. 103–123.

92. Wikesjo, U. M. E., Hanish, O., Sigurdsson, T. J., and Caplanis, N. (1999) Application of rhBMP-2 to alveolar and periodontal defects, in *Tissue Engineering: Applications in Maxillofacial Surgery and Periodontics* (Lynch et al., eds.), Quintessence, Chicago, pp. 269–286.

93. Cochran, D. L., Jones, A., Fontaine, J. D. L., et al. (1999) Periodontal regeneration and localized osseous regeneration in the oral cavity, in *Tissue Engineering: Applications in Maxillofacial Surgery and Periodontics* (Lynch et al., eds.), Quintessence, Chicago, pp. 245–257.

94. Wikesjo, U. M. E., Guglielmoni, P., Promsudthi, A., et al. (1999) Periodontal repair in dogs: effect of rhBMP-2 concentration on regeneration of alveolar bone and periodontal attachment. *J. Clin. Periodontol.* **26,** 392–400.

95. Selvig, K. A., Sorensen, R. G., Wozney, J. M., and Wikesjo, U. M. (2002) Bone repair following recombinant human bone morphogenetic protein-2 stimulated periodontal regeneration. *J. Periodontol.* **73(9),** 1020–1029.

96. Kuboki, Y., Sasaki, M., Saito, A., Takita, H., and Kato, H. (1998) Regeneration of periodontal ligament and cementum by BMP-applied tissue engineering. *Eur. J. Oral Sci.* **106,** S197–S203.

97. King, G. N., King, N., and Hughes, F. J. (1998) Effect of two delivery systems for recombinant human bone morphogenetic protein-2 on periodontal regeneration in vivo. *J. Periodont. Res.* **33,** 226–236.

98. Strayhorn, C. L., Garrett, J. S., Dunn, R. L., Benedict, J. J., and Somerman, M. J. (1999) Growth factors regulate expression of osteoblast-associated genes. *J. Periodontol.* **70,** 1345–1354.

99. Zaman, K. U., Sugaya, T., and Kato, H. (1999) Effect of recombinant human platelet-derived growth factor-BB and bone morphogenetic protein-2 application to demineralized dentin on early periodontal ligament cell response. *J. Periodont. Res.* **34**, 244–250.

100. Stefani, C. M., Machado, M. A., Sallum, E. A., Sallum, A. W., Toledo, S., and Nociti, F. H. Jr. (2000) Platelet-derived growth factor/insulin-like growth factor-1 combination and bone regeneration around implants placed into extraction sockets: a histometric study in dogs. *Implant Dent.* **9(2)**, 126–131.

101. Nociti Junior, F. H., Stefani, C. M., Machado, M. A., Sallum, E. A., Toledo, S., and Sallum, A. W. (2000) Histometric evaluation of bone regeneration around immediate implants partially in contact with bone: a pilot study in dogs. *Implant Dent.* **9(4)**, 321–328.

102. Meraw, S. J., Reeve, C. M., Lohse, C. M., and Sioussat, T. M. (2000) Treatment of peri-implant defects with combination growth factor cement. *J. Periodontol.* **71(1)**, 8–13.

103. Jung, R. E., Glauser, R., Scharer, P., Hammerle, C. H., Sailer, H. F., and Weber, F. E. (2003) Effect of rhBMP-2 on guided bone regeneration in humans. *Clin. Oral Implants Res.* **14(5)**, 556–568.

104. Wikesjo, U. M., Xiropaidis, A. V., Thomson, R. C., Cook, A. D., Selvig, K. A., and Hardwick, W. R. (2003) Periodontal repair in dogs: space-providing ePTFE devices increase rhBMP-2/ACS-induced bone formation. *J. Clin. Periodontol.* **30(8)**, 715–725.

105. Hanisch, O., Sorensen, R. G., Kinoshita, A., Spiekermann, H., Wozney, J. M., and Wikesjo, U. M. (2003) Effect of recombinant human bone morphogenetic protein-2 in dehiscence defects with non-submerged immediate implants: an experimental study in Cynomolgus monkeys. *J. Periodontol.* **74(5)**, 648–657.

106. Wikesjo, U. M., Sorensen, R. G., Kinoshita, A., and Wozney, J. M. (2002) RhBMP-2/alphaBSM induces significant vertical alveolar ridge augmentation and dental implant osseointegration. *Clin. Implant Dent. Rel. Res.* **4(4)**, 174–182.

107. Marukawa, E., Asahina, I., Oda, M., Seto, I., Alam, M., and Enomoto, S. (2002) Functional reconstruction of the non-human primate mandible using recombinant human bone morphogenetic protein-2. *Int. J. Oral Maxillofac. Surg.* **31(3)**, 287–295.

108. Nevins, M., Kirker-Head, C., Nevins, M., et al. (1996) Bone formation in the goat maxillary sinus induced by absorbable collagen sponge implants impregnated with recombinant human bone morphogenetic protein-2. *Int. J. Periodont. Restor. Dent.* **16**, 8–19.

109. Boyne, P. J., Marx, R. E., Nevins, M., et al. (1997) A feasibility study evaluating rhBMP-2/absorbable collagen sponge for maxillary sinus floor augmentation. *Int. J. Periodont. Restor. Dent.* **17**, 10–25.

110. Margolin, M. D., Cogan, A. G., Taylor, M., et al. (1998) Maxillary sinus augmentation in the nonhuman primate. A comparative radiographic and histologic study between recombinant human osteogenic protein-1 and natural bone mineral. *J. Periodontol.* **69**, 911–919.

111. McAllister, B., Margolin, M., Cogan, A., Taylor, M., and Wollins, J. (1998) Residual lateral wall defects following sinus grafting with recombinant human osteogenic protein-1 or Bio-Oss in the chimpanzee. *Int. J. Periodont. Restor. Dent.* **18**, 227–239.

112. Groeneveld, E. H. J., van den Bergh, J. P. A., Holzmann, P., et al. (1999) Histomorphometric analysis of bone formed in human maxillary sinus floor elevations grafted with OP-1 device, demineralized bone matrix or autogenous bone. *Clin. Oral Implants Rest.* **10**, 499–509.

113. Terheyden, H., Jepsen, S., Moller, B., Tucker, M. M., and Rueger, D. C. (1999) Sinus floor augmentation with simultaneous placement of dental implants using a combination of deproteinized xenografts and recombinant human osteogenic protein-1. *Clin. Oral Implants Res.* **10**, 510–521.

114. van den Bergh, J. P., ten Bruggenkate, C. M., Groeneveld, H. H., Burger, E. H., and Tuinzing, D. B. (2000) Recombinant human bone morphogenetic protein-7 in maxillary sinus floor elevation surgery in 3 patients compared to autogenous bone grafts. A clinical pilot study. *J. Clin. Periodontol.* **27(9)**, 627–636.

115. Boyne, P. J. (1999) Studies of the surgical application of osteoconductive and osteoinductive materials, in *Tissue Engineering: Applications in Maxillofacial Surgery and Periodontics* (Lynch et al., eds.), Quintessence, Chicago, pp. 125–130.

116. Howell, T. H., Fiorellini, J., Jones, A., et al. (1997) A feasibility study evaluating rhBMP-2/absorbable collagen sponge device for loca alveolar ridge preservation or augmentation. *Int. J. Periodont. Restor. Dent.* **17**, 125–139.

117. Sanchez, A. R., Sheridan, P. J., and Kupp, L. I. (2003) Is platelet-rich plasma the perfect enhancement factor? A current review. *Int. J. Oral Maxillofac. Implants* **18(1)**, 93–103.

118. Marx, R. E. (1999) Platelet-rich plasma: a source of multiple autologous growth factors for bone grafts, in *Tissue Engineering Applications in Maxillofacial Surgery and Periodontics* (Lynch, S. E., Genco, R. J., and Marx, R. E., eds.), Quintessence, Chicago, pp. 71–82.

119. Bessho, K., Carnes, D. L., Cavin, R., and Ong, J. L. (2002) Experimental studies on bone induction using low-molecular-weight poly (DL-lactide-co-glycolide) as a carrier for recombinant human bone morphogenetic protein-2. *J. Biomed. Mater. Res.* **61(1)**, 61–65.

120. Jiang, D., Dziak, R., Lynch, S. E., and Stephan, E. B. (1999) Modification of an osteoconductive anorganic bovine bone mineral matrix with growth factors. *J. Periodontol.* **70(8)**, 834–839.

121. Lee, Y. M., Park, Y. J., Lee, S. J., et al. (2000) The bone regenerative effect of platelet-derived growth factor-BB delivered with a chitosan/tricalcium phosphate sponge carrier. *J. Periodontol.* **71(3)**, 418–424.

122. Park, Y. J., Lee, Y. M., Lee, J. Y., Seol, Y. J., Chung, C. P., and Lee, S. J. (2000) Controlled release of platelet-derived growth factor-BB from chondroitin sulfate-chitosan sponge for guided bone regeneration. *J. Control. Release* **67(2–3)**, 385–394.
123. Richardson, T. P., Peters, M. C., Ennett, A. B., and Mooney, D. J. (2001) Polymeric system for dual growth factor delivery. *Nat. Biotechnol.* **19(11)**, 1029–1034.
124. Lee, J. Y., Nam, Sil, Im, S. Y., et al. (2002) Enhanced bone formation by controlled growth factor delivery from chitosan-based biomaterials. *J. Control Release* **78(1–3)**, 187–197.
125. Mott, D. A., Mailhot, J., Cuenin, M. F., Sharawy, M., and Borke, J. (2002) Enhancement of osteoblast proliferation in vitro by selective enrichment of demineralized freeze-dried bone allograft with specific growth factors. *J. Oral Implantol.* **28(2)**, 57–66.
126. Stephan, E. B., Renjen, R., Lynch, S. E., and Dziak, R. (2000) Platelet-derived growth factor enhancement of a mineral- collagen bone substitute. *J. Periodontol.* **71(12)**, 1887–1892.
127. Lee, S. J., Park, Y. J., Park, S. N., et al. (2001) Molded porous poly (L-lactide) membranes for guided bone regeneration with enhanced effects by controlled growth factor release. *J. Biomed. Mater. Res.* **55(3)**, 295–303.
128. Lee, J. Y., Nam Sil, Im, S. Y., et al. (2002) Enhanced bone formation by controlled growth factor delivery from chitosan-based biomaterials. *J. Control Release* **78(1–3)**, 187–197.
129. Fujisawa, K., Miyamoto, Y., and Nagayama, M. (2003) Basic fibroblast growth factor and epidermal growth factor reverse impaired ulcer healing of the rabbit oral mucosa. *J. Oral Pathol. Med.* **32(6)**, 358–366.
130. Yeo, S. W., Kim, S. W., Suh, B. D., and Cho, S. H. (2000) Effects of platelet-derived growth factor-AA on the healing process of tympanic membrane perforation. *Am. J. Otolaryngol.* **21(3)**, 153–160.
131. Ma, Y., Zhao, H., and Zhou, X. (2002) Topical treatment with growth factors for tympanic membrane perforations: progress towards clinical application. *Acta Otolaryngol.* **122(6)**, 586–599.